PLANTS

PLANTS
Basic Concepts in Botany

WATSON M. LAETSCH
University of California, Berkeley

LITTLE, BROWN AND COMPANY
BOSTON TORONTO

To my parents, who encouraged and supported my interest in biology, and to Sita, John, and Krishen for their aid and tolerance during the preparation of this book

---

Library of Congress Catalog Card No. 78-64497

Second Printing

Published simultaneously in Canada
by Little, Brown & Company (Canada) Limited

PRINTED IN THE UNITED STATES OF AMERICA

Artists: Douglas Gilbert, Laszlo Meszoly

Cover design: Tonia Noell-Roberts

Cover photo: Mike Mazzaschi, Stock, Boston

Designer: Edgar H. Allard

Title page photo: Stephen Dalton

Lines from "The Hollow Men" by T. S. Eliot on page 58 are reprinted by permission of Harcourt Brace Jovanovich, Inc., and Faber and Faber Ltd.

All illustration credits can be found on pages 491-494.

Preface

Plants is a textbook for students encountering botany for the first time, whether fulfilling liberal arts requirements in the sciences or as prospective science majors. It is unconventional in many ways, perhaps most particularly in its attempts to build on the students' familiarity with plants to introduce them to botanical ideas, systems, and facts. My aim throughout has been to introduce botany as an intellectual discipline closer, more vital to the everyday world we perceive than most students may realize.

My approach in *Plants* derives from the principle that students can best assimilate the new and unfamiliar by working from the concrete to the abstract. It contrasts deliberately with a traditional approach in biology which asks students to understand things they cannot see before they consider familiar objects. In *Plants* the familiar and concrete comes first whenever possible; the "unseen" appears slowly and sparingly. For example, the survey of the plant kingdom begins with organisms the student easily recognizes, the flowering plants, and ends with microbes, organisms more remote from their senses. In a similar way, I treat the whole plant first before taking up cells and organelles; and I describe energy transfer by relating it to hydrogen, a known and demonstrable element. To further bridge the gulf between the concrete and the abstract, I use many recognizable examples and applications whenever possible.

Redundancy is equally important for effective learning of ideas and principles, and I have made an effort to build a measure of repetition into the structure of the text, so that important ideas re-emerge in different contexts. One theme, the origin of diversity and its relation to available energy and matter, I introduce in Chapter 1 and refer to in most other chapters. I also repeatedly bring up transfer of energy and genetic variation, ideas that require frequent exposure to be properly understood. The overall effect is like a spiral, with chapters building on each other and expanding ideas presented earlier. The spiral finally forms a loop as the last chapter brings the reader around to the first chapter, though now with a deeper sense of the complexity of the plant world. However, the parts need not be read in sequence. They can be rearranged to suit different course structures.

Plants deals with botanical facts as well as ideas and processes, but I have not attempted to explain everything there is to know about plants. It is easy to learn facts, but easier still to forget them. Then too, many facts in science change or become irrelevant. Rather than burden the student with an encyclopedia of information about plants, I have used factual information to illustrate general ideas, selecting facts for their interest to students as well as their appropriateness. I also avoid giving the impression that we know all there is to know about plants, frequently identifying controversies and unresolved questions, giving the uncertain and the unknown their due.

The Organization of *Plants*

In developing the organization for *Plants*, I have presented botany to students by linking the discipline as we know it to the plant world as they have encountered it, moving from familiar to abstract and emphasizing the importance of plants to human life. In Part I, I stress the role of plants in providing energy for all living things, a function that has given plants a considerable impact on the shaping of human civilization. This Part includes an examination of the ecosystem in Chapter 1 and discussions of how transformation of energy and transfer of matter work, and how they influence patterns of consumption among living things. In the last chapter, we return to consider the ecosystem in detail, because community structure and adaptations of organisms to their environment will be more fully understood with the chapters on genetics and evolution as background. The second chapter also introduces a main theme of the book: the relationship of plants to human culture, politics, and philosophy.

Part II is the survey of the plant world, its classification and its history, as well as the structural features and relationships of the main plant groups. In Chapters 3 and 4 I describe kinds of plants and in Chapter 5 continue the theme of how people and plants are related by describing plants of special interest and importance to society. Examining these plants, I bring up terms and topics on plant growth, structure, and inheritance, but wait to cover them in detail in subsequent chapters. A brief introduction to cells appears in the Supplement to Chapter 3, providing a background for discussion of specific cell functions in Chapters 6 and 7.

Moving on to the whole plant and plant processes, the five chapters in Part III cover plant structure and function, particularly in the flowering plants. Chapters 6 and 7 are on development, and I have explained features of plants that students may encounter daily. In these chapters, the cell is depicted in its growth along with the rest of the organism, leaving some cellular parts for later concentrated study. (It is not necessary to know about all the cell's organelles to understand basics of cell enlargement.) In Chapter 8 I develop the fine structure of energy transformation and cycling of matter introduced in Chapter 1, explaining all cell metabolism in terms of what the plant needs to grow.

Mineral nutrition of plants and the cycling of specific minerals is the subject in Chapter 9. The nitrogen cycle and the use of nitrogen in plants get full emphasis because the element is so vital to all organisms and because it is a limiting factor in agricultural productivity. This chapter also displays the effect of limitations of energy and minerals on adaptation of plants. Once the importance of mineral nutrition is established, we explore in Chapter 10 the basic anatomy and physiology required to appreciate how vascular plants transport materials. This section goes beyond coverage in previous chapters of specialized cell structure, energy requirements for uptake and movement of materials, and the plant's requirements for minerals.

The major themes of *Plants* are treated explicitly in Part IV. Because evolution occurs by selection of variants in populations, it is important to know how variation works to grasp the idea of natural selection fully. In Chapter 11 I discuss variation by examining sexual reproduction and its significance. Pollination is treated here because this is the way most plants effect sexual reproduction, ensuring and promoting genetic variation. It also balances the complexities of Mendelian and molecular genetics. Genetics is treated historically, because it vividly shows science as an intellectual endeavor. Asexual reproduction (Chapter 12) appears in this Part because it is another reproductive strategy, and tissue culture comes in here rather than in the chapters on development because much current interest in this subject is aroused by its use in plant genetics.

Two chapters focus on evolution. I trace in Chapter 13 the establishment of geological time as well as past and present efforts to study former living organisms. In Chapter 14 I emphasize historical and current efforts to understand how life has evolved, giving many examples illustrating evolutionary phenomena in animals as well as plants. I have not shied away from introducing animals here because they illustrate some concepts more dramatically than do plants, and because it is important to emphasize the interrelationships of plants and animals. My approach to evolution takes me down the historical and natural-history path, rather than the more mathematical track that often confuses beginning students more than it enlightens.

Having established how plants and how organisms adapt, I go on in Chapter 15 to explore the structure of plant communities and the environmental factors shaping them, closely attending also to the human effect in managing ecosystems. This chapter closes the loop I referred to earlier, bringing the reader back to the broad view of the plant world described more generally in Chapter 1.

The text is equipped with varied instructional aids to help students understand the material. Each chapter has a summary and review questions. Many of these relate material and concepts from different chapters and encourage students to integrate their own experiences with discussions in the text. Many chapters end with special sections called Supplements, designed to entice students to look into a selected topic in some detail. Each Supplement is focused on a subject related to material in the preceding chapter, but considered here in more depth.

The text has a glossary at the end, as well as an extensive bibliography including both popular articles and texts, scholarly papers, and monographs. The Supermarket Taxonomy in the Appendix is a survey of species of common plants found in the grocery store, and the families to which they belong. It supplements the discussion of supermarket plants in Chapter 3, providing additional background that students will find interesting. An instructor's manual (*Instructor's Manual to Accompany Plants: Basic Concepts in Botany*, by Mary S. Manteuffel) is also available. It has an overview of every chapter, learning objectives, lecture suggestions, ideas and directions for laboratories and field trips, questions for discussion, and objective questions.

Botany is a visual science, and plants, like so much in biology, are often best presented with pictures rather than words. The numerous illustrations in this text were carefully selected for their value in motivating students and providing information; they form an important instructional medium. Most science books use pictures to illustrate an occasional point in the text, but in many cases I have attempted to use words to illustrate pictures.

Acknowledgments

An old friend, Senior Editor Milton Johnson, Jr., of Little, Brown and Company, started me off on this project and nurtured it through adolescence. His support was essential and is much appreciated. Arthur Wester and Ian Irvine, science editors at Little, Brown, have been encouraging and uncommonly patient, and I wish to thank them as well as the very capable production staff. Kathleen Field revised, rearranged, and taught me a great deal, and I learned much from the incisive editing of David W. Lynch. Janet Beatty Welch guided, prodded, and consoled during the frantic turning of manuscript and illustrations into a book. The quality and variety of illustrations attest to the talents of Tonia Noell-Roberts and Tina Schwinder.

The book owes a great deal to Percy Knauth for his help with the organization and style of the original drafts. His contribution is appreciated. Eric Golanty is to be thanked for his help with many of the Supplements, Lynn Carter for her help with the Supplement to Chapter 1, and Spencer Barnett for his assistance with Chapter 5. The criticism and expertise of Michael Guerin, Mary Manteuffel, and John Gerhart were most useful, and I especially want to thank Lewis Feldman, Robert Ornduff, and Ruth Strand for their suggestions for sources for illustrations as well as for actual photographs. Many reviewers provided valuable comments on various portions of the manuscript; I am grateful for their suggestions. Timely and expert typing is essential, and I thank Helen Diggins and Pat Pope.

I have expressed appreciation to family members elsewhere, but more specificity is only fair. Sita typed for critical deadlines, reviewed manuscript, and helped select photographs; John criticized manuscript and judged photographs, and Krishen helped with illustrations and did not complain loudly when I didn't take him places "because of the book."

Contents

PLANTS

Plants in the

PART I

Biosphere

Few things are so much taken for granted in our everyday environment as plants. They are all around us—weeds defying the inhospitable city by growing out of sidewalks and parking lots, grass and trees in city parks and suburban lots, forests and cultivated fields in the countryside. Plants are everywhere, and we cannot escape dependence on them. From them comes the energy that sustains all living things, including, of course, ourselves.

This dependence is a main topic of Chapter 1. We will see how living and nonliving things are related, how energy is transformed and matter transferred by earth's organisms. Plants form the basis for the transfer of energy required by all living things, and they are in turn affected by them. We will see again and again that the world of plants is wonderfully diverse and that this diversity is related to the energy and matter available to plants.

Plants touch our lives in other ways as well. They have helped shape human culture, politics, and philosophy from the beginning. In Chapter 2 we will follow the history of revolutionary discoveries in agriculture, from the first domestication of plants and origin of civilization to today's agricultural technology. We find that it is a story in which the characteristics of plants, the resources of an area, and human ingenuity are all main characters. These two chapters give the primary ideas that will open the way for a closer look at the world of plants and how we use them.

CHAPTER 1

Plants: The Basis of Life

Consider this statement: plants are the most important organisms on earth. Why? They are the form of life on which all other living things depend, the earth's basic foodstuff. Every living thing on this planet is fueled by the energy that arrives as light from the sun, but animals cannot use sunlight itself as fuel. First it must be transformed to usable energy, and plants are the transformers. They are able to capture the sun's energy and to transform it into a form of fuel we call food.

Think of a place without plants and animals. Once in a while an undersea volcano erupts and throws enough material up above the water's surface to form an island. Soon after this geologic creation, birds and sea mammals may land on the island, but at this point the island is good only as a resting place for migrants, who find their food elsewhere. It is not long, though, before bacteria and other **microbes** (extremely small living beings) arrive, carried on the wind or in the feathers or stomachs of passing birds, to be deposited in rain or in guano on the island. Spores and seeds of plants come in the same ways and begin to take hold in newly formed soil. Only then does the volcanic rock begin to be livable for the first animal colonizers: insects, birds, reptiles, and mammals.

We will look at the earth's living inhabitants in a special way in this chapter—as energy producers and energy consumers, as members of a complex cycle of interdependence. What is the fuel, this energy, that keeps things going? How does it come to us and how is it used? How do plants and animals depend on each other?

1.1 The earth from space. This view of the Mediterranean Sea, Africa, and Antarctica, was taken from the Apollo 17 spacecraft during the last trip to the moon. In comparison to the mass of the earth and the vastness of space, the biosphere is a wafer-thin husk.

THE BIOSPHERE

Think of a basketball with a postage stamp glued to it. Picture next how thin the ink on that stamp is. The part of earth occupied by living things is equal to the thickness of the ink on the stamp on the ball, no more. We are, in a sense, creatures who live in a thin layer of green scum painted on a small rock whirling through space. Everything we earth-dwellers—plants, microbes, fungi, animals—need to sustain life we find within this layer, known as the **biosphere,** that part of the planet where life is found.

Part of the biosphere is solid land. Part of it is water, including all the world's oceans. Its third portion is a part of the atmosphere. Taken together, the biosphere is a thin layer sandwiched between the **geosphere,** the mass of matter from which the earth is formed, and the **atmosphere,** the gaseous envelope that surrounds earth as it spins in space. Very uneven both in depth and in density, the biosphere extends upward to at most about 8 kilometers (5 miles) into the atmosphere. Eight kilometers up and 8 kilometers down (only in oceans)—those are the limits between which every form of life on earth exists.

Matter and Energy Each living thing is a package of matter, made of chemical elements from the geosphere and atmosphere. These elements enter the biosphere and cycle through it in a complex fashion, the process we call living. When a living thing dies, the matter of which it was composed returns to the earth and the air, continuing in the eternal cycle between living and nonliving things. Earth's matter thus forms, for all practical purposes, a **closed system:** it is a *system* because a change in any part will affect all the other parts; it is *closed* because elements outside the system do not affect it significantly. Occasionally a few meteorites survive the fiery trip through the atmosphere and land on earth, but little matter enters the system that was not there before.

Earth's energy, in contrast to its matter, forms an **open system:** energy enters from outside the system, from the sun. In fact, if the only energy the earth had were that found within the biosphere, it would soon be exhausted, for two reasons.

First, energy can be neither created nor destroyed (this is known as the first law of thermodynamics), although it can be—and constantly is—transformed. Second, even though the amount of energy remains the same, the energy in the universe available to do *work* is decreasing (this is the second law of thermodynamics). Why? When wind breaks a tree, when we cut a flower, when our muscles contract, when a cell divides, work is being done. All work requires the expenditure of energy; in fact, we can define energy as the "thing" that enables work to be done. When energy is used to do work, its form is changed, though the total amount remains constant (see Figure 1.3). If you place buckets of water on top of a hill, the water has a certain energy (known as *potential energy:* because of its position, the water has the potential of doing work). If you then pour the water over the hill to fall on a waterwheel below, it makes the wheel turn. The potential energy is now transformed to an active form, which turns the wheel. Some of the energy will be converted into yet another form, heat, which happens whenever work is done. At both times the amount of energy is the same. But heat is a diffuse, disorderly kind of energy, frequently useless in itself for doing work (think of the steam venting from power plants: the heat in the steam represents energy, but it is too dispersed to do work). Thus some energy is always transformed into a useless form, and the amount of useful energy is decreased.

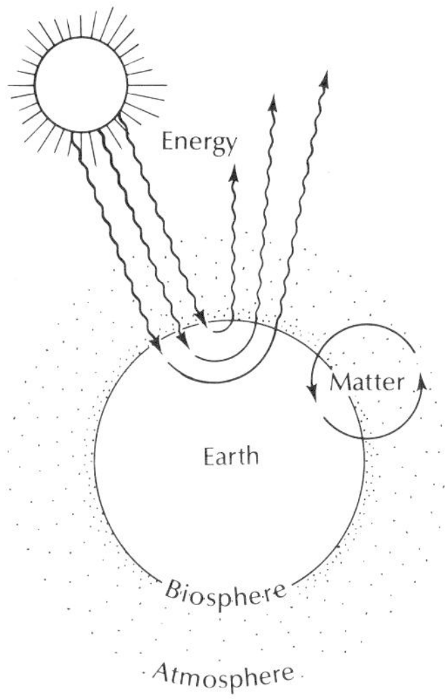

1.2 Energy enters and leaves the biosphere, forming an open system. This energy drives the cycling of matter within the biosphere. Matter does not enter or leave the earth, so matter is cycled within a closed system.

As work is done in the biosphere, most energy is lost as heat. Both this heat and that absorbed by earth and water are reflected from earth's surface as infrared rays, which are absorbed by clouds. On cloudy nights this radiation is reflected back by the cloud cover, giving us warmer nights. Eventually, though, this heat escapes to

1.3 ENERGY: ITS FORMS AND TRANSFORMATIONS

The sun (photo, below) provides all the energy to power life, and much of life's story is the transformation of energy between its various forms. Only a fraction of the radiation from the sun, that part of the spectrum called *visible light* (below), is absorbed by green plants and transformed to chemical energy. Sunlight splits water and chemical bond energy and hydrogen are produced (facing page, bottom). This energy is used to attach hydrogen to carbon dioxide from the atmosphere, and carbohydrates, fats, and proteins are subsequently produced, which provide the fuel for inhabitants of the biosphere. The burning of this fuel to provide energy results when hydrogen is released from carbon and united with oxygen. The force of the hydrogen-oxygen reaction is illustrated by the explosion and burning of the hydrogen-filled zeppelin *Hindenburg* at Lakehurst, N.J., in 1937 (facing page, top left).

Electricity is produced by generators powered by fossil fuel, water power, and recently by nuclear reactors. Hydroelectric power, such as that produced at Hoover Dam (facing page, top right), is transformed solar energy because the sun powers the water cycle. The dam illustrates potential (standing water) and kinetic (falling water) energy (facing page, middle right). This conversion of potential energy to kinetic energy (the kind that does work) affects organisms, as discussed in the Supplement, page 25. Every transfer of energy produces heat. As energy is transferred in the biosphere, heat is lost. This loss eventually equals that which is supplied by the sun.

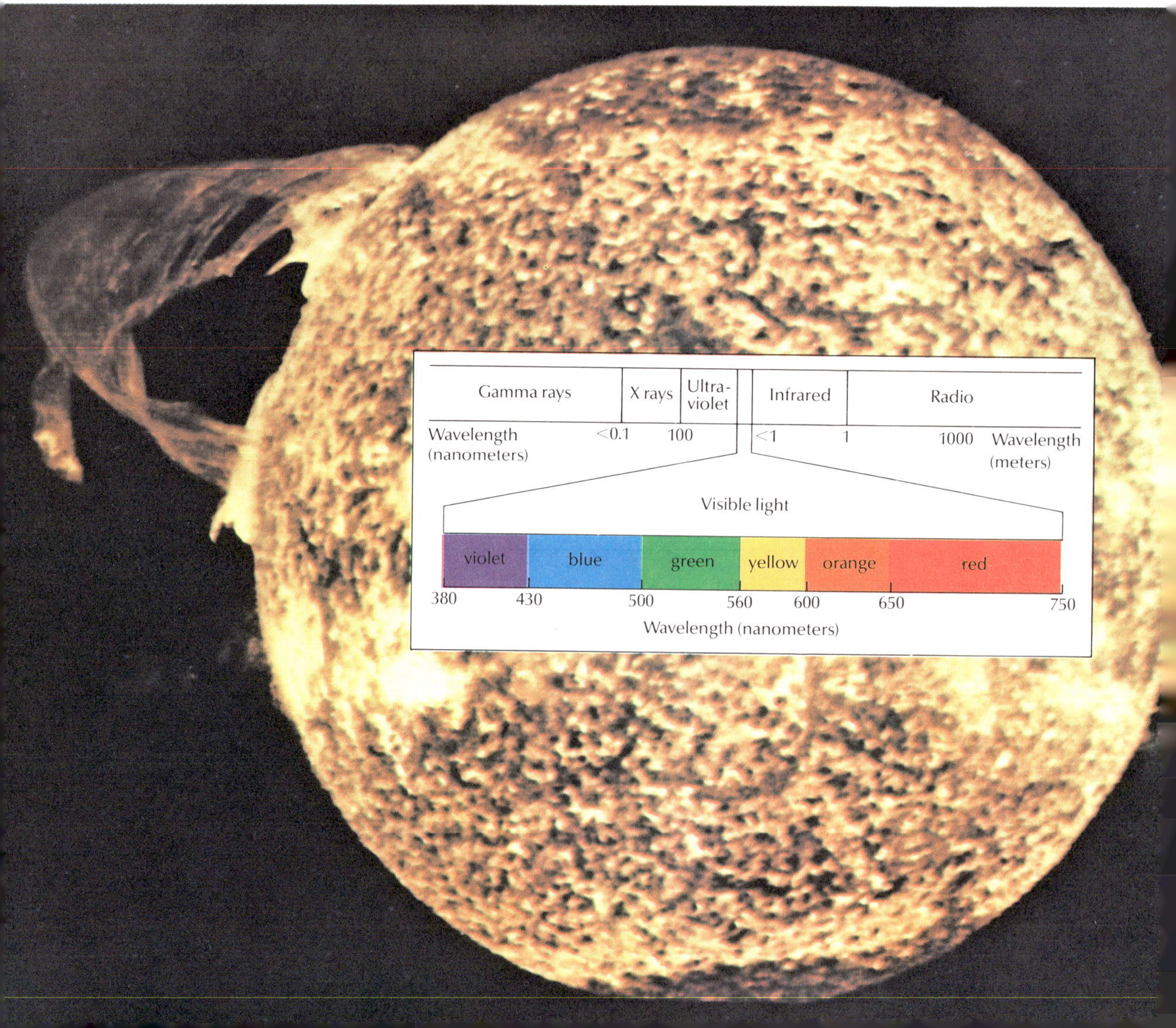

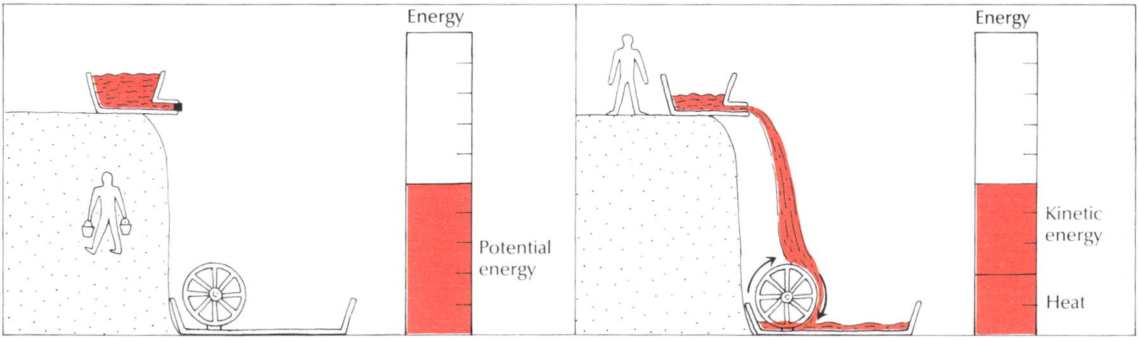
Energy
Potential energy
Energy
Kinetic energy
Heat

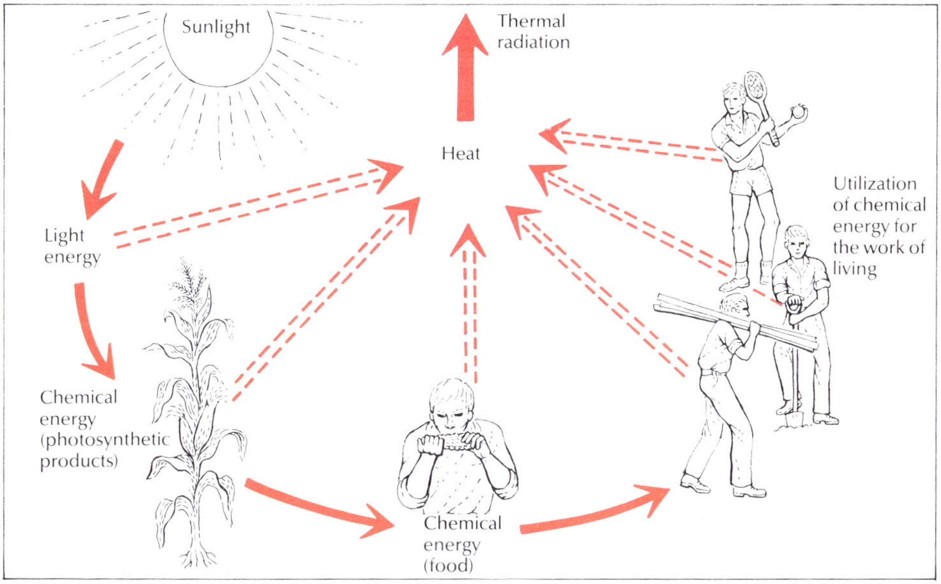
Sunlight
Thermal radiation
Heat
Light energy
Chemical energy (photosynthetic products)
Chemical energy (food)
Utilization of chemical energy for the work of living

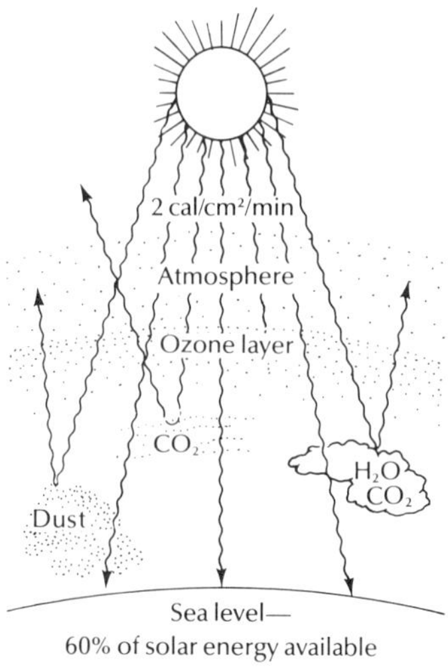

1.4 The atmosphere contains ozone (O_3), water (H_2O), carbon dioxide (CO_2), and dust, which either reflect incoming radiation or absorb it and reradiate energy back to space. As a result, about 40 percent of the sun's radiation received by the atmosphere is intercepted.

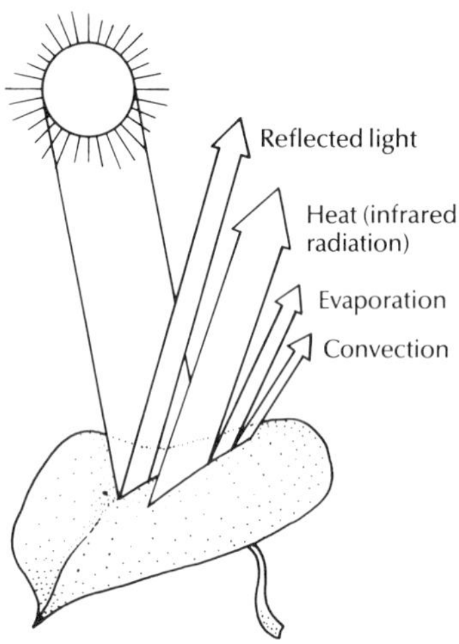

1.5 All organisms are fueled by light energy, which is transformed to chemical energy by photosynthesis. Only a small percentage, about 1 percent, of the radiation striking a leaf's surface is available for photosynthesis.

outer space, lost to earth forever; but as much energy comes in as goes out, so the biosphere's energy system is in a steady state. Nevertheless, the sun will cease to burn eventually, our earth will lose its energy, and the conditions of the second law of thermodynamics will be fulfilled.

The Source of Energy Within the biosphere itself several forms of energy are produced: hydraulic energy, created by water in motion (a river pouring over a dam, storm waves striking a shoreline); electrical energy (lightning); and geothermal energy (underground water converted to steam by hot rock formations). Powerful as these sources of energy can be, they are insignificant compared with the huge flow of energy that comes to earth from the sun.

The sun's energy begins with reactions like that of a hydrogen bomb. Nuclear fusion deep in the sun's core creates radiation, which makes its way to the sun's surface and is then radiated away—most of it as visible light, some as ultraviolet light and infrared light and X rays (see Figure 1.3). This sunlight is the dominant form of energy in our world, one that primitive peoples recognized eons ago as the giver of life. All the energy humans produce in a single year from our many energy sources—coal, oil, hydraulic power, nuclear power—amounts, according to our present estimates, to only two-ten thousandths of the total energy coming to us each day from the sun.

Every minute of every day, every square inch of the earth's surface has beamed toward it an average of 2 calories (a calorie is the energy needed to raise the temperature of 1 gram of water 1°C). But as much as 40 percent of this energy never even reaches the ground because it is reflected back into space by clouds and dust. (Because of this reflection, the earth seen from outer space is as bright as the planet Venus, which we see as the brightest "star.") A small but significant amount of radiation is absorbed by the layer of ozone (O_3) that envelops the globe in the upper reaches of the atmosphere; carbon dioxide (CO_2) absorbs still more.

Even the sunlight that finally reaches a plant is not fully used. About 20 percent is reflected from the leaf's surface, allowing us to see the plant. Some of the solar energy is used for **evaporation** of water from the surfaces of leaves, which helps to cool the leaves and move water through the plant. Some energy is lost by **convection,** the transfer of heat by means of a circulating gas or liquid. (For example, on a warm summer day, the sun-warmed earth is hotter than the air above it and hence transfers heat to it, which sets the air in motion, resulting in a rising column of air in which hawks hover and sailplanes soar. In the same way a sun-warmed leaf transfers heat to the air around it.) The plant does not use some kinds of radiation at all, but it does use visible light. Of the total energy coming from the sun, most plants finally use less than 1 percent.

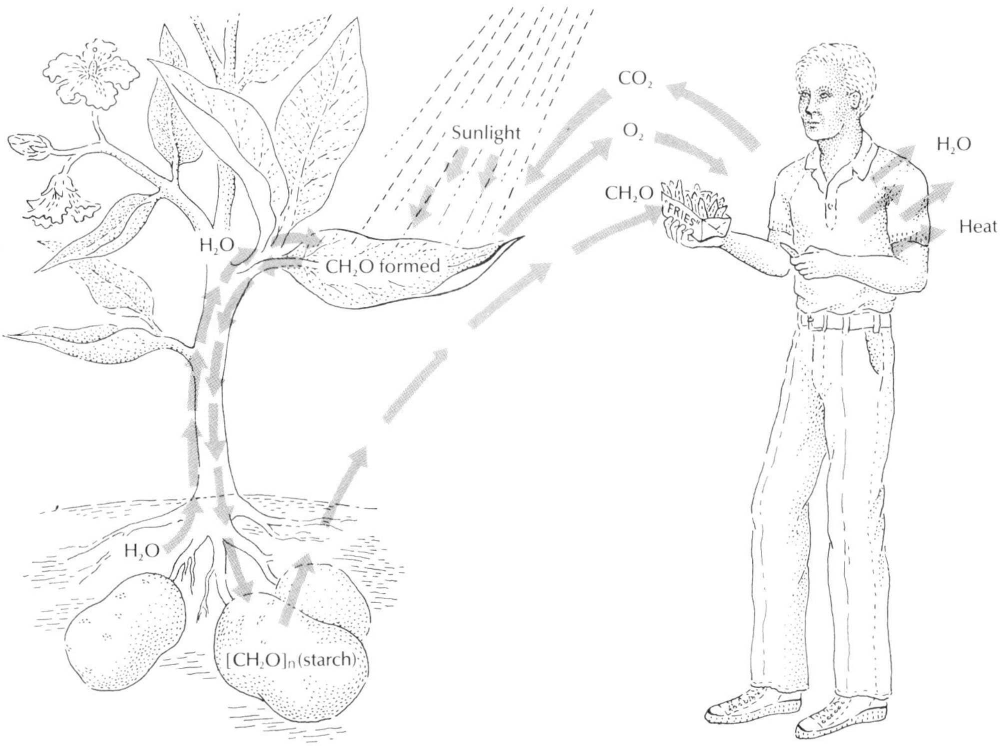

1.6 By means of photosynthesis, plants combine carbon dioxide with hydrogen from water to produce carbohydrates and give off oxygen in the process. A primary consumer such as this college student burns the carbohydrate when hydrogen is removed and combined with oxygen to form water. Chemical energy and heat energy are produced, as well as carbon dioxide, which is made available to plants.

The Transformation of Light Energy Visible light becomes the fuel for **photosynthesis:** with energy provided by the light absorbed by green leaves and stems, the plant breaks down water (H_2O) into oxygen, which is released to the atmosphere, and hydrogen, which it combines with carbon provided by the carbon dioxide in the atmosphere (see Figure 1.6). Chlorophyll, which looks green to us because it reflects green light, mediates this transformation of light: it absorbs the visible light and splits water, providing chemical energy—the hydrogen that fuels the biosphere. (Chlorophyll's role is analogous to that of a windmill: the windmill turns a generator to produce electricity, which can be used to split water and produce hydrogen, which can then be stored for use as a fuel.)

Hydrogen can act as a fuel because when it is **burned** (that is, combined with oxygen, forming water) energy is released. Two examples can illustrate hydrogen's power (for a more detailed explanation, see the Supplement on page 25). A kitchen stove or a home-heating furnace may be fueled with natural gas, that is, methane (CH_4), which contains hydrogen. When CH_4 burns it combines with oxygen:

$$CH_4 + 2O_2 \rightarrow CO_2 + 2H_2O + \text{heat}$$

methane plus oxygen reacts to form carbon dioxide plus water plus heat

The burning of the dirigible *Hindenburg* (shown on page 7) dramatically illustrated hydrogen's immense power. The huge gas bags that gave lift to the airship were filled with hydrogen gas, which is lighter than air. During a landing at Lakehurst, New Jersey, in 1937, an electrical spark or an explosive device ignited the hydrogen, destroying the *Hindenburg* and taking many lives.

Most of the work done in the biosphere results from burning hydrogen by the process known as **respiration.** Using the hydrogen obtained through photosynthesis as a fuel, the plant forms chemical compounds such as sugar and starch, which store energy in the form of combined hydrogen. The starch in a potato, for example, contains hydrogen along with carbon and oxygen, and when you burn it you produce $CO_2 + H_2O +$ energy (calories). When you exercise to burn off the food stored by eating too many french fries, you are burning (chemically) the hydrogen that the potato plant obtained from H_2O when it absorbed light energy.

The entire amount of energy accumulated by plants on earth in the course of a year has been calculated at 10^{18} kilocalories (1 kilocalorie is the energy needed to raise the temperature of 1 kilogram of water 1° C). This figure represents the estimated **gross productivity** of plants—the total amount of energy represented by plant material produced by photosynthesis. The plant uses some of this energy itself to do such work as growing, and about 50 percent of this energy passes off as heat. The amount of energy left in the plant after the tax of respiration is paid is called **net productivity.** Roughly

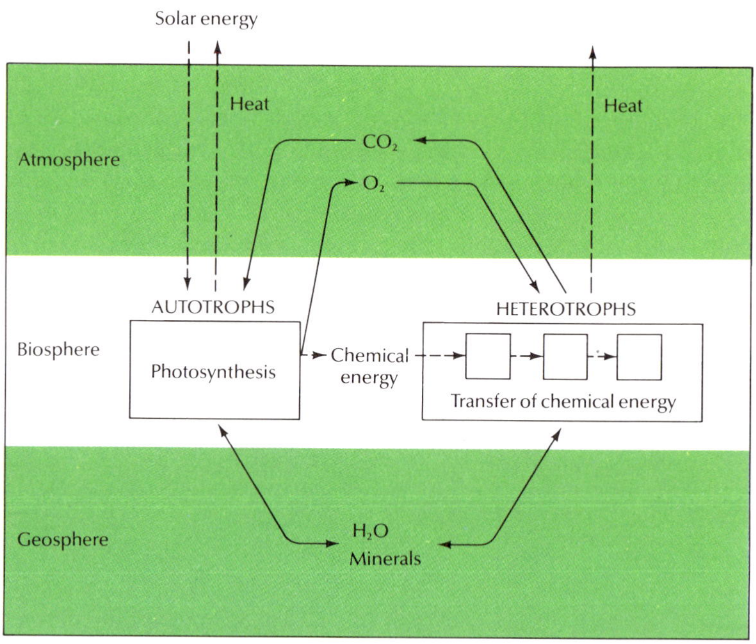

1.7 Solar energy is transformed by autotrophs through photosynthesis into chemical energy, which is transformed from one heterotroph to another. In the process, energy is transformed to heat and leaves the biosphere. In photosynthesis, autotrophs take in carbon dioxide and water and give off oxygen. Heterotrophs take in oxygen and give off carbon dioxide. Minerals cycle from the geosphere and back among both autotrophs and heterotrophs.

50 percent of the energy initially converted from sunlight is available to an organism that eats the plant.

These organisms that consume the plant are known as **heterotrophs** ("other-feeders"), because they get their energy from other living things. Plants, because they capture energy directly from the sun and transform it to the energy they need to live, are known as **autotrophs** ("self-feeders"). Plants are thus the first step in the cycling of energy in the biosphere: taking energy from the sun, they make it available to all other living things.

ENERGY TRANSFER IN THE BIOSPHERE

Lions in the great grasslands of Africa are very fond of zebras. When a lion kills and feasts on zebra flesh, it is filled with new energy, which enables it to live, sire offspring, and hunt more zebras. For their part, the zebras find food in the grasses and other plants. Taking in energy from the plants, the zebras can live and multiply. But the lions limit the number of zebras by hunting and killing them; if they did not kill the zebras, there might be more zebras than grass to support the zebras. The quantity of grass limits the lion population, too, because the number of lions that can survive depends on the quantity of zebras available as food, as energy. And thus there exists, between lions, zebras, and plants, a balance of energy—a balance of life.

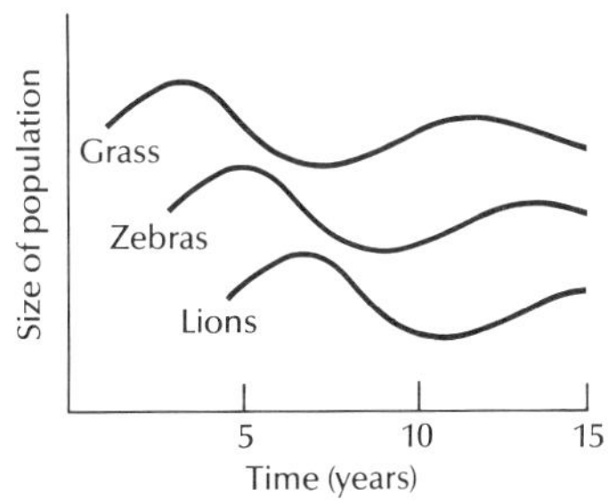

1.8 Grass, zebras, and lions are interdependent, for they form a food chain in which grass is the base. The number of zebras will increase as the grass does, but the grass population can support only so many zebras. The zebra population overshoots the supply of grass and will crash when the grass declines, a crash helped by the large lion population, which has been feeding on the zebras. When the zebra population falls, the lion population will crash.

It was the grass, first of all, that made the balance possible. Living apparently passively in its hospitable environment, the grass drew its energy from sunlight and its matter from soil and air. From grass the zebra took energy and matter, and the lion took both from the zebra. The lion dies, and some of the energy still in its flesh is passed on to jackals and vultures tearing at its carcass and to microbes invading and decomposing it. Finally, some of the matter in the lion's flesh returns to the earth and nourishes the grasses, which, day by day, passively draw their primary energy from the sun.

Food Chains This cycle from grass to lion, from autotroph to carnivore ("flesh eater"), from lowest organism to highest, this **food chain,** transfers energy—food—through many **trophic** ("feeding") levels. Food is supplied from eater to eaten from the bottom link in any food chain, plants.

Every earthly environment has its own kinds of food chains (see Figure 1.9). In prairie country or meadowland, the chain might start with grass seeds, eaten by mice. Snakes eat the mice and then are eaten by owls. A pond food chain could begin with water plants eaten by minnows, which are eaten by bullfrogs, and the bullfrogs by raccoons, which may fall prey to a mountain lion.

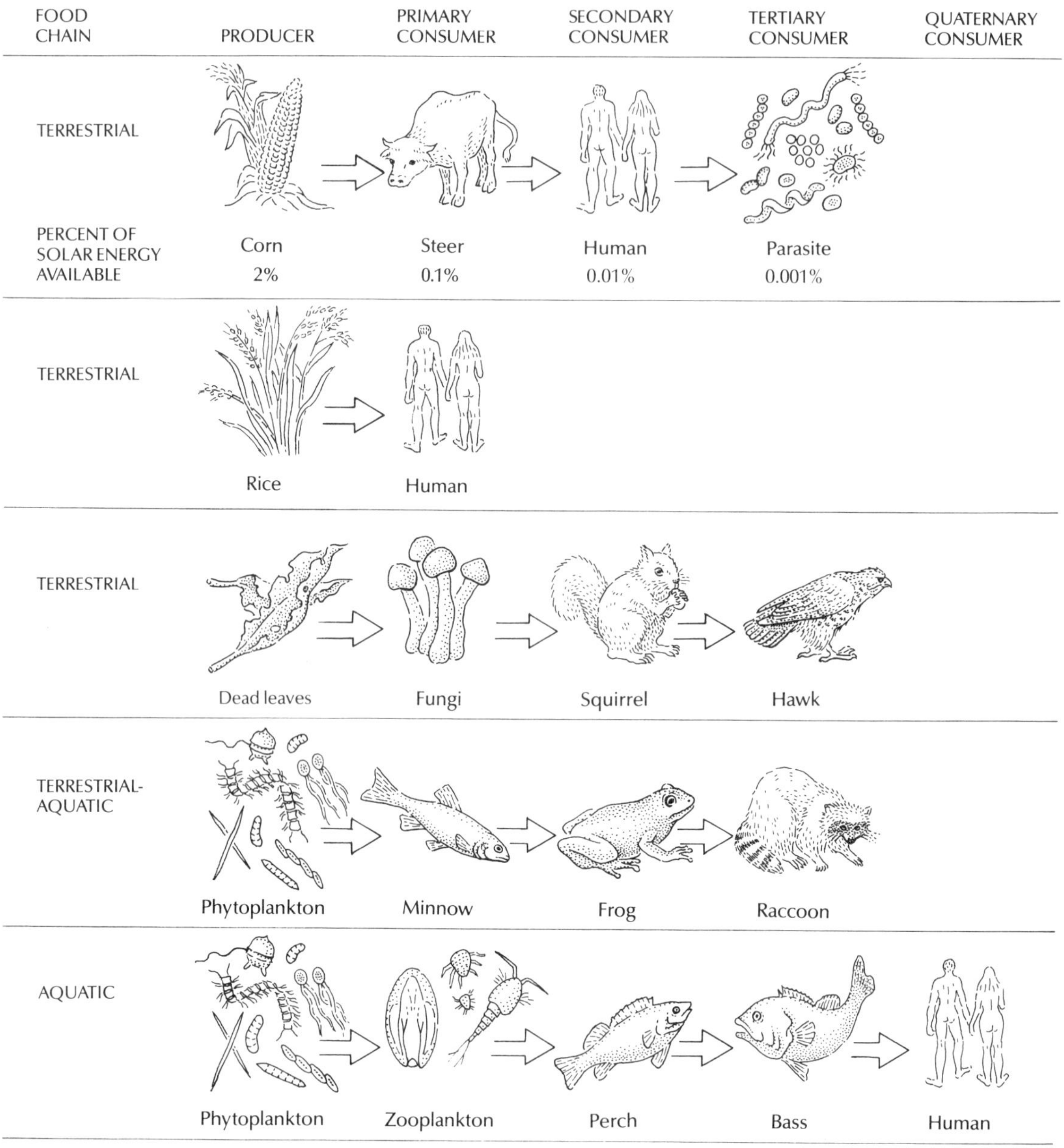

1.9 The number of trophic (feeding) levels is usually limited by the small amount of energy passed on from one level to another. Although three or four levels are usual, some aquatic food chains have many more links.

Here is another example, which I like to call the "collegiate food chain." The sun beams its energy on a corn plant growing in a field, which uses 1–2 percent of that energy for photosynthesis. One percent is passed on by the ears of corn the plant produces to the **primary consumers,** in this case the cattle that eat the corn. Only 0.1 percent or less of the energy in the corn then goes into hamburger and is passed on to the **secondary consumer**—say, a college student.

From the college student only 0.01 percent is passed on to a third trophic level or **tertiary consumer.** Such consumers are rare in this food chain, but let's say that an occasional college student is eaten by a shark or some ferocious animal. (Another tertiary consumer would be an internal parasite, for like most people, college students have internal parasites of various sorts.) The tertiary consumers at this point are benefiting from 0.001 percent of the energy originally beamed down by the sun.

The moral of this tale, of course, is not to show how little of the energy from the sun is acquired by eating a college student but to show why the number of trophic levels in any food chain is limited. With each transfer the energy still available is at least halved and is usually only 10 percent of the energy that was available to the previous trophic level. This loss of energy limits extensions of the food chain. When we eat a can of tuna fish, for example, we are getting only a small fraction of the light energy transformed by the algae starting the food chain that ended in our eating the tuna. And yet there may have been only four basic feeding levels involved: the algae, which were consumed by small animals or **zooplankton,** which were consumed by small fish, which were in turn consumed by the tuna.

Consumers Some two-thirds of all primary energy production occurs on dry land, which occupies only about one-third of the area of the globe. Obviously, little photosynthesis goes on in the oceans as a whole compared with that on the land. (Very high rates of photosynthesis, however, occur in certain parts of oceans.) On land, deserts are the least productive areas; the intensively cultivated fields of modern industrialized agriculture are the most productive. Here people have increased energy production many times by irrigating, fertilizing, and plant breeding. Edges of oceans, estuaries, coral reefs, temperate forests, and many polluted streams and lakes are also very productive, as we will discuss in Chapter 9.

We tend to think of primary consumers as **macroherbivores**—large plant-eating animals like elephants, cattle, horses, and deer—but these occupy a relatively small part of the food chain. Much of the primary productivity is not even available to them. The tops of trees, most of the tree's leaves, and bark, for example, cannot be eaten by these animals. They need habitats with large areas of grass, herbs, and shrubs. But one class, the insects, can find something to eat in almost every habitat (Figure 1.10); insects are major primary consumers on land. Earth's foremost opportunists for at least 300 million years, insects have an estimated one million species, about three times more than all other kinds of animals. They are adapted to so many habitats and have become so specialized in where they live and what they eat that many species can live side by side in a very small area, each species taking what it needs without

1.10 FOOD CHAINS AND CONSUMERS

The role of decomposers such as fungi and bacteria is illustrated in the tropical rain forest (left page). Most of the food energy in trees falls to the forest floor as detritus that provides food for decomposers. The great productivity and luxuriant vegetation of tropical rain forests are due to the rapid decomposition and recycling of the nutrients in this detritus. As a result, the soils of the tropical rain forest have relatively little organic content. The food chain is complex, with many cross-links, and there may be hundreds of species involved. Insects are the principal herbivores. Birds and mammals are also herbivores. Decomposers, such as bits of the fungus illustrated (bottom photo), may be consumed by an insect, such as the caterpillar or beetle shown (top and middle photos). The insects may in turn be consumed by a bird of the tropical rain forest.

competing with another. Insects far outnumber every other type of animal and hence are extremely important as primary consumers. The most common form of insect, numbering perhaps as many as three hundred thousand separate species, is the beetle. By comparison, fish, amphibians, birds, reptiles, and mammals together have fewer than forty-four thousand species.

Part of the success some insects have had comes from the stages they pass through in their development. Changing from larva to adult forms, they live in a different environment and eat different foods at each stage. Caterpillars are an example. As caterpillars they feed on the leaves of a host plant; anyone who has ever seen a tree infested by tent caterpillars can testify to their enormous numbers. Then, through metamorphosis, caterpillars develop into completely different types of organisms, ones with wings: butterflies and moths. In this phase of their lives, they can leave the host plant and fly to other types of plants, even changing their diet from leaves to the nectar in plant blossoms. Because larvae and adults have different diets, the organism's adaptability to varying habitats is increased.

Extremely important consumers in food chains are **microherbivores.** These are microscopic creatures such as protozoans and other small animals that feed on **detritus,** the product of wearing away or disintegration: leaves that dry up in the autumn and fall to the ground; trees that grow old, die, and topple; and bits and pieces of animal carcasses. Microbes like bacteria and fungi (described on pages 117–124) also feed on these things and are in turn food for the microherbivores. Were it not for these scavengers, collectively called **decomposers,** forests would long ago have choked on the products of their own growth, maturation, and death.

Very little live plant material is directly available to herbivores on the ground, as a rain forest shows (see Figure 1.10). The crown, branches, and leaves of the tree hold 7 percent of the energy; the trunk has 19 percent, and litterlike falling leaves and broken twigs make up 50 percent. Decomposers capture most of the energy made available by trees in a rain forest.

In estuaries, the broad marshy tidal wetland along the mouths of rivers, dead plant material falls into the water and is consumed by microbes and small animals such as crabs. These in turn are eaten by larger animals, such as fish and water birds. These again fall prey to still larger fish and animals, as the food chain progresses from many microfeeders to a few macrofeeders.

Symbiosis Some organisms find energy by ingesting other organisms, but there are other forms of harvesting energy: symbiotic associations. **Symbiosis** is the intimate living together of two or more organisms, and it may take several forms. **Parasitism** is one: one organism benefits and the other is harmed but usually not killed. The flea, for example, is a parasite of the wolf. **Commen-**

salism is another form: one organism benefits and the other is not harmed. **Epiphytes,** plants of the tropical rain forests living on the branches of trees, are an example. Thanks to the tree's height, they have easy access to the sun, but they do no harm to the tree. **Mutualism** is a symbiotic association in which each organism gets some benefit. The cow, for example, lives from birth to death in close association with microbes and protozoans in one of its stomachs, the rumen (see Figure 1.11).

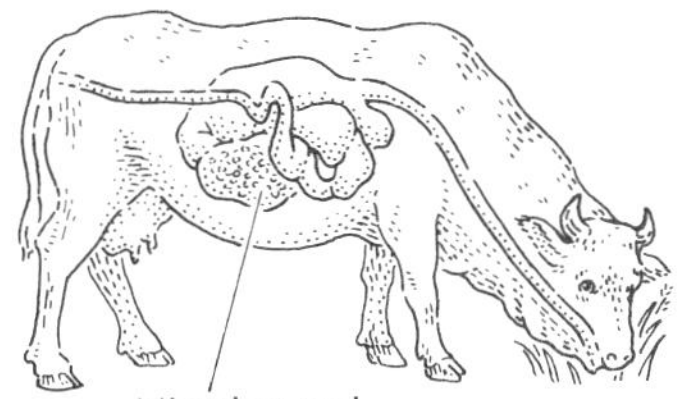

1.11 The cow is a mobile fermentation tank that has a symbiotic relationship with microbes and protozoans living in its rumen. The cow harvests and grinds up grass, which is digested by the microbes, and the cow lives on the microbes and the compounds they produce.

When a cow grabs a big bite of hay and starts chewing it, it seems that the cow is a primary consumer, eating the hay and getting the energy stored in it. Surprisingly, the cow is actually a secondary consumer. The primary consumers are the microbes and protozoans in the rumen. It is there that the microbes and protozoans break down the hay, feeding themselves in the process and leaving for the cow the products of their digestion. From these, not from the hay itself, the cow derives its quota of energy. In effect, what feeds the cow is not the hay but the microbes and the compounds produced from the hay by these primary consumers. Termites feed in much the same way. They chew and ingest wood; microbes in their gut digest the wood; and the termites feed on the microbes.

A LARGER VIEW

We have seen that the transformation of energy can take surprising forms, but it is always accomplished by essentially the same system: food chains. The many food chains in nature, however, are not independent of each other.

Food Webs Several interacting food chains usually form a complex feeding system, a **food web.** A primary producer of energy such as a tree passes on energy in several ways: as detritus that falls to the forest floor and provides food for primary consumers like microfeeders, and as foliage for primary consumers like deer or insects (see Figure 1.12).

The microfeeders, which consume detritus, are in turn consumed by invertebrates such as insects or worms, which in turn are consumed by vertebrates such as birds, fish, and raccoons.

The deer that feeds on tree leaves passes on some energy in the form of dung falling on the forest floor, there to be consumed by microbes, worms, and insects. Most of its energy, however, is passed on to predators, such as the mountain lion, which kills and eats it.

The lion, too, leaves dung on the forest floor, which is food for microbes, worms, and insects. When the mountain lion dies, its carcass may be consumed by a tertiary consumer, a vulture, which also leaves dung, which provides food for microbes, which then

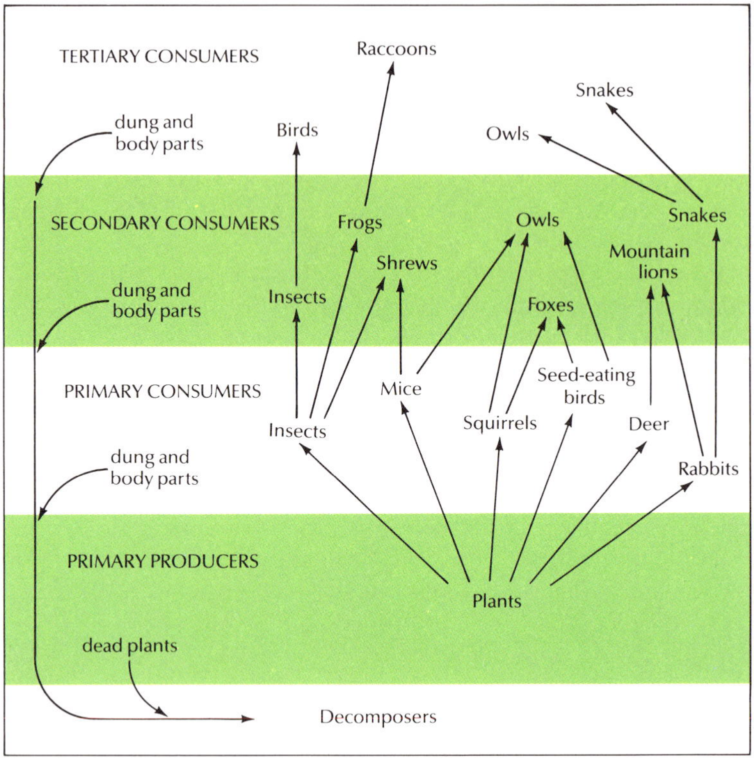

1.12 An ecosystem has a series of trophic (feeding) levels. Each level is sustained by the one below it. These food levels, often called food chains, are really complex webs. Many animals occupy more than one trophic level. Decomposers have a critical position in the food cycle. They make elements such as nitrogen temporarily captured in organisms available to primary producers. In turn, they die and fuel other decomposers or they are eaten by primary consumers in the form of small animals.

provide food for microscopic animals, which are eaten in their turn by worms and insects. Thus we have, from plant to consumer, a complex interweaving of energy transfers: a food web.

Each organism in a food web is involved in a number of relationships, not only with other living things in the web but also with the nonliving—for example, the soil in which plants grow. The whole system serves primarily to transfer energy from one living thing to another, but it also moves inorganic materials, such as minerals, from a nonliving environment (earth and air) through the bodies of living organisms and finally back to the nonliving environment from which they came.

Ecosystems This complex of living things interacting with their nonliving environment is known as an ecosystem. Ecosystems may be extremely large—the entire biosphere is an ecosystem—or they may be very, very small, like a puddle of water in a forest. Whatever its size, an **ecosystem** is a system of interdependence that functions by the transformation and transfer of energy and of matter (see Figure 1.13).

Every ecosystem is composed of ecological **niches,** which are geographic places or habitats, but they are also more than that. A niche is the total of all climatic, environmental, and biological factors that make it possible for a plant or animal to "make a living," to survive. A niche is not just a geographical definition but a functional one as well: how an organism makes its living is just as important in defining its niche as where. A flea lives by sucking the blood of a wolf, wherever the wolf may be and whatever the environmental conditions. A few square inches of the wolf's hide defines the flea's niche in geographical terms; the geographical niche of the wolf may cover tens of square miles in summer heat or winter snow. The flea's and the wolf's ecological niches are the complex way they function in a particular habitat.

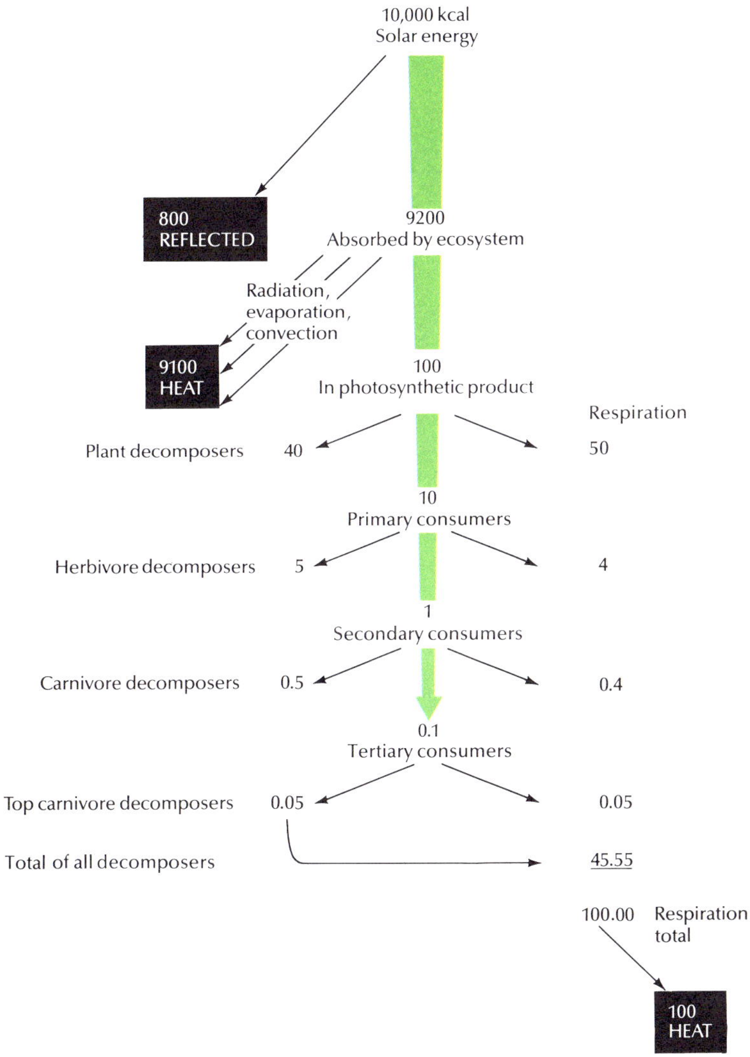

1.13 The loss of energy at each trophic level is about 90 percent. This limits the length of the food chain and the size of populations at each trophic level.

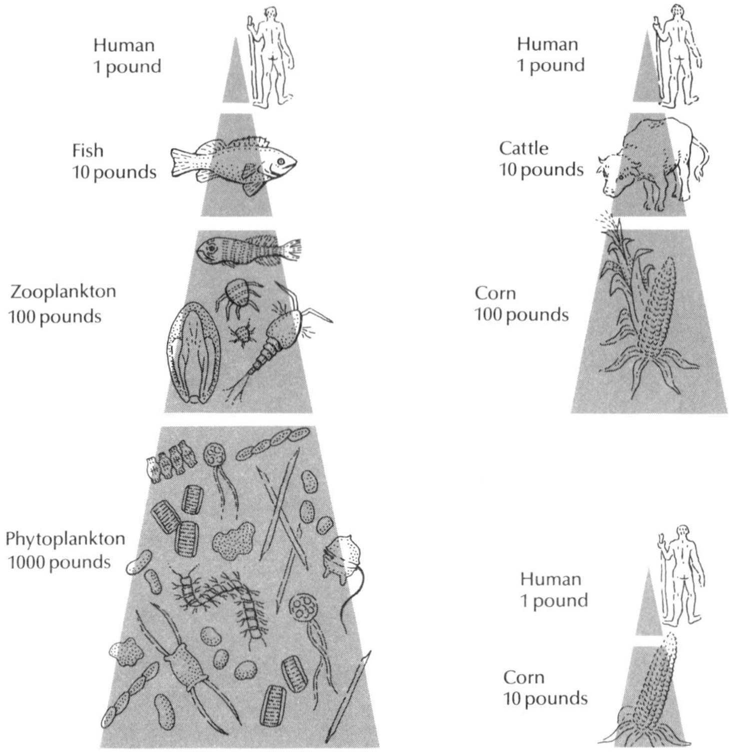

1.14 Energy transfers between trophic levels can be described by food pyramids. The taller the pyramid, the broader the base, and the more primary production required. The human–corn pyramid represents a relatively efficient energy transfer, and shows clearly why poor countries often favor vegetarianism.

Food Pyramids Humanity's position within ecosystems becomes clear when we look at food chains in yet another way: in terms of the amount of energy lost at each trophic (feeding) level. The collegiate food chain can again serve as our example.

The farmer produces about 100,000 pounds of corn so that the cattle grower and feedlot operator can send to the slaughterhouse about 10,000 pounds of beef. These 10,000 pounds of beef, in turn, sustain 1,000 pounds of college students—five or six people. Thus the **food pyramid** grows quickly from an immensely broad base to a narrow peak. From that perch a few students look at the 50 tons of corn it took to feed them.

In a different, far shorter food chain, people, not cattle, could be the primary consumers. The 100,000 tons of corn could be made into cornmeal, bread, or tortillas. Then it would be enough food (but nutritionally poor in comparison with meat) for 10,000 pounds of people—fifty or sixty of them. People of poor countries feel beef is a luxury simply because they cannot afford to lose the energy it takes to produce it.

A food pyramid gives a very practical way of looking at how resources are used, and it leads us to divide the world into two categories of consumers: those who eat meat and those who do not.

The difference is clear in our own population of native Americans before the middle of the nineteenth century.

Indians in the Great Plains were primarily meat eaters, preying on buffalo and moving with the herds. These animals roamed in incredible numbers, feeding on the seasonal prairie grasses, which were thick and plentiful. With only thousands of Indians hunting millions of buffalo on almost limitless pasturelands, the Plains Indians could afford the high loss of energy in the grass–buffalo–people food chain.

Some distance to the west and south were another people, the Pueblo Indians. They lived in settled communities and were mostly vegetarian. Not that they had no meat—they domesticated and ate turkeys and, when they could, they supplemented their vegetable diet with jackrabbit, antelope, and prairie dog. But meat was too scarce to be a practical staple of their diet. These Indians therefore raised corn, beans, and melons, not to feed animals that they would then eat but to feed themselves. They were the primary consumers.

Still today some have meat and others do not. Most industrial nations and countries with small populations eat meat because they can afford the wasted energy. Most poor nations are vegetarian. "Bread is the staff of life," the Bible says; it was written a few thousand years ago in lands where meat was eaten only by the privileged. In any of our markets, the imbalance shows clearly in prices. Bread is still the staff of life because it is usually the cheapest of foods: it can be produced directly from grain, with little added to it, providing the most energy for the least effort.

How great is the energy loss in the corn-to-beef-to-people chain? The United States consumes more beef than any other country. We plant food crops on about half our land that is usable for agriculture. Eighty-nine percent of our corn crop and 87 percent of our oat crop, or 30 million tons of protein a year, we feed to livestock. In fact, the protein in grain fed to livestock is too dilute to feed humans economically; the animals concentrate it and make it tasty for us. But by eating the livestock, people obtain only about 2 million tons of this protein, a loss of 28 million tons in the food chain. That amount would make up 90 percent of the world's deficit in protein. It would give every human being 12 grams (about ½ ounce) of protein every day, which is about a quarter of the human daily requirement of protein.

The system sounds wasteful, and it certainly is, but for some good biological reasons, the human being is not entirely a primary consumer (see the Supplement on page 61). Like some modern primates (monkeys, apes, and people), the earliest people may have been vegetarians, but there came a time when they began to supplement their diet of plants with eggs, insects, and an occasional small animal they were able to catch. If it did nothing else for them, adding these foods helped them exploit the environment more fully, and that is always part of the reason that one animal succeeds better than another; it may also have helped the human animal become highly adaptable. Because we are adapted for eating meat, we require proteins and vitamins that are difficult to obtain from plant food alone.

Through most of human existence, people have lived on the land, aware of how their lives depended on plants. In the far-off past they ate wild grasses and grains, only learning to cultivate them much later. Then they learned to refine the grain into flour and bake it into bread. Bread and plants, then, were clearly indispensable for survival. But in the last half century, people have moved in great numbers from the land to the city, the greatest migration in history. "Home" today is far less likely to be down on the farm than an apartment in a crowded city or a split-level in a suburban development. A whole generation has grown up never having seen a harvest taken in. Plastic-packaged and uniform in bins of unvarying chain supermarkets, in endless shopping centers sometimes set in formerly productive farmland, food for most Americans has lost the dust of the land and every trace of its origin. Plants no longer seem necessary.

Yet the loss of energy at every step in the chain leading to the food in the supermarket has never diminished, nor will it ever diminish. This is a fact that must be faced and understood by urban society. The long history of food plants and the factors involved in current agricultural productivity are the topics of the next chapter.

Summary

The biosphere, that part of the earth occupied by living things, is a thin layer sandwiched between the atmosphere and the geosphere. Plants are the basis of life in the biosphere because they transform solar energy to chemical energy.

All living things are composed of matter that cycles through the biosphere and between it and the atmosphere and geosphere, but the earth is a closed system with respect to matter. Matter neither enters nor leaves the earth in appreciable amounts. In contrast, the earth is an open system with respect to energy. Energy comes into the biosphere from the sun; then it is transformed by plants to chemical energy and cycled through the biosphere. In the process it is transformed to heat, which is radiated and reflected back into space, lost to the biosphere. This transformation and loss of energy to living organisms is described by the first and second laws of thermodynamics.

About 40 percent of the sun's radiation is reflected and absorbed by clouds, particles, and gases in the atmosphere. Much of the light that does strike a leaf is reflected, used in the evaporation of water from the leaf's surface, or lost as a result of convection; and plants can transform only visible light, which is a small portion of the light spectrum. Plants commonly use only 1 percent or less of solar radiation.

Photosynthesis is the process by which plants transform light energy to chemical energy. The green pigment chlorophyll absorbs light energy and uses it to split water into oxygen and hydrogen. The oxygen is lost to the atmosphere, but the hydrogen is combined with carbon, which is obtained by plants from carbon dioxide in the atmosphere. The burning of hydrogen-containing compounds in the process called respiration produces the energy required by all organisms.

Organisms that can transform solar energy to chemical energy are called self-feeders, or autotrophs. Only green plants and certain bacteria and blue green algae are autotrophs. Organisms that must consume other organisms to live are called other-feeders, heterotrophs. Animals, fungi, and most microbes are heterotrophs.

The transfer of energy and matter from autotrophs to and between heterotrophs results in food chains. Each transfer is referred to as a trophic (feeding) level. About 90 percent of the energy is lost with each transfer to another trophic level; it is used to maintain the organisms at that level. The first trophic level consists of primary producers (plants), which are consumed by primary consumers. The chain extends to a second-, third-, and sometimes a fourth-level consumer.

Land plants account for much of the earth's primary production, and most of the primary consumption is done by invertebrates, particularly insects, and by decomposers such as bacteria, fungi, and protozoans. In many ecosystems most of the energy and matter captured by plants is recycled by decomposers.

A common and complex form of energy harvesting is symbiosis, the association or relationship of dissimilar organisms. Types of symbiosis include parasitism, commensalism, and mutualism.

Food chains usually interlock, forming a complex tangle called a food web. In fact, the biosphere's transformation and transfer of energy and cycling of matter take place in interdependent systems, called ecosystems, that involve the nonliving as well as the living environment. An ecosystem consists of ecological niches, which include both the space in which an organism lives as well as the means by which it lives.

Yet another way of analyzing relationships in the biosphere is to construct food pyramids, which describe the relationship between the number of organisms or the amount of energy available at each trophic level. Rich and poor nations of the world can be identified by the shapes of their food pyramids.

Review Questions

1. Why is the biosphere described as a closed system with respect to matter and an open system with respect to energy?
2. People and plants have similar problems in collecting solar energy efficiently. Why? What makes it possible for plants to use solar energy as their sole energy source?
3. In the future our society may rely on a "hydrogen economy." Plants already do this. Explain.

4. What is the basic difference between autotrophs and heterotrophs?
5. Contrast food chains and food webs. Why are trophic levels limited?
6. Select an organism familiar to you and define its ecological niche.
7. Weeds and people can be said to have a symbiotic relationship. Why?
8. The corn–cattle–human food pyramid could be altered by feeding cattle only grass and other forage rather than grain. Assuming current consumption of beef did not change, what would such a pyramid look like? What effect would it have on land use?

Chemical Energy: Burning Hydrogen

In a food chain, the energy transferred is chemical; it is stored and transferred in chemical compounds. What this means may be clearer if we look at one molecule, H_2O, and especially at hydrogen.

With a continuous supply of energy, water molecules, H_2O, can be broken up into molecules of H_2 and O_2. For example, if you provide an electric current to water, the longer the current runs, the more H_2 and O_2 you get and the less H_2O is left. The small amount of energy in a spark or a match is sufficient to induce the H_2 and O_2 to recombine forming water. When they do so, energy will be released. Why?

The two hydrogen atoms in an H_2 molecule or the two oxygen atoms in an O_2 molecule are a definite distance from each other, drawn close together by the attractions between electrons and nuclei (see Figure 1), yet

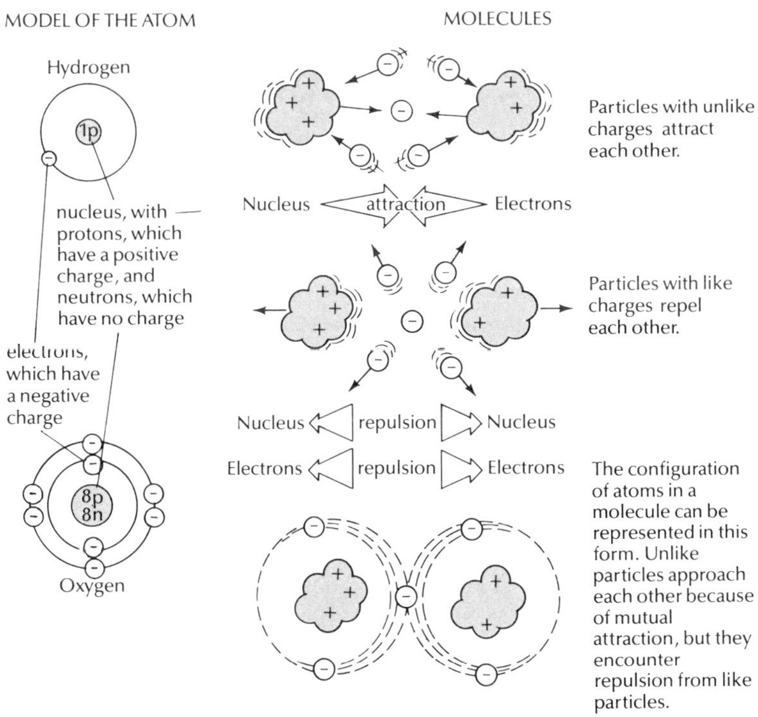

1 Atoms consist of charged and uncharged particles. The forces of attraction and repulsion within and between atoms result in chemical bonds that connect atoms together to form molecules.

held a small distance apart by the repulsion between the two nuclei and between the electrons on each atom. It would take energy to push the two nuclei closer together, because the repulsive forces increase greatly as the nuclei approach each other. And pulling the two atoms apart takes energy (because the electrons and nuclei are attracted to each other), just as it takes energy to pull apart two magnets that are touching each other at opposite poles. Chemists call this tendency of two atoms to stay close to each other a chemical **bond.**

The atoms in the H_2 and O_2 molecules have to separate (their bonds must be broken) before they are rearranged to form water, and this requires some energy. But the energy it takes to begin this reaction is small. A spark is sufficient to disrupt a small fraction of the H-H bonds and O-O bonds, and the energy released by this disruption is enough to disrupt more H_2 and O_2 molecules and ensure that the reaction keeps going, forming H_2O, without any more energy coming from the outside.

To split H_2O into H_2 and O_2, however, energy must be supplied continuously; when they recombine to form water, the same amount of energy is released (for an analogy, see page 5). It is therefore useful to think of H_2O as being at a lower level of energy than H_2 and O_2. Figure 2 shows the relative levels. A "downhill" reaction (such as the formation of water), releases energy; an "uphill" reaction (such as the splitting of water) requires the sustained input of energy.

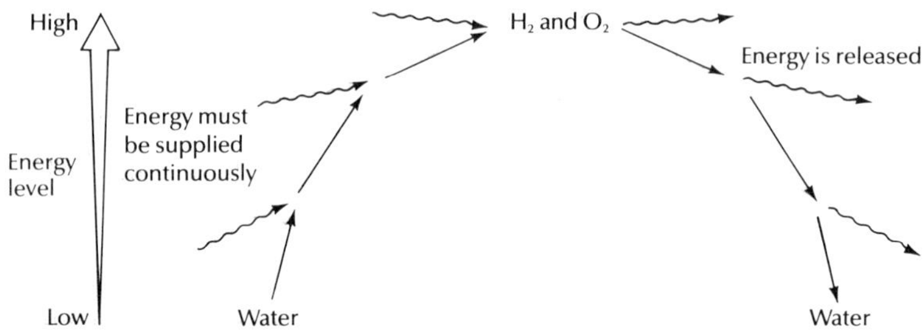

2 Energy relationships in the splitting and regeneration of water.

What properties determine whether a compound will be relatively high-energy or low-energy when compared with other compounds? One major determinant is how easily the atoms in the compound can form strong bonds; that is, how closely nuclei and electrons can approach each other and therefore experience strong mutual attraction. Anything that weakens bonding will raise the compound's relative energy level; the stronger the bonds, the more stable the compound and the lower its relative energy. Three oxygen atoms close together, each having a negative charge and repelling the other two oxygen atoms, help make the molecule shown in Figure 3 a compound with relatively high energy.

A second factor fixing the level of energy is the relative amount of disorder or randomness in the spatial arrangement of atoms in the compound. A jumbled pile of lumber, bricks, pipes, wires, and nails is more disordered or random than a house. In general, energy must be added to turn a random arrangement of things into a more highly ordered arrangement, to turn a pile of building materials into a house. It also takes energy to convert atoms in a relatively random or disordered arrangement into a

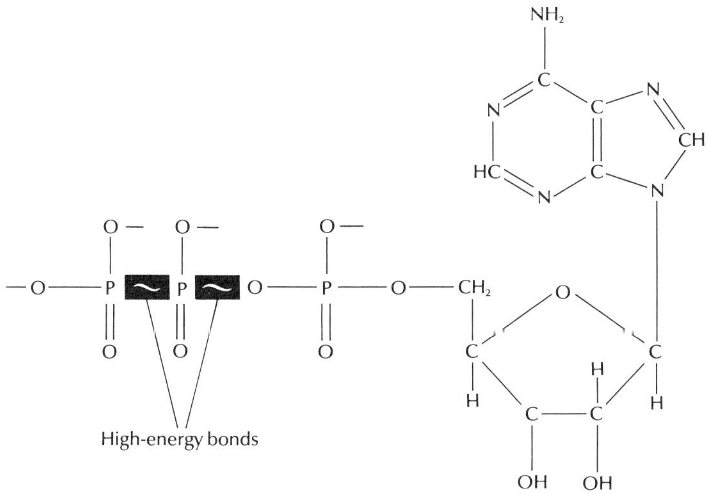

3 Molecular structure of adenosine tri-phosphate (ATP), the energy broker of cells.

more highly ordered arrangement. Energy is required to make this reaction (diagrammed in Figure 4) go forward:

6 carbon dioxide and 6 water molecules	$\rightarrow$	glucose and 6 oxygen

This reaction summarizes what happens in photosynthesis. Just as it takes energy to convert a pile of lumber into a complicated structure like a house

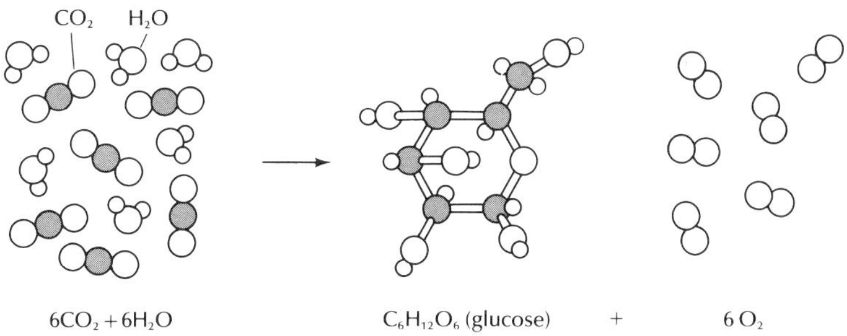

4 The construction of more complex molecules from simple molecules requires energy.

(some energy goes into arranging the lumber and some into connecting the pieces), it takes energy to convert simple building blocks—CO_2 and water—into a relatively complicated molecule, glucose. Plants capture and use sunlight as the energy to run this reaction.

When H_2 and O_2 combine ($2\ H_2 + O_2 = 2H_2O$), disorder decreases because two molecules are produced from three molecules, which results in fewer positions for the molecules in space and hence a more orderly arrangement. The hydrogen-to-oxygen bonds in water, however, are sufficiently stronger than the hydrogen-to-hydrogen and oxygen-to-oxygen bonds in hydrogen and oxygen gas respectively to make water lower in energy, overall. Therefore when H_2 and O_2 combine to form water, energy is released.

Living organisms make use of this relative difference in energy between

5 Photosynthesis uses light energy to make elements into compounds such as carbohydrates. As these compounds are passed along the food chain, the bonds uniting elements are broken, and the energy released is either transformed to sustain life at each trophic level or is given off as heat. At the end of the food chain, the chemical bond energy resulting from photosynthesis is dissipated as heat and the components of the photosynthetic reaction are regenerated.

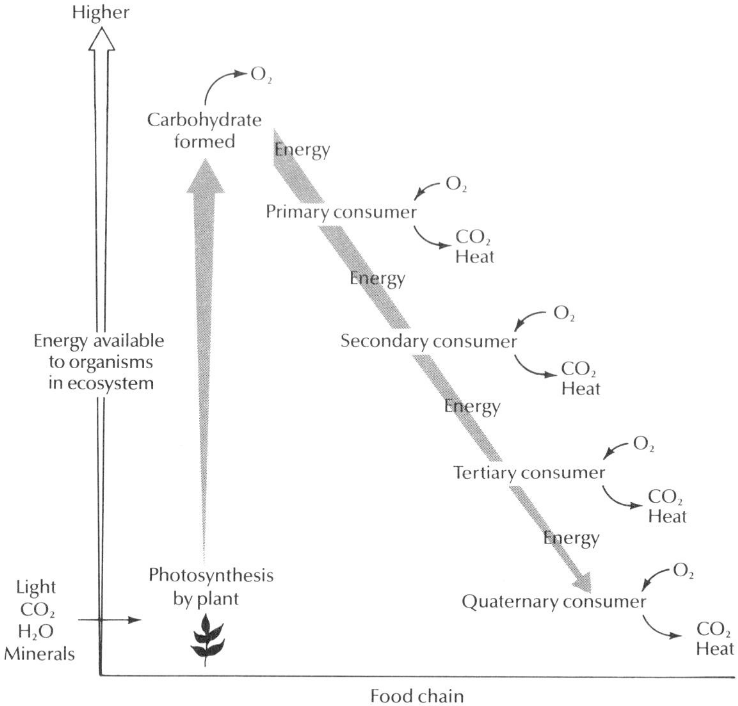

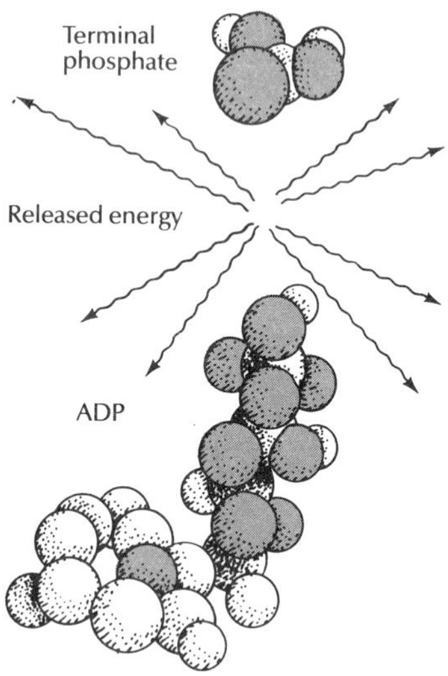

6 When the terminal phosphate group in adenosine tri-phosphate (ATP) is detached by enzymatic action, energy is released as a result of breaking the chemical bond. This energy is used to make other chemical bonds involved in the work of the cell. Photosynthesis and the burning of carbohydrates provide energy to bond the phosphate group with adenosine di-phosphate (ADP), thus creating a storage bank of chemical energy.

water and hydrogen and oxygen. Hydrogen atoms are removed from organic compounds such as the sugar glucose and combined with oxygen from the atmosphere. Water is thus formed and energy released. Carbon dioxide is all that is left of the sugar, once all its hydrogen atoms have been removed. Just as water is at a relatively lower level of energy than H_2 and O_2, so too water and carbon dioxide are at a lower level than glucose and O_2.

Outside the cell, H_2 and O_2 combine with flashing rapidity once a reaction gets started, and great quantities of energy are released almost all at once, as an explosion or a burst of flame. Energy released this fast would be of little use to an organism even if the organism survived the explosion or fire. Instead, cells combine oxygen with the hydrogen from organic molecules in a series of steps, and the energy is released not all at once but piecemeal. Cells use energy released in this way to make an uphill reaction take place: this reaction is the formation of ATP (adenosine tri-phosphate) from ADP (adenosine di-phosphate) and phosphate. When the cell needs energy for any purpose, the ATP splits up into ADP and phosphate again, releasing the same amount of energy that was required to form ATP from ADP and phosphate in the first place (see Figure 6). Thus the ATP acts as a middleman, capturing and transferring the energy originally released when oxygen combined gradually with the hydrogen of organic fuel molecules to form water.

CHAPTER 2

Agricultural Revolutions

Ten thousand years ago plants were involved in the start of a quiet revolution that changed the world. For about three thousand years it went on, but once this new technology thrived, human life was forever altered. No longer wild creatures living in caves and taking their food where they could find it, people had learned how to use plants, and they became civilized. Further revolutions in agriculture followed.

FIRST REVOLUTION: DOMESTICATING PLANTS

It began probably between 10,000 and 8000 B.C. In the Fertile Crescent between the Tigris and Euphrates rivers of what is now Iraq, in the Tehuacán Valley of Mexico, by the Yellow River in modern China, people took the seeds of a wild plant, a grain or cereal no doubt, and cultivated it in primitive gardens. They domesticated plants.

Early humans, before the first agricultural revolution, had subsisted mainly on meat and wild plants such as grains and fruits. Some of their vegetable diet came from the grains they gathered from grass plants, and primitive tools indicate that they ground these into flour, making thin cakes like tortillas to be dried in the sun and then baked, perhaps, on stones heated by fires. Many hunted for meat.

How do we know what people ate all those millennia ago? First of all, we can collect the pollen of plants that lived at a particular time in the past (see Chapter 13). In addition, many substances leave a trace in the fecal matter people excrete. Much of this material disappears, absorbed into the earth. Sometimes, however, it is fossilized in little piles called **coprolites.** Studying these shows us, sometimes very clearly, what people ate in ancient times because traces of indigestible matter in the food, particularly seed and plant fibers, may be found. From these and other fossils like those found in the camp garbage heap, scientists reconstruct prehistory.

The Earliest Agriculture From the work of many scholars, archaeologists, botanists, and other scientists, we can reconstruct with fair accuracy how the plant revolution, at least the one in the Middle East, most probably got under way.

The Fertile Crescent of the Middle East is a land of varied habitats and was an early abode of human beings along the Tigris and Euphrates rivers. To the east lie the rugged Zagros Mountains, unsuitable for permanent settlement but offering good hunting and a cool summer for nomadic bands. West of its foothills, dry and windswept steppes dip gradually to the rich flood plains along the rivers. It seems likely that the base camps of wandering tribes lay in the foothills, but the men and women who lived there went far afield. They dwelt in the fertile bottomlands during the cooler seasons, and in the burning summers they migrated east to the mountains. Whenever they traveled they subsisted on game and the

2.1 The foothills of the Zagros Mountains in modern Iraq are the site of some of the earliest agricultural settlements. The climate and scenery have changed little in the past 8000 years.

edible plants they found. Their food grew wild, each crop maturing in its season, and it was varied and plentiful because of the varied habitats that lay between rivers and hills.

The opportunity, an invitation almost, to bring the sources of food to where they wanted them, rather than always migrating to where the food was, must have gradually become obvious to the nomads as they watched how plants reproduced themselves and grew. Undoubtedly, as the women gathered the wild grain, they brought it back to the encampment, spilling some seeds, throwing some onto a garbage heap after a meal, or otherwise accidentally scattering them near the camp. Some of these seeds, variants hardier than the parent plants, in time would hold and grow. The women, seeing this, may well have reasoned that these plants came from the grain they had brought home, and they might experimentally scratch or dig a few holes in the ground and plant more. Seeing plants thrive when supplied ample water, they would bring water to them, at first in gourds, then in ditches or canals as their planted fields spread. And so, slowly, generations may have come to realize that they could stay in one place and grow the plants they needed by their homes.

The earliest evidence for agriculture is from the Zagros Mountains, but perhaps this is because the climate is dry, preserving remains better than elsewhere, and because many scientists look there for evidence of early farming. But half a world away, in the Tehuacán Valley south of where Mexico City stands today, other people were trying similar experiments with other plants (see Figure 2.2). In Mexico maize (corn) was probably among the first seeds planted.

Another birthplace for agriculture was around the middle reaches of the Yellow River in northern China, the center for the domestication of the cereal grains we call millet. These are annual grasses that are grown in semiarid regions, now a staple food for millions in Asia and Africa.

Scientists also have evidence that **root-crop agriculture** in the humid tropics is very ancient. In this type of agriculture, the starchy underground stems and roots are the food. Root-crop plants are mostly species in three families—taro, cassava, and yams (not our yam, which is a sweet potato). These are easy to raise, have high yields, and are the staple and sometimes only food of people in many parts of the tropics.

Fossil remains are less likely to turn up in the tropics, however, so our record of early agriculture is somewhat biased. Decay is rapid there, and the difficulty of finding sites suitable for exploration in places like the jungles of Southeast Asia has made them less attractive to archaeologists than places like the arid Mideast, where signs of ancient settlements litter the landscape. Thus most of our evidence about early agriculture comes from temperate regions.

2.2 Sites of early agriculture have been studied (a.) in the mountains ringing the Fertile Crescent in the Middle East, (b.) along the Yellow River in China, and (c.) in the Tehuacán region in Mexico. These were the centers for the domestication of barley and wheat, millets, and maize, respectively. Sites in the Middle East are at least 10,000 years old (dating from 8000 B.C.).

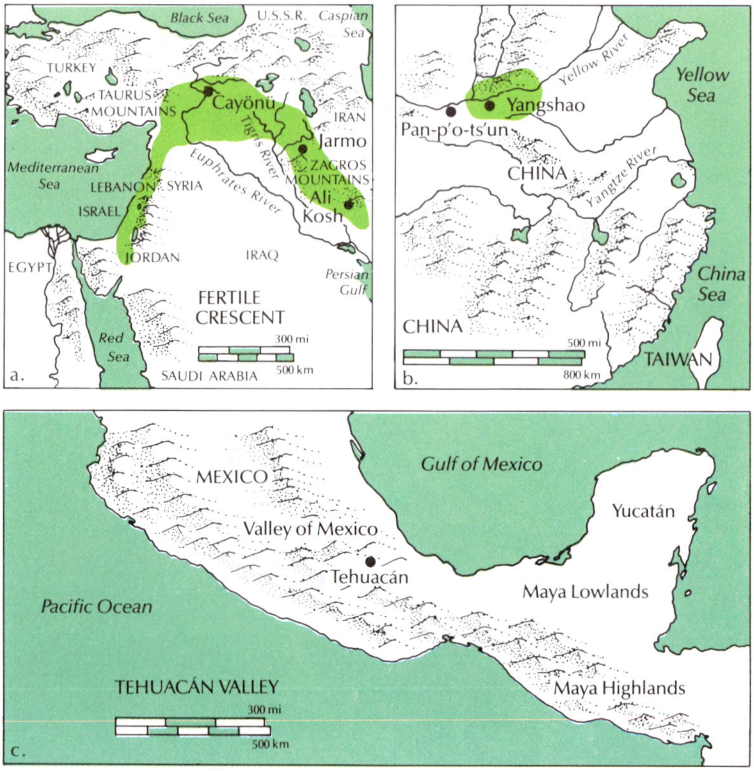

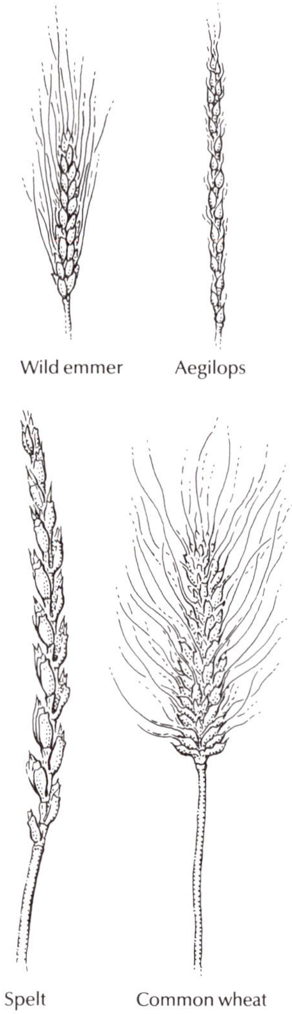

The first **economic plants,** as useful plants are called today, were hardy plants to begin with; they flourished in the new and favorable environment created for them by human care. Competing vegetation was cleaned out so that the planted crops could grow better. As time went by a domesticated crop was regularly grown in cultivated fields, and other plants were added—figs, dates, olives, grapes. Just how these and other plants first cultivated were transformed into their domesticated versions we will never know (see Figure 2.3). Some of the change probably was natural and accidental, some probably was deliberate as cultivators learned more about what they were doing. Either way, our debt to these primitive first farmers is huge. They domesticated most of the food plants we eat today, laying the basis for agriculture as it developed right up to modern times.

2.3 Wheat was domesticated in the Middle East. Spelt was once common in Europe and is a hybrid of the wild grasses emmer and Aegilops. Common wheat is believed to have resulted from a cross between Aegilops and early domesticated wheat.

From Plant Domestication to Civilization By the time the first agricultural evolution had run its course, the Fertile Crescent was producing several types of millet, field peas, lentils, vetch, chick-peas, horsebeans, and flax. Among the fruits were wine grapes, olives, dates, apples, figs, cherries, and pears. In Mexico our familiar tomatoes, peanuts, and lima beans were grown. By 1500 B.C. the wild maize had crossed with other wild grasses, creating new and hardy forms that were cultivated from then on (see Figure 2.4). The Egyptians cultivated cucumbers, artichokes, lentils, garlic, leeks, onions, lettuce, mint, endive, chicory, radishes, and a variety of melons, as well as dates, grapes, olives, figs, bananas, lemons, and pomegranates. Their ability to grow all these crops, however, rested on other changes, part of the wide-ranging social and cultural effects of the domestication of plants.

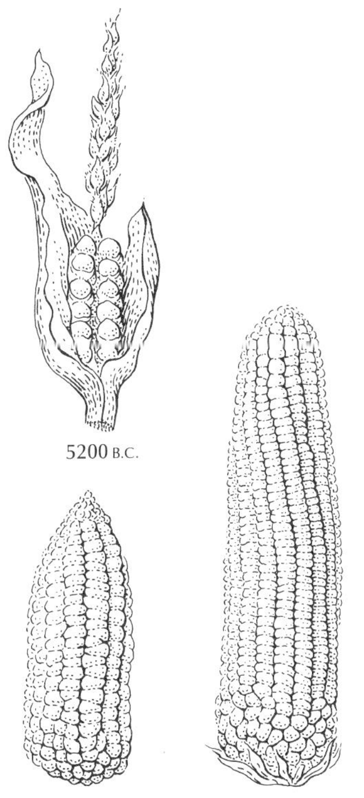

2.4 The extinct wild maize of Mexico had only eight rows of kernels and dates from 5200 B.C. By 500 B.C. ears of maize grown in Mexico were four times longer (10 centimeters, or 4 inches) and had about eleven rows of kernels. A contemporary variety of Mexican maize is 20 centimeters long (8 inches) and has seventeen rows of kernels.

Before plants were domesticated, people were hunters and gatherers, having to move to find enough food to survive. But if they grew their food, they could settle down. When people stop wandering, their society and their culture change radically (see Figure 2.7 on page 36). No society can develop further than its food supply allows, and now populations could increase. They probably devised ways of treating diseases; they sharpened their hunting skills and produced better weapons, lessening the dangers of the hunt. There was more time to think now, to observe, and as people studied the world around them, they began to understand something of it, increasing their efficiency still more. Huntsmen lost their first place in the tribes as a new priestly class, armed with knowledge beyond that of most men, came to power.

One of the activities for which these early people now had more time was creating art, and on the walls of caves, in frescoes on their buildings, and scratched into fragments of bone, we find records of what they did and how they lived. Religion flourished, too, and with it the construction of religious edifices: temples, monuments, and elaborate grave sites. The rising priestly class brought an entirely new group into their social structure, men who wielded power because of their knowledge. And with religion, festivals appeared and the times of sowing seeds and harvesting crops were given solemn significance in complex rituals designed to placate and appeal to the gods. Half a world apart, these events took similar courses in Mexico and the Middle East: in both places the occult sciences flourished, astrology above all, as we can see in constructions like the Aztec temples in Mexico and the monumental pyramids of Egypt, oriented to the heavens.

The Egyptians also formed the technological base for a productive agriculture: preparation of the land, including plowing with oxen; irrigation by ditches and canals; and pruning. Besides the food crops already mentioned, in order to serve various industries, they grew castor beans, date palms, and papyrus, from which they produced

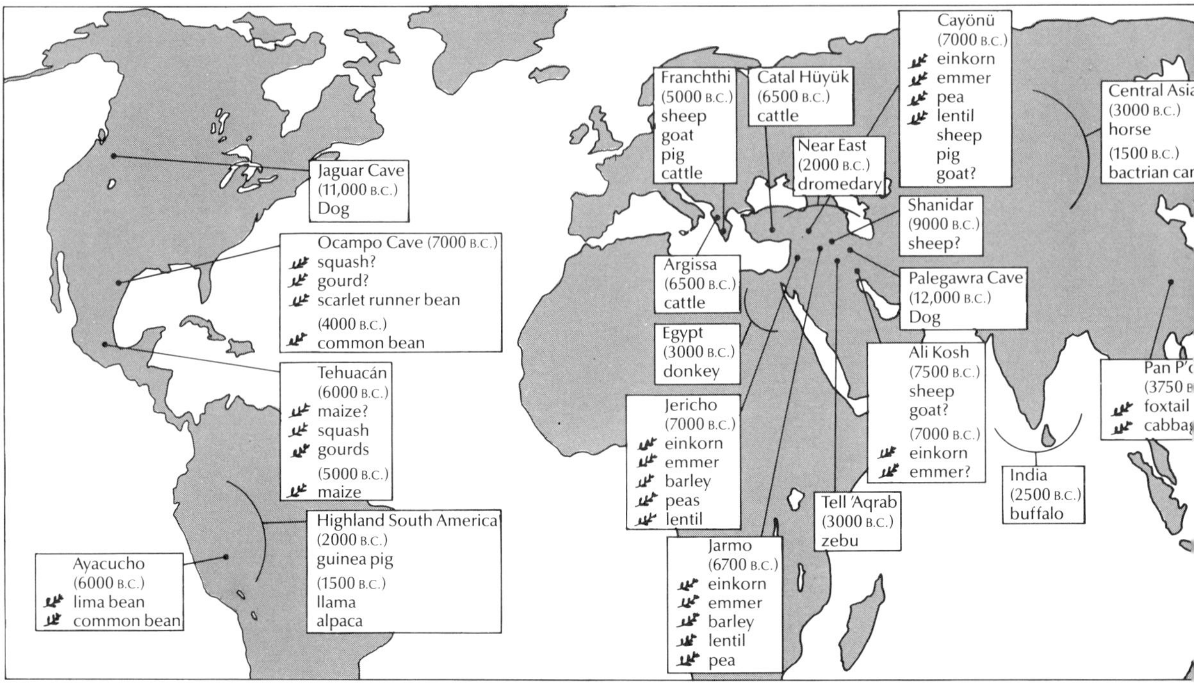

2.5 There were many early centers for the domestication of plants and animals. The Middle East has yielded much information, partially as a result of the relative ease and long period of exploration and the dry climate favoring preservation of artifacts. The Middle East is also the location of the oldest sites to date. Missing from this map are sites for domestication of tropical root crops; far less is known about tropical root crops than about crops from more temperate zones.

oil, fiber, and a form of paper for written records. They also pioneered in baking bread and fermenting beer and wine, drying food, and making pottery to cook it in. In Mexico the people changed character and produced the highly advanced civilizations whose traces we find today in the ruins of the Mayans and the Aztecs.

Studying Plants Another phase of this plant revolution was launched around 1000 B.C. in Egypt. People systematically examined plants in an attempt to understand them, and devised ways by which useful plants might be improved. In a very primitive way, this was the beginning of the scientific study of plants, which today we call *botany;* it has continued to this day and will certainly go on far into the future.

By the time these things started happening, the world was already heavily dependent on cultivated plants. In ancient Mesopotamia some 16,000 square kilometers (10,000 square miles) were under cultivation, feeding a dense population with a sophisticated irrigation system built with bricks whose joints were sealed with asphalt. The search for medicinal herbs was advancing; a list compiled in Egypt can be seen today in carvings on the temple walls at Karnak. The Egyptians now had a spice and perfume industry, and they were pioneers in the art that was to become an obsession everywhere: planting and care of stylized, formal gardens.

2.6 About 3400 years ago, Egyptian pharaoh Thutmose III ordered carvings of many plants on the walls of his temple at Karnak. This panel contains many carvings of pomegranates, perhaps the first depiction of this fruit in Egypt.

It was at this time, too, that special-purpose plants began to be dispersed throughout the world. Flax was one of the first: cultivated in Egypt as a source of textile products, it spread to Europe, where it was grown as far north as Scandinavia and the British Isles. In Greece, Aristotle (384–322 B.C.) wrote about plants, and some of his records have survived. His most famous student, Theophrastus of Eresos, studied classification, propagation, forestry and horticulture, geographic botany, pharmacology, viticulture (grape growing), plant disease, savors, and odors. For good reason he is known as "Father of Botany." His ideas influenced the science for many centuries and inspired such men of ancient Rome as Pliny the Elder, Cato, and Columella, who left to posterity a large work on botanical subjects. By the time that Christ was born, plants had been studied quite extensively.

In the Middle East and Egypt, agriculture had flourished for thousands of years by the time of the Roman Empire. The Romans colonized North Africa and grew good crops there. Except for the banks of major rivers like the Nile, much of this region is now desert. What happened? The intensive agriculture had been sustained by sophisticated irrigation and probably by wise management of the land. Systems of irrigation crumbled in war; prosperity ended in the Fertile Crescent when the irrigation works on the Tigris and Euphrates rivers were destroyed by Mongol hordes in the mid-

2.7 PLANTS AND EARLY CIVILIZATION

These Sumerian clay tablets (below, left) are among the earliest known written records (3000 B.C.). They depict ears of grain, and the circles and impressions probably indicate numerical tabulations of the crops. The map on the 1300 B.C. clay tablet (below, right) is of the fields and irrigation system of the ancient city of Nippur in modern Iraq. Each field is identified with the owner's name. The large "V" and branches from it are irrigation canals, and the field within the narrow portion of the "V" belongs to the king. An Indian-Spanish artist made drawings (facing page, top) of the seasonal activity of Inca farmers in the 1580's as a gift to the king of Spain. The August drawing depicts the Inca emperor and members of the court in a symbolic ceremony in which they dig holes in a sacred field with digging sticks. Women bow, and the empress offers corn beer to the men. In the drawing representing September, a farmer digs holes with a digging stick and a woman plants corn seeds. Inca women did the planting because of their symbolic association with fertility. Indian farmers of the Western Hemisphere made holes in the ground for seed rather than turning the soil with plows, which was typical of farming in Eurasia. Plowless planting, increasingly used in U.S. agriculture, is similar in many respects to this ancient practice of the digging stick. Agriculture was a highly developed science in Egypt, and yearly flooding by the Nile River guaranteed good harvests. This scene of a nobleman and his wife annointing a harvest of produce, bread, and meat (facing page, bottom) is based on tomb paintings. The scenes at the right illustrate gathering, threshing, and winnowing of grain.

dle of the thirteenth century. Waterways in many places filled with silt following deforestation and overgrazing. Fields in some regions were lost to cultivation because waterborne salts accumulated by years of irrigation poisoned the soils. (Think of the white crust that forms on the surface soil in flowerpots after months of watering.) Agriculture in the cradles of civilization fell into a long sleep; later agricultural revolutions were made in Western Europe and the New World.

Pause in Progress Throughout the Dark Ages following the Fall of Rome, agriculture was looked upon as a lowly occupation, despite the obvious reality that no fortified town, no knight's castle, no monastery or learned monks could exist without it. The one place in Europe where plants were systematically cultivated and developed was the monastery garden. In the quiet and seclusion behind their walls, medieval monks grew the plants they needed for food and bred the flowers they cherished as ornaments and offerings to God. Many kept detailed records of what they grew and learned, and the gardener (*Hortulanus* or *Gardinarius*) became a respected officer of the monastic hierarchy. Here vegetable and fruit strains were preserved, and techniques such as grafting and pruning were improved.

European civilization progressed little in plant technology for several hundred years. From the Fall of Rome to the Renaissance in Italy, the science of agriculture and horticulture scarcely advanced beyond techniques that Greeks, Romans, and Egyptians had known and practiced. Most of the material written in these years was based on Roman scholar Pliny's *Natural History* and was as full of errors in observation and interpretation as that work.

Slowly, in the thirteenth and fourteenth centuries, the restricted feudal life in fortified castles and towns gave way to growing commerce as trading caravans wound their way from place to place, finally linking even such distant places as the Middle East and northern Europe. In this more open atmosphere, seeds began to travel from place to place. Meat became a bigger part of the human diet, increasing the demand for condiments and spices and spreading herbs throughout the civilized world. Plants of all kinds traveled far from their origin, becoming familiar in places that had never known them. The science of botany, too, began to flourish, and much that was learned was set down in books like the *Maison Rustique,* a French work published in the sixteenth century by Charles Estienne. In it we read of practices such as fertilization, grafting, pruning, breeding, dwarfing, transplanting, control of insects, girdling fruit trees to promote their flowering, new methods of harvesting, processing, and cooking, and ways of using plants as medicines.

SECOND REVOLUTION: PLANT MIGRATIONS AND AGRICULTURAL IMPROVEMENTS

The fifteenth and sixteenth centuries were an age of exploration, when ships built and equipped by such farsighted monarchs as Prince Henry the Navigator of Portugal ventured out on the uncharted oceans to discover whole new continents. Plants now were spread worldwide (see Figure 2.8). Many new ones—maize, beans, tomatoes, and potatoes—were introduced to Europe by explorers who touched American soil. Some, like the potato and tomato, were then reexported to the Far East, which in turn supplied such flavorings as cumin, turmeric, black pepper, and that instant best-seller, tea. As trade in these plant commodities spread and flourished, plants began to influence the world economy and even politics, and nations competed fiercely for trade routes and markets. Transplanting the rubber tree from Brazil to Asia via England was to affect relationships of nations right down to the present.

Soon followed the age of colonization. As settlers migrated from continent to continent across the seas, plants went with them, and they sent new plants back to the mother country. Sugar came to the New World and became an economic staple in the West Indies.

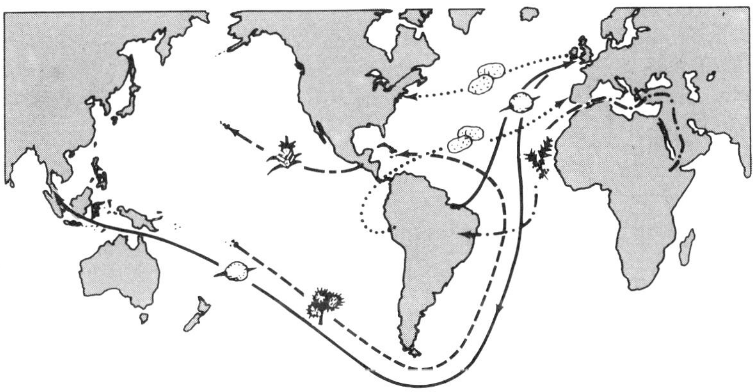

2.8 Production centers for many important economic plants are located far from each plant's original home.

— *Rubber* seeds, smuggled out of Brazil in 1875 by the British, became the basis for the rubber industries of Ceylon, Malaya, and Indonesia.

– – *Breadfruit* was discovered in the South Pacific in 1595 and later brought to the West Indies by Captain Bligh to feed slaves.

—-*Pineapples* originated in tropical America and were exported to other tropical areas.

.... *Potatoes* were taken by the Spanish from South America to Europe in the middle of the sixteenth century. They were reintroduced to the Western Hemisphere and North America by the middle of the seventeenth century.

.—. *Coffee* was spread from Africa to Asia to the New World, where most of today's coffee is grown.

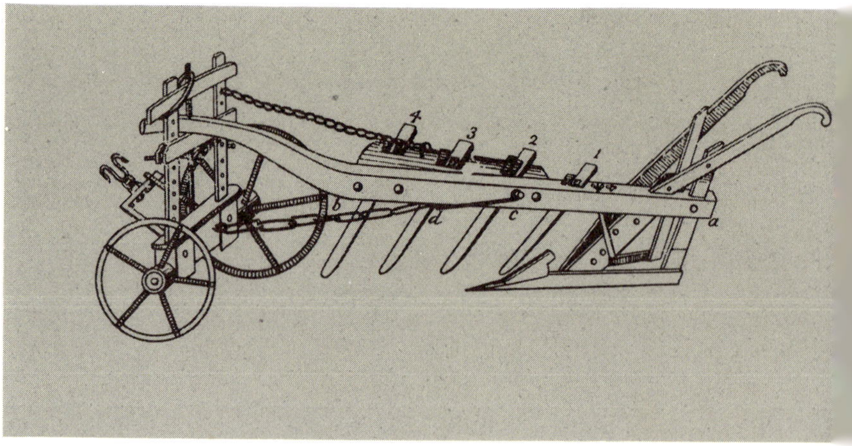

2.9 Agricultural technology developed slowly over most of the ten thousand years of plant domestication. An ancient Egyptian peasant, an Inca farmer, and a European peasant of the Middle Ages had much in common. This technology began to change rapidly in the eighteenth century and the rate of change continues to accelerate. The Aztecs constructed a sophisticated system of plant beds called "chinampas" (top, left) by digging canals through marshes and stacking the excavated material and topsoil between retaining walls. The willow trees at the edges stabilized the chinampas, which resembled the floating gardens that can be seen today near Mexico City. Terracing is as old as agriculture and permits the cultivation of steep slopes. Terraces in Machu Pichu, Peru, (top, right) were built by the Incas about 1440 A.D. Some terraces, like those in Pisac, Peru, are still being used today. The mural of agricultural practice in ancient Egypt (right) demonstrates that still-used techniques for planting, reaping, and sowing grain were well established thousands of years ago. The eighteenth century saw many new agricultural practices. Jethro Tull, a well-known innovator and inventor, developed the deep-cutting steel plow (bottom, right).

Pineapples migrated from Central America to Asia, Africa, and the South Pacific. One attempt at wholesale transfer of plants to a new habitat was the voyage of the British Navy's ship *Bounty*, sent to Tahiti to load a cargo of a thousand young breadfruit plants that could be grown as cheap food for slaves in the Caribbean Islands. The *Bounty's* crew interrupted the botanical experiment with mutiny, as we know from the classic book by Charles Nordhoff and James Hall, as well as the popular movie renditions.

Some of these introductions of plants to new environments did not have their full effect until many years later. Strawberries, native to California and eastern North America were not domesticated in Europe until the nineteenth century. The tomato, which some thought poisonous, did not come into its own until early in this century, by which time it had spread all across the globe. Soybeans were imported to Europe from China in the seventeenth century and then into the United States at the beginning of the nineteenth century. They have been an important crop here only since the end of World War II. But the hunt for new varieties of useful plants has never slackened, especially now as a growing world population demands more and more food and fiber. Intensive searches are also being made for plants useful to medicine.

Down through all the centuries since human beings planted the first crop, some kinds of food plants have always been more valued by one population than by others. These are the staple foods that are a people's basic diet. They may be plants native to the area, or they may have been imported and then proved indispensable. History shows how and why each of these has pushed the world's agricultural food supply to its enormous economic importance, which we will examine further in Chapter 5.

Improvements in agriculture since the Renaissance show a long slow rise through the sixteenth, seventeenth, and eighteenth centuries. Tribes of native Americans knew as much about irrigation in pre-Columbian times as Europeans did in the eighteenth century; the Incas in Peru built terraces as efficient as any in the Alps or the Apennines of later centuries, and it is probable that many Western Hemisphere Indians knew as much about the structure and function of the plant as most Old World scientists until Robert Hooke and others laid bare the structure of plant cells late in the seventeenth century by using microscopes. In all that time there was no great incentive to improve farming methods or add to knowledge handed down by peasant fathers to their sons. Populations were still small; the great, hungry, sprawling cities had not yet evolved; and few understood exhaustion of the soil by overcultivation.

But by late in the eighteenth century, pressures were beginning to be felt. A few large cities were asserting demands for food, colonization was opening vast new areas of the world to human habitation and plant cultivation, and new farms and plantations were growing

2.10 Grain harvesting machines such as McCormick's reaper could cut, rake, and bind the grain faster, more cheaply, and more efficiently than the hand harvesting crews they replaced. Harvesting machines made possible wheat production on the vast prairies of the West.

beyond one family's ability to till the soil, sow the seeds, and harvest the crop. Improved varieties of farm animals, better cultivation, and an expanding labor market all contributed to new agricultural practices, particularly where the Netherlands and Belgium, England, and Scotland are today. These changes in agriculture were precursors to the revolution that really shook the world: the Industrial Revolution of the nineteenth century.

THIRD REVOLUTION: MACHINES REPLACE ANIMALS

At this time agricultural efficiency took a sudden leap. Through the nineteenth century, both science and industry helped make possible improvements in methods of preparing the soil, planting, caring for, and reaping crops. Machines like Cyrus McCormick's reaper, invented in 1831, enabled one man with a horse to do the work of many, and in less time. Eli Whitney's cotton gin, which separated the lint from the seeds in the cotton fields of the American South greatly increased the production of raw cotton simply by speeding up this one time-consuming step in getting cotton from the fields to the textile mills. Almost instantly pressure was put on the fields to produce more cotton faster.

The mobile steam engine appeared. Not only did a wide-flung network of railroads suddenly provide far easier access to markets, but steam engines came to the fields themselves—in tractors that could operate faster, more efficiently, and more cheaply than draft animals. The internal combustion engine also appeared, rendering steam tractors obsolete by its power, mobility, and speed. The combine was invented, to reap, thresh, and haul grain all in one operation, pouring its load into trucks that could drive into fields where the combine worked and from there haul grain to the railroad. From a small-time operation occupying families and a few hired laborers, farming became a big business requiring large investments, with commensurate high profits luring the farmer on.

Science grew in close parallel to the Industrial Revolution. In 1840 Justus von Liebig, a German, formulated his **law of the minimum,** the principle explaining exhaustion of the soil. The yield of any crop, said Liebig, is irreversibly limited by the minimum quantity of any essential mineral element (such as nitrogen, potassium, or phosphorous) in the soil. Research on fertilizers suddenly grew urgent as the implications in Liebig's findings hit home. In midcentury the discovery was made that plant diseases may be caused by specific organisms invading a crop plant. All of a sudden the farmer could take active steps to protect the crop instead of annually invoking the aid of the gods to keep plants healthy.

In the 1860s Gregor Mendel published discoveries about the laws governing inheritance of characteristics in plants, on which the science of genetics was to be built (Chapter 11). He found that units of inheritance—we now call them **genes**—were transmitted from one generation to another in the male and female gametes or sex cells, of plants. No one grasped the full significance of Mendel's work in his own time, but many years later botanists and zoologists were able to produce desired traits in plants and animals by systematic and selective crossbreeding.

FOURTH REVOLUTION: GENETICS AND CHEMISTRY

The United States could not have today's high standard of living without both rich soil and favorable climate, which are ours mostly by chance, and high technology (see Figure 2.11). These advantages have helped make agriculture our most important industry, just as it is in much of the rest of the world. It is clear that the need for food and fiber grows along with population, urbanization, and flourishing technology. Without the fourth revolution in agriculture, we would be even further from meeting these needs.

Improving on Nature When Mendel's work finally bore fruit early in this century, agriculture was revolutionized once again. Primitive people and plants themselves—by trial and error or

2.11 AMERICAN AGRICULTURE: RESOURCES AND TECHNOLOGY

The world's main croplands occupy only about 10 percent of the earth's surface (below). When combined with pasture lands, this amounts to only 30 percent of the globe's surface. Europe and the eastern United States are well endowed with rich land. The great river basins of India and Southeast Asia are the main crop areas of the tropics. The fertile soil of the United States is in the valleys of the Ohio and Mississippi rivers. Much of the West is mountainous or arid plains and desert (bottom). The amount of rain varies greatly across the country, and this moisture gradient is the primary determinant of plant distribution. Ample moisture and rich soil are the combination which makes the Midwest the world's breadbasket. The plains of Kansas, the Dakotas, eastern Colorado, and eastern Washington produce most of our wheat. This combine harvesting wheat in southeastern Washington (right) typifies the scale and mechanization of U.S. agriculture. The dramatic increase in agricultural productivity in the United States since 1945 is illustrated by the graph (facing page, bottom left). At the turn of the century, one farmer grew only enough food to support five others. Now our food is grown by only 5 percent of the population. Hybrid crops, mechanization, new types of nitrogen fertilizers, the wide use of new herbicides and pesticides, and the availability of cheap energy in the form of fossil fuels account for much of this productivity. This increase in productivity is seen in one crop, corn (facing page, right). The great harvests of hybrid corn require mechanized equipment and the hybrids respond to increased fertilizer application.

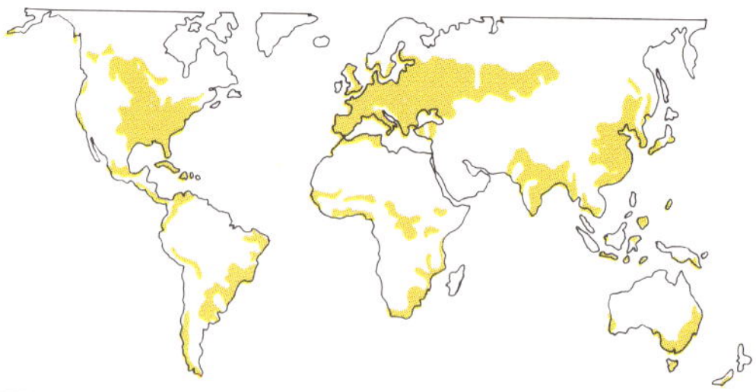

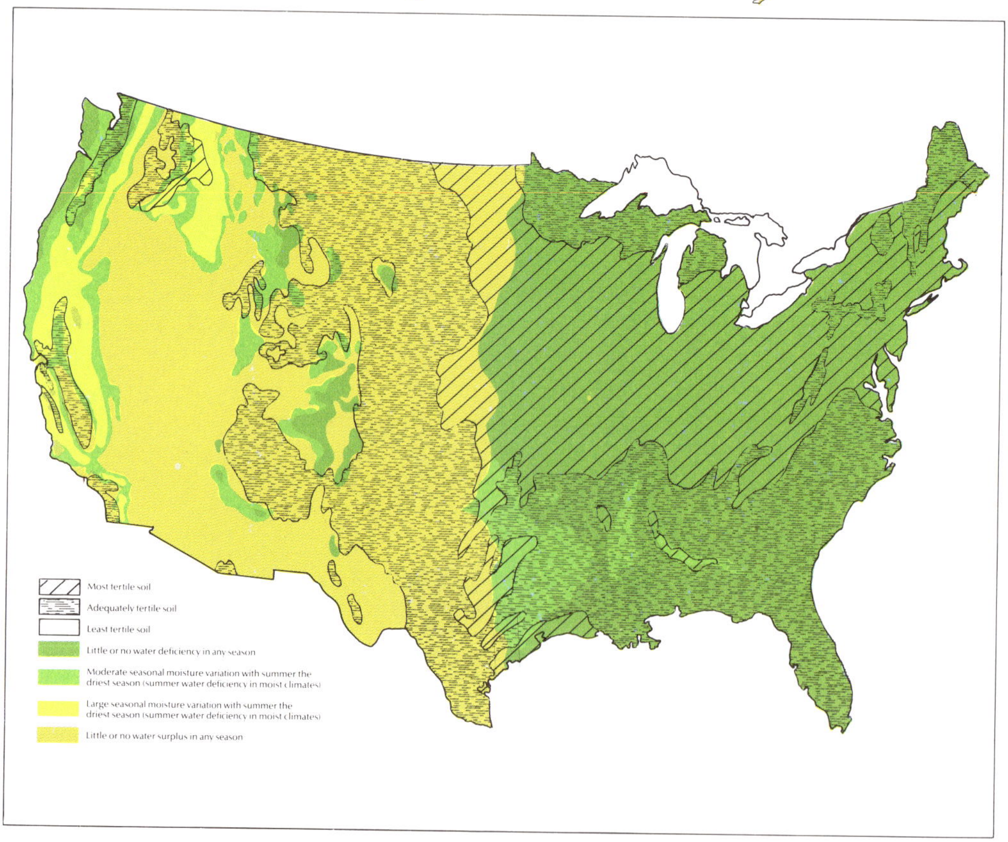

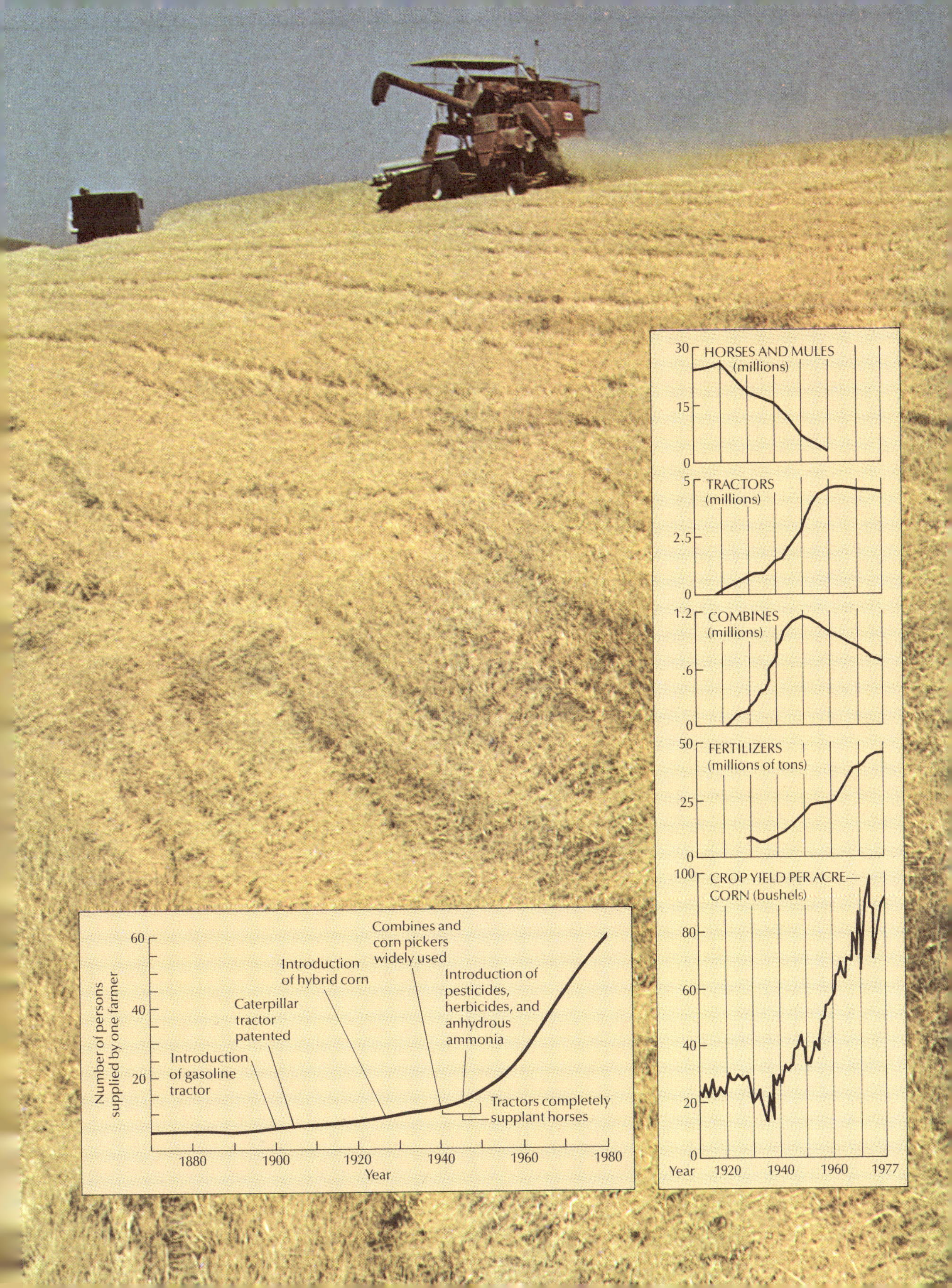
HORSES AND MULES
(millions)
30
15
0
TRACTORS
(millions)
5
2.5
0
COMBINES
(millions)
1.2
.6
0
FERTILIZERS
(millions of tons)
50
25
0
CROP YIELD PER ACRE—
CORN (bushels)
100
80
60
40
20
0
Year
1920
1940
1960
1977
Number of persons supplied by one farmer
60
40
20
Introduction of gasoline tractor
Caterpillar tractor patented
Introduction of hybrid corn
Combines and corn pickers widely used
Introduction of pesticides, herbicides, and anhydrous ammonia
Tractors completely supplant horses
1880
1900
1920
1940
1960
1980
Year

by chance—had **hybridized** ("interbred") the first wild grains millennia ago; now it could be undertaken scientifically, and the result was hybrid plants of many varieties with much greater yield and resistence to pests and disease. Ultimately, hybrid corn plants would be bred that were not only hardier and more disease resistant than their ancestors but yielded 25 percent more corn at harvest time.

In the 1930s hybrid corn started a "green revolution" in the United States compared to which agricultural efforts of only a decade earlier seemed primitive. Hybrid wheats and corn grew such staggering harvests that it began to seem possible for America to feed the world. Plant breeding affected other crop plants no less profoundly. Whereas crop failures caused by disease had been common throughout the history of agriculture, such wholesale disasters as the Irish potato famine and the catastrophic American wheat failures before the 1940s were now less likely because resistance to disease could be bred into plants. New diseases, when they did invade, could be combated by developing resistant strains of crop plants, which is a continuous project in America, particularly for wheat and corn.

Giving Plants Nutrients Liebig's mid-nineteenth-century studies on soil exhaustion led nearly a century later to great discoveries about plant chemistry and plant nutrition. In this century a new field, mineral nutrition, opened up, and it became apparent how important **trace elements**—chemicals present in very small quantities—are in feeding plants. Soil lacking zinc, for example, was found to cause growth defects in plants. A chain of discoveries followed, and finally chemical fertilizers could be precisely tailored to soils.

One mineral that plants require in large amounts is nitrogen, a deficiency of which is one cause of soil exhaustion. Nitrogen-containing fertilizers have always been in short supply. Nitrogen is so necessary to agriculture that the guano trade of the late eighteenth century was supported entirely by the accumulated droppings (guano) of huge populations of sea birds on a few islands off the coast of Chile. The mountains of guano accumulated through untold millennia were mined and shipped in sailing vessels halfway around the world to be used in wheat and corn fields.

During World War I nitrogen-containing fertilizers and explosives were in critically short supply in Germany. Under pressure to provide substitutes, German chemists discovered how to make them synthetically by combining atmospheric nitrogen (N_2) with hydrogen from natural gas (CH_4) to make ammonia (NH_3); the process is known as the Haber-Bosch method. A whole new industry was formed, producing large amounts of cheap nitrogen, later to help in revitalizing exhausted soils throughout the world. As always, a chain of events followed. The higher-yielding hybrid crop plants needed greater amounts of nitrogen, which meant using more natural gas to produce it. And today high-priced natural gas increases the cost of fertilizer, which is finally reflected in more costly food all over the world.

Controlling Weeds and Pests Another offspring of the new science of plant chemistry was the herbicides or weed killers used increasingly after World War II. They are more and more popular around the world because they decrease the labor and cost of removing weeds. Weed killers grew out of research on plant **hormones** (substances secreted in one part of an organism that stimulate growth in another; see Chapter 7). Synthesized plant hormones are also important to agriculture. Chemical compounds can be applied to delay or speed up ripening of fruits, to maintain dormancy in such staple foods as potatoes so that they can be safely stored for long periods, and to increase sugar storage in sugar cane. Chemical inducement of flowering in some plants is also a major goal in chemical plant research (see Chapter 7).

Throughout agricultural history insects have been a plague and not infrequently a catastrophe. Birds and rodents, too, have always taken their toll. India, one of the most populous countries, would be much closer to self-sufficiency in food production if it could reduce the 25 percent loss of foodstuffs to animal pests after the harvest. All agricultural countries profited, therefore, by the effective and selective pesticides discovered during and after World War II. Since 1945 far less food has been lost to pests throughout Europe and America and somewhat less in other food-producing countries. It is now known that indiscriminate use of pesticides may influence ecological balances; yet pesticides are applied increasingly and are hard to dispense with in food production.

But pesticides, too, have helped swell the cost of living everywhere. Like herbicides, they are synthesized from fossil fuels, which, as they grow scarcer and more expensive, inevitably push costs of the food basket higher.

The effects of indiscriminate use of chemical pesticides like DDT, widely publicized as in Rachel Carson's *A Silent Spring,* have turned research in recent years to seek other ways of controlling pests. One of the most promising is **biological control,** which is particularly desirable where insect pests have been imported but their natural predators have been left behind. Biological pest control can work in several ways:

1. Predators may be introduced to keep pest populations under natural control, like lady beetles and lacewings, which prey on aphids in many crops (see Table 2.1). Such parasites as the calcid wasp can be used to control a number of major pest species.

2. Normal breeding of an insect population can be drastically reduced by introducing large numbers of sterile males. In some insects the likelihood of fertilization decreases after the first mating. Mating with sterile males, therefore, will markedly decrease the production of fertilized eggs and should reduce the population.

3. Hormones may be used to regulate insect life cycles or lure them to traps. Juvenile hormones prevent some species from completing their life cycles by inhibiting metamorphosis from larval to adult stages.

4. Specific weeds can be controlled by inducing insect hordes to eat them.

Some very promising results have been achieved with biological pest control. In California alone about $300 million in crops have been saved by biological control projects. There is still much to be done, and it may be that the ultimate method will mix biological and chemical controls, a combination that economic entomologists call **integrated pest control.**

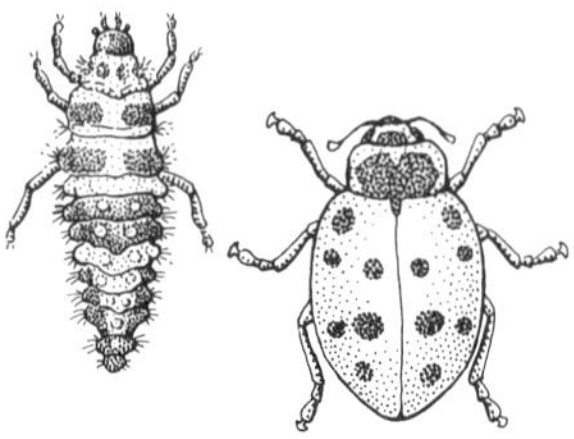

Ladybug (larva and adult)

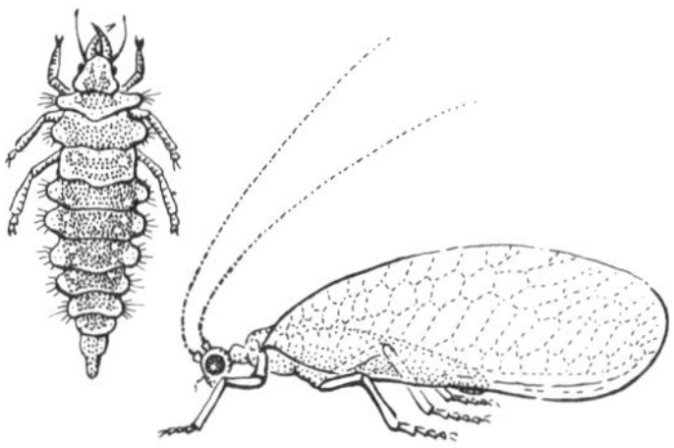

Lacewing (larva and adult)

Table 2.1 Examples of Natural Controls

Natural Control	Victims
LADYBUGS Also known as the lady beetle and ladybird, the ladybug can be any of hundreds of different beetles, the most familiar of which is the spotted *Hippodamia convergens* illustrated here. Even in the larval stage, they devour as many as twenty or thirty aphids a day apiece; ladybugs eat twice as much when they are adults. They are most active in hot weather and will reproduce several times in a single growing season if there are enough pests to feed them all.	Aphids Leaf worms Mealy bugs Mites Scale insects
LACEWING LARVAE The adult lacewing is about ½ to ¾ inch long with gauzelike wings and large glittering eyes. It is the little larvae, called aphid lions, that are most useful as pest killers. In their two-week larval period they feed almost constantly on aphids and other insects—including other lacewing larvae—by seizing them in tong-shaped jaws and sucking out their body fluids. As adults, only a few varieties of lacewings continue to eat insects. The rest subsist on aphid honeydew and flower nectar.	Aphids Leaf hoppers Mites Thrips White flies

PLANTS IN OUR FUTURE

One lesson appears in all these efforts to revolutionize agriculture: we must struggle to live on a planet whose resources are diminishing under pressure from constantly growing populations. For a dramatic proof compare medicine and botany. Medical research in the last hundred years has enabled humanity to stamp out or control a number of diseases that have periodically cut down human populations. Bubonic plague, the black plague of medieval times, has all but disappeared. Smallpox, another wholesale killer, is thought to be totally eradicated. Sanitary engineering and medicine have brought cholera, typhoid, and typhus under effective control in most of the world. The two-edged DDT has just about wiped out malaria in tropical areas where once it was an ever-present menace, though the disease may be struggling toward rebirth. Poliomyelitis, or infantile paralysis, is no longer cause for dread among growing children or adults. And the list could go on and on.

Natural Control	Victims
PRAYING MANTISES A relative of the grasshopper, the praying mantis gets its name from the way it sits with front legs held up as if in prayer. Actually the powerful legs are used to seize and hold insects. The mantis is a predator from birth, feeding almost immediately on aphids and small flies. The adults, up to five inches long in some species, are indiscriminate gluttons that tackle the largest insects—including each other—and have been known to attack frogs and lizards.	Aphids Beetles Caterpillars
MILKY DISEASE This slow-working bacterial disease attacks the larvae of Japanese beetles, turning their normally clear blood to milky white and halting their development into adult beetles. Death may take weeks or months. The disease is introduced by a dust containing spores of either *Bacillus popilliae* or *Bacillus lentimorbus* mixed with talc and chalk to prevent the spores from being killed by sunlight.	Larvae of Japanese beetles

Praying mantis

Milky disease

Source: Adapted from James Underwood Crockett and the Editors of Time-Life Books, *Perennials*, page 53.

2.12 This scene in Zaire is repeated periodically in the less developed world. Obtaining enough calories, let alone a balanced diet, is the basic goal of many in the world, and it is probable that increasing numbers will participate in this struggle.

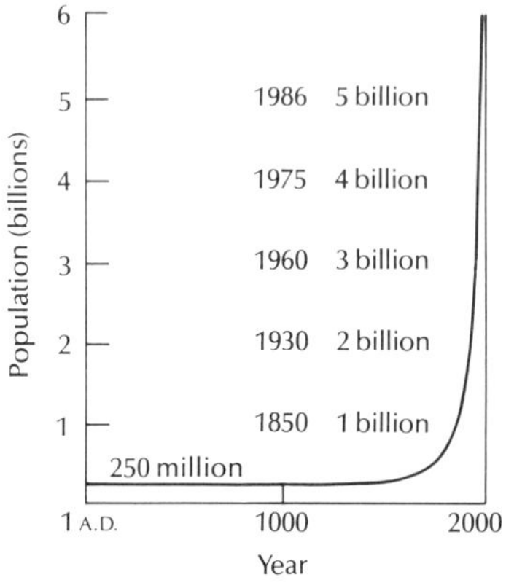

2.13 The increase in the world's population was barely detectable until 300 years ago. The increase in this century alone has been spectacular. Everything we know about populations of organisms states that the slope of this curve must soon flatten out. This has happened in industrialized nations, but when and how it will happen in the rest of the world is not known.

Many of these things came about because of the explosive increase in medical research and knowledge in the nineteenth century. And because the human death rate was so sharply controlled in so short a time, the world's human population has increased explosively. Today, considered as a world population, humans live on the brink of starvation. Why? One reason is that medical research has so far outstripped botanical research that the productivity of the plant world has trouble keeping up with the continuing expansion of the human population (see Figure 2.13).

Research in plant chemistry, plant genetics, and the economies of crop plants has made it possible to speak theoretically of a green revolution that could sweep the world's agricultural countries, increasing production of food crops so greatly that world food supplies could keep up with expanding population. The green revolution really means exporting the genetic and chemical revolutions from the highly developed, industrial nations of North America and Western Europe to the less developed nations of India, Asia, Africa, South America (and, agriculturally, the Soviet Union), which account for the greatest increases in population but at the same time are the least advanced in methods of increasing agricultural production.

Theoretically the present production of food throughout the world could be increased by about twenty times, which is more than enough to offset any food shortages now foreseeable. That is theory, however. The practical problems in the way of achieving this enor-

Table 2.2 Increase in Food Production, Population, and per Capita Food Production, 1954–73

	Total Increase 1954–73 (percent)	Annual Rate of Increase (percent)
FOOD PRODUCTION		
World	69	2.8
Developed countries	65	2.7
Developing countries	75	3.0
POPULATION		
World	44	2.0
Developed countries	22	1.0
Developing countries	61	2.5
PER CAPITA FOOD PRODUCTION		
World	17	0.8
Developed countries	33	1.5
Developing countries	8	0.4

Source: From "The World Food Situation and Prospects to 1985," *Foreign Economic Report*; 98 (U.S. Department of Agriculture), page 12.

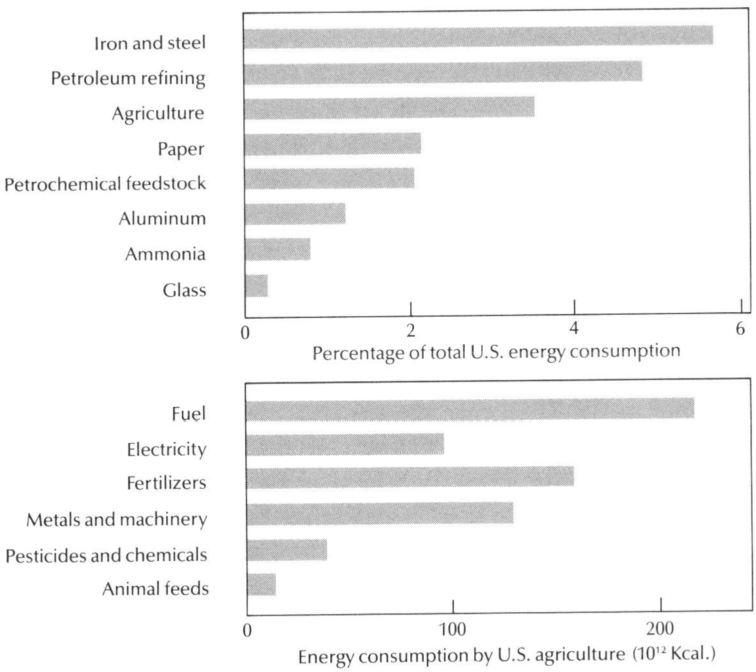

2.14 Energy expenditure for agriculture in the United States is small when compared with transportation (25 percent) and space heating (18 percent). Savings in fuel consumption and less costly methods for producing nitrogen fertilizer would significantly affect energy expenditures for agriculture.

mous increase in food production are immense, to be solved only by researchers and food producers and by the nations themselves. And we must see to it that knowledge gained is not left to the few who are already far ahead in the race to increase their production of food but is made available to all.

First, agriculture must depend less on energy. Yields have often been increased by heavy expenditures of energy, but energy from fossil fuels is growing scarce. It is absolutely necessary, therefore, that we either use less energy to produce food or improve agricultural yields with the same amount of energy.

How can this saving be achieved? Research has a lot to do with part of the answer. Research is needed, for one thing, on ways to produce nitrogenous fertilizers. By the end of this century, 200×10^6 metric tons of fertilizer will be required annually throughout the world to feed an anticipated population of some 7 billion people. The present world production, thanks mainly to the Haber-Bosch method developed in Germany, is 40×10^6 metric tons, with a value of some $8 billion. The explosive demand for nitrogenous fertilizers may be judged when we consider that just twenty-five years ago the world got along with 3.5×10^6 metric tons. How such an increase can be achieved no one can say.

Two agricultural innovations are noticeably changing the American landscape and increasing crop yields: **center-pivot irrigation** and **plowless planting** (see Figure 2.15). In the first a long pipe with sprinkler nozzles is mounted on wheels and attached at one end to a water supply. The pipe rotates around the central axis and irrigates all the soil inside the circle. The circles usually cover 133 acres but some units can irrigate a circle a mile in diameter. The unit rotates automatically, eliminating the labor usually needed for irrigation. The water supplied to fields can be carefully regulated, and fertilizers and weed killers can also be included in the irrigation water. The rate of water flow, the concentration of fertilizer, and the speed of rotation can be controlled automatically and suited to the climate and to the age and type of crop. Sandy soil, which does not hold enough water and is not normally cultivated, can produce excellent crops using this type of irrigation with its uniform and repeated application of small amounts of water. The view from an airplane over western farmland has changed greatly in the last few years, from the familiar pattern of squares and rectangles to the circles of center-pivot irrigation.

Even greater technological change is coming to planting and cultivating, the traditional tilling of soil. Crops are being planted without plowing the soil. Traditionally, ground would be plowed and remains of the previous crop turned under. Then harrowing would break the clumps of soil, and when the crop was planted, the fields were tilled several times more to destroy weeds. This sequence required many miles of tractor travel, using a lot of fuel and labor.

2.15 Minimum tillage techniques are being used with increasing frequency. Harvesting and planting (foreground) are going on simultaneously in this field (left). Shallow furrows are scratched in the stubble-covered field, fertilizer and seeds are deposited, and the furrows are covered by the wheels in the back. This method requires fewer tractor miles and minimizes soil erosion and water evaporation. Center pivot irrigation (below) is changing the pattern of the rural landscape. The self-moving pipe radiates from a central water source and provides the right amount of water and fertilizer for the type of soil and kind and age of crop.

Table 2.3 Present Population and Cultivated Land on Each Continent, Compared with Potentially Arable Land

Continent	Population in 1965 (millions of persons)	Potentially Arable Land (billions of acres)	Acres of Cultivated[a] Land per Person	Ratio of Cultivated[a] to Potentially Arable Land (percent)
Africa	310	1.81	1.3	22
Asia	1855	1.55	0.7	83
Australia and New Zealand	14	0.38	2.9	2
Europe	445	0.43	0.9	88
North America	255	1.15	2.3	51
South America	197	1.68	1.0	11
U.S.S.R.	234	0.88	2.4	64
Total	3310	7.88		

Source: From Paul R. Ehrlich and Anne H. Ehrlich, *Population, Resources, Environment: Issues in Human Ecology*, 2d ed., Table 5-2. Reprinted by permission of W. H. Freeman and Company. Copyright © 1972. After the President's Science Advisory Committee, 1967, *The World Food Problem*, vol. II, Table 7-9.

[a]Arable land and land under permanent crops. The definition of cultivated area used here includes land under crops, temporary fallow, temporary meadows, for mowing or pasture, market and kitchen gardens, fruit trees, vines, shrubs, and rubber plantations. Within this definition there are said to be wide variations among reporting countries. The land actually harvested during any particular year is about one-half to two-thirds of the total cultivated land.

The exposed soil was susceptible to erosion and loss of water by runoff, and the surface dried quickly.

Plowless planting, in contrast, is done by a machine that makes a small furrow, applies fertilizer, plants the seed, and covers it, all in one operation. The new crop, planted as the former crop is harvested, grows up among the stalks or straw of the previous crop, which act as a mulch. Weeds are controlled by herbicides instead of by tilling with a cultivator. Plowless planting uses only 25 percent as much fuel and can increase yields. It also saves labor, soil, and water. Thirty years from now 50 percent or more of our cropland may use plowless planting and most of our crops will be grown with much less tilling. Perhaps you will have to take your grandchildren to a museum to see a plow.

Another world problem is the loss of **arable** land—fit for crop growing—to cities. Many new developments at the edges of today's

Table 2.4 Estimate of Relative Percentages of Feed Nutrients Converted to Edible Products of Animal Species

Animal Product	Energy Conversion (percent)	Protein Conversion (percent)	Gross Edible Product Output as Percent of Feed Intake
Milk	20	30	90
Beef	8	15	10
Lamb	6	10	7
Pork	15	20	30
Eggs	15	20	33
Chicken (broilers)	10	25	45
Turkey	10	20	29

Source: From R. E. Hodgson, "Place of Animals in World Agriculture," *Journal of Dairy Science* 54, no. 3, page 442. Reprinted by permission.

huge sprawling cities destroy arable land. Indiscriminate development means loss of potentially high-yielding acreage for grain or vegetable crops. As an example, much of California's Santa Clara Valley, extremely rich agricultural land, has fallen victim to urban sprawl. The world population must learn to treasure its crop-growing land as one of its most vital natural resources and stop squandering it for the comforts of modern living. We must consider the possibility that in years to come we may have to expand the arable land on this planet four times, which is possible in theory if urbanization can be controlled (see Table 2.3).

To solve our critical food supply problem, we must take several steps. First, we must alter our place on the food chain. For most of the world's population in the less developed countries, eating less red meat would not change the diet greatly. But for beef eaters in the more affluent nations like those in the United States, Canada, Europe, Argentina, New Zealand, and Australia, the change would be drastic. We have seen the enormous waste in eating beef, a loss of about 99.98 percent between the energy beamed to earth by the sun and the energy derived by one of us eating a steak. This waste of energy can be decreased by moving down the food chain and eating plants or by using more efficient converters of energy such as hogs, fish, and chicken. Also, the amount of grain-fed meat will diminish in favor of grass-fed meat. Most food that cattle and sheep eat is forage like grass, clover, and alfalfa, which cannot be digested by people. Converting this unavailable matter and energy to meat and dairy products is an excellent way to use this vast source of energy.

2.16 This chart shows why potatoes and range-fed beef will be a major factor in our future diets, while shrimp from distant places, tuna fish, and tender, well-marbled beef will be luxuries.

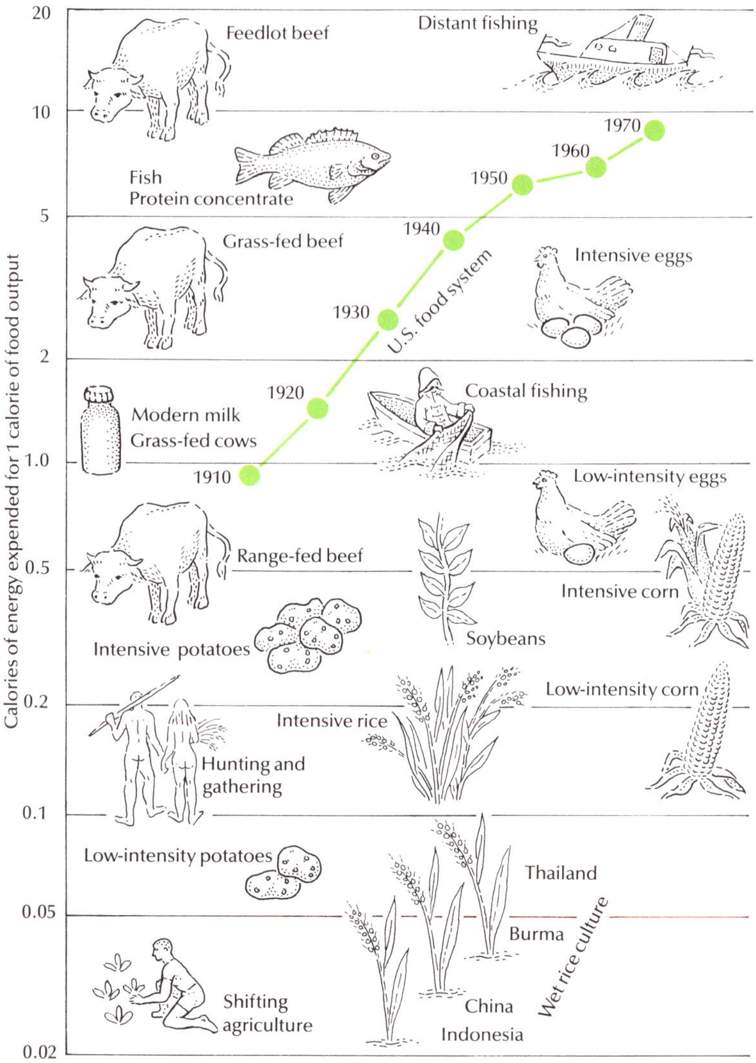

We must improve plant varieties for both higher yields and greater resistance to animal pests and diseases (see Figure 2.18). This is one of the more promising possibilities, and research is booming.

We must try to alter the plant's **metabolism**—all the chemical processes that supply energy and assimilate new material—to increase its efficiency as a converter of energy. Under natural conditions plants convert less than 2 percent of the sun's energy into chemical energy. Theoretically, making photosynthesis more efficient could increase the product substantially. The **respiration rate** of plants—the rate at which chemical compounds are broken down to release energy to do work—might also be decreased. Respiration results in the loss of CO_2, and the effect is less food for primary

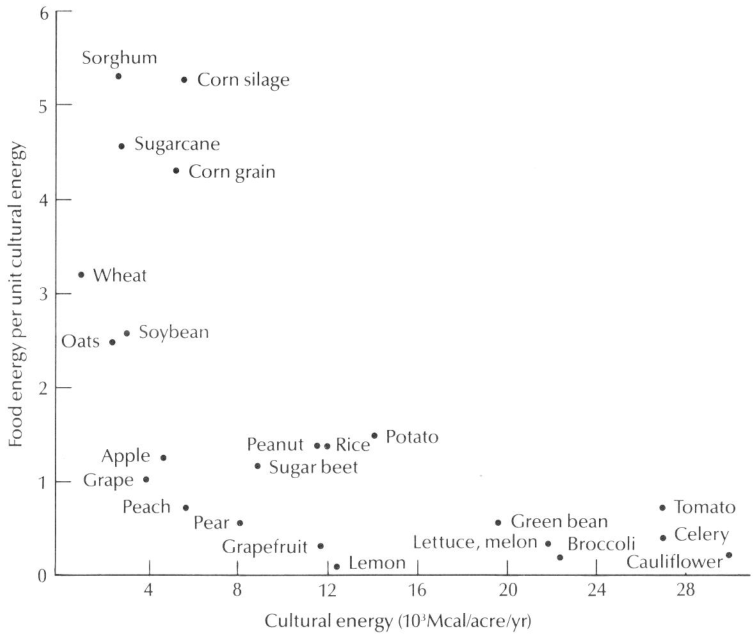

2.17 Vegetables will be expensive as long as energy is expensive, and bread and sugar will be cheap in comparison. See Figure 2.14 to understand why vegetables will increase in price every time there is an increase in the price of oil.

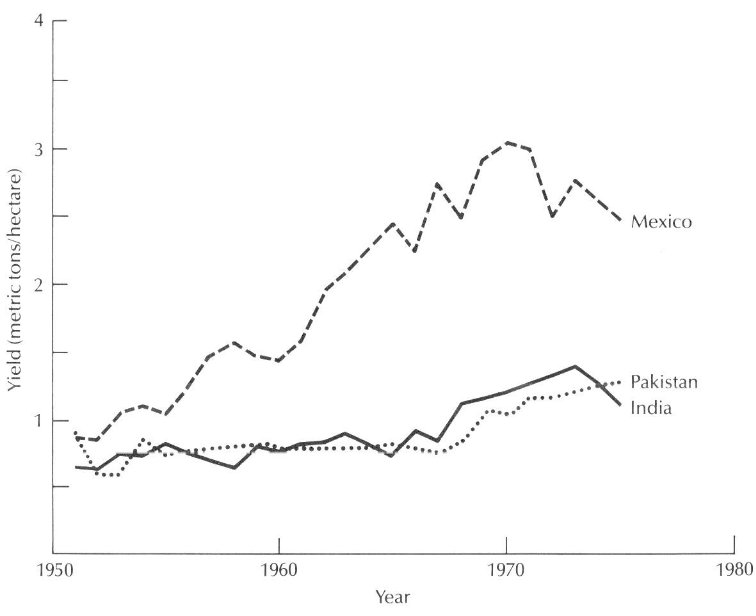

2.18 High-yielding wheat varieties developed in Mexico resulted in a tripling of wheat production in that country. Introduction of these varieties into India and Pakistan also resulted in substantial gains. Decreased yields in India were due to the failure of the monsoon, but several years of good weather have resulted in increased wheat production.

consumers. Still another possibility is to fix nitrogen more efficiently.

Research may lead us down strange and fascinating paths. The leaves of plants are still a vast, almost completely untapped source of energy: few are now harvested as food for people. We do have a machine (called the "mechanical cow") that can harvest and convert the leaves of any plant to edible protein. Like soybean protein, it can be added to other foods, or it can be flavored and textured and eaten directly.

Plants might also be used some day as a source of energy for the petrochemical industry. All our fossil fuels are derived from plants that hundreds of thousands of years ago were covered over, compressed, and eventually fossilized by the tremendous forces that shaped our present world. This trapped energy today drives our cars, heats our homes, propels our ships and trains, and fuels our industry. The same plant hydrocarbons that we now use in fossilized form might be put to work directly to increase our supplies of energy and to feed our chemical industry. For example, plant hydrocarbons can be converted to alcohol by microbes and used directly as fuel or to synthesize a great variety of chemicals.

These are a few of the fascinations and challenges of botany today. Studying plants and their relationships to each other, their structure, their chemistry, and their efficiency as converters of energy is not just an interesting thing to do. It is one of the most vital fields of effort, critical to continued human survival.

All the revolutionary changes in agriculture were achieved by research. Even those long-ago primitive peoples, observing the cause and effect of scattering seeds, were conducting applied botanical research. Present and future feeding of the world's peoples demands vigorous and well-supported basic and applied research in the plant sciences.

The alternative is clear. For the time being we have succeeded in controlling the awesome destructive powers of the atom. We have not yet controlled the equally awesome, if far less dramatic, threat of starvation. Starvation is now and potentially the lot of vast numbers of people on this planet, a challenge to every plant scientist. If we do not avert it, then T. S. Eliot will indeed have proven to have been prophetic:

> This is the way the world ends,
> Not with a bang, but a whimper.

Summary

The first agricultural revolution is known to have taken place at about the same time in different parts of the world. Important centers were the Fertile Crescent in modern Iraq, parts of China, and central Mexico. The earliest evidence for domesticated plants, dating from 10,000 B.C., comes from sites in the Zagros Mountains in Iraq, but there is evidence that root-crop agriculture of the humid tropics is also very ancient. Wheat was domesticated in the Middle East, millet in China, corn in Central America, and yams and cassava in the humid tropics. By 3000 B.C. most of our domesticated plants were widely cultivated. The domestication of plants provided leisure and wealth, influenced religion, stimulated science, and created new technologies.

Agriculture in the Middle East and North Africa became stagnant in post-Roman times as a result of deforestation and overgrazing, the accumulation of salt in the soil, and disruption of irrigation systems by warfare. During most of the Middle Ages, European agriculture varied little from that of classical times.

The second agricultural revolution started in the sixteenth century and was induced by voyages of exploration and the discovery and dissemination of new crops. Botany as a science developed during this period. Improved varieties of crop plants and domesticated animals as well as improved farming methods led to better crop yields, which meant that a larger population could be fed, allowing for the labor force required for the Industrial Revolution.

The third revolution in agriculture resulted from the industrial revolution, when machines supplemented and eventually replaced the muscle power of people and beasts of burden. The fossil fuel that powered steam and, eventually, internal combustion engines replaced the plant food required to fuel animal power, so more of the crop could be used to produce meat or to feed people directly. A new understanding of the mineral nutrients required by plants led to increased use of fertilizer and, after World War I, to the synthesis and availability of nitrogen fertilizer. Botanical research provided new insights into plant pathogens and provided the basis for the science of plant breeding.

Application of new knowledge in genetics and chemistry produced the fourth and latest agricultural revolution. Hybrid corn was introduced in 1926 and within a few years America's corn belt developed the capacity to function as the world's breadbasket. Scientific plant and animal breeding led to dramatic increases in the agricultural productivity of the Western world. The introduction of herbicides, pesticides, antibiotics, and new fertilizer technology after World War II changed our agriculture and permitted a small rural population to produce food and fiber for an urban society. This new technology is now being exported to other parts of the globe.

Contemporary agriculture, mechanized and chemically based, depends on petroleum, which is expensive. Extensive use of pesticides can be detrimental to the environment, so a major area of research concerns biological control of pests, such as the introduction of predators and interference with the life cycles of pests.

Continuing supplies of plentiful and moderately priced food and fiber will depend on research of the genetics and metabolism of plants and pests, new and less energy-intensive methods for synthesizing nitrogen fertilizer, and new methods of irrigation and cultivation. We have the technology to increase food production and keep pace with population increases for a generation or two, but the long-term race with starvation will be a close one. To win we must increase our basic understanding of plants as well as develop the social and political skills for putting new knowledge into practice on a world-wide scale.

Review Questions

1. Where were the first centers of plant domestication? What basic food plants were associated with these centers?
2. How did the voyages of discovery stimulate agriculture and the study of plants?
3. During the first half of the nineteenth century, the United States was a rural community consisting of self-sustaining farms. By the end of the century, the rural population was migrating to cities. How did agricultural developments influence this migration?
4. A "green revolution" resulting in greatly increased yields started in the United States in the 1930s and accelerated after World War II. What were the main factors in this "revolution"?
5. What is meant by biological control? Provide some examples.
6. Outline several areas of agricultural research that will be emphasized during the next twenty years.

Foods

Our domesticated plants are a product of many centuries of human intervention, and many species are more or less dependent on human care. We of course depend on these plants to harvest the sun's energy and to provide many nutrients, drugs, and fiber. Plants are particularly important because they provide **food** (technically, any compound that an organism uses to get energy or to make or repair its tissues) for us and other animals. Our bodies and our food consist mainly of three kinds of compounds: carbohydrates, fats, and proteins (see Table 1).

The plant makes the simple carbohydrate glucose as a product of photosynthesis. With the energy it thus obtains and with minerals it gets from the soil, it can synthesize all compounds it needs—other sugars, amino acids, proteins, fats, and so on. Plants combine glucose molecules into the more complex carbohydrates cellulose, and starch. By taking in and then decomposing these substances (as well as plant protein), animals obtain much of their energy, which they in turn often store in fats.

We need not only energy but also particular nutrients such as amino acids, vitamins, and minerals. But animals are unable to synthesize many compounds essential to their lives; they must obtain these in their diet. Perhaps most important of these are amino acids—nitrogen-containing compounds that are linked together to form proteins (see Figure 2). Given

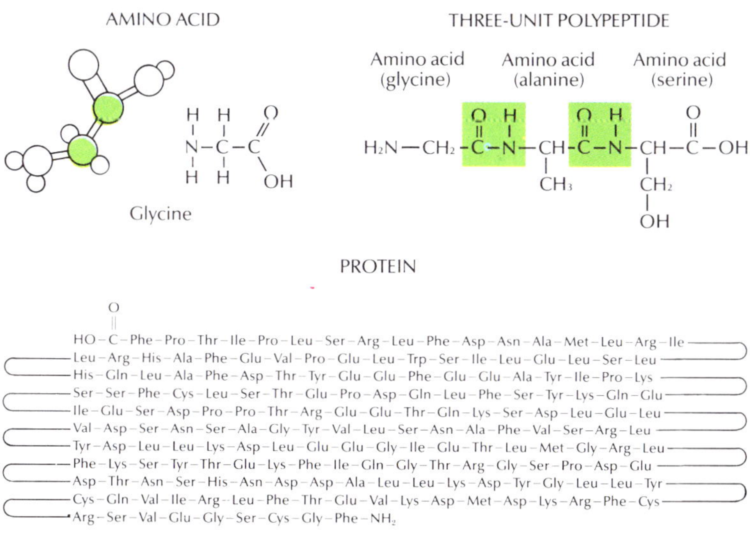

1 Molecular structures of a carbohydrate (glucose) and a fat.

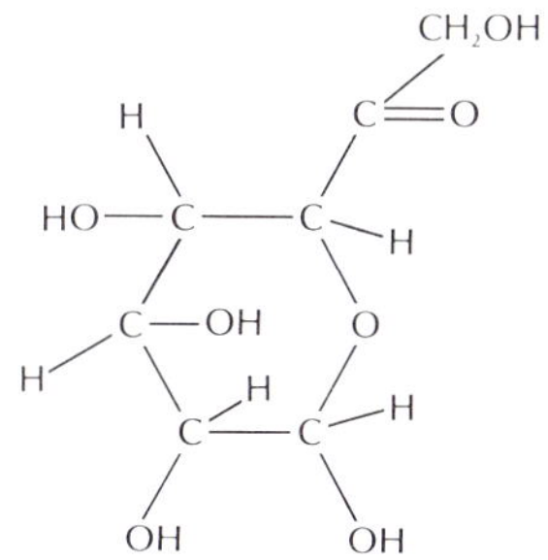

A carbohydrate (glucose)

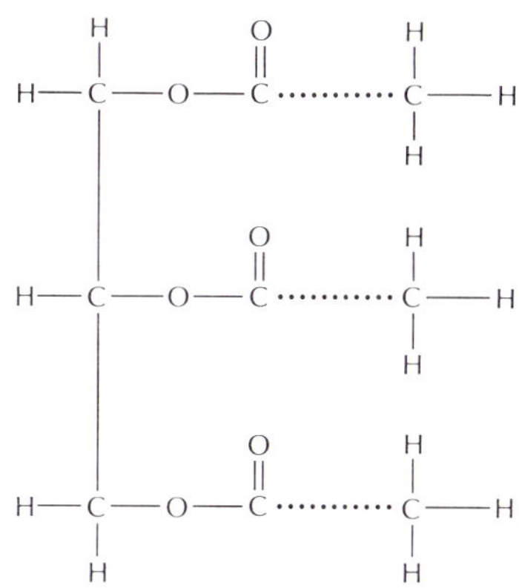

A fat

2 Proteins are built of nitrogen-containing amino acids. Strings of amino acids are called polypeptides and proteins are constructed of folded and interconnected polypeptides.

Table 1 Characteristics of the Principal Classes of Foods

Food	Description	Main Functions in Plants	Main Functions in Humans
Carbohydrates	Compounds containing carbon, hydrogen, and oxygen, with two hydrogens for every one oxygen (CH_2O). Important kinds of carbohydrates are: *Sugars:* Simple sugars are known as monosaccharides. Examples are the 6-carbon sugar glucose and fructose. Monosaccharides may be joined to form 12-carbon disaccharides, such as sucrose and lactose (milk sugar). *Polysaccharides:* Composed of many joined monosaccharides. Examples are cellulose and starch, which are formed of strings of glucose molecules.	Supply energy; furnish the carbon skeleton from which fats and proteins are made. Cellulose builds the plant's structural framework. Starch is the carbohydrate most commonly serving as a reserve food.	Supply energy; furnish the carbon skeleton from which fats and proteins are made. Stored in the liver and muscle as glycogen.
Fats	Compounds that, like the carbohydrates, contain carbon, hydrogen, and oxygen, but with a high hydrogen/oxygen ratio, which is why fats are light and float on water.	A primary component of all cell membranes. Stored as reserve foods, especially in seeds and fruits.	A primary component of all cell membranes. Stored as reserve food in various body regions.
Proteins	Very large, complex molecules made up of chains of amino acids.	A primary component of all membranes. All enzymes are proteins. A storage compound in some seeds, especially legumes.	Provides not only the basis of the cell's living substance but also much of its structure. Skin, hair, and muscle tissue are largely protein; much of our bones are too. All enzymes are proteins.

simple carbohydrates and combined nitrogen, plants can synthesize all the amino acids that make up their proteins, but humans can make only about half the required amino acids. To get the rest we take in proteins in our diet, our digestive systems break down the proteins into their component amino acids, and then the amino acids are recombined into the particular proteins humans need.

The proteins found in plants, however, often either lack an adequate amount of a needed amino acid or are difficult for our digestive systems to handle. As a result, we obtain much protein from other animals in the form of meat, dairy products, and eggs. The animals have already synthesized proteins that contain amino acids in the proportions which we too need, and their protein is easily digestible. But, as we saw in Chapter 1, energy is meanwhile "lost" in the food chain. On the other hand, humans as well as many other animals cannot assimilate much plant material; herbivores convert food we cannot use into food that we can.

From plants we also obtain minerals such as calcium (from dark green vegetables) for bone growth, iron (from whole grains and leafy vegetables) for healthy blood, potassium (from fruits) for our nervous system. Animals, too, supply many of these nutritional needs.

Both plants and animals require **vitamins,** relatively small organic molecules essential for the proper functioning of living things. The function of most vitamins is to promote biochemical reactions. Unlike carbohydrates, fats, and proteins, which are used in large quantities, vitamins are needed in very small amounts. Only plants can synthesize most vitamins, however. If a person's diet is deficient in one or another vitamin, a disease may result: for example, scurvy from the lack of vitamin C (ascorbic acid), beri-beri from the lack of vitamin B_1 (thiamine), or pellagra from the lack of niacin. We obtain most vitamins directly from plants. Plants, however, have little or no vitamin B_{12}, a deficiency of which can lead to pernicious anemia. This vitamin must be obtained from animal products such as milk.

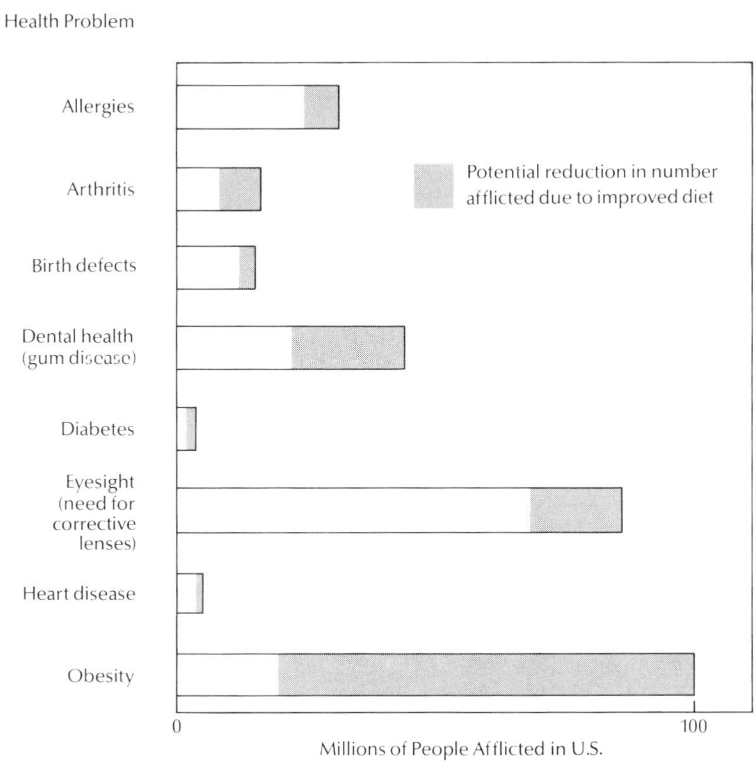

3 This graph indicates how major health problems of the United States population can be decreased by a better diet. An improved diet means a balanced diet of the basic food groups and an appropriate caloric intake. The number of people afflicted with each problem is known, but the effect of diet on each problem is an estimate.

Kinds and

PART II

Uses of Plants

Once we can see the plant world around us as something more than green wallpaper—a pleasant but unobtrusive backdrop for the events of daily life—its incredible diversity becomes striking. How is order to be found in this apparent chaos? How are these many and varied plants related?

To answer these questions Chapter 3 begins with a look at classification, a basic concept of biology, which provides an introduction to the kingdom of plants, their structures and "life-styles." In all this diversity flowering plants are the most familiar and most important to people, and the most numerous. In what ways are they alike? Not in where they live; their habitats are richly varied. However, they do have similarities in basic structure, as we will see. Although flowering plants are emphasized here and throughout the book, in Chapter 4 we look too at plants without flowers, especially those you are most likely to see—pines and fir trees, ferns and mosses—as well as the less familiar algae and, briefly, some organisms that are neither plant nor animal. If these short descriptions fire your curiosity, it is hoped that you will seek additional information from the reading list at the end of the text.

Only a relatively small number of the many kinds of plants are cultivated by humans for food, fiber, pleasure, or industry. Examining some of these vital economic plants (Chapter 5), with facts about their growth, structure, and uses, completes this part of our introduction to an awareness of plants. With these first two sections of Plants *as background, we can then focus on a microscopic view of plants.*

CHAPTER 3

Classifying the Green World: Plants with Flowers

How much do we know about the food we eat, especially the plants that make up so much of our diet? Did you know, for example, that beets and swiss chard belong to the same group of related organisms, the same **family?** The beet family (its scientific name is Chenopodiaceae) also includes some of the salt-tolerant plants in saltwater marshes by the sea as well as some common garden weeds. Potatoes and tomatoes belong to the same family (Solanaceae) as the eggplant, and their green parts—leaves and stems—are poisonous. In this family too is the deadly nightshade. And poison hemlock, made famous by the death of Socrates, is in the same family (Umbelliferae) as celery, carrots, and parsley.

Perhaps these close relationships among apparently different plants seem surprising. The diversity of plants is extraordinary (think of all the colors, the shapes, the sizes in a formal summer flower garden or in the woods through the seasons); but as we examine plants, we come to recognize that they also have many similarities. It is with these similarities that we classify them and bring order to the mass of plant life.

The science of classification is **taxonomy,** and applying it to plants is a job that is likely never to be finished. Approximately three hundred twenty-five thousand kinds of green plants have been described, and new ones are constantly being added. The entire living world has between 2 million and 3 million kinds of organisms, but even set against that huge throng, the number of

3.1 A supermarket can be a laboratory for studying plant classification. Apparently diverse, all these supermarket plants belong to the same family, the Cruciferae. Cabbage, cauliflower, Brussels sprouts, broccoli, and kohlrabi belong to the same species, *Brassica oleracea,* and the turnips are also members of the genus *Brassica.*

kinds of plants is impressive. In this chapter we will see how taxonomists order this tremendous diversity and then look in detail at the similarities and differences within one large group of plants, the angiosperms.

CLASSIFICATION

People have always been interested in classifying plants. Millennia ago primitive people recognized that a few plants were good to eat, that others were valuable as medicines, that still others were poisonous, and that some seemed good as potions for love or magic. Because of their importance, classification systems have been refined throughout history. These systems have sometimes taken odd forms. During the Middle Ages, for example, some plants were classified by their similarity in shape to parts of the human body. The mandrake has forked roots that seem to resemble human legs; it was thought to have magical powers and was used as an aphrodisiac. Liverworts were named because of their liverlike shape, which suggested their usefulness as a drug for liver ailments.

The practical importance of classification is very great, for information—about plants or anything else—is useful only if it is retrievable. Think of a large library with its thousand of books storing the world's knowledge. Obviously, that knowledge would be of little use if we did not know where to find a specific book when we needed it. For that reason classification systems have been devised for libraries. The Dewey decimal system is one; the Library of Congress system is another. They are different, but both systems tell us where to find the information we need.

The same idea applies to systems for guiding us about the living world: different, equally useful classifications are possible. These have always been influenced by philosophical systems and have reflected the cultural bias of those who devised them. A nation of farmers would find a way of organizing the living world that was different from that devised by a nation of fishers; a bright lion probably would classify the animal world differently than would a person. But the purpose is the same: to organize diversity so that it becomes understandable, to explain relationships so that the working of the living world can be understood. In fact, the whole intellectual base of biology is concerned with explaining the relationships among organisms. The history of taxonomy shows how classification grew from a sketchy practical system for identifying useful plants into a theoretical science concerned with all forms of plant life.

The Early History of Classification The Greek philosopher Aristotle was one pioneer of taxonomy; his system of classification lasted two thousand years. He separated all living things into a plant kingdom and an animal kingdom according to whether the organism could move and feel or was rooted and could not feel. But his system came to a dead end with many questions unanswered.

Classification studies made great progress when Europeans began to explore the world beyond their seacoasts during the sixteenth and seventeenth centuries. The known world was suddenly much enlarged as they sailed to distant, unknown places. Their voyages, which spanned a period of over three hundred years, continuing well into the nineteenth century, were enormously stimulating to biology in general and to the science of classification in particular, because they showed how large and diversified the world really is.

Many voyagers brought back specimens of plants and animals, starting the collections that ultimately became our museums of natural history. People began to study these strange new examples of life, and to make detailed drawings of them. Descriptions of living things were far more accurate than earlier, but it was not until the nineteenth century that satisfactory explanations were given for how these diverse organisms originated.

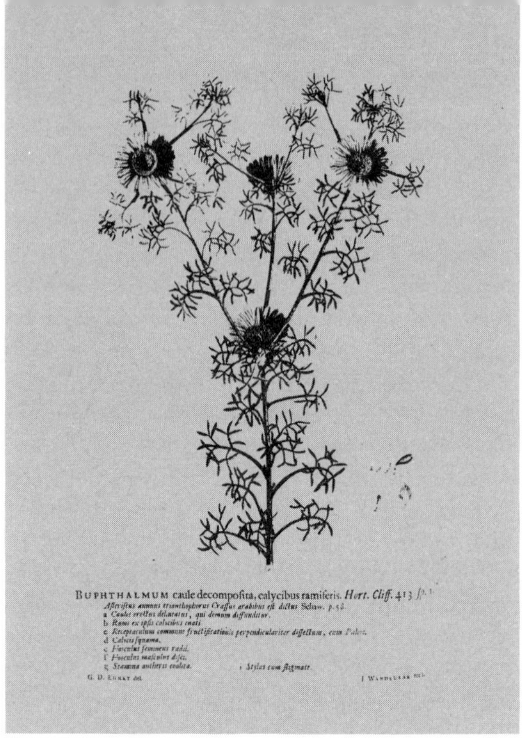

3.2 The Doctrine of Signatures is illustrated by the drawing of scorpion grass (left) in John Gerard's *Herball* (1597). The seed pods are considered to resemble scorpion tails, and the *Herball* states: "the leaves of Scorpion grass applied to the place, is a present remedy against the sting of Scorpions: and likewise boiled in wine and drunk, prevaileth against the said bitings, as also of adders, snakes, and such venomous beasts." By Linnaeus's time, much progress had been made in classifying plants. After Linnaeus obtained his medical degree in Holland, he spent 1733–1735 classifying plants on the country estate of a rich Amsterdam banker, George Clifford. This collection of plants was described in a book published in 1737 titled *Hortus Cliffortianus* (literally, Clifford's garden), one of Linnaeus's first works on the classification of plants. A page from this book is shown on the right.

Linnaeus and the Plant Kingdom Modern classification of plants has its roots in the relationship between botany and medicine. Medical doctors in times gone by had to know their plants because so many were a source of medicines. During the late Middle Ages botanical gardens were started at universities because they were needed as laboratories of pharmacology, and the first botany professors were teachers in medical schools. In fact, it was an eighteenth-century medical doctor who produced a better basis for systematic classification of plants, which grew into the system we now use. He was a Swedish doctor named Carolus Linnaeus (1707–1778).

Linnaeus was physician to the Swedish Admiralty and professor of medicine at the University of Uppsala before botany completely claimed him. Then he wrought a botanical and indeed a biological revolution. In 1753 he published his *Species Plantarum,* using a very simple logical method of classification. He based it on his observation that the number of stamens (the male sexual organs) in a flower is always similar in the same kind of plant. Linnaeus was a disciple of Aristotle, who had taught that reproduction was the "essential nature of plants," and out of that doctrine he developed his classification system. As Linnaeus's system of classification became widely known, organisms were sent to him from all over the world. He gave many common plants and animals the formal names we know today. (In the Appendix on page 465, you will find some results of his work.)

Linnaeus's system of naming plants was actually based on one invented a hundred years earlier, but it was more consistent and easier to use and soon gained wide acceptance. One reason was his use of **binomial,** or double-named, system. We are known as *Homo sapiens,* and the garden bean as *Phaseolus vulgaris.* The second word (which is not capitalized) tells what kind of plant it is, the **species;** the first word tells what group of related species, what **genus,** the plant belongs to. (Both words are italicized.) For some plants the name of the person who first described the species is acknowledged by placing either an initial (for Linnaeus) or an abbreviation of the person's name after the specific name. We find the garden bean listed as *Phaseolus vulgaris* L., informing us that Linnaeus first described and named it according to the binomial system; *Aesculus californica* Nutt. indicates that Nuttall was the first to describe the California buckeye. These rules of nomenclature are maintained and when necessary modified by the International Association of Plant Taxonomists.

Every organism has many different common names in many different languages, and some organisms with different scientific names have the same common name. In the United States "corn" refers to *Zea mays,* but in Europe "corn" refers to other grains such as rye in the genus *Secale.* Latin is used for scientific names because until recently it was the common language of all educated people; its use continues to provide a worldwide standard and reduces the confusion of common names.

From Kingdom to Species In Linnaeus's system similar species were classified in similar genera (*genera* is the plural of *genus*), similar genera in similar families, and similar families in orders. We go on to lump different orders into classes, the classes into divisions, and the divisions into the plant kingdom. (The same system applies to animals, but the word *phylum* came to be used for *division* in the animal system.)

Here is an example of how the Linnaean system of classification works, beginning at the most specific level, comparing people and garden beans:

Common Name:	Human	Bean
Species:	*sapiens*	*vulgaris*
Genus:	*Homo*	*Phaseolus*
Family:	Hominidae	Leguminosae
Order:	Primate	Rosales
Class:	Mammalia	Dicotyledoneae
Phylum/Division:	Chordata	Anthophyta
Kingdom:	Animal	Plant

Present classification systems most commonly divide all organisms into three to five **kingdoms** according to how they obtain nutrients or according to their cell structure (see the Supplement on page 96). For example, fungi were long thought to be part of the plant kingdom just because they grow in the earth like plants. But looking at fungi more closely, we see that in several ways they differ fundamentally from plants. For example, they have quite a different way of "making a living"—they do not draw their energy from sunlight, as plants do. Modern classifications therefore separate the fungi from the plant kingdom (see Table 3.1). Important characteristics of all members of the plant kingdom are that they have autotrophic growth (Chapter 1) and cells with cellulose walls (Chapter 6).

The category just below the kingdom, the **division,** still arouses much learned argument, for there is no universal agreement on how organisms are divided into divisions. The ambiguity is reflected in the variability of names. For many years it was common to place all seed plants into one division called the Spermatophyta. This division had two subdivisions, the Angiospermae and the Gymnospermae. It is now common to place all the flowering plants (the old angiosperms) into a single division called the Anthophyta (literally, "flower plants") and to subdivide the Gymnospermae into a number of separate divisions. Because *angiosperm* is a fairly common word, we will use it to indicate the flowering plants.

All the other taxonomic categories (collectively called **taxa**) are also inventions of the human brain, so their limits are relative and not absolute. As with divisions, some authorities will split classes, orders, and families into a number of lower taxonomic categories, and others will combine a number of taxa into a higher category.

A unit of classification called *species* will often be mentioned. It is fundamental in the classification system. Trained people will organize a population of anything into different sets in order to understand that population. Look around at the human race and you will see an enormous diversity of faces, shapes, ways of vocalizing, and so on. All are different but all are assigned to the same species.

Why? What is a species? It is often defined as a population of similar organisms that normally mate with each other. If population *A* does not interbreed with population *B*, then *A* and *B* belong to different species. *Homo sapiens* is a species; the white oak tree *(Quercus alba)* is too, and the sweet cherry tree *(Prunus avium)*. The definition is not very satisfactory, however, because different species of plants can interbreed and form fertile hybrids (see page 337). And there are other complications, even when only animals are considered.

Before Charles Darwin (1809–1882) published his milestone work, *On the Origin of Species* (1859), in which he presented the first reasonable theory explaining how all the enormous diversity on earth evolved, most people believed that species were naturally occurring objects, like chemical elements. They were thought to be

Table 3.1 A Classification of Organisms

KINGDOM MONERA
The cells of Monerans do not have a nuclear membrane or organelles (see page 96). Most are unicellular, although some do form colonies. They include the blue green algae (Cyanophyta) and bacteria.

KINGDOM PROTISTA
The cells of these organisms have nuclear membranes and organelles. Most are unicellular; some are colonial. Examples include protozoans (single-celled animal types) and certain green, motile unicellular organisms, such as *Euglena* and dinoflagellates, commonly classified with algae.

KINGDOM FUNGI
Fungal cells have nuclei with membranes and other organelles. The cells are frequently organized in filaments. Most fungi do not move and they obtain nutrients by absorption. Main divisions include:

DIVISION MYXOMYCOTA
Known as slime molds, many members of this group consist of a multinucleate mass capable of amoeboid movement. They are common in dead vegetable matter.

DIVISION EUMYCOTA
The true fungi, these organisms are frequently filamentous and their cells have walls. They consist of several classes. These include:

Class Phycomycetes
Water molds, some mildews, and the common bread mold are among this group of organisms.

Class Ascomycetes
Yeasts, powdery mildews, and molds that synthesize antibiotics are examples.

Class Basidiomycetes
Major agents of decay, these organisms produce spores on basidia. Among them are the mushrooms, rusts, and smuts.

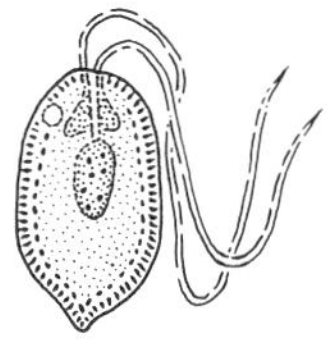
Green alga

KINGDOM PLANTAE
Plant cells have nuclei with membranes, organelles, cellulose cell walls, and photosynthetic pigments. Most are multicellular and nonmoving. Major divisions include:

Moss

ALGAE: DIVISIONS CHLOROPHYTA (GREEN ALGAE), CHRYSOPHYTA (GOLDEN ALGAE), PHAEOPHYTA (BROWN ALGAE), RHODOPHYTA (RED ALGAE), AND XANTHOPHYTA (YELLOW ALGAE)

DIVISION BRYOPHYTA
Land plants without highly specialized structures for conducting water, minerals, and nutrients, the bryophytes have no true leaves or roots. This group includes the mosses and liverworts.

(Table 3.1 continued on next page)

(Table 3.1 continued from page 73)

Club moss

DIVISION PSILOPHYTA

Primitive vascular plants (see page 108), these plants have neither seeds nor true leaves and roots.

DIVISION LYCOPODOPHYTA

An ancient group of vascular plants with true leaves and roots. Most members of this division are extinct, but club mosses and quillworts are living representatives.

DIVISION ARTHROPHYTA

Another ancient group of vascular plants represented by one living genus, *Equisetum,* the horsetails.

Fern

DIVISION PTEROPHYTA

The ferns. These are vascular plants with prominent leaves, which usually bear sporangia (see page 107) on the lower surfaces.

DIVISION PTERIDOSPHERMA

The seed ferns are an enigmatic group of extinct plants frequently invoked in discussions on the evolution of seed plants because they demonstrate the first occurrence of seeds in the fossil record.

Cycad

DIVISION CYCADOPHYTA

An ancient group of seed-bearing plants, many of whose members are now extinct.

DIVISION GINKGOPHYTA

Woody seed plants extinct except for the cultivated tree *Ginkgo.*

DIVISION GNETOPHYTA

This group of seed plants has three orders of one genus each. The genera *Ephedra* and *Welwitschia* are found in certain arid regions while *Gnetum* consists of tropical vines. The ancestry and relationship of these genera are uncertain.

Conifer

DIVISION CONIFEROPHYTA

The conifers are the major group of living gymnosperms. Their seeds are borne in cones.

DIVISION ANTHOPHYTA

Vascular plants with flowers, the most familiar kind of plant, this group includes two major classes:

Class Monocotyledonae
The seeds have one cotyledon and the flower parts are generally in groups of three or multiples of three. The stems do not have secondary growth (see Chapter 6).

Flowering plant

Class Dicotyledonae
The seeds have two cotyledons and the flower parts are usually in groups or multiples of four or five. The stems have secondary growth.

Kingdom Animalia

something God or Nature or Divine Providence just created on earth and left to live here; the forms of living things, their similarities and differences, had existed from the beginning. Then Charles Darwin, after returning to England from his voyage on the H.M.S. *Beagle,* proposed that species had evolved through **natural selection.** That is, any plant or animal variant that was more successful than others around it in dealing with its environment usually left the most offspring to reproduce in their turn. To put it differently, some variants of a species usually coped better with their environment; they were better *adapted.* These variants produced more offspring than other less successful ones. If some varieties were isolated, cut off from reproducing with populations of other varieties, then in time, because of continuous selection among the variants, they would become different from other members of the species (see Figure 3.3). Eventually they might be considered a separate species.

What this means, and what Darwin emphasized, is that distinguishing species from varieties is frequently difficult. It is easy to distinguish species such as human beings and chimpanzees, because intermediate forms between the two are extinct. If the ancestors of *Homo sapiens* and chimps were alive, however, it would be extremely difficult to differentiate between the two species. Even though the idea of species is useful to help biologists organize diversity, it remains a relatively arbitrary construction of the human mind.

Modern Classification The Linnaean system is artificial. Certain characteristics were selected as the base of the system, like the number of stamens in a flower, but other characteristics that are reasonably constant, such as synthesis of certain chemicals, might have served equally well. As in classifying the books in a library, this system was as "right" as another might have been.

We still *identify* plants using arbitrary systems like Linnaeus's number of stamens in a flower, shape and alternation of leaves, and so on. Identification, however, is not the same as classification. To identify we need only divide objects into sets and subsets according to established criteria. Determining the criteria, however (which is what classification is), comes close to being a philosophical problem. Linnaeus chose his system of establishing criteria because of his devotion to Aristotelian principles, and he, as well as others before Darwin, classified living things according to the **morphology** ("form") of selected organs. Current classification schemes, however, are designed to reflect the genetic relationships of organisms. (We will examine these in Chapter 11.) The modern taxonomist is searching for evidence that will help to establish relationships. Plants, like people, are not necessarily related because they happen to look alike.

Why does the task of classifying organisms go on, apparently never to end? One reason is that new information is coming in from

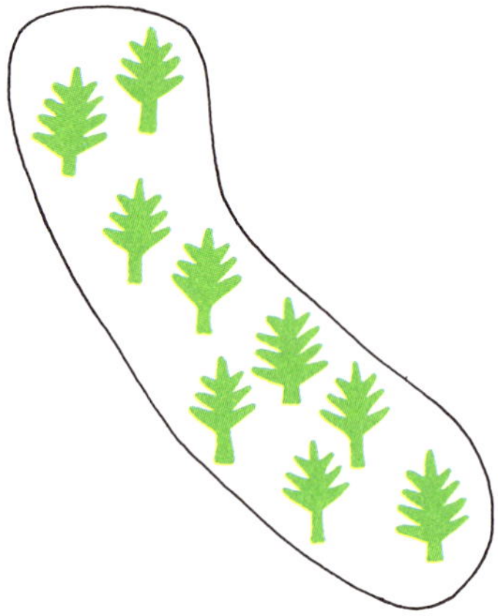

1. A single population in a homogeneous environment.

2. Changes in parts of the environment and dispersal of plants to new environments lead to change within the population.

3. Further change and dispersal produce geographic isolation of some offspring of the ancestral population. Later some of these isolated populations undergo changes in genes that control reproductive mechanisms.

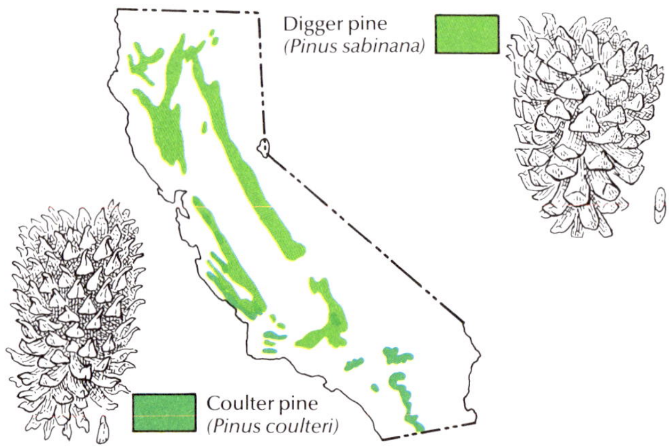

4. Further changes in the environment permit the geographically isolated populations to exist together again in the same region. They now remain distinct because they are reproductively isolated. They are considered to be two separate species.

3.3 Formation of new species (speciation) often involves isolation. At first, one plant species is distributed throughout an area. As the environment changes, some plants are dispersed to new regions and become geographically isolated. In time, some plants undergo genetic changes that make it impossible for them to interbreed with other populations. Although geographical barriers are broken down and the previously isolated populations again grow in the same region, they cannot interbreed, and they are designated separate species.

all branches of biology, constantly clarifying relationships among organisms. And unknown organisms are being discovered all the time. Just a few years ago a whole phylum of wormlike organisms was found of which we had known nothing. Classifying is a continuous process, too, because all around us new species are evolving and others becoming extinct. If we stopped classifying organisms, we would eventually face a strange new world. And without such classification the kingdom of flowering plants would be a confused if beautiful jungle. Once we can find our way with a classification system as guide, we are ready to look closely at the plant and its parts.

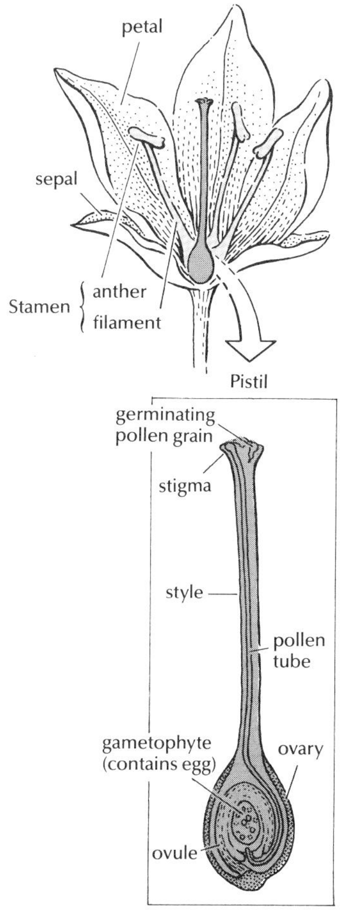

3.4 This lily flower has both female (pistil) and male (stamen) sex organs. The single pistil with its terminal stigma is surrounded by six stamens with their pollen-bearing anthers, and these are surrounded by six petals. Many flowers are unisexual and possess either pistils or stamens. In pollination, pollen grains from the anthers are transferred to the stigma where they produce pollen tubes. Pollen tubes grow through the tissue of the style and into the ovules. Two sperm cells are released within the gametophyte and one fertilizes the egg.

FLOWERING PLANTS: SIMILARITIES

We know that most varieties of beings are animals, that most animals are insects, and that most insects are beetles. With respect to plants, we know that most plant species (about 85 percent of them or three hundred families) are flowering plants, which belong to a category of plants called **angiosperms** ("enclosed seeds"). By examining their structure we will see both why these plants are considered one taxon and why they have been so successful.

The Basic Structure of Angiosperms Some of the similarities among the angiosperms are obvious. It is easy to see that an apple tree, a pear tree, and a cherry tree have a similar structure (morphology). There are the parts above ground—the trunks or stems, leaves and flowers, and fruits. Then there are the parts below ground, the roots. These different structures, called **organs,** have similar functions in most of the many species of plants, although their outward appearance may vary considerably. For example, many plants have underground stems, and there are even some, like the peanut (also called groundnut), that have fruits that develop underground.

Other similarities are less obvious. Plants are classified primarily by their reproductive structures, and in angiosperms sexual reproduction occurs in the flowers, which are often the most conspicuous and most beautiful feature of a plant. Flowers are actually groups of modified leaves. These have evolved over millions of years into forms, sizes, and colors of endless variety.

In many flowers male and female reproductive organs are close beside each other (see Figure 3.4). The male organs, the **stamens,** are commonly long stemlike filaments, each with a structure at its end called the **anther.** This organ produces pollen grains, releasing them when ripe by rupturing.

Inside the circle of stamens are one or more female reproductive organs, the **pistils.** Each pistil has an ovary and ovules. The **ovary** is an enlarged base that may look like the lower part of a vase. It will later form much of the fruit. Inside the ovary are the **ovules,** which will develop into seeds. In many flowers a slender stalk, the **style,** grows upward from the ovary. At its end is an often flattish, sticky, hairy structure, the **stigma,** or pollen catcher.

The flower's outermost parts are the most leaflike; these are the **sepals.** Usually green, they enclose and protect the flower bud. Collectively they are known as the **calyx.** Next to the sepals are the petals, which form the **corolla.** These too may show their leafy derivation, although they are often brightly colored. It is usually the petals that advertise the flower's presence to pollen-carrying insects and to birds that come in search of nectar and pollen.

There are many variations on this basic design, yet the elements are generally much as described here. Stamens and pistils may have a complex structure, but the stigma always receives the pollen and, being sticky and sometimes hairy, it can catch and hold the pollen.

Pollination Pollen, you must have gathered, has something to do with reproduction. Even if we do not always recognize it, pollen is a common sight. In the early summer you might see its yellowish specks floating like golden dust through breezeless air in the sunlit forest or even in a city park. You can see it on the flower itself, on anthers or stigma. Landing on a flower, an insect can pick up some pollen from the anthers. Landing on another flower, the pollen rubs off the bug, sticking to the stigma and held there by the gooey surface.

The pollen is produced in the anthers. Within the pollen grain's thick walls is a two-celled structure, the **gametophyte.** Gametophytes produce sex cells, which are called **gametes.** Male gametes are produced in pollen, female gametes in ovules. (Compare the cross-sections of a pollen grain and an ovule in Figure 3.6.) Reproduction of many plants requires that pollen be transferred from the stigma to the anther, a process known as **pollination.**

A good many plants among the angiosperms are **self-pollinating:** anther and stigma are close together in the same flower, and pollen can be easily transferred from one to the other. It may even be moved from anther to stigma by an insect pushing its way into the flower. Many other plants must be **cross-pollinated:** pollen from one flower must be placed on the stigma of another flower. The plant cannot move around to bring pollen and stigma together. It needs something to do the walking for it, so to speak, and much of the angiosperms' success is due to their highly effective systems for ensuring that pollination does occur.

3.5 Honey bees and flowering plants have a symbiotic relationship. Bees obtain nectar and pollen and many plants depend on bees for pollination. Attracted by the yellow color of the crocus, this bee is landing to obtain more pollen. As the bee gathers pollen, it is stuffed into pollen baskets on its hind legs, clearly visible here. In the process of foraging among stamens, the bee will deposit pollen on the stigmas of many flowers.

Some flowers, like those of grasses and many trees, are **wind-pollinated.** Clouds of windblown pollen may be carried miles to alight at random on stigmas. Many other flowers rely on insects, birds, and even bats for pollination. As a result, the evolution of flowers closely parallels that of some animals, particularly the insects: they evolved together. (We will discuss this coevolution further in Chapter 14.)

The gaudy colors and scents and sweet nectar of many flowers have just one function: to attract insects and nectar-eating birds. Each insect and bird will respond most often to one color or combination in a flower, as the hummingbird does to red. Attracted to the flower by color (or nectar or smell), it deposits on the stigma any pollen taken (unintentionally) from another plant. Light as dust, the pollen sticks to the insect's body or the bird's feathers. As the bird or insect searches for the nectar, more pollen sticks to it, to be left on the next flower it stops at.

Fertilization and the Life Cycle A **pollen tube** will begin to grow from the pollen grain soon after it is deposited on the

3.6 ANGIOSPERMS: LIFE CYCLE

The life cycle of all plants has two phases, the sporophyte and the gametophyte. In flowering plants, the gametophyte stage takes place within the tissues of the reproductive organs of the flower. The gametophyte which will produce an egg cell (female gamete) develops within the ovule(s) inside the ovary at the base of each pistil (bottom). Gametophytes develop from specialized cells called *spores*. The gametophytes which produce sperm (male gametes) are produced within anthers, the terminal portions of stamens. During pollination, the pollen grains (male gametophytes) are deposited on the stigma. A pollen tube emerges from a pore and grows through the pistil's style (below). One cell of the two-celled gametophyte divides and forms two sperm cells. The pollen tube penetrates the tissue of the ovule and discharges the sperm cells into the female gametophyte. The nucleus of one cell unites with the nucleus of the egg cell. The product of this fertilization is the zygote which initiates the sporophyte generation (facing page, top). The nucleus of the other sperm cell fuses with the two nuclei of the gametophyte's central cell to form the cell which will produce endosperm. This double fertilization is only found in the life cycle of flowering plants. Seeds of flowering plants consist of an embryo which develops from the zygote endosperm and the seed coat derived from tissue of the ovule. The ovary forms tissue of the fruit and encloses the seed. Only angiosperms have seeds enclosed within a fruit. The genetic basis of the flowering plant life cycle is provided in Chapter 12. This photograph of a pollen grain (below, left) made with a scanning electron microscope illustrates the highly sculptured cell walls of pollen. The morphology of pollen grains is characteristic of different kinds of plants and is an important taxonomic feature. Pollen walls are made of a complex material extremely resistant to destruction, and this means that vegetation of former periods can be readily identified because of the abundant deposits of pollen that have become fossilized.

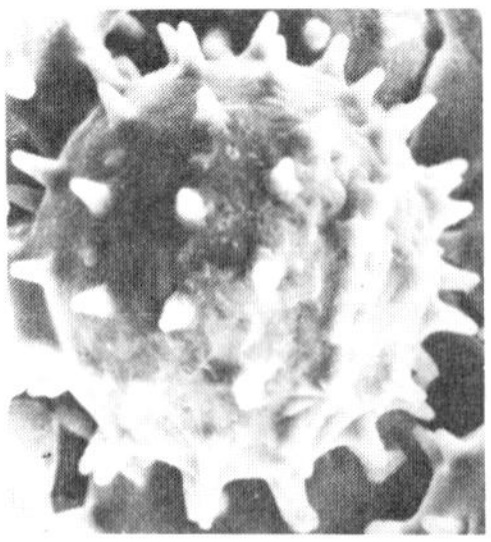

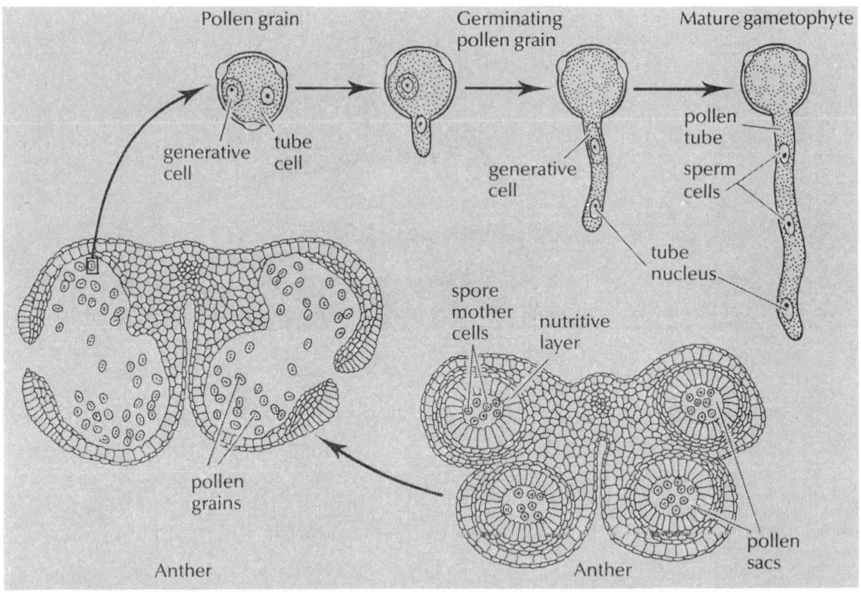

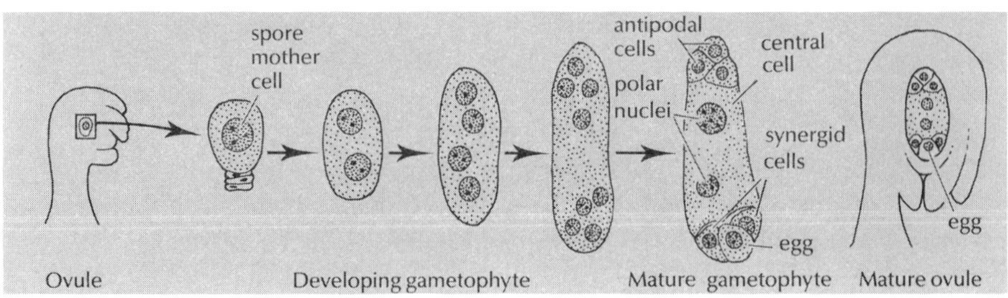

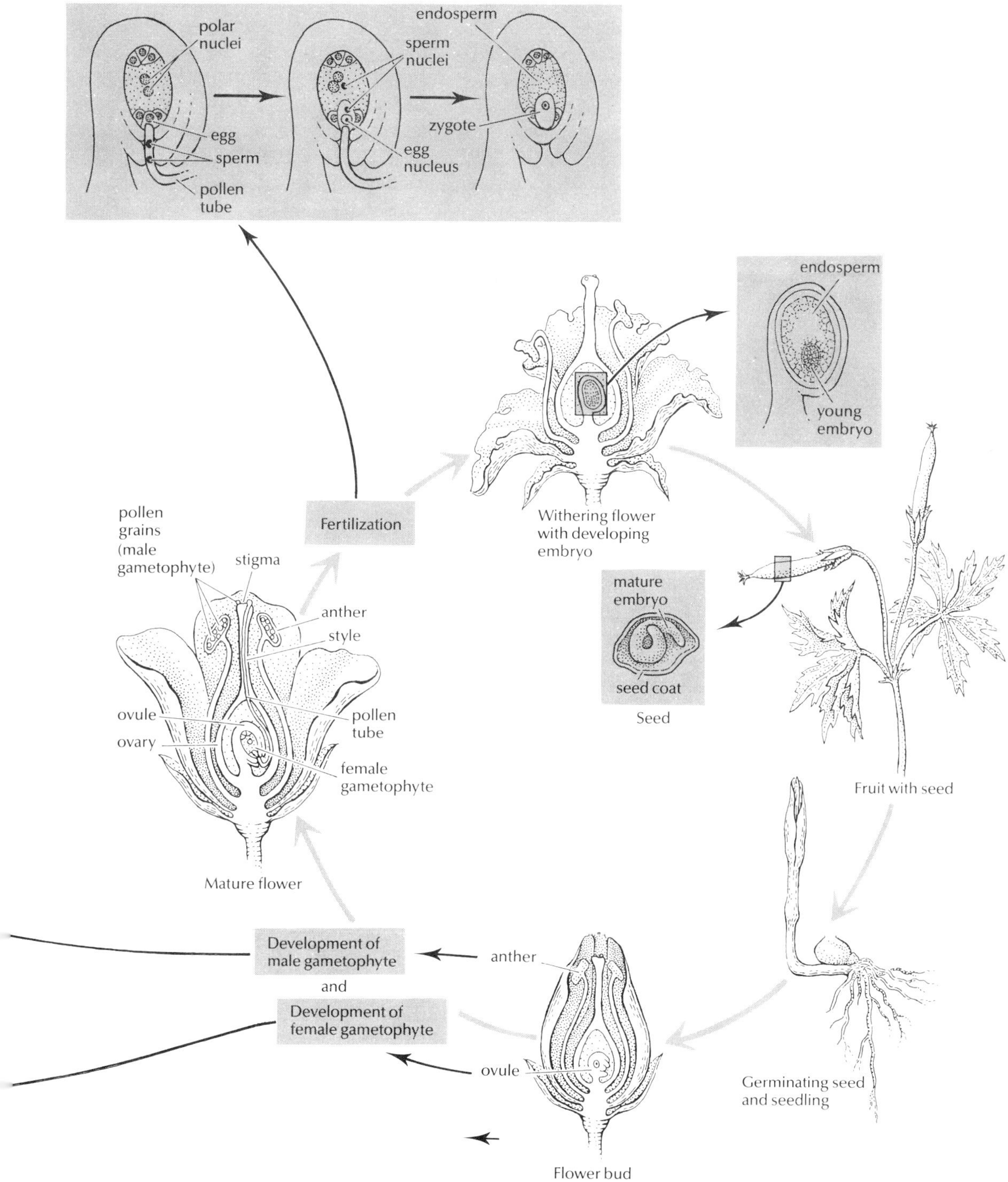
polar
nuclei
endosperm
sperm
nuclei
egg
sperm
pollen
tube
egg
nucleus
zygote
endosperm
young
embryo
Withering flower
with developing
embryo
Fertilization
pollen
grains
(male
gametophyte)
stigma
anther
style
ovule
ovary
pollen
tube
female
gametophyte
Mature flower
mature
embryo
seed coat
Seed
Fruit with seed
Germinating seed
and seedling
Development of
male gametophyte
and
Development of
female gametophyte
anther
ovule
Flower bud

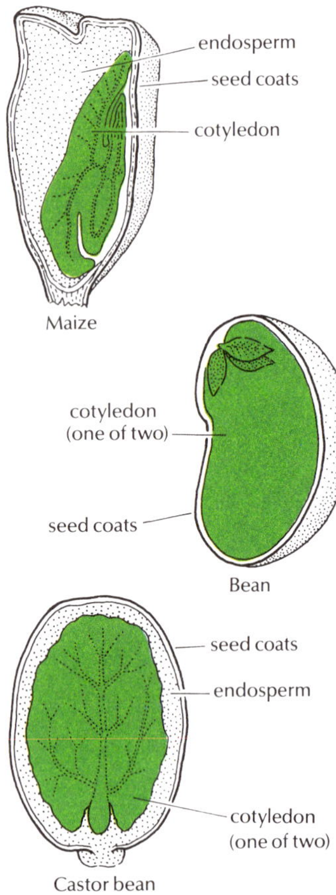

3.7 Three types of seeds are illustrated here. Maize is representative of grasses. It has one cotyledon and abundant starchy endosperm. Both the common bean and castor bean have two cotyledons. The common bean has no endosperm in the mature seed; the cotyledons serve as the site of stored food. The mature castor bean still has endosperm, in addition to large cotyledons.

stigma, expanding within a few hours to many times the size of the pollen grain from which it came. Down into the style and ovule it penetrates, finally entering the female gametophyte. Inside the tube are two male gametes, which are discharged in the female gametophyte in the ovule. One male gamete nucleus then enters the **egg cell,** which is a cell of the female gametophyte. [The **nucleus,** a prominent structure within the cell, holds the genes (see page 96).] The egg cell nucleus and the male gamete nucleus fuse, and **fertilization** has occurred. Fertilization takes place a few hours to a few days after pollination. The fertilized egg, called the **zygote,** produces an **embryo,** which will develop into the plant.

Meanwhile, the nucleus from the second male gamete of the pollen tube fuses with nuclei from another cell in the female gametophyte, forming a cell that later will produce the **endosperm.** The endosperm develops rapidly and provides stored food for the growth of the zygote into a seedling (see Figure 3.7). (When we eat corn, we eat it for the endosperm, the biggest part of the corn kernel.)

In many plants, grasses being one example, embryo and endosperm grow side by side, gradually absorbing the other cells inside the ovule which surrounded the female gametophyte. When their growth is completed, a **seed** has been formed. The wall of the ovule has become the **seed coat,** like the brown papery skin around a peanut. And the ovary, greatly changed, becomes the carrier of the seeds; it is called the **fruit.** Just as insects and birds are indispensable in the pollination of many plants, fruits are required for seed dispersal. Here again plants get around despite their eternal handicap—their inability to move.

Only plants produce gametes from special bodies known as gametophytes, and these give the plant two separate reproductive phases, which are called the **alternation of generations** (see Figure 3.8). In one of these phases or generations, the gametophyte makes sex cells, the gametes, and male and female gametes fuse, as we have seen, to produce a zygote, which then gives rise to an embryo, which develops as a **sporophyte.** The sporophyte produces specialized cells, the **spores,** which in turn will produce gametophytes. One alternate generation, then, is the sporophyte; trees, grass, and rose plants are all sporophytes. Spores of some plants like ferns are visible on the underside of leaves, but in angiosperms they are enclosed in anthers and ovules. Pollen grains develop from the

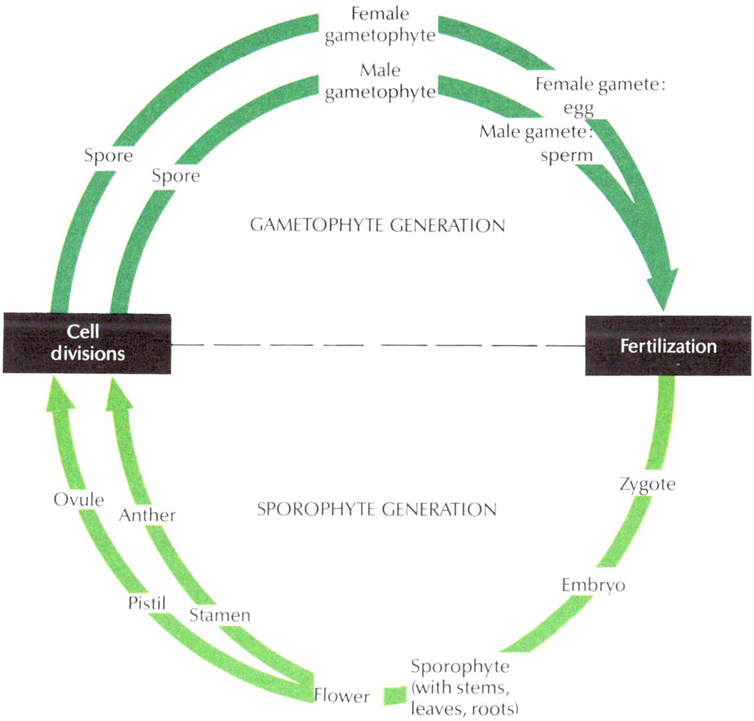

3.8 The life cycle of all plants consists of two stages determined by the final product of the organs involved in sexual reproduction. The sporophyte produces specialized cells called spores, and the plant called the gametophyte produces gametes. Unlike plants such as mosses and ferns, gametophytes of flowering plants are hidden within the tissues of anthers and ovules in the flower. In the course of their lives, gametophytes produce sex cells and witness fertilization, which yields the zygote that initiates a new sporophyte generation.

spores. The gametophytes, the other generation, are tiny structures in the ovule and pollen grains of flowering plants, but in some nonflowering plants such as moss and many algae (to be discussed in Chapter 4), the gametophytes are the dominant generation.

TAXONOMY IN THE SUPERMARKET

We recognize the angiosperms as a taxon because of these very basic similarities in their structure and life cycle. Built on these similarities, however, are infinite variations.

There are, first of all, two major classes of angiosperms: Monocotyledoneae and Dicotyledoneae (**monocots** and **dicots,** for short). The **cotyledons** are structures within the seed containing stored nutrients that support the growth of the seedling. Each half of a pea or bean seed is a cotyledon. Some plants have one cotyledon (monocots), others have two (dicots). Onions, lilies, corn plants, and palm trees are monocots; apple and oak trees, rose bushes, and African violets are dicots. You can tell the two groups apart by the features of their leaves, flowers, and seeds (see Figure 3.10 on page 86).

You do not have to go far to see both monocots and dicots and examine more closely angiosperms and their variability. The nearest supermarket, in fact, will do. Looking at all the fruits and vegetables neatly classified in the bins before us, we can start by thinking of plant families. The family, we have seen, is one of the larger taxonomic categories, and there are about three hundred different families of flowering plants. Most of the plants we eat or use for fiber, drugs, and other purposes are found within seventy of the three hundred, and most of our food plants are covered by ten families or even fewer. Around the world people make use of about a thousand species of plants for food; but in North America we choose to eat only a fraction of these.

Dicots Here in the supermarket are a great many familiar plants, and we can compare these with other plants of the same family that are less familiar. We will begin with families that are dicots. (For a more detailed look at classification of supermarket plants, see the Appendix on page 465.)

Roses *(Rosa),* always a favorite flower of the human race, are in the family Rosaceae, which also provides most of our common temperate fruits. Strawberries, raspberries, loganberries, and blackberries are in the same family as roses, as are apples, cherries, pears, peaches, and plums.

The cherry *(Prunus)* provides a good example of the structure of a fruit. Its flesh is the swollen ovary, and the pit is the hardened inner walls of the ovary. The seed is inside the pit and consists of the seedcoat (outer layers of the ovule), the endosperm, and the embryo. A fruit always develops, at least partially, from a flower.

The strawberry *(Fragaria)* is more complex, being not just a single fruit but an aggregate of many fruits. The fruits are all those small seedlike structures on the surface of the berry that stick between your teeth. The flesh is not an enlarged ovary but the swollen base of the flower. Earlier we mentioned fruits doing the delivery work for plants in seed dispersal. Imagine how widely the seeds of the strawberry are distributed by the animals that eat them.

Now for some vegetables. The name *vegetable* is not precise. It can mean a fruit such as string beans or tomatoes, a root such as beets, a stem as in potatoes, or a leafy shoot such as a cabbage. The vegetables cabbage, cauliflower, Brussels sprouts, kale, collards, and broccoli all belong to the same species; all are derived from the wild cabbage *(Brassica oleracea)* of the Middle East. Although the leaves and stems of these plants differ, their flowers are almost identical and closely resemble those of the wild cabbage. We can easily examine the flowers of this group; the part of the broccoli plant we eat, for instance, is the flower head, or **inflorescence.** The cabbages belong to the family Cruciferae, the mustard family. And we know more plants in this family: radishes, turnips, rutabaga, and mustard.

A family of trailing vines is very important in the supermarket. These are the Cucurbitaceae, whose Latin name, meaning "gourd," hints at what they are: cucumbers, watermelons, cantaloupes, pumpkins, squashes, and gourds, collectively known as the melons and the squashes. All are closely related; pumpkins and summer squash are the same species *(Cucurbita pepo)* and the winter squashes *(C. maxima)* are close relatives. In fact, canned pumpkin is often made from squash fruits.

The variability that leads to such extreme differences as we saw in the cabbages and which is likely to keep taxonomists busy always, is striking in still other supermarket species. Swiss chard and sugar beets belong to the same species *(Beta vulgaris)*, and beets belong to the same family as spinach. These are the Chenopodiaceae (Latin for "goosefoot"), the family that includes many plants in saltwater marshes and around the rims of salt ponds in desert areas. In this same family are some plants we think of as weedy species, among them lamb's-quarters and Russian thistle.

It is easy enough to see that daisies and sunflowers are related, though their sizes are so different, but it would be hard to guess that lettuce is a species in the sunflower family too. Along with daisies, artichokes, thistles, and wormwood, it belongs to the Compositae, one of the three largest families of flowering plants.

3.9 Lamb's-quarters (*Chenopodium album* L.) is one of our most common weeds, and like so many of them, it originated in Eurasia. As a member of the Chenopodiaceae, it is related to spinach and beets.

Probably the most familiar family of plants in the supermarket is that of the bean, the Leguminosae. The legumes are the most important plants in the human diet after the grasses. Besides all the forms of beans we eat, the legume family includes peanuts, soybeans (now the biggest cash crop in America), and alfalfa, the most important source of hay.

It is a rare day when we do not eat potatoes *(Solanum tuberosum)* and tomatoes *(Lycopersicon esculentum)* in some form. Both belong to the Solanaceae, the nightshade family. Eggplants *(S. melongena)* are also in this family, the vegetable loved by many but loathed by some (like so many "vegetables," it is in fact a fruit). If you buy a pack of cigarettes or a can of tobacco on the way out of the supermarket, you will be broadening your acquaintance with the nightshade family, because tobacco is a member. Tobacco *(Nicotiana)* contains nicotine, a poison, and potent poisons are one characteristic of this family. Beware of the green parts of the potato plant and of the tomato plant. Before they became popular as a food, tomatoes were called "love apples" and considered poisonous, which they are, except for the ripe fruit, the part we eat.

The carrot and celery family, or Umbelliferae, also has poisonous plants. Celery, carrots, and parsley are umbels, as is poison hemlock, which Socrates drank to end his life. The umbels have a branched umbrellalike flower stalk and are easy to recognize by their flowers. Parsley often has flower stalks, and many of us have probably seen the wild carrot, or Queen Anne's lace, which decorates the nation's highways so abundantly.

The sweet potato *(Ipomoea batatas)* is really a morning glory, one of the Convolvulaceae. This is a family both loved and hated by gardeners. Bindweed, a nasty and common weed, is a wild morning

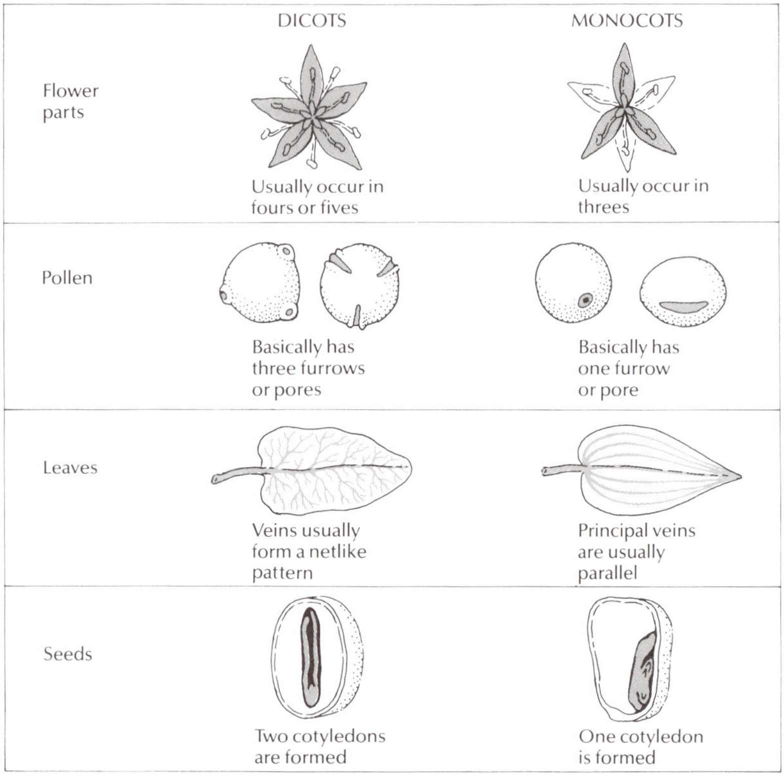
DICOTS
MONOCOTS
Flower parts
Usually occur in fours or fives
Usually occur in threes
Pollen
Basically has three furrows or pores
Basically has one furrow or pore
Leaves
Veins usually form a netlike pattern
Principal veins are usually parallel
Seeds
Two cotyledons are formed
One cotyledon is formed

3.10 MONOCOTS AND DICOTS

Angiosperms form a division with two classes, monocotyledons and dicotyledons. The date palm (facing page, top left), Turk's cap lily (facing page, bottom left), and Pink Ladyslipper (facing page, bottom right) or moccasin flower, belong to different families of monocots. Even though palms and many bamboos, which are also monocots, are as big as trees, their stems do not increase in thickness like dicots. Monocot flower parts are usually in threes or sixes such as the lily. The monocots evolved later than the first dicots, and the orchid family to which the ladyslipper belongs has the most specialized flowers of all families. The dicots have many woody species, including most of the trees in the world. *Fuschia* (below, left) and Columbine (below, right) belong to dicot families. As illustrated by the *Fuschia*, dicot flower parts are usually in combinations of five. An easy way to tell most monocots and dicots apart is to look at their leaves. Monocots have leaves with parallel veins, as illustrated on the corn leaf (bottom, center photo). Dicot leaves have veins which form a network, as illustrated by the tuliptree leaf (bottom right). Veins are the vascular tissue; leaves are the plumbing system which transports water, minerals, and food.

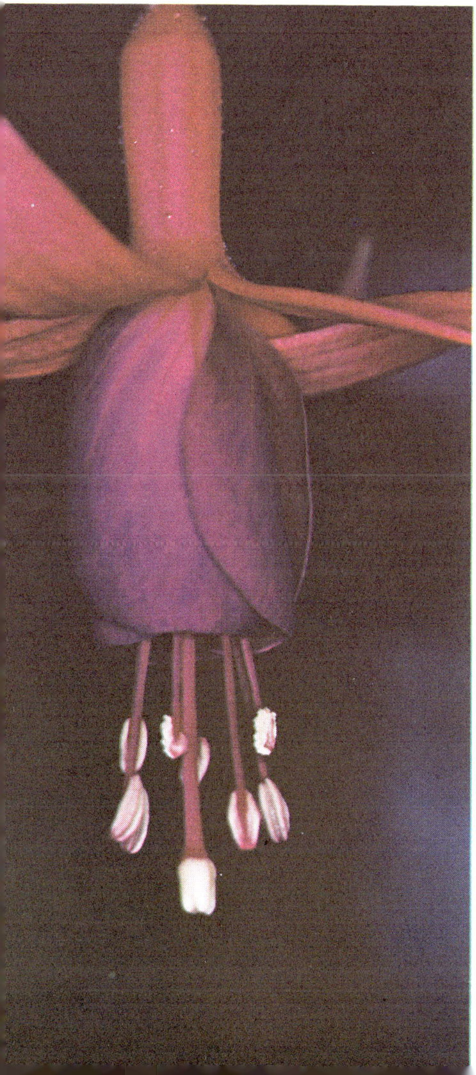

glory. The yam *(Dioscorea)* is not in this family, though many people call a sweet potato variety a "yam." In fact, yams are in another family *(Dioscoreaceae)* and are a primary source of food in the humid tropics (Chapter 5).

All supermarkets sell tea, and in some states they also sell flowering shrubs called camellias. Tea *(Camellia sinensis)* is really a camellia of which we use only the young leaves. They belong to the family Theaceae (tea family).

Poinsettia plants *(Euphorbia pulcherrima)* have become a traditional Christmas decoration in American homes because of their bright red "flowers." These spectacular spreading blooms are not flowers at all but modified leaves called **bracts.** The flowers proper are small, yellow, unobtrusive little clusters in the center of the bracts. Poinsettias are in the same family as the rubber tree, the Euphorbiaceae, or spurge family. Pick a leaf from a poinsettia plant and you will see a white sticky sap drip from the leaf stalk. This *latex* is characteristic of the family. Rubber trees produce it in great abundance: exposed to air, it hardens, and when heated, it becomes resilient and quite tough.

Other Euphorbiaceae include many cactuslike house plants and some common weeds. The cassava *(Manihot),* main food plant of the humid tropics, is the source of tapioca. Castor oil, that cure-all forced down the throats of generations of youngsters, is produced by a member of this family *(Ricinus communis).* It is, by the way, an industrial lubricant usually found in quantity around an automobile race track. The seed from which the oil is extracted is itself quite poisonous.

Monocots We have discussed some of the most important dicots in the supermarket. Now a look at some of the monocots will illustrate differences between the two groups of angiosperms.

The largest family of monocots is the Gramineae, or grasses. From this family come all the flour and cereal products, cornstarch, beer, whiskey, and much of the sugar that we use. Corn, wheat, oats, sugarcane, sorghum, bamboo, and crabgrass all belong to the grasses. The grasses have evolved recently compared to many other plant families. Their flowers are specialized for wind-pollination, and because animal pollinators are not involved, the flowers are small and not very colorful. So inconspicuous are they that many people do not realize that grasses have flowers.

Pineapples *(Ananas comosus)* are monocots and are a fruit derived from the whole flower stalk. They belong to a large group of tropical plants, the Bromeliaceae. Many bromeliads are epiphytes, plants that grow upon other plants, and many do not depend on roots for water and nutrients because they absorb these directly through their leaves. Spanish moss *(Tillandsia usneoides)* is a common epiphyte of the American Southeast, where it drapes tree branches and creates a distinctive landscape. Spanish moss is a relative of the pineapple; it is not a moss at all but a perfectly good angiosperm.

Easter lilies and garlic are appreciated by humanity for very different reasons, but both are members of the lily family, the Liliaceae. Onions and asparagus, among others, are equally popular members of this family. Members of the Liliaceae like garlic, onions, and shallots make many other food plants palatable and appealing.

The plant family with the most species is very poorly represented in the supermarket: the orchid family, or Orchidaceae. It has only one plant commonly used for food in North America, but we take it for granted. That is vanilla *(Vanilla)* a flavoring without which many foods would be almost tasteless. Originally taken from the fruit of a tropical orchid, its flavor has been synthesized by chemists, so that most of the vanilla we use nowadays is artificial. Orchids have the most specialized and the most wondrous of all blossoms, so appealing that collecting them is a highly specialized and absorbing hobby for those who can afford it (one of the best-known orchid collectors was Rex Stout's fat, fastidious, fictional detective, Nero Wolfe).

A GLIMPSE OF VARIATIONS: HABITAT AND STRUCTURE

This tour of the supermarket certainly does not exhaust the variations among angiosperms, nor does it explain them. Examining the environments in which these diverse plants arise provides clues to the source of this wondrous variability. Angiosperms literally blanket the earth in all habitats except for deep water and regions of perpetual snow. With these differences in environment, we see variations in the structures of plants.

One variation is size, and the range is enormous. The smallest of all flowering plants are those commonly called duckweed (Lemnaceae), often confused with floating algae. Duckweeds have only one or at most a few leaves, but they do produce a flower. They float on freshwater ponds, streams, and lakes and, if growing conditions are good, may cover the surface completely. During the growing season they reproduce not by *sexual reproduction,* as we discussed on page 79, but by **vegetative reproduction:** a small bud breaks off the parent plant and forms a full-sized miniature plant. (We will discuss this kind of reproduction further in Chapter 12.)

A common species of duckweed is *Lemna minor* (see Figure 3.11). *Minor* is a good description, for its leaves are only a fraction of an inch across. Yet it grows in such quantity that seen through the trees beside quiet lakes of the American north woods, it can give the impression of a green meadow.

From the insignificant-looking duckweed, we glance up to one of the largest flowering plants, the eucalyptus tree *(Eucalyptus)* of the tropical forests in Australia. A large eucalyptus may reach more than 90 meters (300 feet) into the sky, close in height to the largest trees in the world, the giant redwoods of California. For all this huge difference in size, *Lemna* and *Eucalyptus* have the same basic structure, and their reproductive systems, too, are very similar.

Another important variation in the angiosperms is life span. Many **herbaceous** ("nonwoody") plants live, flower, and produce seed in one growing season, then die. They are the **annuals,** including such familiar plants as dandelions, petunias, and peas. Other species, the **biennials,** will live for two growing seasons. Beets, cabbage, carrots, and foxglove are all biennials. They flower and produce seed in their second year. Then there are the woody shrubs and trees that live for many years, the **perennials.** Most of them flower each year, but some, like bamboo, flower infrequently. The aboveground parts of some perennials die back each winter, but new shoots will develop from the base of the stem or from underground parts; potatoes and tulips are examples. Similar reproductive strategies are seen in animals. Many invertebrates live a brief span and produce thousands of offspring, while humans and elephants live a long time and produce few offspring.

Terrestrial Plants The best place to examine plant diversity is in the tropical rain forest (shown in Figure 1.10). Rainfall is plentiful here and this, combined with a great deal of sunshine, produces an environment favorable for the growth of many kinds of plants, from epiphytes to tall trees. As as result, the average number of species per square mile is many times greater in a rain forest than in a New England or Midwest forest in America, where a temperate climate and pronounced seasonal changes—hot summers, cold winters—limit the number of plant species. In fact, diversity is greater in the tropical rain forest than anywhere else on earth.

Plants of the arctic and regions of high altitude are faced with severe problems, and their forms differ drastically from those of the tropical forest. For one thing they are small, so they require relatively little moisture and nutrients. They remain close to the ground and are firmly anchored, an adaptation permitting them to cope with the strong winds that prevail in arctic and high alpine regions.

Aquatic Plants Marine angiosperms have their own environmental problems. Water is abundant, because they live in it, but they must cope with salt (how they adapt we will see on page 288).

Freshwater aquatic plants also show interesting adaptations to the extremes of life in the water. The chief problem of these **hydrophytes** ("water plants") is availability of light and air. Plants like the

3.11 Duckweed *(Lemna minor)* is one of the smallest of the flowering plants (above, right). It consists of one to five leaves, each with a single root. Three tiny flowers are produced from a cleft in the margin of the leaf. The smallest flowering plants are members of the genus *Wolffia,* which also belongs to the monocot duckweed family (Lemnaceae). Duckweed can form a carpet on still waters in both temperate and tropical regions. This duckweed (above, left) is on a pond in the Botanical Gardens in Calcutta, India. This garden played an important role in the plant explorations and introductions associated with the British Empire.

3.12 The bumblebee flower (*Pedicularis lanata,* family Scrophulariaceae) at left occurs in the tundra of northern Alaska. At this latitude, the south side of the plant receives the most sunlight and is warmest. As a result, flowers on that side open first. Note the relative size of the floral and vegetative portions of the plant. In regions with limited growing season, annual plants must have a rapid life cycle.

3.13 Water lilies and water hyacinths possess different adaptations for survival in an aquatic environment. The water lily is firmly rooted in the earth and sends up leaves and flowers to the surface on long stalks. The leaves are like a diver's platform providing air conduits for what lives underwater. The water hyacinth, however, is like a life raft at the mercy of wind and currents.

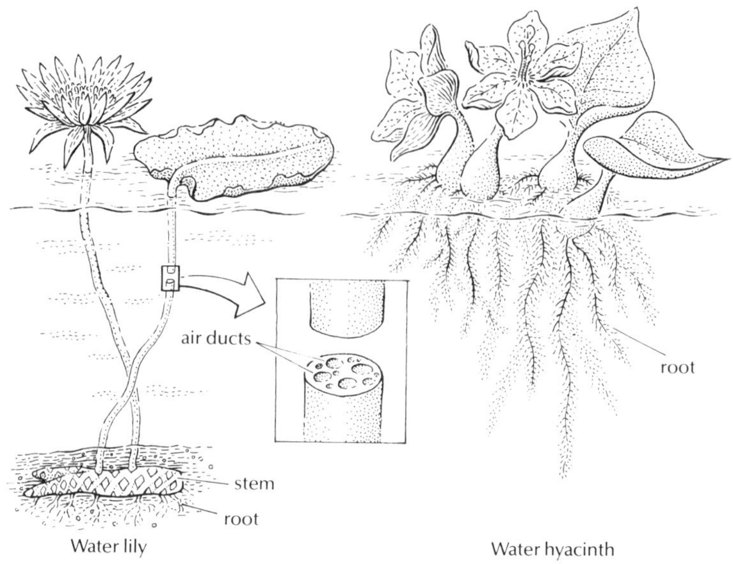

duckweed float on the surface and get their air directly through the leaves. The water hyacinth *(Eichhornia)* is another floating plant and a very mobile one (see Figure 3.13). Water hyacinths reproduce so rapidly (by vegetative means) that they are a major pest in many areas, including our Southeast, where they spread in such profusion that they clog canals and other waterways and are difficult and expensive to control.

Other plants, less fully adapted to life in the water, retain some characteristics of plant life on land, such as roots that draw nutrients from the bottom soil. One is the highly specialized water lily *(Nymphaea)*, shown in Figure 3.13, which makes the best of two worlds. Its roots grow into the pond bottom like any plant on land and its long stems reach up to the surface. There, broad leaves absorb air and light. An elaborate system of channels pipes air down to the roots.

Plants that are completely submerged are able to photosynthesize at considerable depths and can also absorb gases dissolved in the water. The minerals they require come from the water as well as through roots branching into the bottom soil.

Meat Eaters and Parasites A fascinating adaptation to extreme conditions is that of **insectivorous** ("insect-eating") plants, which live where the soil is poor in nitrogen and therefore must find other ways of getting that nutrient, which plants require in large amounts for growth. A wide variety of adaptations helps them to trap insects. Some, like the sundew, catch them as if with flypaper, using sticky secretions; some lure insects with nectarlike secretions into deep pitchers formed by modified leaves (see Figure 3.14); and some, like the Venus's-flytrap, catch their prey with a trap that snaps shut. All these elaborate specializations allow the plant to obtain nitrogen, for these plants live in bogs and other places where it is scarce. Insects have it, and the plants are adapted to catch and feed on insects. (This highly interesting aspect of plant adaptation will be discussed in more detail in Chapter 9.)

Then there are plants that live by growing on and in other plants, the true parasites. One of these is mistletoe *(Viscum album),* hung in doorways at Christmas. This plant, which grows on the branches of certain trees in temperate and tropical regions, can photosynthesize to some extent. But mistletoe mainly lives off its host, which in the end can be destroyed by it. The mistletoe puts out rootlike processes called **haustoria** along its stem to grow through the bark and tap the tree's sap, providing the parasite with food, minerals, and water at the host's expense.

Another parasitic adaptation is seen in the snow plant *(Sarcodes sanguinea),* a fleshy bright red native of the high Sierras in California. Its common name derives from its early appearance in the spring, before the snow has melted. The snow plant is a root parasite, tapping the roots of trees and stealing nourishment from them. In midwestern and eastern forests, the root parasite Indian pipe *(Monotropa uniflora)* is common.

We will look at these and many other plant adaptations as we examine the structures, functions, and evolution of the flowering plants in more detail in later chapters. But first we will look at another part of the green world: plants without flowers.

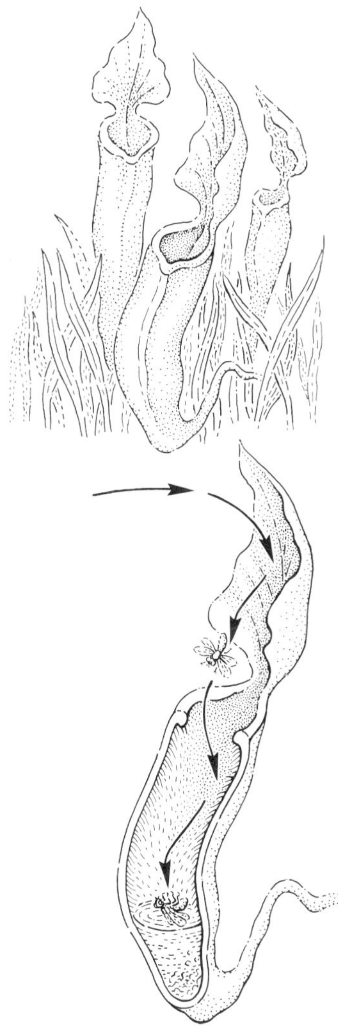

3.14 Carnivorous plants occupy two trophic levels: primary producers and primary consumers. Like other carnivorous plants, the pitcher plant (right) lives in nitrogen-poor soils. The unfortunate bug entering the one-way chamber formed by modified leaves will provide the plant with nitrogen fertilizer.

3.15 The parasitic mistletoe (left) obtains water, minerals, and some food from the host plant by means of rootlike structures called haustoria that penetrate the host tissues. Unlike some other angiosperm parasites, mistletoe is green and conducts photosynthesis.

Summary

The history of taxonomy goes back at least as far as Aristotle's writings, but the classification system devised by the eighteenth-century physician Carl Linnaeus is a more direct ancestor of today's classifications. Linnaeus consistently applied a binomial system for naming plants, and he developed a more comprehensive classification than his predecessors had, a scheme based on similarities in reproductive structure.

Taxonomists still use a binomial system derived from Linnaeus's achievement, but they attempt to devise classifications based not just on similarities of form, as Linnaeus's was, but on the natural, genetic relationships among organisms. Our understanding of these relationships was much advanced by Charles Darwin, who proposed that natural selection was the principal explanation for the evolution of organisms. The differences among organisms are not fixed for all time, and, it follows, the distinctions among groups of organisms, which classification systems reflect, are neither fixed nor absolute. There is thus no one "correct" classification system, nor can the task of classifying end. Although classification schemes are to some extent arbitrary inventions, they provide an invaluable guide through the diversity of the plant kingdom.

Most plants are flowering plants—angiosperms. They share a similar structure, which includes a trunk or stem, roots, leaves, flowers, and fruits, and a similar life cycle.

Angiosperms have reproductive structures called flowers. The anther at the end of the stamen produces pollen, which contains the male gametophyte, which produces the male sex cells (gametes). During pollination pollen grains are deposited on the stigma at the tip of the style; both stigma and style are part of the female reproductive organ, the pistil. Some plants depend on the wind to ensure that pollination occurs, others are pollinated by insects, bats, or birds, and some are self-pollinating.

After pollination, a pollen tube containing the male gametophyte grows from the grain and eventually passes down the style into the female gametophyte inside the ovule. Fertilization then occurs: one male gamete nucleus fuses with the nucleus of the egg cell in the female gametophyte. The fertilized egg will produce the embryo, which will develop into the mature plant.

The nucleus of the other male gamete fuses with other nuclei of the female gametophyte, eventually producing the endosperm, which will provide food for the embryo and the immature plant. Embryo and endosperm develop to form the angiosperm's seed; it is enclosed in a fruit, which develops from the flower's ovary.

Angiosperms, like all other plants, have an alternation of generations. In angiosperms, the gametophyte generation produces gametes, which fuse during fertilization and form a zygote, which produces the sporophyte. The sporophyte produces spores from which the gametophytes arise.

There are about three hundred families of angiosperms but only two major classes are recognized: monocotyledons, whose seeds have one cotyledon, and dicotyledons, which have two. Other differences among these two classes include the pattern of veins in the leaves and the number of parts in the flowers.

Size is a more obvious variation among flowering plants. Yet another is their life span: angiosperms may be annuals, biennials, or perennials. They are diverse in habitat too, being found in arctic as well as tropical climates, in water as well as on land. Although all photosynthesize, some such as the Venus's-flytrap and sundew obtain nitrogen by capturing insects, and some are parasites, such as mistletoe and snow plants and obtain most or all of their nutrition from the host plant.

Review Questions

1. Why are the contributions of Linnaeus considered so significant in the history of botany?
2. What are some differences between classification and identification?
3. Once it is decided that a new book is about philosophy, history, or political science, it can be placed within a library cataloging system. Compare the problems of book classifiers with those faced by plant classifiers.
4. What are the primary characteristics of angiosperms?
5. Most of our edible plants are flowering plants. What connection might this have with pollination and fruit dispersal mechanisms of many flowering plants?
6. Describe the "alternation of generations" as it applies to flowering plants.
7. List any questions you have about adaptations of flowering plants. Save your list and read it again after you have finished the book.

SUPPLEMENT

Cells

1 Cells of blue green algae (below, left) and bacteria are characterized by the absence of membrane-bound structures called organelles. These include the nucleus (N), chloroplasts (C), and mitochondria (M). The absence of a cellulose cell wall (CW) and chloroplasts is one of the major differences between an animal cell (below, right) and a plant cell (facing page). They possess many of the same kinds of organelles. The soft tissues of plants are composed of thin-walled cells called parenchyma cells. Plant cells are characterized by a rigid cell wall, a watery vacuole (V), and, if green, chloroplasts.

Cells are the building blocks of life, the structural units that distinguish rocks from gorillas and orchids. All of the life-sustaining processes (respiration, energy production, reproduction, etc.) take place because of cells. The two basic kinds of cells are prokaryotic and eukaryotic. **Prokaryotic cells** have a relatively unspecialized internal structure and are unique to unicellular organisms such as bacteria (see Chapter 4). All plants and animals are made up of **eukaryotic cells,** which are different from prokaryotic cells in that they contain a nucleus (see Figure 1).

All cells contain **protoplasm,** the material that is life. In plant and animal cells, the protoplasm is bounded by the **plasmalemma,** which regulates the flow of nutrients into the cell and wastes out of it. Protoplasm consists of specialized structures called **organelles** and a viscous mass called **cytoplasm.** The **nucleus,** which is an organelle, is the cell's control center. It

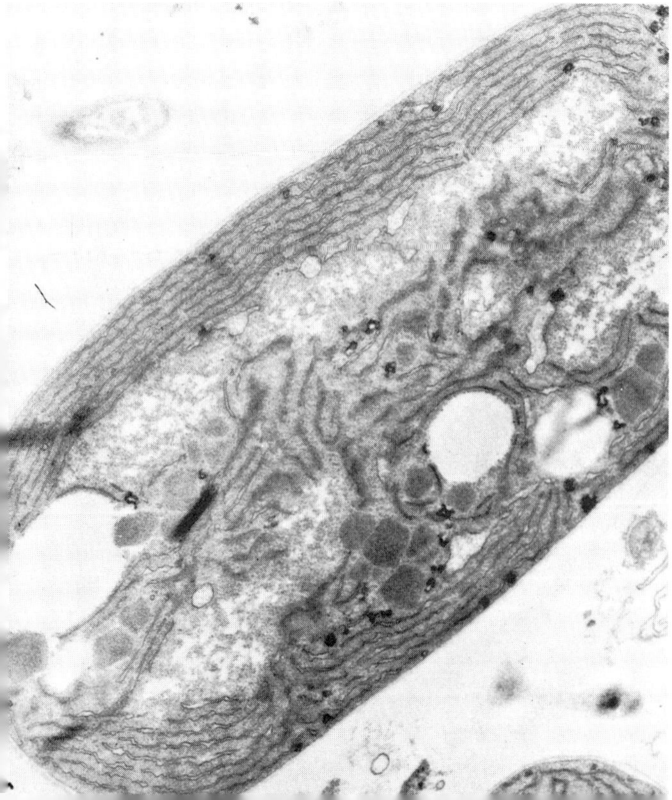

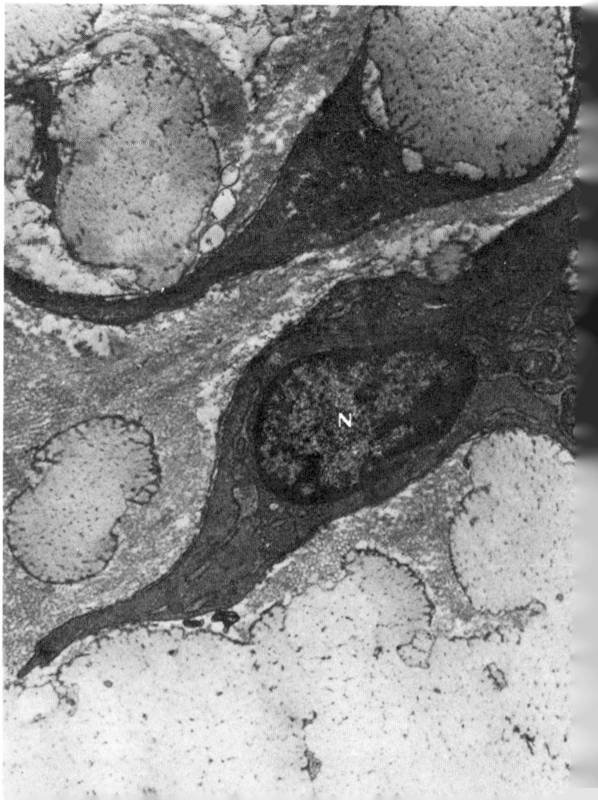

contains the genetic material that specifies how the cell will develop and function. Two **nuclear membranes** surround the nucleus. Other organelles include **mitochondria,** rod-shaped bodies vital to energy production; **Golgi bodies,** membranous structures involved in the synthesis of complex chemical compounds; and the **endoplasmic reticulum,** a complex network of membranes that aid in cell metabolism (see Figure 2). The cytoplasm contains many small structures called **ribosomes,** where protein synthesis takes place (see Chapter 11). Cells reproduce by dividing to form two identical cells from one.

Unlike animal cells, plant cells are surrounded by a rigid, nonliving wall that is made up primarily of the carbohydrate cellulose (see Figure 3). The wall has a lot to do with the way a plant grows. It also provides structural support (see pages 200–201). Extending through pores in certain regions of the cell wall are strands of cytoplasm that connect individual cells.

Another difference between plant and animal cells is that plant cells contain discus-shaped organelles called **plastids;** green plastids, or **chloroplasts,** are the site of photosynthesis. The most conspicuous element in mature plant cells is the **vacuole.** It is filled with a watery sap and is separated from the cytoplasm by the **tonoplast.**

There are many variations on the basic structure of a cell that allow cells to carry on specialized functions (see Figure 4). The shapes of cells vary considerably, as do the composition and thickness of the cell wall. Some specialized plant cells store food. Others transport it. Still others provide structural support. The list of functions could go on and on.

Some plant cells are as large as 200 millimeters (8 inches) in length, but most are too small to be seen without a microscope. Our knowledge of the internal structure of a cell is limited largely to what we can see with an

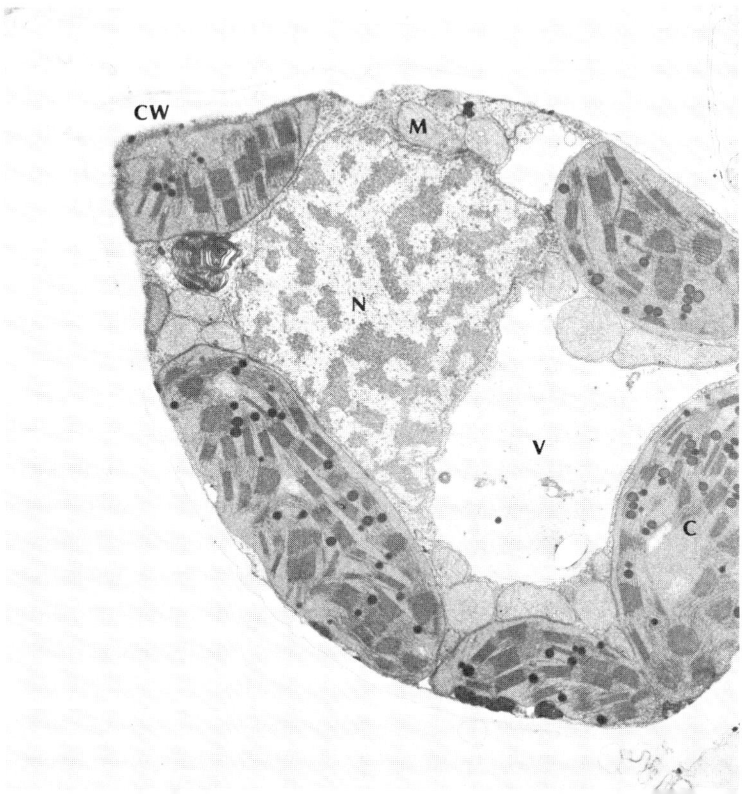

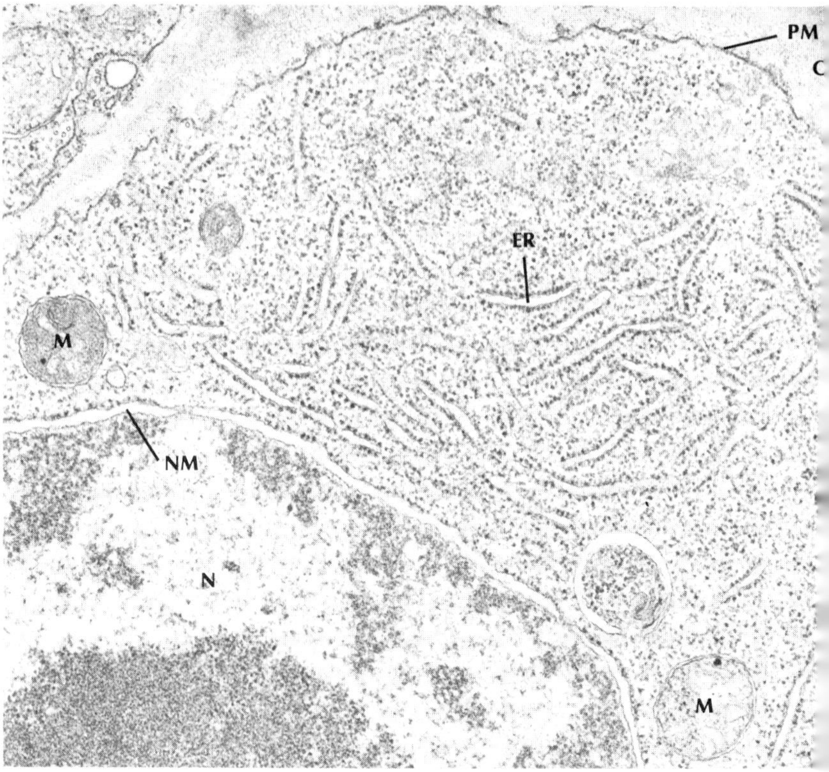

2 The endoplasmic reticulum (ER) is a complex system of membranes within the cells of plants, fungi, and animals that is involved in many synthetic activities in cells. In this plant cell, nucleus (N), plasmalemma (PM), mitochondria (M), and nuclear membrane (NM) are all clearly visible.

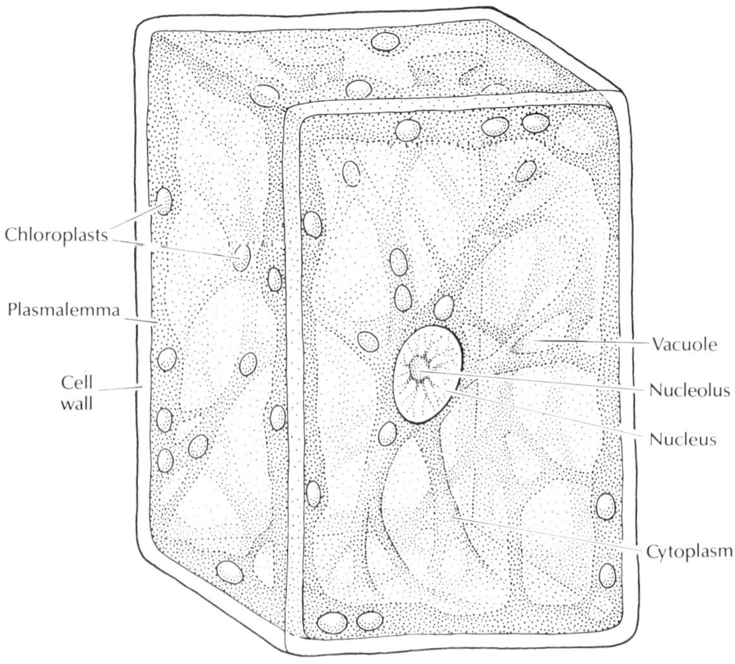

3 This schematic representation of a cell shows it as a box filled with strands of cytoplasm traversing the watery vacuole. The cytoplasmic strands are constantly changing, so a true representation of this cell would show a new pattern every second.

electron microscope, an invention that has enabled us to view the organelles (the resolution of the best electron microscopes is about 3 angstroms, or 0.0003 millimeters). With an electron microscope it is possible to distinguish the structural features of membranes and ribosomes and even to discern large molecules.

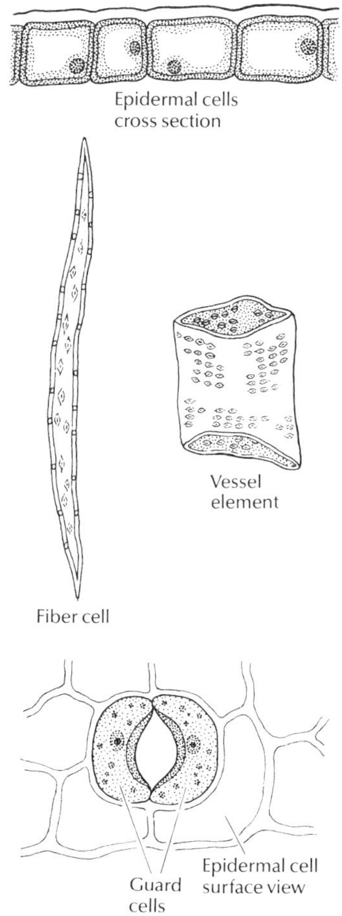

4 Plant cells are less diverse in both structure and function than animal cells, but there is a lot of variety. The cells illustrated range from those in the epidermal layer of leaves to vessel elements that make up wood.

CHAPTER 4

The Green World: Plants without Flowers

Angiosperm, we have seen, means "enclosed seed," and indeed all flowering plants have their seeds enclosed in a fruit. There are other plants, such as the familiar pine trees, whose seeds are not so enclosed. These are the **gymnosperms,** or "naked seed" plants, which have no flowers. Still others—the ferns and their relatives—have neither seeds nor flowers. All these plants are similar in that they are **vascular;** that is, they have cells that are specialized to conduct water, minerals, and nutrients between the plant's organs. Other plants, we will see, have neither flowers nor seeds nor a vascular system.

THE GYMNOSPERMS

The term *gymnosperm* illustrates the relativity of taxonomic nomenclature. It has been used to designate a subclass, a class, and a subdivision. As used here, it will designate a group of divisions with the common character of having seeds borne in cones.

The gymnosperms are a more ancient group than the angiosperms. In fact, primitive gymnospermlike plants might have been the ancestors from which, eons ago, the angiosperms evolved. Pines, fir trees, spruces, redwoods—all these are familiar gymnosperms today. They, and all gymnosperms, are like angiosperms in being vascular plants and having seeds, but we can distinguish them from angiosperms by the way they reproduce.

4.1 Pollen is discharged in abundance from male cones of pine trees, in this case a Jack pine (*Pinus banksiana*).

Insect pollination, so common in the angiosperms, is absent in gymnosperms. Instead, they characteristically depend on wind-borne pollination, which is bound to be very much a hit-a-few, miss-a-lot affair. It succeeds by sheer extravagance. Gymnosperms produce so much pollen that some of it inevitably lands where it will fertilize an egg. Walk in a pine forest in late spring or early summer and you will see pollen mantling everything in gold. Even on lakes and ponds a golden carpet can cover the water.

The best-known gymnosperms, the **conifers,** such as pines and fir trees, carry their seeds on scales, the shelflike structures of which the cone is constructed. The seeds develop on the surface of these scales and are not enclosed within cone tissue (see Figure 4.2). The cones contribute to the conifers' success. Until it matures, opens, and lets the seeds drop free, the cone gives the seeds almost armor-plated protection. Some pine cones, including those of the jack pine

4.2 CONIFERS

Conifers are an order of gymnosperms, and include the Pinaceae, the most common gymnosperm family and the one usually referred to when people speak of conifers. All conifers bear reproductive organs and gametophyte generations in structures called *cones.* Examples of different cones and foliage are provided by white fir *(Abies concolor)* (below, left), Sargent's weeping hemlock *(Tsuga canandensis,* variety pendula) (facing page, top), and Eastern white pine *(Pinus strobus)* (below, right). Like all plant life cycles, those of conifers (see the life cycle of pine, facing page) have sporophyte and gametophyte generations. Sporophytes (the trees) bear two types of cones. Male cones are small and short-lived and produce winged pollen (male gametophytes). Female gametophytes develop in ovules on the surfaces of cone scales of the female cones. A pollen tube penetrates ovule tissue and fertilization is accomplished when a sperm nucleus unites with an egg cell nucleus. Unlike flowering plants, there is no double fertilization, and conifer seed endosperm is derived from the female gametophyte. After seeds mature, scales of the cone separate and the naked seeds fall out. They are not enclosed by a fruit.

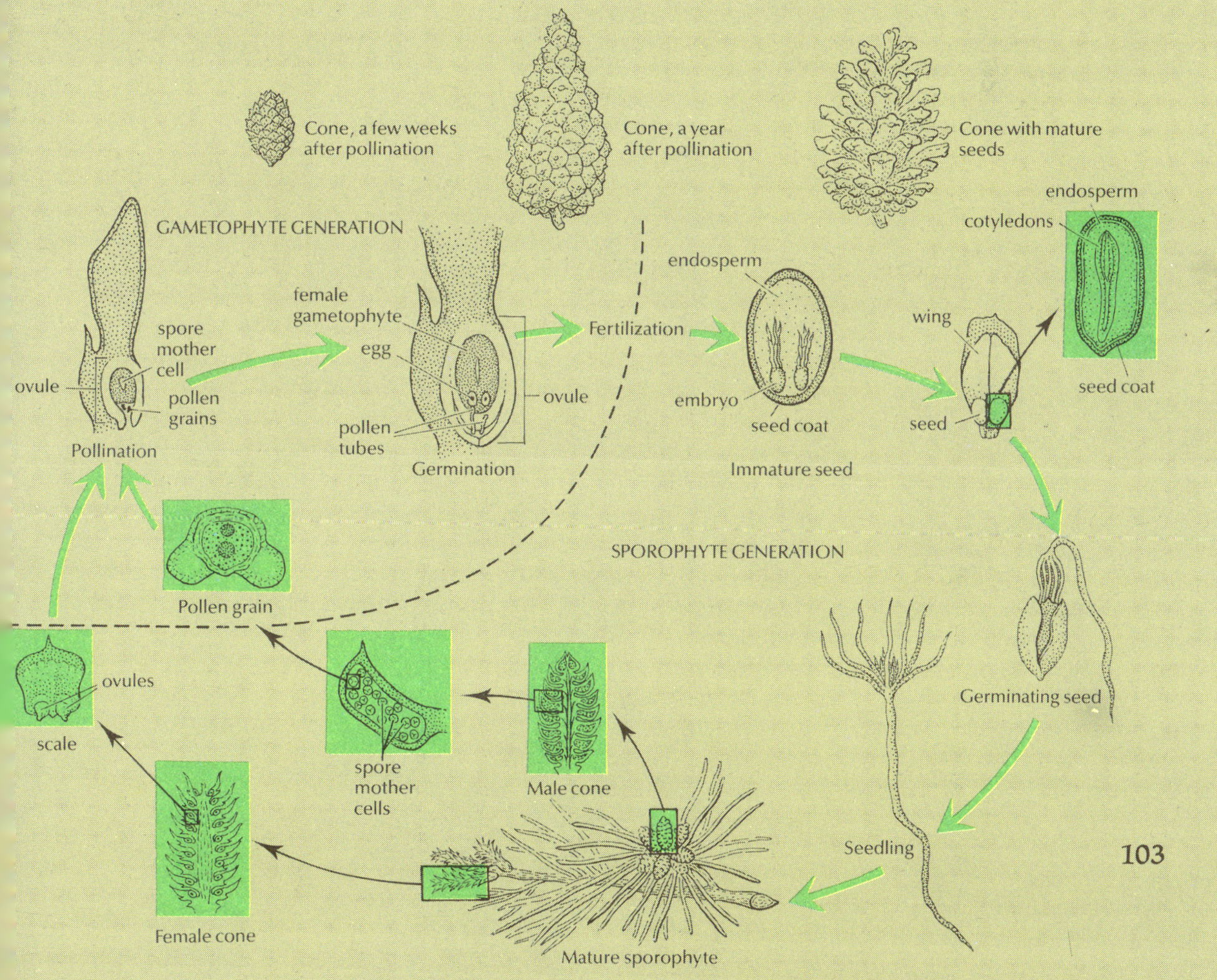
Cone, a few weeks after pollination
Cone, a year after pollination
Cone with mature seeds
GAMETOPHYTE GENERATION
spore mother cell
ovule
pollen grains
Pollination
female gametophyte
egg
ovule
pollen tubes
Germination
Fertilization
endosperm
embryo
seed coat
Immature seed
endosperm
cotyledons
wing
seed
seed coat
SPOROPHYTE GENERATION
Pollen grain
Germinating seed
ovules
scale
spore mother cells
Male cone
Seedling
Female cone
Mature sporophyte

in the upper Midwest, protect the seeds even through fires. The fire's heat causes the cones to open, and the seeds drop to the ground. As a result, the jack pine is among the first plants to take hold and grow again in the ruins left by fire.

Another difference between the two groups is that gymnosperms do not have the double fertilization characteristic of angiosperms. The embryo's development begins by fertilization of the egg by sperm, but the endosperm of a gymnosperm seed develops from the female gametophyte alone, not from the union of another male gamete with a female one, as in the angiosperms.

The Cone Bearers Among the gymnosperms the most familiar division is the Coniferophyta and its most familiar family is the Pinaceae. These are the evergreens, with their cones and needle-shaped leaves: pines, firs, spruces, hemlocks, and junipers all belong to this group. The cone is really a stem with modified branches on the surface of which seeds can develop. There are both male cones, producing pollen, and female cones, producing seeds. The female cones are produced on short lateral branches and are larger than the male cones. Each "branch" that makes up the cone has scales; these produce the female gametophyte.

Each spring the male cones are produced in clusters near the tips of branches; they will wither and fall off after releasing pollen. The pollen of conifers, like that of angiosperms, is the structure in which the male gametophyte is formed. This pollen is adapted for air transport by winglike extensions of the pollen grain's wall. At the same time that pollen is released, the scales of the female cone separate slightly, and pollen blows into the cracks. The cones subsequently close, capturing the pollen in a moist chamber.

Although there are great differences, reproduction in pines is similar to that in other conifers in its main features. In pines at least a year will elapse after pollination while the male gametophyte produces a pollen tube with sperm. The tube slowly grows toward the female gametophyte developing on the surface of the cone's scales. Fertilization of an egg cell by a sperm from the pollen tube will take place during the summer, as long as fourteen months after pollination. This extended fertilization period is in contrast to that of angiosperms, for which fertilization usually follows rapidly after pollination.

Three years after the cones formed, the pine's seeds are ready to be dispersed. The scales of the cone separate; the seeds, which have been resting on the surface of the appressed scales, are now separated from the scales and fall.

Many conifers have large, nutritious, and tasty seeds. These seeds were important in the diet of native Americans in California and the Southwest, and the seeds of the pinyon pine (pine nuts) are still a highly prized food. Conifer seeds illustrate the principle that the size of the seed has nothing to do with the size of the plant. Giant redwoods have seeds about as big as those of a large tomato; the 10-meter (30-foot) digger pine has massive cones with seeds the size of small almonds.

Members of the redwood family, Taxodiaceae, are also conifers, and the coastal redwoods *(Sequoia sempervirens)* grow taller than any other tree. Of these the highest reaches 112 meters (364 feet) into the sky, about as tall as a thirty-six story building. The most massive, and among the oldest, of plants are the Sierra redwoods *(Sequoiadendron giganteum),* or "Big Trees," of California.

The bald cypress *(Taxodium distichum)* of the southeastern United States also belongs to the redwood family, as does the rare dawn redwood *(Metasequoia glyptostroides)* of China, a tree that differs from

4.3 The Sierran redwoods, or "Big Trees" *(Sequoiadendron giganteum),* are restricted to the Sierra Nevada range of California and rank with bristlecone pine as the oldest tree species in the world. As their name implies, Big Trees are the most massive of trees, but are not as tall as their cousin, the coastal redwood (*Sequoia sempervirens*).

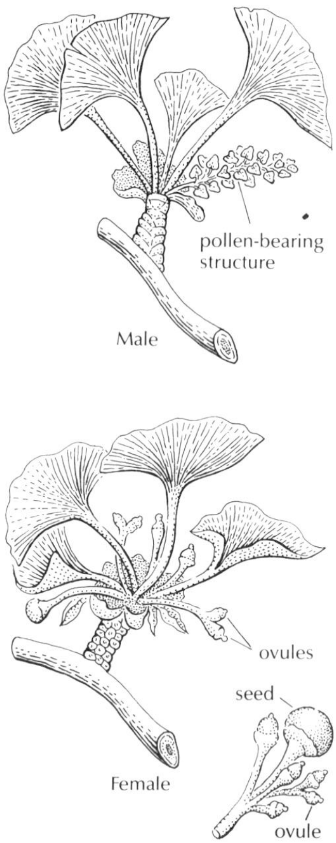

4.4 Ginkgo, like other gymnosperms, has seeds that are not enclosed in fruits. The female gametophyte is borne within the naked ovules. After fertilization the ovules develop into seeds.

the American redwoods but resembles the bald cypress in shedding its leaves each year. By contrast to most **deciduous** or ("leaf-shedding") trees, however, the dawn redwood does not lose the leaf alone; the short shoot to which the leaf is attached goes with it, breaking off at the base where it is attached to the branch.

Redwoods invite interest not just by their size but also because they appear to be in the final stages of a history that goes back some 50 million years. They were very common then in temperate forests, but an increasingly rigorous, dry, cold climate gradually confined them to the isolated pockets where they exist today. The dawn redwood may already be extinct in nature; the California redwood, in terms of geologic time, probably does not have very long to go before it, too, joins the host of other plants that, unable to adapt to changed conditions, die out.

Living Fossils: Ginkgo Trees and Cycads So far we have a picture of a gymnosperm as a cone-bearing tree with needle-like leaves, but not all gymnosperms are like this. There is, for example, the ginkgo tree, or maidenhair, which has broad leaves and looks like a large angiosperm (see Figure 4.4). Like other gymnosperms, however, its seeds are not enclosed in fruits. In fact, the seeds of the ginkgo, which superficially look like fruits, are borne at the ends of short thin stems. Already extinct in the wild, the ginkgo tree *(Ginkgo biloba)* is the only living species of its division, the Ginkogophyta. Its fossil record has been traced back 300 million years. This "living fossil" is planted in many places as an ornamental tree. You may know it from the streets of your town or city; the seeds produced by female trees smell like rancid butter. It has been saved from total extinction because of its appeal to humans.

The Cycadophyta are another very old gymnosperm group. Fairly common in semitropical and tropical areas, cycads are often mistaken for large ferns or small palm trees. They were very widespread in the Mesozoic forests, 130 to 200 million years ago, helping to give these forests their characteristic look. One genus, *Zamia,* is native to Florida. Today only a few species of cycads remain, and these are restricted to warm and humid places. Many live as ornamental plantings in greenhouses, but, like redwoods and gingkos, they are probably headed for extinction in the wild. They will not vanish tomorrow, but in the sense of geological time, the future of these relics of former ages is not bright.

Ginkgo and cycads have a reproductive cycle similar to other gymnosperms, but they differ in a very interesting way. Like other gymnosperms, they produce pollen, which is blown to the structures bearing the female reproductive organs, and pollen grains then produce pollen tubes, which penetrate through the tissues of the ovule and reach the female gametophyte. But unlike sperm in angiosperms and other gymnosperms, the two sperm cells in these pollen tubes can move. They have many whiplike appendages

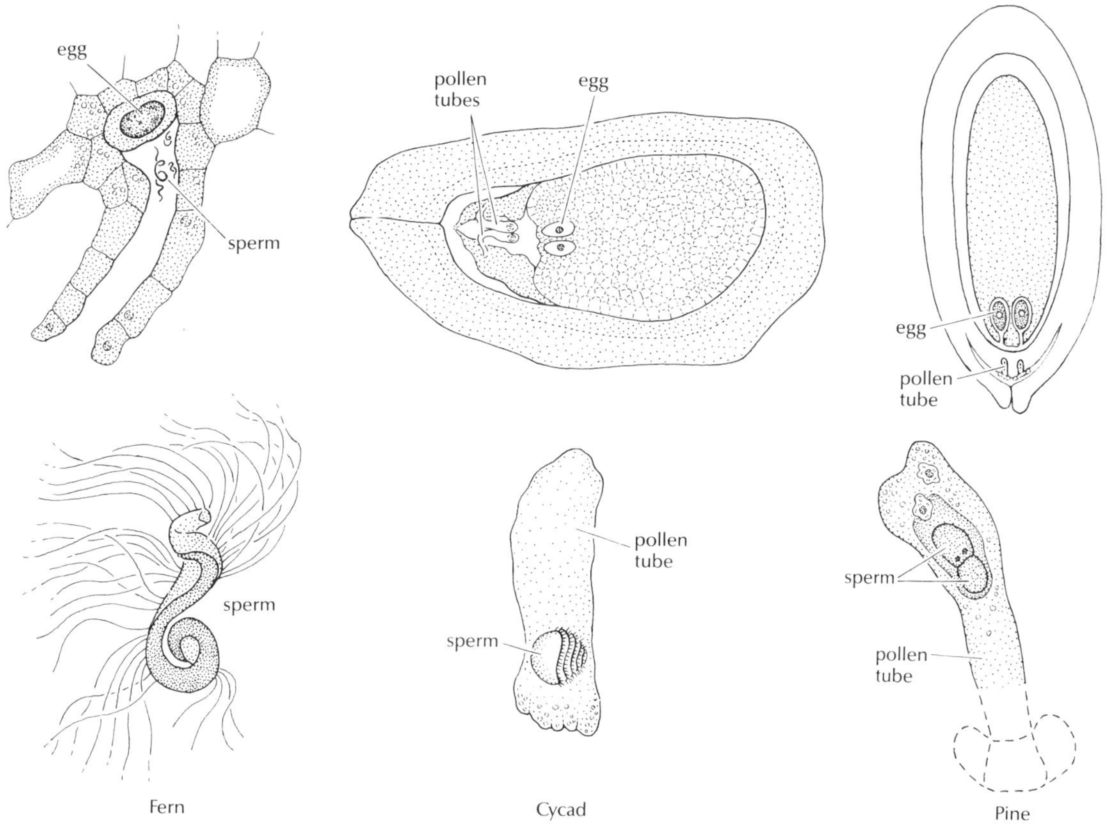

4.5 Plant sperm cells take many forms. Ferns and other seedless plants have free-swimming sperm. Cycads have flagellated sperm, which are discharged from a pollen tube. Pines, like the flowering plants, have nonmotile sperm, which are discharged directly into the cells being fertilized.

called **flagella,** and these flagella propel the sperm to the egg cells after they are discharged from the pollen tube.

Ginkgo, cycads, and the more ancient vascular plants to be discussed next all have motile sperm (see Figure 4.5). When motile sperm are launched from a pollen tube, they can only exist in a watery environment. The very specialized pollen tubes of conifers and angiosperms deliver the sperm cells to their targets within the liquid-filled tubes; the sperm are not exposed to the external environment. This is an adaptation ensuring fertilization in an arid environment. As plants evolved on land, the dry air was a major force affecting their evolution. Studies of reproductive structures in ancient and modern plant groups clearly show adaptation to the arid conditions that characterize the environment in which the aerial parts of land plants live.

LAND PLANTS WITHOUT SEEDS

Ferns Besides the angiosperms and gymnosperms, a commonly observed group of land plants are the ferns (division Pterophyta). Walking through a woodland with a canopy of gymnosperms or angiosperms overhead, we may see close to the ground the ferns' graceful-looking, upward-sweeping "branches." (The branches are really leaves.) Numerous as they are in temperate-zone landscape, such as damp forest floors and the edges of ponds and streams, ferns are much more diverse in the tropics. However much of nature's beauty you may have seen, a whole forest of palmlike tree ferns, with trunks 10 meters (30 feet) high, is impressive.

Ferns have specialized vascular cells much like angiosperms and gymnosperms, but they have no seeds at all. If you look on the underside of a fern leaf, you will find clusters **(sori)** of small round structures called **sporangia** (see Figure 4.8). The placement and form of the clusters mark the species of fern. In the sporangia are many separated cells with thick walls, the **spores**. Sporangia propel spores into the air much like a handful of steel shot would be launched from a sling. When fern spores fall on moist ground, they produce a green, sheetlike, multicelled organism, the gametophyte. It is **free-living** (needing no nourishment from the fern itself) and so small—usually less than a centimeter wide—that it escapes general notice. This gametophyte produces both male and female gametes in specialized structures known as **antheridia** and **archegonia**. Fern sperm are motile, swimming through the water drops on the surface of the gametophyte by means of many beating flagella and entering the egg-containing archegonia. When the egg is fertilized by the sperm, it produces an embryo that initially draws its nourishment from the surrounding gametophyte tissue. Eventually the embryo grows into a mature sporophyte, the familiar fern plant. The gametophyte withers and dies.

Thus the fern's alternation of generations is basically the same as that of angiosperms and gymnosperms: fertilization produces a sporophyte, and the sporophyte produces spores, which form the gamete-producing gametophyte. Unlike the spores of angiosperms and gymnosperms, however, the fern's spores separate from the sporophyte and produce a free-living gametophyte.

From tundra to tropics, from deserts to tropical rain forests, the fern has adapted to extremely varied habitats. Some ferns are even aquatic, adapted completely to life in the water. Like duckweed, some of these float on the surface; others sink their roots into the bottom mud and float their leaves at the surface.

All told, there are about ten thousand species of ferns, most of which exist in the tropics. The fern's part in the human economy, however, is almost limited to satisfying aesthetic appetites—as an ornamental plant it rivals flowering species. One exception is the orchid grower's unfortunate use of the fibrous trunks of tree ferns to hang their orchids on. For this one use tree ferns are being destroyed wholesale in some of their native habitats.

Remnants of the Earliest Forests The ancestors of ferns living today died out millions of years ago. We know of them because they grew in vast forests that were later compressed under geologic formations to make deposits of oil and coal, the trapped energy of these ancient plants. In the veins of coal we have an extensive fossil record of the ancient ferns, for the fossils consist of impressions showing many details of these plants long dead, most long since extinct. With the aid of these fossils, we can identify living descendants of these ancient land plants: the club mosses, horsetails, and whisk ferns (or, respectively, the divisions Lycopodophyta, Anthrophyta, and Psilophyta).

Few species of these three quite different groups live on. They are not very important in the plant world, although they are beloved by most botany texts. Like all the plants we have discussed so far, they are vascular plants because they have cells that are specialized to conduct water, minerals, and nutrients throughout their roots, stems, and branches. But the psilopsids—the whisk ferns, rarest of the three groups—are different from the others. They have no roots, and they either lack leaves entirely or have small leaves that look like scales. The fossil record suggests that they resemble early land plants (see the Supplement on page 382); some botanists even regard them as "living fossils." The steps of their life cycle are similar to that of the ferns, although there are specific differences in the form and location of their reproductive structures.

4.6 This photograph shows both vegetative and spore-bearing shoots of *Equisetum telmatei*. The vegetative shoots in the background indicate why they are called horsetails. Spores are produced in the tips of the shoots in the foreground and they will produce gametophytes. Ancestors of these curious plants formed the coal forests of 250 million years ago.

Club mosses (lycopods) are most familiar as the miniature evergreens in Christmas decorations. For many millions of years, they and the psilopsids, horsetails, and ferns were the principal dry-land vegetation on earth. The horsetails, consisting of a single genus, *Equisetum,* favor the damp areas along the banks of streams and ponds. Strange-looking plants, they have a naked stalk that produces circles of green branches all along its length. The outer cells of the stems hold deposits of silica, a major component of both beach sand and window glass, which makes the stems rough and hard. They were once known as "scouring rushes," the forerunners of our pot scourers.

Mosses and Liverworts Not all plants are vascular. Among the nonvascular land plants, those without specialized tissue for conducting food and water, the most common is moss. With fourteen to fifteen thousand species, mosses are found wherever there are moist conditions. They are very adaptable and can be found on the tundra, on the damp floors of deep forests, on rocks along the shores of streams, lakes, or ponds, on rocky overhangs where rainwater flows, even on many trees. They are small plants, seldom more than 15 centimeters (6 inches) tall, and they do not have the specialized conducting tissues of the vascular plants. Nor do they have true stems, roots, or leaves. Their cells absorb moisture through the air and directly from the ground, and these plants become **dormant**—living processes are reduced to a low level—when the environment becomes dry. (See Figure 4.9 on page 112.)

4.7 *Psilotum nudum* is one of the most primitive land plants. The whisk ferns have no true roots or leaves. They are thought to resemble very early land plants.

4.8 FERNS

Ferns are very diverse in form and size. The most spectacular are tree ferns. This 60-foot specimen grows in Tahiti (facing page). An important taxonomic feature of ferns is the arrangement of sporangia on the underside of many leaves. Spores are borne in sporangia which are clustered in distinctive groups called *sori* (singular, *sorus*). In the illustration of the undersurface of a fertile leaf of a polypody fern, (inset, facing page), each of the dots is a sorus. When spores fall on moist ground, they produce small, green, sheetlike gametophytes. Specialized organs on the gametophytes produce male and female gametes. (See the life cycle of fern, below). After fertilization, young sporophytes are nourished by gametophytes until the first leaves are produced and they begin to photosynthesize.

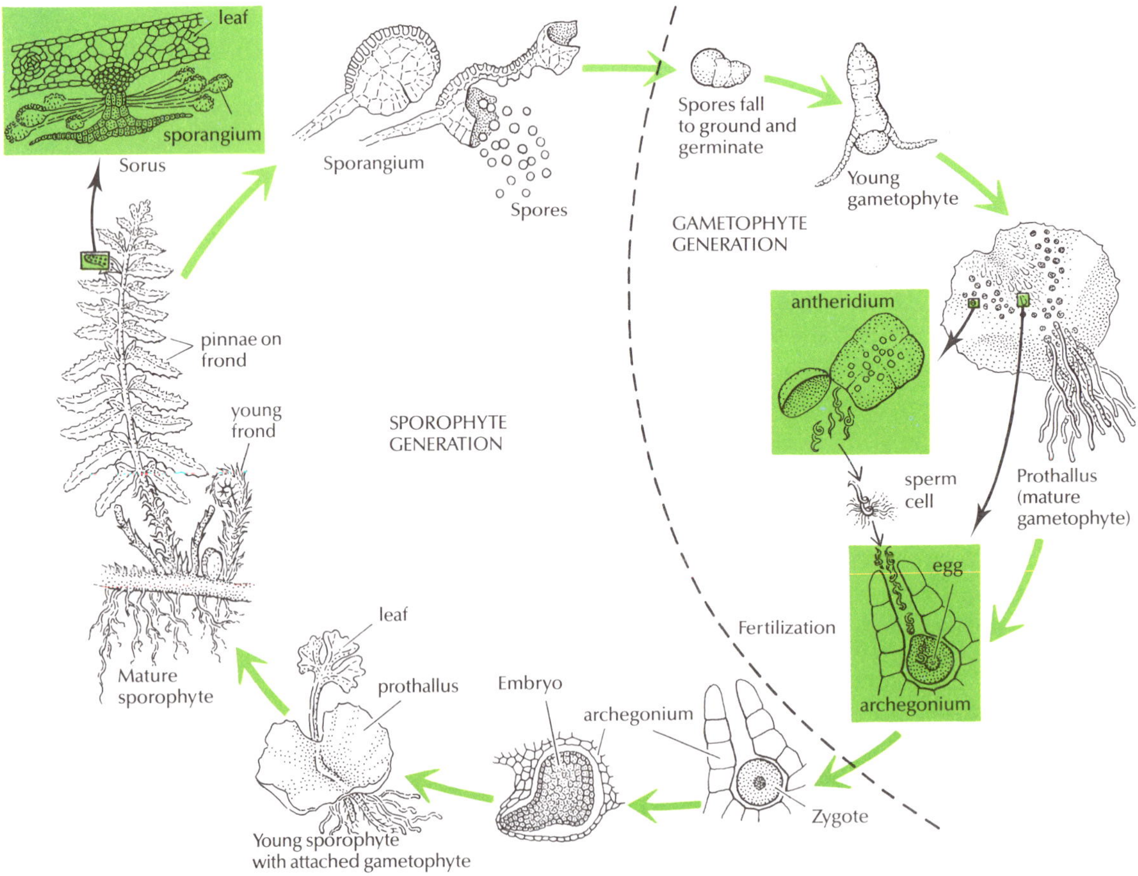

4.9 MOSSES AND LIVERWORTS

Although distinct groups, mosses and liverworts are both bryophytes and they exhibit similar life cycles. In the life cycle of moss (below), spores germinate and subsequent division gives rise to young gametophyte plants called *protonemas*. Rhizoids soon grow out of the protonemas and budlike structures form on some cells. The green leafy gametophyte develops from these buds. In mosses, both archegonia and antheridia may be present in adjacent plants or they may be widely distributed. Egg cells are produced in archegonia. Sperm develop within antheridia. Fertilization takes place and the zygote divides to produce a sporophyte plant that remains attached to the leafy gametophyte (see the photo of moss, facing page, top). The sporophyte consists of a stalk and a capsule full of spores. The caplike structure covering the end of the capsule is called the *calyptra*. Spores from the capsule fall to the ground and produce gametophytes. A single sporophyte plant may produce hundreds of thousands of spores, each of which can potentially develop into a new gametophyte. The liverwort illustrated (facing page, bottom right) is in the genus *Marchantia* and its flat, leaflike gametophytes are common on damp, shaded soil and rocks.

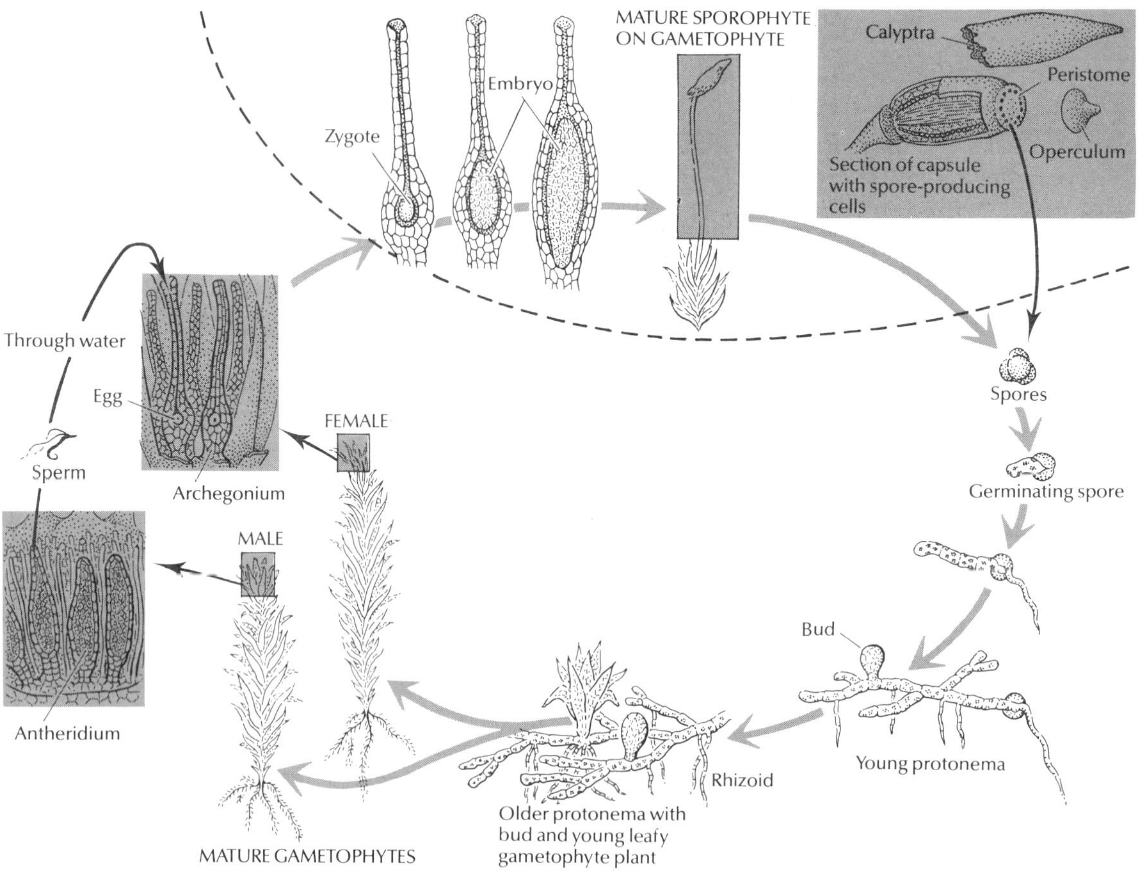

Some Differences Between Mosses and Liverworts

Mosses	Liverworts
Gametophytes usually erect and leafy, with leaves in ranks on central stalk	Gametophytes generally flat, with notched, plate-like structure
Leaves usually have midrib-like region	Leaves, if any, lack midrib-like region
Young gametophytes are filamentous protonema	Lack protonema stage
Multicellular rhizoids with chloroplasts	Unicellular rhizoids without chloroplasts
Complex, long-lived sporophytes	Less complex, short-lived sporophytes
Young sporophytes are green	Sporophytes are not green
Sporophytes have peristome	Sporophytes lack peristome
Common in Arctic regions	Not common in Arctic regions

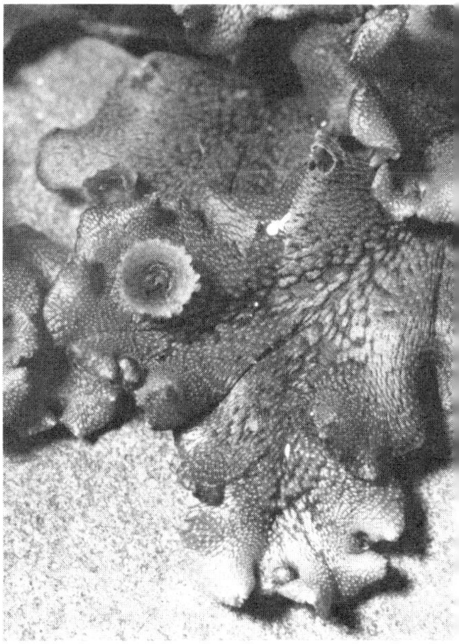

Most often the part of mosses we notice is a gametophyte, not a sporophyte. The dense green gametophytes have leafy stalks. At the tips of these, male and female gametes are produced in specialized structures (antheridia and archegonia). After the male and female gametes fuse and an embryo grows, the sporophyte forms. It obtains its nutrients most of the time from the gametophyte. Thin brown stalks, the sporophytes have capsules at their tips; in these the spores are produced. If they fall on a moist place, the spores will produce a filamentous gametophyte that later forms the familiar moss plant.

In the same damp and shady places grow relatives of the mosses, the liverworts. (Both mosses and liverworts are members of the division Bryophyta and are called bryophytes.) So named because of their supposed resemblance to the liver, they are frequently green platelike structures growing flat on the ground. These are gametophytes; the sporophytes are even less obvious and depend on the gametophyte for nutrients. There are about nine thousand species of liverwort, and they are found in moist habitats throughout the world.

AQUATIC PLANTS: THE ALGAE

Some vascular plants live in water, but most of the aquatic plants are nonvascular and are called algae—a group consisting of almost twenty thousand species. Some algae are relatively simple forms and are probably much like the first plants to appear on earth. Others, the giant kelp of the ocean, grow almost as tall as the tallest trees and possess a variety of specialized cells. Algae are extremely important because, as we saw in Chapter 1, they are the base of the food chain in oceans and lakes.

All algae have chlorophyll, and they are all capable of photosynthesis, but not all are green. In some species another color masks the characteristic green of the chlorophyll. The dominant color is the key for identifying algae: we speak of brown (Phaeophyta), red (Rhodophyta), and green (Chlorophyta), indicating the different divisions. In their often complicated life cycles, algae have an alternation of generations such as we have seen in other plants. The way the gametophyte and the sporophyte are produced, however, and the dominance of each generation can vary greatly from one group of algae to another.

Brown Algae Many algae are what we usually call "seaweed," the aquatic plants along the seacoast. Among the most common and easiest to identify is rockweed (the genus *Fucus*) with its many small air bladders floating it to the surface. The giant kelps such as *Laminaria* and *Macrocystis* are found farther offshore. Both

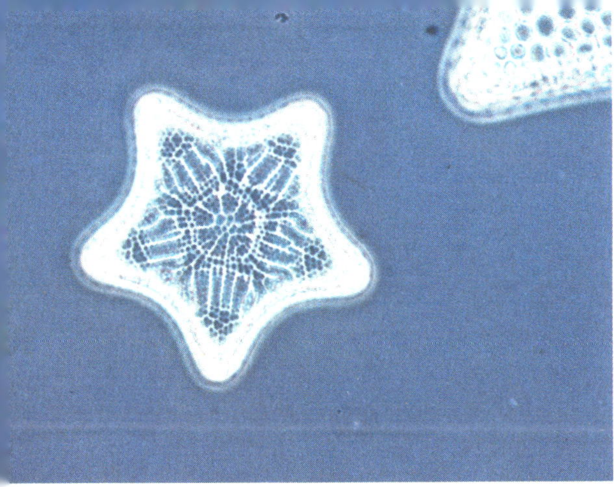

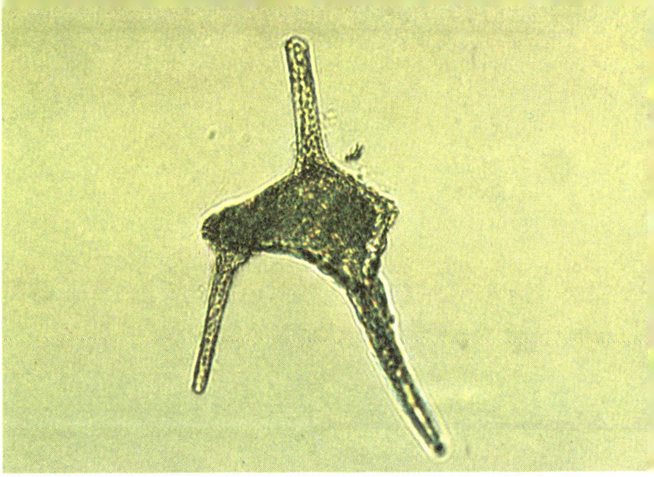

4.10 Diatoms (top left) are among the most beautiful forms of microscopic life. Their cell walls contain silica, and deposits of cell walls (diatomeaceous earth) have many uses. Dinoflagellates (top right) are important components of plankton, so important in oceanic food chains. The unicellular species moves about by means of two flagella. *Pelagophycus* (above) is one of the giant kelps. The brown alga has a long stalk called a stripe, which reaches from the sea bottom where it is anchored by a holdfast to the surface. The fronds are held near the surface by a hollow bladder. A tidal pool (left) on the Massachusetts coast contains green, brown and red algae. These algae are grazed by a multitude of aquatic herbivores, and form the basis of the aquatic food chain.

Table 4.1 **Major Groups of Algae**

Common Name	Division	Main Habitats
Green algae	Chlorophyta	Chiefly freshwater, but some marine forms are common
Brown algae	Phaeophyta	Marine with a few exceptions and mostly in temperate climates
Red algae	Rhodophyta	Chiefly marine, widely distributed, but most species in tropical or subtropical climates
Golden algae (including diatoms)	Chrysophyta	Chiefly freshwater but many marine species
Yellow-green algae	Xanthophyta	Chiefly freshwater with relatively few marine forms
Dinoflagellates and cryptomonads	Pyrrophyta	Chiefly marine and free-floating, but many freshwater species
Euglenids	Euglenophyta	Chiefly freshwater with relatively few marine forms

Source: Adapted from Knut Norstog and Robert W. Long, *Plant Biology* (Philadelphia: W. B. Saunders, 1976), page 247. © 1976 by W. B. Saunders Company. Reprinted by permission.

rockweed and kelp have firm anchoring systems, or **holdfasts,** as anyone knows who has ever tried to pull a clump of rockweed off a rock. The holdfast that moors kelp to rock is sturdy enough to steady a 100-meter (300-foot) plant. Both rockweeds and kelps have a specific brown pigment, which gives them their name, brown algae. Brown algae, particularly the giant kelps, may be used as food. Algin, a stabilizing agent, is derived from brown algae.

Red Algae With more species than the brown algae, the red algae are the dominant seaweed of warm seas, often intermixed with brown algae along the shore. They are the most important of the seaweeds used for food. The red alga *Porphyra* is cultivated by the Japanese, for whom seaweed is an important part of the diet, and it is their most valuable maritime product. Agar, the gelatinous material used to grow cultures of microorganisms, is extracted from red algae. Similar compounds derived from red algae are used in everything from insect sprays to ice cream. They make shaving lather foam, paint spread, and provide the smooth texture in inexpensive ice cream.

Phytoplankton Of great importance to fish life in the oceans are the microscopic algae known as phytoplankton. They consist of representatives of a number of divisions such as green

4.11 These Japanese are harvesting the red alga nori (*Porphyra tenera*), which grows on nets suspended in the intertidal zone. Nori is a common element in the Japanese diet.

algae, diatoms, and dinoflagellates. They form the base of the food chain in open waters. Among the phytoplankton, **diatoms** are particularly noteworthy: they are distinguished by their cell walls, which are made of silicon, the same substance that makes horsetails rough. Under the microscope diatoms show exceptionally beautiful, endlessly varied shapes. Their crystalline cell walls make them look like engraved crystal boxes. Diatomaceous earth, derived directly from the remains of diatoms, is found in many places and is used as a filter in scientific research, as insulation for buildings, as an abrasive, and in perfume to prevent quick evaporation.

Groups of microscopic algae are frequently visible as green scum on lakes and ponds in summer. If nutrients are available, they will multiply with extraordinary rapidity. Farm ponds turn green with algae faster than others because of manure and other fertilizers draining into the pond from surrounding fields (Chapter 9). Fish ponds are often fertilized on purpose to produce a bloom of green algae that will support fish higher on the food chain. In addition to living in ocean, lakes, and ponds, green algae can also be seen on damp soil, on tree trunks, and even on mountain snowbanks.

PLANTLIKE ORGANISMS

There is a large group of organisms that are very different from green plants, although they often grow, plantlike, in the ground. These are the fungi: the mushrooms, toadstools, breadmolds, mildews, rusts, smuts, and yeasts. Although they have tradi-

4.12 FUNGI

Unlike plants, fungi are heterotrophs and are considered here as a separate kingdom of organisms. They are a large and varied group and a few representatives of important classes are given here. The phycomycetes include the water molds of which this species of *Saprolegnia* (below, right) is an example. It is growing on wood in this case, but it is also common on dead aquatic animals and some relatives are parasites of aquatic animals. The common black bread mold, *Rhizopus nigricans* (top right, facing page), belongs to another order of phycomycetes. The sac fungi or ascomycetes include the powdery mildews illustrated here (bottom right, facing page). Members of the class basidomycetes are the most commonly observed fungi because they include mushrooms, toadstools, bracket fungi, and rusts. Bracket fungi usually appear on dead wood and their varied colors and shapes, such as in this *Polyporus versicolor* (bottom left, facing page) add to the forest scene. Rusts are major diseases of plants, and it is a rare plant which is free of infection. These bigleaf maple leaves are infected with black speckled leaf spot (top left, facing page). The mushroom (below, left) illustrates some general features of the life cycle of basidomycetes. Spores germinate and produce a mycelium consisting of cells with single nuclei. Primary mycelia of different mating types come into contact and exchange nuclei. The resultant cells divide to form a secondary mycelium consisting of cells with two nuclei. Eventually this mycelium will form a complex spore-producing structure such as a mushroom. Before spore formation, nuclei in each cell fuse and undergo meiosis (Chapter 11).

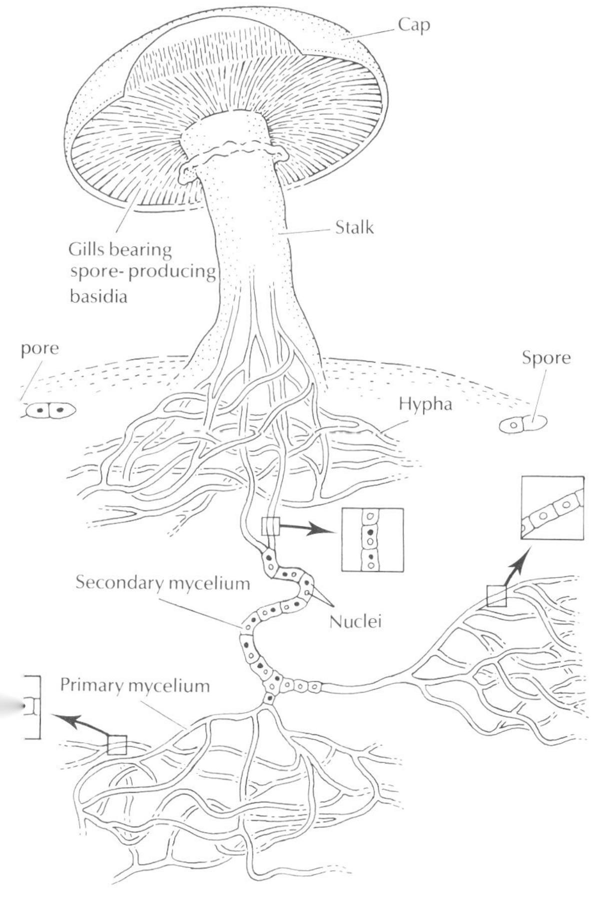

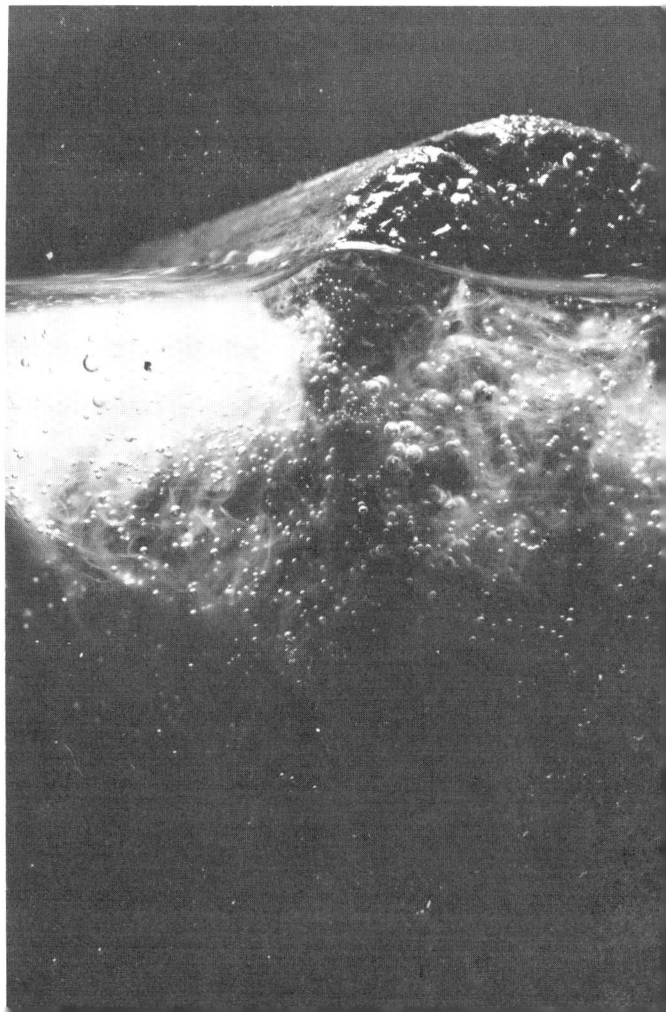

tionally been discussed in botany books because of their superficial similarity to plants, many biologists now consider fungi so different that they assign them a kingdom entirely their own.

Characteristics of Fungi Unlike plants, fungi do not have chlorophyll; as a result, they cannot carry on photosynthesis. They get their nutrients instead by absorbing organic materials, living or dead, plants or animals. (Thus they are heterotrophs, not autotrophs.) The mushrooms growing on a dead log are assimilating the carbon compounds the tree produced by photosynthesis. A fish floating dead in the water will be covered with fuzz, an aquatic fungus; the fungus is absorbing the nutrients in the fish tissues. Bins containing manure are used to grow commercial mushrooms because the manure contains organic carbon and nitrogen compounds, which the mushrooms break down and absorb. When fungi get their food from dead organisms, they are called **saprophytes;** when they feast on live organisms, they are called **parasites.**

Fungi differ from green plants in other ways. Their cell walls usually contain **chitin,** a compound related to cellulose. The exoskeleton of insects is also made of chitin, but it is never found in green plants. Because we are emphasizing green plants, and particularly flowering plants, our coverage of fungi will be brief. They are a major group of organisms, however, consisting of perhaps two hundred thousand species and, as we saw in Chapter 1, they are important decomposers in ecosystems, playing an essential role in cycling matter and energy.

Whether breadmold *(Rhizopus)* or mushroom *(Basidiomycetes),* most fungi are thin filaments called **hyphae** (one filament is a *hypha*). A colony of hyphae is a **mycelium.** Yeasts usually are single cells that form new cells by pinching off small segments of the parent cell. Many fungi reproduce by forming spores that are resistant to desiccation ("drying out") and are blown about freely in the air. Fungal spores are almost everywhere. If they land on something that is damp and has some minerals and organic compounds, they will quickly produce a hypha, which in short order will form a mycelium. Usually we are aware of a fungus only when it forms the spore-bearing structures called sporangia, because the spores are often black, greenish, blue, or pink.

Mushrooms are spore-bearing structures. They look complex and solid but are really a mass of tangled hyphae. Think of them as sculptures made of tangled string. The stalk and cap we see are only the tip of a mycelial iceberg, because in the ground or rotten log beneath, the mushrooms are connected with widespread mycelia. Beneath the mushroom cap, gills contain spore-bearing structures **(basidia)** that produce millions of spores, which will fall on the ground or blow to new places. You can see this spore production by cutting a mushroom cap from its stalk and placing it bottomside down on paper. Remove the cap in a few days and you will see the pattern of the gills outlined by spores deposited on the paper. The color of the spores helps to identify the species—an important and often difficult task if you are deciding whether a mushroom is edible.

The life cycles of some fungi consist of simple **asexual reproduction** (no union of gametes), such as fragmentation of the hyphae. In other fungi very complex sequences of sexual reproduction are combined with asexual reproduction involving different types of cells, called spores, that are resistant to desiccation. An example of a complex life cycle is that of the fungus *Puccinia graminis,* which is responsible for black wheat rust, a major disease of wheat. At one time the formidable task of memorizing the stages in the life cycle of *P. graminis* was a prime hurdle in passing elementary botany and biology courses. This fungus has five kinds of spores in its reproduction and two different but specific host plants: wheat and the American barberry, a small-thorned bush with oblong red berries. Part of the life cycle must be accomplished in the stems and leaves of wheat; another part must occur in barberry leaves. Attempts to fight the rust by eradicating barberry plants are longstanding but not particularly successful. The race between the constant natural production of new varieties of the rust and the breeding of new wheat varieties resistant to the fungus is never-ending.

Fungi, Plants, and People Anyone dealing with plants cannot be ignorant of the fungi; they cause many diseases of plants. Rare is the plant that has no fungus infection. It may be as minimal as the fungus-induced athlete's foot of humans—not killing the plant, just "nibbling" around the edges. But some fungus diseases can kill plants quickly. Dutch elm disease is a fungus infection. Because of it, streets of our towns and cities once shaded by arching ranks of gigantic American elm trees are now bared. The wheat rust fungus has caused massive human starvation, as has the fungus *(Phytophthora infestans)* responsible for potato blight. Production of coffee trees is cut by a fungus disease, and any gardener knows how damaging mildew can be to roses.

The amount of economic destruction done by plant diseases is incredible. The world could have far more food if losses caused by fungi could be decreased. A worldwide army of plant pathologists is working constantly to understand and control fungal infections of plants, and many plant breeders are trying to improve the resistance of crop plants to these infections.

All the relations between plants and fungi discussed so far are parasitic, but fungi do have other relations to plants. Some of them are **mycorrhizal associations**—a fungus and the roots of a plant are related in a way that is beneficial to both. The roots of the plant have a fungus on their surface, and this fungus penetrates the root tissues. The fungus helps the root to absorb minerals, and the fungus obtains organic compounds from the root.

Another more familiar symbiotic relationship between fungi and plants is the lichen, formed by a fungus and an alga together. Two organisms in one, the lichen is a symbiotic organism with a morphology different from either of the partners. The fungus forms a tangled mass of filaments, and the algal cells are held within this meshwork (see Figure 4.13). From its algal partner the fungus obtains photosynthetic products; biologists are not certain what the alga gains from the relationship, but perhaps it obtains minerals more readily because of fungal action on the substrate of rock or bark. Lichens grow on rocks, walls, and tree trunks and are an important food for caribou and reindeer in the arctic tundra.

Like green plants, mushrooms provide food, hallucinogens, medicinal drugs, and poison. Some mushrooms, like the white or death angel, have a deadly poison; many others have poisons of various potencies. People react differently to the poisons, especially the

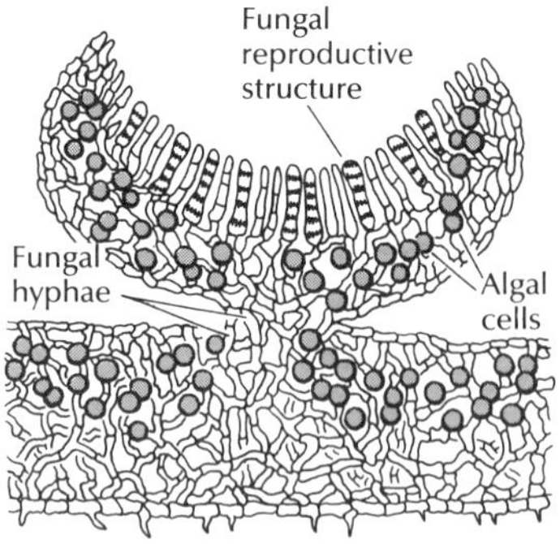

4.13 Lichens are made of both fungi and algae. Algal cells are embedded in the tangled fungal filaments.

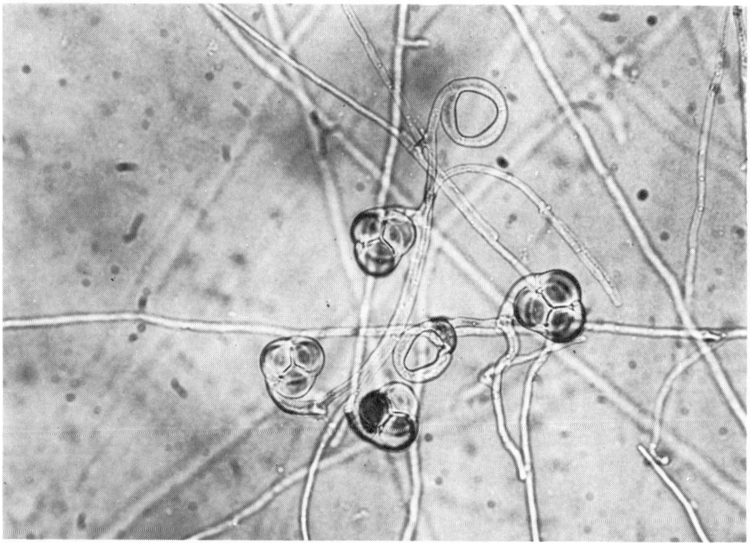

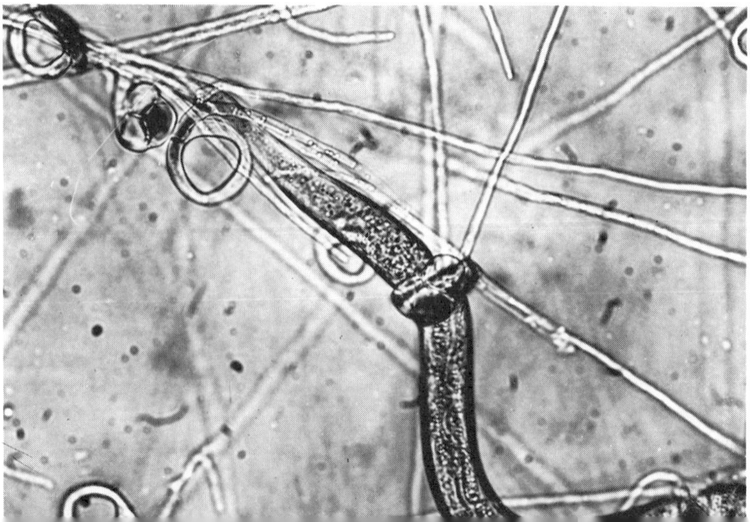

4.14 Animal-catching plants, such as the Venus's-flytrap, are well-known, but microscopic soil fungi also catch animals. In this case, a microscopic worm called a nematode is the victim. The loops formed by the fungal hypha swell when touched, and the lassoed nematode is consumed by fungal hyphae, which invade its body.

milder kinds. A good rule is to stick to mushrooms you are positive are harmless.

Yeast, a ubiquitous group of fungi, uses sugar as food, breaking it down and producing energy, carbon dioxide, and alcohol (among other compounds) in a process known as **fermentation** (see Chapter 8). To yeast we owe the existence of wine, beer, bread, spirits, some cheeses, and a host of other products of fermentation.

Yeasts are also responsible for serious infections. Many skin diseases such as ringworm are caused by fungi, and they are also behind some respiratory infections that can be fatal. But fungi are also the source of some lifesaving drugs, such as the antibiotic penicillin, which was first isolated from the fungus *Penicillium chrysogenum* in the 1930s. Recognized or not, members of the genus *Penicillium* are seen by almost everyone as colonies with greenish blue spores on overripe oranges and cantaloupes. *Penicillium* is also consumed in great quantities by lovers of blue cheeses like Roquefort and Stilton. It is hard to imagine how much human suffering has been eliminated by antibiotics such as penicillin, streptomycin, and Terramycin that we now take for granted.

Blue Green Algae Still another group of organisms is often considered with plants, although they are not related to them. To make matters more confusing, they are called blue green algae, although they are not related to the algae we have been discussing. Their cell structure is very different from that of both plants and animals but very similar to that of bacteria. (See the Supplement on pages 128–131 for a description of bacteria.) For this reason blue green algae have been placed in a kingdom together with the bacteria, the Monera.

Both bacteria and blue green algae lack a nucleus, the prominent structure in the cells of plants, fungi, and animals that contains the gene-bearing chromosomes (Chapter 11). They have genes, but the genetic material is not carried in a specialized structure; it is free in the cell's internal medium, the **cytoplasm.** They also lack the other

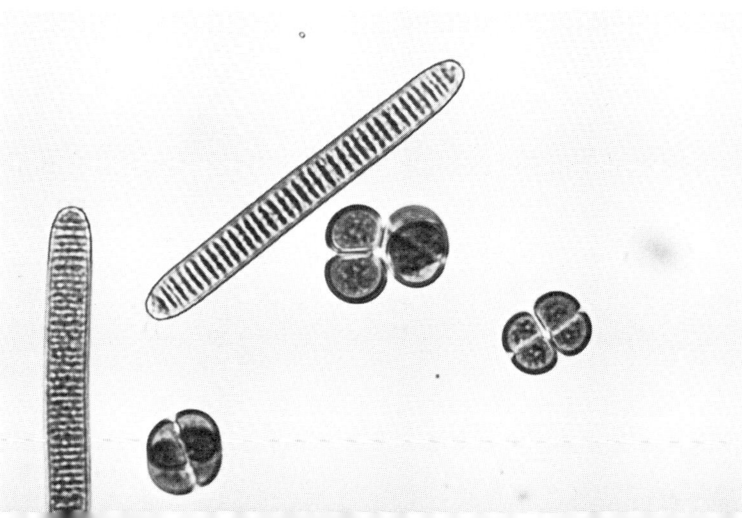

4.15 Blue green algae consist of single and multicellular forms. As its name implies, the filamentous *Oscillatoria* moves back and forth. The cells of *Chroococcus* divide in two planes and a disklike colony is formed. Blue greens are characterized by a mucilaginous sheath around the cells.

structures possessed by cells of organisms in the other kingdoms (see page 96). Blue green algae and a few kinds of bacteria have chlorophyll, and they can and do photosynthesize. Their chlorophyll, however, is not in **chloroplasts,** the green bodies in plant cells where photosynthesis occurs; it is in membranes in the cytoplasm. In addition to chlorophyll, blue green algae often contain blue pigments, and this combination produces the blue green color. Sexual reproduction is well known in bacteria but has not been clearly demonstrated in blue green algae; reproduction is by cell division.

Blue green algae can be as small as microscopic single cells and filaments, or they can be aggregations of filaments forming balls larger than grapefruit. Many have a sticky sheath outside the cell wall with which they may adhere to other algae. If algae are abundant in the water, you can run slimy films of blue greens through your fingers. They may become very abundant in water rich in minerals. They can produce odors and change the taste of fresh water and, when growing rapidly, can ruin a body of water for swimmers.

Blue green algae, like bacteria and fungi, exist in symbiotic relationships with green plants. Many blue greens (and some bacteria) are able to **fix nitrogen:** they are able to take the nitrogen from the atmosphere and form nitrogen compounds. The aquatic fern *Azolla,* often mistaken for the duckweed *(Lemna),* harbors a blue green in pouches on the underside of its small leaves. The blue green provides the fern with nitrogen. In another symbiotic pair, a cycad has the blue green in its roots and obtains nitrogen from it. This ability to fix nitrogen, which can produce masses of blue greens in lakes and ponds, is a bounty for rice farmers of Asia: blue greens in rice paddies provide nitrogen to the rice. In a number of countries, inoculating the flooded fields with blue greens is common practice in order to improve the yield and avoid buying expensive nitrogen fertilizer.

SOME UNANSWERED QUESTIONS

This brief tour of much of the plant kingdom has hit only the high spots; much of importance has, by necessity, been omitted. Only divisions with living representatives have been included, but a fuller understanding of the relationship of these higher taxonomic categories depends on studying extinct taxa known to us only as fossils. (A somewhat mysterious group called the seed ferns is one example.) If intermediate groups still existed, our categories of flowering plants, conifers, and ferns would be far less distinct and classification would be much more difficult.

A close study of living and extinct plant taxa clearly demonstrates

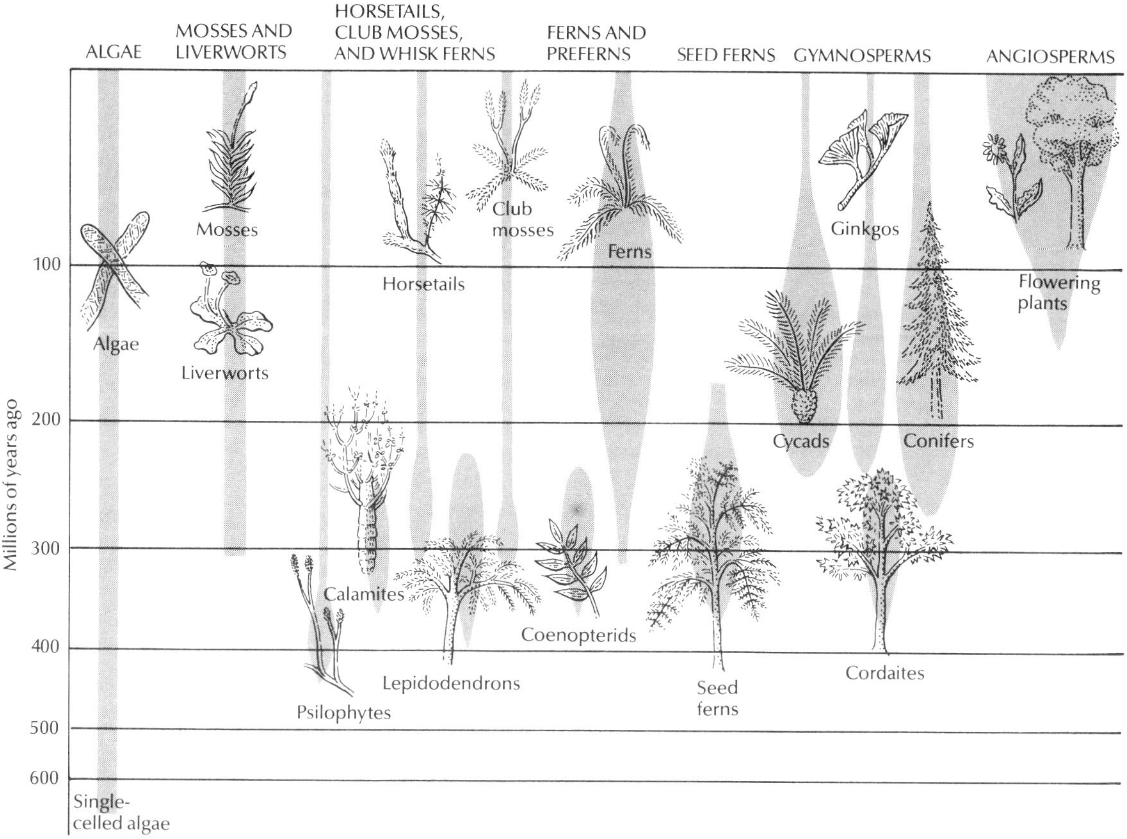

4.16 The main groups of living plants and selected extinct taxa are shown. The width of each bar represents the estimated number of species existing at that time.

many relationships, but although we can construct the outlines of an evolutionary sequence, many links are difficult to establish. For example, we think the first land plants were probably derived from the green algae (although not the existing ones), but it is difficult to construct a specific family tree for the various groups of vascular plants. The origin of the flowering plants is a particularly knotty problem. Contemporary taxa of ancient groups such as bryophytes and ferns are far removed from the taxa that were the direct ancestors of the angiosperms. Clearly, much inspired study and research is needed to trace the ancestral roots of flowering plants.

In comparison with the age of the most recently evolved division of plants, those with flowers, humans have existed for only an instant. Our ancestors evolved in a landscape dominated by flowering plants, and they quickly made use of them and, indeed, influenced the evolution of many plants. In fact, a strong argument can be made that human culture could not have developed without flowering plants. How we use angiosperms will be investigated in the next chapter.

Summary

The gymnosperms are an ancient group of plants and it is probably from ancestors of some of the living conifers that angiosperms evolved. Like the angiosperms, they are vascular plants (that is, they have specialized structures to conduct water, minerals, and nutrients), and they have seeds. Their seeds, however, are not enclosed as are those of angiosperms; they are not pollinated by insects but usually by wind; and the endosperm of their seeds develops solely from the female gametophyte.

Conifers are the most familiar kind of gymnosperm, and most have a life cycle much like that of the pine. Three years may elapse after pollination before seeds are ready to be dispersed.

Most familiar gymnosperms bear cones and needlelike leaves, but the ginkgos and cycads, though gymnosperms, look quite different. The ginkgos are now extinct in the wild, and the cycads appear to be headed for the same fate. Unlike other gymnosperms, ginkgos and cycads have motile sperm with flagella.

Ferns (Pterophyta), too, are vascular plants, but without seeds. On their leaves are sporangia in which are spores. When the spores are flung to the ground, they grow into green, sheetlike gametophytes. Male gametes produced by the gametophyte are motile. They swim through water on the surface of the gametophyte and enter the female gametophyte and fertilize eggs. As the resulting embryo grows into a mature sporophyte, the gametophyte, which has nourished it, dies.

The club mosses, horsetails, and whisk ferns (Lycopodophyta, Anthrophyta, and Psilophyta) are among the first groups of land plants and living species are few in number. These three groups are also vascular, but the psilopsids have no roots. Like the ferns they do not have seeds.

Mosses and their relatives, the liverworts, are nonvascular plants with no seeds. Both groups (Bryophyta) live in moist places and absorb water directly into their cells from the air and ground. In these plants the gametophyte generation is dominant.

Most of the plants that live in water are algae, a group of nonvascular plants that are relatively simple in structure. Although all algae have chlorophyll, many are not green, and algae are commonly identified by their dominant color. Giant kelps and rockweed are brown algae; red algae are important seaweeds and a few are used for food. In summer, microscopic green algae often form a green scum over lakes and ponds. Microscopic algae known as phytoplankton are the base of the aquatic food chain. One kind of phytoplankton in particular, the diatoms, has many industrial uses.

Fungi do not photosynthesize and, although they grow in the ground and cannot move, we do not consider them as plants. Most fungi are filaments called hyphae, which form colonies known as mycelia, and their cell walls contain chitin. Some reproduce by fragmentation of hyphae, but many reproduce by forming spores, which are spread by the wind.

Many fungi are saprophytes, obtaining their nutrients from dead organisms; many others are parasites. Some form mycorrhizal associations with the roots of plants, benefiting both plant and fungus. Lichens illustrate yet another relationship: here the fungus forms a mass of filaments that holds algal cells.

Infections from fungi are widespread among plants and people; Dutch elm disease, wheat rust, and potato blight are familiar examples. The uses of fungi are many, too, including drugs such as penicillin and yeasts for fermentation.

Unlike both plants and fungi, blue green algae have no nucleus, but they do contain chlorophyll, as well as blue pigments. Blue greens also form symbiotic associations with some plants in which the blue green provides nitrogen to the plant.

Precise evolutionary links within and among these groups cannot be definitely established, and the evolution of the angiosperms is especially difficult to trace.

Review Questions

1. What are the basic differences between the reproductive cycles of angiosperms and gymnosperms?
2. How is the endosperm of a pine tree seed different from the endosperm of a corn plant seed?
3. What features of the more ancient groups of vascular land plants demonstrate their limited adaptation to living on dry land? Contrast ways in which gymnosperms and angiosperms are more adapted to this environment.
4. How are mosses and liverworts similar to algae?
5. The algae consist of fewer species than the flowering plants—in spite of the oceans covering almost three-quarters of the earth's surface. Give reasons for this phenomenon. Review the question and your answers after you have completed the book.
6. Read the labels on food packages, cosmetics, and toilet articles and hunt for evidence that algae affect your daily life.
7. How do fungi and blue green algae interact with plants? List specific relationships.
8. "Have you thanked a green plant today?" is a common bumper sticker. List the effects of fungi on your daily life and decide if they qualify for a similar bumper sticker.
9. Suppose that some newly discovered organisms are suspected of being bacteria. How would you test this hypothesis?

SUPPLEMENT

Bacteria and Viruses

Bacteria are microscopic unicellular organisms that are structurally less complex than the cells of plants and animals. Like blue green algae, they are prokaryotes—cells that lack a discrete membrane-bound nucleus and other organelles typifying the eukaryotic cells of most other organisms. The genetic material of bacteria lies within the cytoplasm (see Figure 1). Like plant cells, they are surrounded by a rigid cell wall, but the bacterial cell wall is comprised not of cellulose but of complexes of different carbohydrates and proteins. A gelatinous layer, the capsule, often lies outside the wall. Many types of bacteria possess flagella, tail-like structures that produce mobility.

Unlike plants, bacteria usually reproduce by **binary fission,** forming two individual bacteria. Some bacteria produce spores, which are more resistant to extremes of moisture and temperature, and some do exchange genes. Under the right conditions, bacteria divide in a matter of minutes, so their reproductive capacity is impressive indeed.

Bacteria are classified into several groups, including myxobacteria (one-celled bacteria that glide), spirochetes (one-celled, helical bacteria), and eubacteria, or true bacteria. Eubacteria may be spherical **(cocci),** cylindrical **(bacilli),** or helical **(spirilla)** (see Figure 2). The eubacteria are also classified according to whether they can be stained by a purple stain called the gram stain: if they can be, they are called **gram positive;** if not, they are

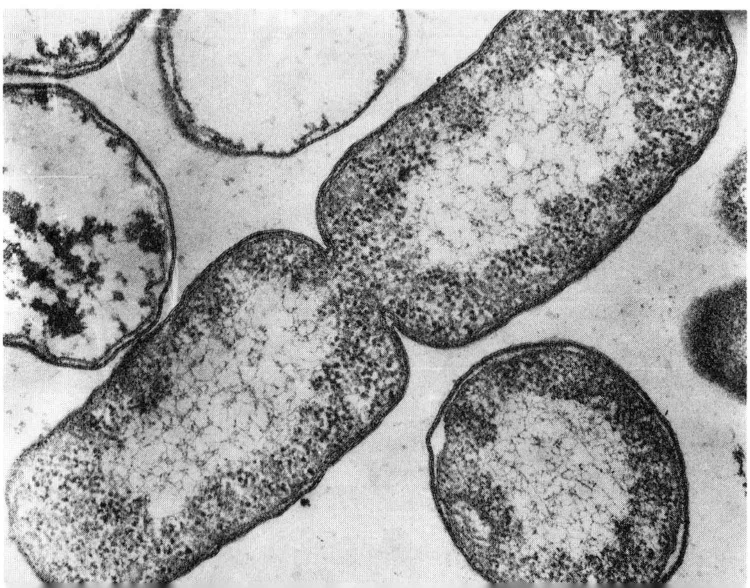

1 An *Escherichia coli* bacterial cell dividing by fission—a process of constriction that yields two cells. These sections of bacterial cells were photographed with an electron microscope. Bacteria do not have nuclei and the fibrous material in the central part of the cell is DNA, which controls the cell's genetics (see Chapter 11).

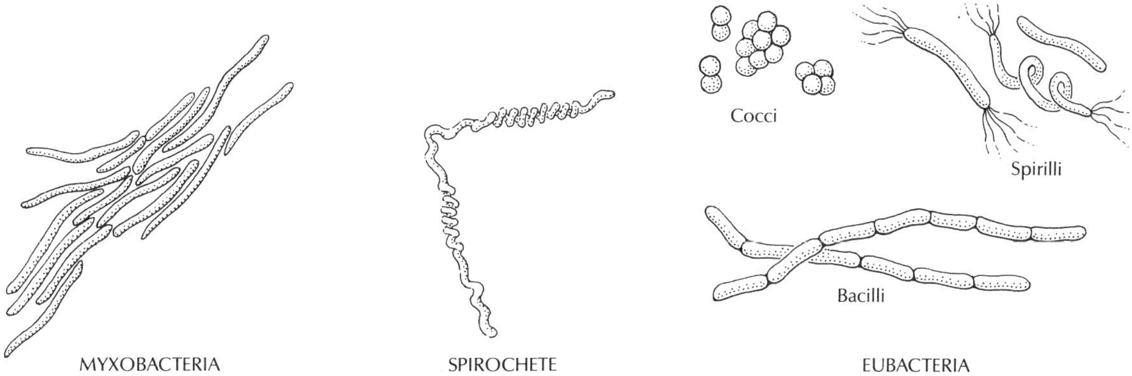

2 Bacteria are classified according to several groups, based on cell structure and type of movement.

labeled **gram negative**. This classification reflects differences in the structure of the cell wall and, as a result, in resistance to antibiotics.

The ubiquitous bacteria are a versatile group with representatives in almost every kind of environment, from polar regions to hot springs to temperate climates. Many species live in the soil or on plants and animals, and even in environments with no oxygen. Like any organism, bacteria can live only in an environment that provides nutrients for growth and reproduction; the remarkable thing about bacteria is that as a group they have diversified to such an extent that they can use almost anything as a source of nutrients. Unlike plants, most bacteria do not photosynthesize; they are either parasites or saprophytes (deriving their food from dead organic matter). A few species are **chemosynthetic;** that is, they are autotrophs deriving energy to synthesize their own food not from other organisms or from the sun but from chemical reactions involving compounds of such elements as sulfur and iron.

Bacteria have numerous effects on our world; most are beneficial. As decomposers, they are essential in the cycling of nutrients through the biosphere. We use some bacteria to treat sewage, others to produce pharmaceuticals, food supplements, yogurt, cheese, and vinegar. Bacteria in the human digestive tract produce several vitamins. But bacteria also cause major diseases in plants and animals (including crown gall, fire blight, and soft rot in plants; pneumonia, diphtheria, and gonorrhea in humans).

Many bacteria produce antibiotics used to treat infectious diseases. The major bacterial sources of antibiotics are *Bacilli* (polymixins and bacitracin) and *Actinomyces* (tetracycline, chloramphenicol, polyemes, aminoglycans). Most antibiotics eliminate bacterial infections by inhibiting synthesis of proteins, thus stopping growth and reproduction of the infecting organisms. (Penicillin is an exception. This antibiotic is produced by a fungus, *Penicillium,* and it works by preventing formation of the cell wall.) Extensive use of antibiotics, however, has led to the development of strains of pathogens that are resistant to presently available antibiotics.

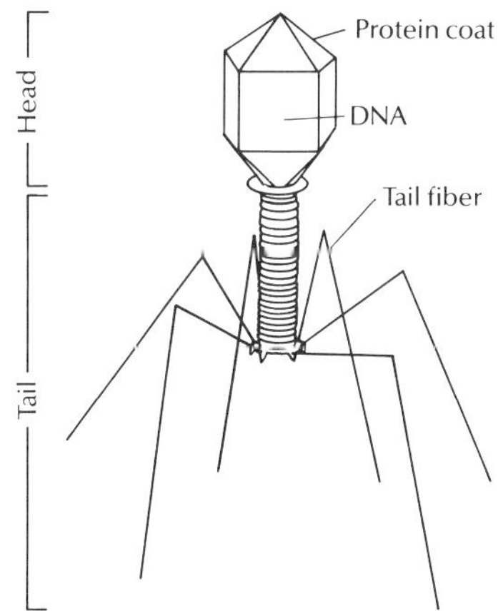

3 Bacteriophages are viruses that attack bacteria. They have a head composed of protein and packed with DNA, a hollow tail, and fibers, which act as a sort of landing gear on the surface of a bacterial cell.

Bacteria are sometimes classified with viruses as members of one kingdom, the Moneran kingdom. Viruses, however, are not capable of existing independently; they are active only when they have infected a host cell. Isolated, a virus will not grow or reproduce, but it is capable of infecting and then growing in plant, animal, and bacterial cells.

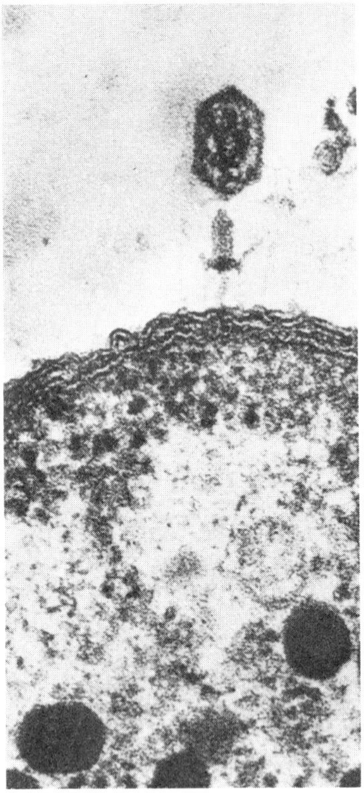

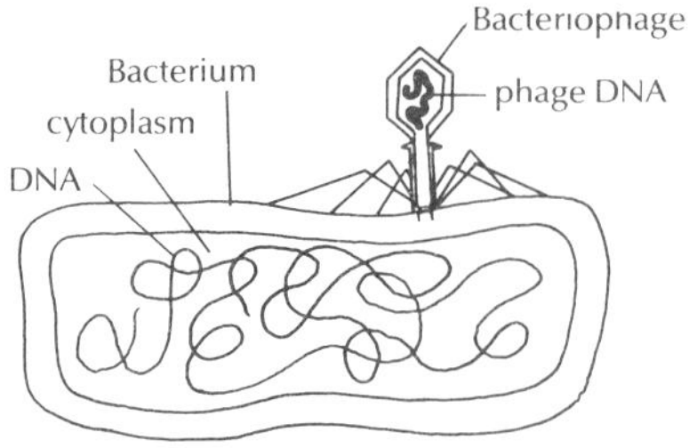

1. The bacteriophage attaches to the bacterium by its tail fibers.

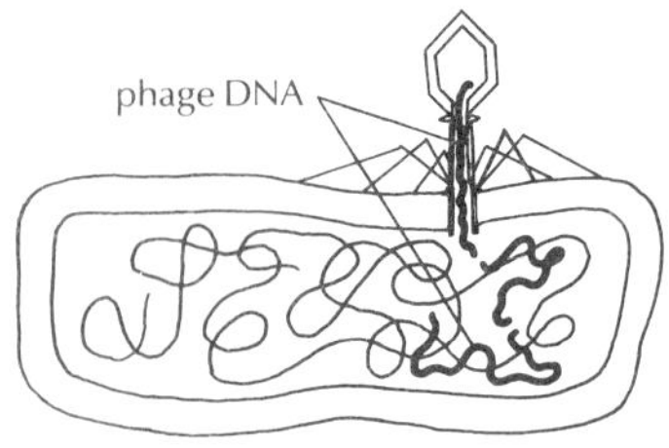

2. The bacterium wall is perforated by an enzyme and the phage injects the genetic material (DNA) into the cytoplasm.

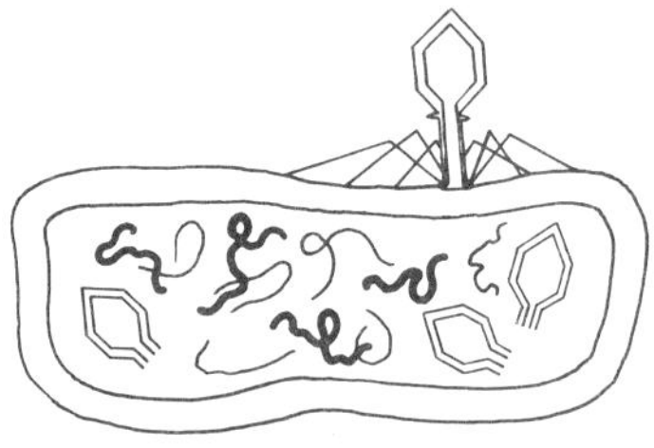

3. The bacterium's DNA begins to disintegrate; the phage's DNA synthesizes millions of new virus particles.

4. The bacterium ruptures, releasing the new phage particles.

4 Photo (above, left) shows a bacteriophage on the surface of an *Escherichia coli* bacteria cell. The virus injects DNA into the bacterial host cell (above, right). This genetic material directs the assembly of new virus particles, which are manufactured from host cell materials.

A virus is composed of a small piece of genetic information—a strand of nucleic acid (see Chapter 11) encased in a protective protein coat. When viruses infect cells they insert their nucleic acid into the host cell. They then take over the cell's metabolic machinery to replicate themselves, filling the host cell with hundreds and sometimes thousands of new virus particles. When the host cell bursts, the released particles can infect other cells (see Figure 4).

Viruses are usually classified either by the disease they produce (for example, poliovirus) or the kind of cells they infect (a **bacteriophage** is any virus that infects bacterial cells). Another scheme classifies by the chemical nature of the virus—whether it contains DNA or RNA and what constitutes its protein coat.

So small they can be seen only with an electron microscope, viruses are responsible for many diseases (see Figure 5). In humans they produce colds, meningitis, measles, mumps, rabies, and other diseases. Major food crops such as potatoes, tomatoes, wheat, oats, rice, and corn are susceptible to infection by specific viruses. Sometimes the viral infection kills the plant, but more often it only weakens the plant and causes lower crop yields. Perhaps the best-studied virus is the tobacco mosaic virus, which can infect over seventy species of plants. One way of combating viral infections of plants is by breeding virus-resistant varieties (see page 339).

5 The tobacco mosaic virus in *Nicotiana,* causing blistered and distorted leaves. This was the first virus to be isolated and studied in detail. Although it is a major plant disease (it can even spread from infected cigarette tobacco), it has been an important research tool for virologists.

CHAPTER 5

People's Plants

The world has more than three hundred thousand species of green plants; we have learned to use only about five thousand of them. Food, fiber, drugs, and industrial uses account for large quantities of these species, and ornamental applications a few more. Still, those thousands of plants are a lot to think about here, and we will consider only a fair sample of them as we take an introductory look at "economic botany." Our goal here is to understand some of the plants that keep our cultures running.

WHEAT

The genus *Triticum* (wheat), foremost of domesticated plants and grown more widely than any other, was among the first wild varieties to be systematically cultivated. A member of the Gramineae (grass) family, it commonly grows in regions that have a cold season and 75 centimeters (30 inches) of rainfall or less every year. Its potential habitats spread far around the planet, but they do not include lowland tropical areas where maize, rice, or root crops are traditionally grown instead.

After a hundred centuries of cultivating, adapting, and hybridizing, three types of wheat make up the world crop today. All three are crosses between cultivated wheat and wild grasses. What can be traced of the evolution of these forms leads us to believe that the steps involved were not conscious acts by farmers but were accidents, the result of wheat coming into contact with wild relatives.

5.1 Temperate-zone human society and domesticated wheat have a symbiotic relationship in which neither could prosper without the other. Wheat is one of the first domesticated crops and is king of the world's cereals. The United States is the world's breadbasket and this wheat shown in a Kansas field might be consumed at any locale between Topeka and Vladivostok.

The human contribution, not to be underestimated, was to select the most vigorous offspring that resulted from this long period of hybridization and cultivate them.

How did this selection come about? It was probably quite simple. As women gathered the wheat at harvest time, they must have noticed that some kinds had larger more edible grains that, unlike those of most wild wheats, did not easily separate from the stalk. This retention of grain was suicide when the plants grew in the wild: they could not propagate nearly as fast or as easily as plants with shattering heads of grain. But the trait was a boon for humanity, because the ripe heads of grain could be harvested intact. It is not too far-fetched to assume further that when grain was brought to the village, some seeds were spilled and grew there. The next logical step would be the deliberate sowing of some grains.

Archaeological sites in the Near East, where the first farmers may have lived, yield evidence that wheat was cultivated there around 7000 B.C. (Chapter 2). The same evidence shows that wheat was not first used for making bread. Instead the kernels were heated and the grains chewed. Sometimes grains were ground, mixed with water, and left to ferment, showing how far back in history people first learned about the effects of alcohol.

The wheat plant, originating in the Old World, has spread worldwide. Two basic types of wheat are grown today: spring and winter wheats. Spring wheats are planted in the spring and harvested in the fall; winter wheats, planted in the fall, are harvested in spring or early summer. Wheat therefore is available year-round in the various parts of the world. The spring varieties are grown in colder regions, whereas winter wheats are cultivated in regions of favorable autumn rainfall and less severe winters. Winter wheats grow more slowly but give higher yields than spring varieties.

Like all plants, wheat has many diseases, and the fungal diseases that have plagued farmers since the beginnings of agriculture are particularly important. The principal of these is wheat rust *(Puccinia graminis),* which, as we saw in Chapter 4, is especially dangerous to wheat.

MAIZE

Indian corn or maize *(Zea mays)* is to the New World what wheat was to the Old: the staple crop, the source of food. It is called *Indian* because Europeans first saw it in native American gardens when they discovered America. Today it has spread worldwide and, after wheat, is the most widely grown of cereal plants, cultivated throughout the tropics and in many temperate countries as well. Half the world supply is produced in the United States, where it takes up twice as much land as any other crop.

Maize is a double-use food crop. In the tropics it is food for humans, baked either as bread or in thin cakes like the tortillas of Mexico. In temperate regions it is mostly used as feed for livestock.

Maize, like wheat, is in the grass family. It is an annual, some varieties of which grow up to 3 meters (10 feet) high, far taller than the familiar grasses. It has a solid stem and prop roots at the base that help support it. Its flowers are unisexual, with the two sexes occurring in different places on the same plant (see Figure 5.2). This arrangement is known as **monoecy** (the adjective is *monoecious*). The familiar tassel is the male inflorescence (flower cluster), which bears stamens and appears on the terminal shoot. The ear or cob is a modified lateral branch that produces pistils, which are the female organs. (Elongated pistils of the female flower make up cornsilk, traditionally smoked by small boys out behind the barn.) After fertilization the grains become wedge-shaped fruits, arranged in closely packed rows along the length of the cob. These corn fruits—the kernels—are the source of human food.

Today many varieties of maize are grown throughout the world. One of the most adaptable of the important food plants, corn is

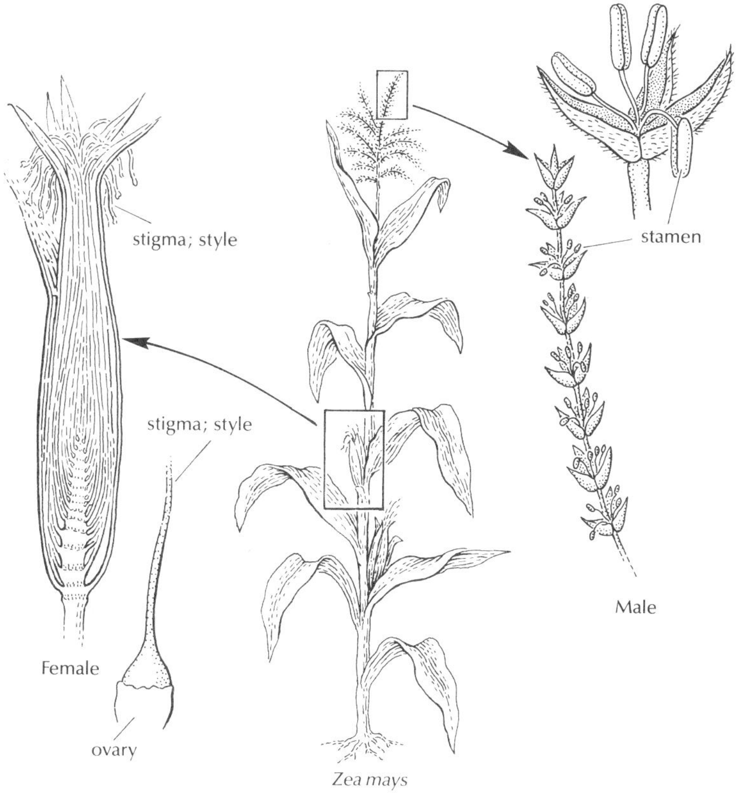

5.2 Corn flowers are unisexual but each plant has both sexes. Corn is therefore a monoecious plant, which facilitates crossbreeding.

much hybridized. The modern hybrid types, produced by crossing two inbred strains, are very high-yielding and have revolutionized agriculture in the American corn belt. These are very different from the South American pod corns, believed to be the most primitive of today's cultivated varieties. Other primitive varieties still in existence are those used by native Americans, who have preserved some kinds of corn because of their religious significance.

A very familiar variety of corn surprisingly different from corn on the cob is popcorn. This type has small hard grains that expand explosively as internal moisture turns to steam when heated and produce the light, white, fluffy mass that moviegoers crunch. It is about the only modern variety with small cobs; generally scientists have sought to enlarge the cobs and individual grains to increase the yield at harvest time.

Truly wild maize is now unknown, so the origin of this extremely useful food plant is a matter of controversy. The search for its ancestry has involved botanists in careful study of primitive varieties still living and of archaeological material that might yield clues. On the basis of the evidence found so far, it is now generally believed that wild maize grew in Mexico until about 5000 B.C. (Chapter 2). At about that time it probably began to be cultivated, which led

5.3 These Indian women are planting rice seedlings raised in seed beds. This back-breaking, labor-intensive practice is followed in regions where there is plentiful labor and low wages. In contrast, the rice paddies of California's Sacramento Valley are sowed with seedlings cast from airplanes.

to its hybridization. Like wheat, it may well have been domesticated when people began to take care of wild maize strains, gradually interspersing them with primitive domestic strains. The wild maize gradually gave way to domesticated varieties and the wild varieties gradually became extinct.

RICE

More people eat rice than any other staple food. More wheat may be grown, but half the world's population lives on rice. Southeast Asia has cultivated it since ancient times and still produces 90 percent of the world's supply. China and India are the largest producers by far, but the United States is the largest exporter. California is the most mechanized of all rice-growing regions: the crops are planted and fertilized and pests are controlled by airplane.

The two species of rice are *Oryza sativa* and *Oryza glabberima.* The first is by far the most prevalent. *Oryza glabberima* grows in Africa, where it was cultivated long before the first white settlers came, but today it is restricted to a few regions in West Africa. Both species occur naturally in swamps, although some varieties of *Oryza sativa,* so-called upland rice, are grown without such permanent flooding.

Rice, like the other important cereals, is in the Gramineae (grass) family. It has hollow stems and erect basal branches called **tillers** that rise from the lower parts of the plant. The number of tillers differs with varieties and growing conditions. The flowers are at the ends of the tillers, and the mature fruits, or grains, are tightly packed inside two dry bracts (husks). These are removed during threshing. When rice is milled to produce the white polished grains often used as a food, the outer layer of the grain is lost. Because this layer contains much of the grain's proteins and vitamins, populations in many parts of the Orient, where rice is the staple food, suffer diseases caused by vitamin deficiency.

Rice is probably the most varied food plant; thousands of cultivated varieties differ in both form and physiology. Each variety is a local type, developed through the centuries to suit local taste and growing conditions.

Most people are familiar with the wild rice grown in the United States. Despite its common name it is not a true rice, although it is in the same family. This plant, *Zizania aquatica,* is a wild species of grass, and the grain is collected from plants growing around the periphery of lakes. The native Americans knew and valued it long before European settlers came; it was and still is a luxury rather than a staple food.

SOYBEANS

Legumes include many food plants (see our supermarket taxonomy, page 465) and are a large source of protein for many peoples. This family also supplies the United States with one of its most valuable cash crops, the soybean (*Glycine max*), which has become a bastion of our economy in a very short time. Although corn is the largest crop in the United States (soybeans and wheat are runners-up), soybeans are the clear winner in export value and are an important part of the total income generated by United States exports. Our agricultural exports sustain our economy in the face of increasing imports of oil, and soybeans are the most profitable exportable agricultural commodity.

The United States grows a far larger crop of soybeans than any other country, supplying most of the stock for world export. Brazil is growing large crops and might soon compete for the international market.

The soybean's importance is all the more remarkable because it is a recent crop in this country. It was cultivated in China at least a

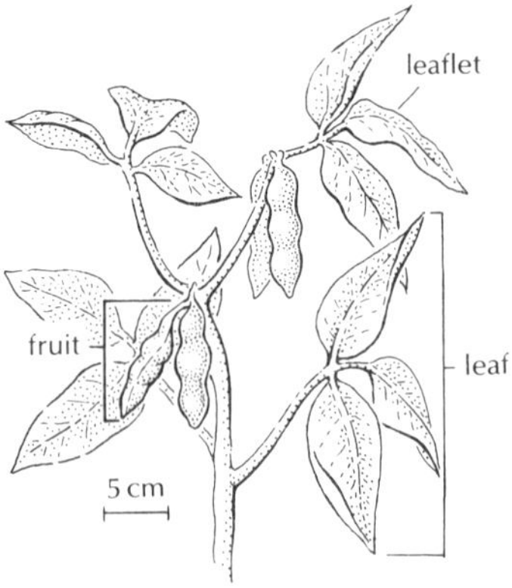

5.4 The soybean has been widely cultivated in this country only during the last fifty years. It is now the biggest cash crop in the United States and a vital world source of protein and fat.

thousand years before Christ, but it was not included in the migrations that spread so many other food plants to Europe following the voyages of discovery. It reached Europe in the seventeenth century and was brought to the United States at the beginning of the nineteenth century. Only a million acres of soybeans were planted here in 1930, but 10 million acres were harvested in 1943. The rapid increase was caused by the shortage of butter in World War II and its replacement by margarine. Margarine is made from plant fats, and soybean oil is a primary ingredient. The oil, composing about 20 percent of the weight, is extracted from the bean. About 45 percent of the weight is protein, so that after oil is extracted an enriched protein meal is left. Both fat and protein are valuable products, which led to rapidly increasing cultivation of soybeans, up to 56 million acres in 1973.

To the lasting distress of dairy farmers, margarine sales are double those of butter, and 75 percent of our margarine is made from soybean oil. It is also used in cooking and salad oils; in fact, half the consumable fat in this country is derived from soybeans. It is the world's largest source of edible fat.

Soybean protein goes into much livestock feed, a reason for the growing demand in many countries where people are eating more meat. People are eating more soybean protein, too, and not just in soy sauce. Always a staple in China, Japan, and Southeast Asia, we now consume great amounts as a filler or as a substitute for meat. Anyone living in a university dormitory or similar institution probably eats sausage, ravioli, and fish sticks fabricated primarily, if not totally, from soybean protein. The pure protein is odorless and tasteless, but it can be textured and flavored to mimic fish, bacon, turkey, and many other meat products. It is also made into a milk substitute. This protein is cheaper than meat and has no fat, so that its consumption will increase (remember the food chain). Instead of having animals concentrate plant protein, we can let the soybean plant do the work and eliminate a trophic level. Its protein is not as nutritious as animal protein, but that is not a serious matter in a balanced diet.

This plant that does so much is an annual beanlike bush about 1 meter (3 feet) high, in this country raised mostly in the Midwest. Several pealike seeds grow in hairy pods. Soybeans and many other legumes have root **nodules** that fix nitrogen, with which the crop supplies some of its own nitrogen fertilizer (see Chapter 9).

SUGARCANE

As the most inexpensive, energy-giving food per unit of land, surgarcane is very important as a food and export crop of tropical countries. From *Saccharum officinarum* about two-thirds of the world's sugar is made. The economies of Cuba and other countries depend almost completely on exporting their sugar, making it significant in international diplomacy. This sugarcane is not only an important economic plant but also a powerful political factor.

A grass growing about 5 meters (15 feet) high does not sound like sugarcane, but close up sugarcane does look much like the grasses. It is a perennial with solid jointed stems. These hold the sugar-rich juices that make up 85 to 88 percent of their weight. Because its mode of photosynthesis is very efficient, growth is quick. A new planting will mature and be ready for harvesting within two years.

Sugarcane probably first grew in Southwest Asia, but it is no longer known as a wild species. In its long history of cultivation, it has been hybridized with both cultivated and wild species. It is grown throughout the tropics and subtropics, often on immense plantations (see Figure 5.6). Stem cuttings (see page 353) are the usual way of planting.

In many countries, to make harvesting easier, the canes are burned to get rid of leaves just before harvesting. The burning also helps control pests. The juices of the cane, after shredders and presses extract them from the stem, are boiled down to a thick syrup, which is then crystallized. This is raw sugar, which is usually refined by washing and recrystallization to remove impurities such as brown pigments. Many think that raw sugar, because it is brown, is healthier than refined white sugar. However, the brown color is there not because of vitamins or other nourishing substances lost in the refining process but because when the stems are broken in shredders and presses, a chemical reaction produces a brown pigment. White sugar is almost pure sucrose and contains only traces of minerals. Brown sugar lacks vitamins and is very poor in minerals.

Sugar refining has one chief by-product, molasses, which is used to make alcohol, especially rum. The molasses is fermented by adding a yeast, and the fermented product is then distilled. The distillate is a clear liquid to which caramel (burnt sugar) is added for dark rum.

5.5 Sugar cane, one of the oldest domesticated plants and perhaps the most rapidly growing plant, does not exist as a wild species. A perennial grass, it is usually harvested at two years of age. Fields are burned before harvesting to remove leaves and to facilitate the harvest. As the growth rate of the stem decreases, the sucrose increases in the stem storage tissue. New stalks are produced by lateral buds at the base of older stalks.

SUGAR BEET

Among the very few plants that have been domesticated in modern times is *Beta vulgaris,* the sugar beet, which two hundred years ago was a wild species but today produces about a third of the world's sugar; it is a leading economic plant and hence is important also in politics.

The domesticated sugar beet is derived from the wild species *Beta maritima,* a common seashore plant in Europe. Both wild and domesticated species are of the family Chenopodiaceae (goosefoot family), which includes spinach and many common weeds. Like other species of the family, the wild sugar beet is adapted to saline

5.6 Although mechanical cane cutters are taking over the harvesting of sugar cane in some parts of the world, much of it is still cut by hand, as is the case in this field in Trinidad. This East Indian woman will cut each cane and lop off the leaves. The canes will be transported to a sugar mill where they will be crushed and the released juice will be boiled down to produce raw sugar and molasses.

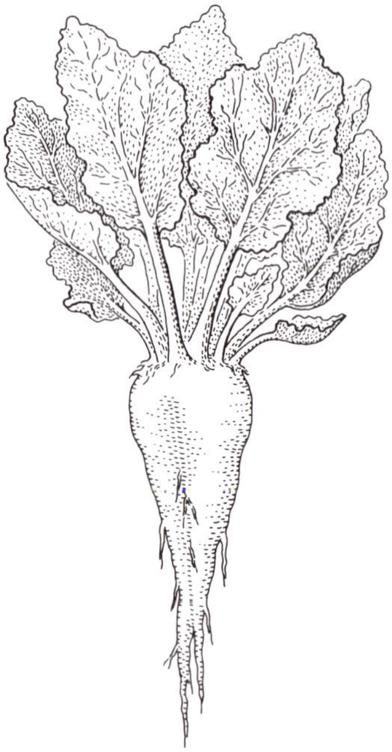

5.7 Sucrose is stored in the root of the sugar beet plant. Domesticated only about 200 years ago, this cheap source of sugar might have altered the history of colonialism had it been available to Europeans 200 years earlier.

habitats such as salt marshes. Originally cultivated for its leaves, it was eaten as a vegetable in some European countries. Germans had the idea of using the swollen taproot as a source of sugar in the eighteenth century and built a factory for extracting it in 1801. Napoleon was an early advocate of the sugar beet industry, which began to flourish only in the 1840s.

Now at least four distinct varieties of the sugar beet are cultivated for the roots and for the leaves as animal feed. All four were developed in Europe and North America in the past century. Sugar beets were cultivated most intensively during World Wars I and II, when the supply of cane sugar from the tropics was interrupted. In both Europe and North America, sugar beets were planted extensively to make up for the missing imports, and the plant has held its important place.

The large, often conical roots can weigh up to 3 kilograms (6 pounds). Mechanical harvesters remove them from the ground. They are then shredded and soaked in hot water. The water removes most of the sugar, which is then purified to produce pure sucrose. Here, too, molasses is a by-product of refining, and much of it is added to the solid waste and used as cattle fodder. Chemically beet sugar is exactly the same as cane sugar and is equally effective as a sweetener.

The sugar beet was domesticated from a temperate-zone species, and so far no attempt to adapt it to tropical regions on a large scale has succeeded. The wild root yields only about 2 percent sugar, but consistent artificial selection has raised the content to 20 percent sugar—a dramatic indication of how man-induced evolution can lead to rapid change.

POTATO

A paradoxical plant is the potato *(Solanum tuberosum).* The part that we eat is not the root but a swollen underground stem, a **tuber.** Its relatives are improbable: the tomato, which grows above ground on a vine, the chili pepper, the eggplant, the tobacco, and the deadly nightshade. (All these plants are in the nightshade family Solanaceae.) In spite of lowly origins, the potato is the most important vegetable in the world; it is an excellent source of starch and of calories, protein, iron, phosphorus, calcium, and vitamins B and C. People can live on this plant if they have no other food source, and some, such as the Irish, have done just that. Most of a human being's need for minerals, vitamins, and calories can be satisfied by 3.5 kilograms (7 pounds) of potatoes a day, but that is a lot of potatoes. American children and teenagers are living proof that people can practically survive on potato chips and french fries. Potatoes are also used to feed stock and for industrial starches, spirits, and industrial alcohol.

What sort of a plant is this economic miracle? It is a branched perennial that grows up to about 1 meter (3 feet) high and produces swollen stem tubers underground. It is reproduced vegetatively (without seeds) by cutting parts of the tubers with at least one bud ("eye") apiece and planting.

Cultivation began in the Andes Mountains of South America where, with maize, it was the staple diet of the Incas, that very advanced culture nearly exterminated by the Spaniards in the sixteenth century. Potatoes are still grown in these mountain regions, but by far the greatest amount is now produced in countries far away. People have spread them all over the world.

Today Germany and the Soviet Union are the primary producers; in the United States the leading producers are Idaho, California, Maine, and Long Island. Although it is a native of the Western Hemisphere, the potato reached the United States by way of Europe. Introduced there in the middle of the sixteenth century by Spanish conquistadors, it was brought to North America in the first half of the seventeenth century. It is also widely grown and eaten in Asia and is much used in Indian cuisine.

In volume and value, potatoes are far greater than any other crop. In favorable conditions, as in Idaho, yields of 35 tons an acre are not unusual, and 100 tons an acre are considered possible. Potatoes give more protein per acre than any other plant we eat because of the large yield. A very large number of people, probably most of the population of the temperate world, make potatoes their staple food.

Potatoes have also changed history, particularly by increasing the population in the United States. In the nineteenth century a fungal disease ravaged potato crops in Europe, setting off great famines followed by large migrations. The disease was the potato blight,

THE POTATO FAMINE IN IRELAND, 1846

5.8 The tubers of the potato plant are enlarged underground stems called *rhizomes* (left). The eyes are nodes, and the lateral bud forming each eye produces a new plant. When a fungus wiped out the potato crop in Ireland in the 1840s, potatoes were the main source of food. People gathered outside workhouses to beg for food, as shown at right. Millions died or emigrated.

caused by the fungus *Phytophthora infestans,* which blackens the leaves of the plants and rots the tubers. In 1845 and 1846 the disease was epidemic. It did its worst work in Ireland, where potatoes were the main food. More than a million Irish died of starvation when the crops failed, and a million more migrated to the New World. Irish immigrants in America have so touched the development of political, religious, and law enforcement institutions that we can say that *Phytophthora infestans* decisively affected United States history.

The sweet potato *(Ipomoea batatas),* despite its name, is no relation to the white potato. It is in the morning glory family (*Convolvulaceae*) and is a vine whose swollen roots we eat—in this case true roots, not underground stems. It too is native to South America.

CASSAVA

Also called manioc, cassava *(Manihot esculenta)* is a shrubby perennial of the family Euphorbiaceae (spurge). It has a large starchy root, which is to the people of the tropics what the potato is to people in temperate zones. Cassava was domesticated in ancient times in the lowlands of eastern South America, and like the potato it spread all over the world after the voyages of Columbus.

The root is almost tasteless and low in food value, but because it is an inexpensive source of calories, millions make it their chief food. It is boiled and eaten or ground into flour and baked or used as porridge. Variations are many, each region having its own style. The

5.9 A market scene in Port-of-Spain, Trinidad, displays typical root crops. The brown elongated cassava roots in the foreground are surrounded by taro roots and sweet potatoes. The cassava plant, like the potato, is native to South America, and is the basic food in the humid tropics. The cassava, with its spindly stalks, seems an unlikely candidate to possess such large roots.

tapioca we use for pudding is made from cassava roots by heating starchy pellets on a metal surface. Cassava is also used to make alcoholic drinks. But some varieties of cassava are very poisonous; they contain cyanide compounds. The juice from their ground roots must be pressed out or the roots boiled before they can be eaten.

Cassava is easy to cultivate and high in productivity. Varieties are numerous, but plant breeders have given the plant little attention. Interest is gaining in producing improved varieties, however. It grows so well and in such abundance that yields of improved varieties might well lead to more general cultivation.

Although cassava is perhaps the most important crop in the root-crop system of tropical agriculture, yams and taros are useful, too. *Taro* is the name for a number of species in different genera of the family Araceae (arum), and yams are members of the genus *Dioscorea* in the Dioscoreaceae. Both are native to Asia and have spread through the tropical world. (The yam, like other plants, has important uses other than food; for example a birth control pill was indirectly derived from *Dioscorea*.)

BANANA

Would you have guessed that the foremost commercial fruit is the banana? It is much more than the breakfast or dessert item that so many fruits are, since it is a staple in many countries. When eaten raw, the mature fruits are sweet and delicious, but the

5.10 The banana flower has a long central axis on which flowers are borne in clusters. These produce the "hands" of banana fruit. The flowers are produced on the large apical bud.

unripe fruits and, more commonly, the related plantain can be cooked to provide a starchy food nutritionally similar to the potato.

The banana, which has several species in the genus *Musa,* is a perennial with a short underground **rhizome** (a rootstalk) that produces both roots and shoots. All edible bananas derive from only two species in the genus *Musa,* both native to Southeast Asia, from which they have spread widely. The banana is cultivated in plantations and in gardens. A plant is started using **suckers,** shoots arising from the rootstalk. Seven to nine months later the plant will flower, but another few months are needed for the fruit to mature. The flowers grow in compact groups on a large stalk, and in many types the fruit contains no seeds. The fruit itself is white and pulpy, with a tissue of large cells filled with starch and protected by a thick skin. During the ripening much of the starch is converted to sugar.

Banana fruits for export are usually harvested before they are fully ripe and shipped at low temperatures, with carefully controlled humidity and balance between oxygen and carbon dioxide, to slow the ripening. Once at the destination, the fruits are allowed to ripen. Bananas continue to ripen even after you buy them.

Every banana-producing region has many varieties for local consumption, and these are usually sweeter than those exported. Today the large tonnages of bananas produced in Asia and Africa are mostly consumed there. The world export trade gets its produce from the countries of Central America and the West Indies, which produce more than three-quarters of the world's export supply. In the Old World, bananas for export are grown in the Canary Islands.

Besides being the most important commercial fruit, the banana is useful in many other ways. Its green leaves make plates for meals where it is grown and, when dried, thatching for roofs. Bananas are doubly serviceable as shade plants for young cacao and coffee trees, a singular way of using one crop plant as a help in cultivating another.

COFFEE

As you read this, maybe you have just had or are looking forward to soon having a cup of coffee. Coffee has woven itself so tightly into the fabric of human life that, in the form of coffee breaks, it is even a factor in labor contracts. In times of war when supplies were short, people have lied, stolen, degraded themselves, even risked imprisonment or worse for the sake of coffee, buying it on the black market. As a drug, not a food plant, it offers no nourishment, and yet it is one of the world's most sought-after plant products.

5.11 Coffee trees are fairly small and produce clusters of lovely, fragrant, white flowers. The fruit is picked by hand and dried in the sun. The outer pulpy part is removed and the remaining beans are roasted.

This plant that has gripped mankind is an evergreen shrub that occurs naturally as an undergrowth species in the tropical forests of Africa. Its family is Rubiaceae, with the genus name *Coffea.* The two main species are *Coffea arabica* and *Coffee canephora.* The first is the more popular, making up about 75 percent of the coffee grown worldwide. *Cinchona,* a major source of quinine, and the ornamental gardenia are close relatives of the coffee plant.

Coffee trees (actually they are more shrubs than trees) are relatively fast-growing, bearing fruits three to four years after planting. The fruits take another seven to nine months to mature. It is from these that the beverage used by at least a third of the world's people is produced. The mature fruits, or berries, are harvested by hand and are processed by one of two methods. The dry method is used in most of Brazil's coffee-producing regions: the fruits are spread out to dry in the sun for fifteen to twenty-five days and then hulled. The wet method calls for pulping the berries after picking to remove the outer layer and part of the fleshy inner layer of the fruit. The pulped fruit is then fermented in tanks, washed, and sun-dried for eight to ten days. The dry skin around the beans is removed by milling and polishing, leaving shiny blue or grayish blue beans. The characteristic brown color is produced by roasting.

Coffee drinking started in Arabia during the fifteenth century, and it is still very popular there. Native to the Old World, most coffee today is grown in South America, where Brazil produces 75 percent of the world's supply. Amazingly, much of the world's coffee is descended from a single coffee plant obtained in Java in 1706 and sent to Amsterdam. Seeds were later sent to the Dutch colony of Surinam. Surinam coffee later spread throughout the New World.

Being descended from just one plant has limited the variability of much of the New World's coffee. Unlike wheat or maize, coffee varieties are very few, a dangerous limitation increasing the plant's susceptibility to disease. The greater the number of varieties, the greater the chance that some of them will be resistant to a particular disease. In recent years a fungus disease known as coffee rust has spread from the Old World to the New; having done deep harm to coffee crops in Ceylon and many parts of India, it now threatens the New World's crops.

CACAO

Chocolate is produced from the seeds of the cacao tree *(Theobroma cacao)* in the family Sterculiaceae, which also has the cola tree, source of caffeine for cola drinks. The word *chocolate* comes from a Mayan word for a drink made from cacao seed. Cacao is a tropical crop, restricted to lowland areas where the temperature never falls below 15° C (60° F). Like coffee, trees of the genus *Theo-*

5.12 Cacao plant with fruits. Cacao has flowers formed from bud-producing regions of the trunk. The cacao pods are cut by hand and the large seeds extracted. Chocolate is produced from the fermented and dried seeds. West Africa is the largest supplier of cacao, but the cacao tree is native to the humid tropical regions of Central and South America.

broma occur naturally as undergrowth in the tropical rain forests of northern South America. Many plantations grow the cacao under shade trees, often rubber trees *(Hevea brasiliensis)*, in conditions like those of their native habitat.

Although the genus *Theobroma* has twenty species, only one, *Theobroma cacao*, is grown on a commercial scale. Cacao is relatively small, seldom growing more than 5 to 7 meters (15 or 20 feet) high. It has an unusual growth form, with upright and horizontal types of branches. Like many other tropical trees, it grows in periodic flushes, rather than producing leaves continuously. But the cacao tree is striking: its flowers grow not on separate flowering shoots but directly on the trunk. This **cauliflory** is common in tropical trees with heavy fruits (see Figure 5.12). Fruits appear six months after the flower is fertilized in large colorful pods with up to forty seeds. The seeds take up a quarter of the pod's weight.

Chocolate, like coffee and tea, does not have its familiar flavor right off the tree. The seeds, or cacao beans as they are known, are removed from the pods and fermented for as long as twelve days. Fermenting kills the embryo in each seed and releases **enzymes** (compounds that can promote chemical reactions without themselves being changed by the reaction), which induce chemical changes. The chocolate flavor is made by drying and roasting the fermented beans. Fermentation leaves a fatty substance, cocoa butter, which is extracted from the seeds and used for making chocolate and the drink cocoa. An **alkaloid** (a complex nitrogen-containing compound) called theobromine is also obtained from the seeds and converted to caffeine for use in cola drinks.

The cacao tree was first domesticated in Central America, where Mayas and Aztecs made chocolate by pounding the seeds and mixing them with maize and *Capsicum* peppers. Cacao was so highly valued in these civilizations that it was used for money. Today, although Central America and the West Indies still have large plantations, about two-thirds of the world's cacao is grown in West Africa.

TEA

The most popular beverage on earth is probably tea; certainly it is one of the most ancient. The Chinese first used it as medicine and have cultivated it for two to three thousand years. Half the world's people drink tea regularly, and it is the most common beverage all through Asia and the Far East. Along the north coast of Africa and southward up the Nile Valley, a strongly brewed tea mixed with sugar, herbs, and spices is popular, and in Tibet the leaves are pounded, mixed with dough and fats, and eaten as a paste. Like coffee, tea is a drug plant. Most of the drug component is caffeine.

Tea came to Europe in 1610, introduced by the Dutch from the Far East. It took so little time to become popular that the East India Company of England was founded to exploit the plant, eventually helping to establish the British Empire in India. In the heyday of the Empire, the British spread their educational and legal systems through the world along with the message that a hot cup of tea was essential to human welfare. Tea is inseparable from the British character even now, and the well-being they find in it has a real cause; the 2 to 5 percent caffeine in its leaves (coffee has 1.5 percent).

Camellia sinensis, the tea plant, probably originated in Burma and spread to China and Assam. It is a small tree, kept shrubby for cultivation by pruning. It is normally reproduced by seeds, sown in a nursery, and transplanted to the field after six to eighteen months. The trees can be harvested after three to four years of growth. The apical bud, the one at the tip of a branch, and the first two leaves of the shoot are picked by hand, using thumb and forefinger. The bud and the youngest leaves have the highest caffeine content and make the best tea. Tea leaves grow out in **flushes,** not one by one, and may be harvested every seven to ten days in warmer climates, less frequently in colder regions. After a number of flushes have been produced, the tree becomes dormant. It is then severely pruned, manured, and new shoots are produced.

5.13 These women in Java are plucking the tips of branches of tea plants. The young leaves are dried and processed to produce tea. Tea is grown on well-drained soils throughout Southeast Asia and other tropical regions, but India and Ceylon are the major producers.

Commercial tea is grown from two major types of *Camellia sinensis:* the high-yielding, large-leaved types of India and the narrow-leaved Chinese types that are most resistant to cold and other adverse conditions. Between these two types many hybrids are cultivated, particularly in India. Most tea is grown in China, but most is also consumed there, so the bulk of the supply for the rest of the world comes from India and Ceylon.

Black tea, the largest part of the world's consumption, is manufactured by a highly technical process in factories on the tea plantations. The leaves are heated in a strong air current. The result is more than drying but less than roasting. Caffeine is liberated and the flavor and color of the tea is developed. Green tea is made by steaming, rolling, and drying the leaves. The dried tea is graded and sorted by color and fineness of particles, and then different types are blended to form varieties that suit the tastes and preferences of consumers over the world.

TOBACCO

No plant in history has caused more addiction than tobacco. The dried cured leaves of *Nicotiana* species are smoked or chewed everywhere, except among a few isolated mountain or island populations. A native of South America, the tobacco plant was used long before Columbus's discovery of the New World in 1492. Early explorers brought it with them to the Old World, and the smoking habit has never stopped spreading regardless of politicians, clergy of all faiths, physicians who now recognize that it causes cancer, and sheer common sense. Ironically, it was first introduced to Europe for its medicinal qualities.

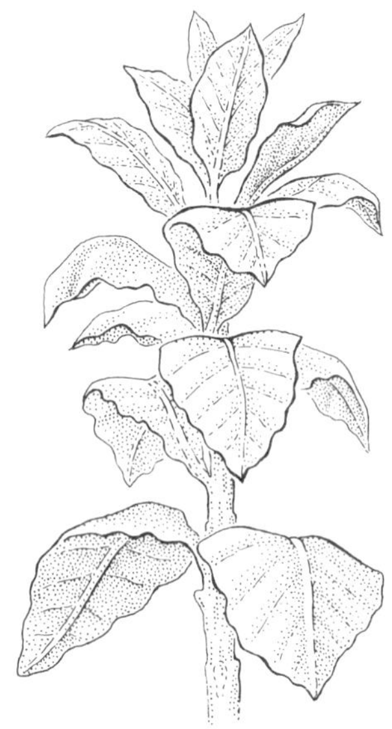

5.14 The tobacco plant, grown worldwide, is the source of much human pleasure and misery.

Cultivated tobacco comes from two species: *Nicotiana rustica* and *Nicotiana tabacum.* Both are in the family Solanaceae (nightshade), to which potatoes and tomatoes also belong. *Nicotiana tabacum* is by far the more important of the two and feeds most of the world's appetite for cigars, cigarettes, snuff, and pipe tobacco. *Nicotiana rustica* occurs in Mexico, and though seldom smoked, it is a valuable commercial plant because its nicotine-rich leaves supply most of the world's demand for commercial nicotine, which is important as an insecticide and in drugs.

Like many long-domesticated plants, such as wheat and maize, neither species has been reliably recorded as growing in the wild. *Nicotiana tabacum* appears to be a hybrid offspring of two wild species and may have originated in Argentina.

The tobacco plant is an annual, growing 3 to 4 meters (9 to 12 feet) high. Its plantations are established from seed. Cultivated plants are drastically modified by removing the tips of growing shoots, which stimulates the growth of **axillary buds** (buds located at the base of each leaf), and these are then also removed. The plants then grow abnormally large leaves, possibly half again the size of plants allowed to grow freely without having their buds pinched off.

Tobacco leaves are allowed to mature through most of the plant's growth cycle, the leaves being harvested only when they are old and are greenish gold from loss of the pigment chlorophyll. Experienced workers can tell by the color when the moment for harvesting has come. Depending on the quality of the tobacco required, whole plants or selected leaves are harvested and then cured by subjecting them to varying temperatures and humidity. The cured leaves are then put in piles to ferment for as long as six weeks.

On plantations tobacco is grown in well-drained soils. It is remarkably tolerant of climate and temperature, thriving in both upland and lowland regions in the tropics and as far north as Connecticut and Scandinavia. It is very sensitive to soil conditions, on the other hand, and the same type grown in different soils can give radically different aromas. That is why the cigar smoker's timeless favorite, Havana tobacco, cannot be reproduced outside Cuba, even when the same varieties and techniques are used.

Tobacco has affected both politics and economics in the New World. The Jamestown colony in Virginia could count on success when John Rolfe began cultivating tobacco there in 1612. Even in those days the demand for tobacco was endless—it was grown in the streets of Jamestown. Tobacco extracts large quantities of nitrogen from the soil, as planters in the southern colonies found when their soil was exhausted (the soils of George Washington's Mount Vernon plantation had used up their fertility by the time he died in 1799). The fertile soils of the east depleted, people demanded new lands west of the Appalachian Mountains, and pressure on the western frontier grew. Pressure for the hand labor necessary to tend the tobacco was a prime cause of slavery on southern plantations.

SPICES

People have cultivated and used spices since the ancient civilizations of Egypt, China, Greece, and Rome. The richly assorted flavors have always been especially sought in countries where the staple diet is monotonous starchy grains, roots, or meat that, when dried to preserve it, takes on an unpleasant flavor. Spices are much used throughout the temperate world, too, and a good many explorations in the fifteenth and sixteenth centuries were undertaken by European navigators spurred by the profits to be won finding new sources of spices. Also, traditional land routes to spice markets in the East had fallen victim to power politics. The eastern empires of both the Portuguese and the Dutch were founded on spices; in fact,

5.15 During the sixteenth and seventeenth centuries, a voyage to the Spice Islands had an aura of mystery and romance similar to our space flights. Those returning from the Molucca Islands brought spices, dyes, and rare woods. The imagined productivity of the Spice Islands is depicted in this illustration of a nutmeg tree with nutmegs the size of beach balls.

the spice trade in that period had something of the mystique and profits peculiar to the oil trade today.

Most plants from which spices are made grow in the wetter parts of the tropics. Strangely, these native areas are mainly islands, like the Moluccas, a group in the Indian Ocean south of the Philippines and west of New Guinea. Here was the first source of clove and nutmeg, for which the group became known as the "Spice Islands."

Spices owe their flavor and aroma to the presence of essential oils, which are derived from parts of varied plants. These are organic oils with relatively small molecules. They are volatile: like perfume, they evaporate and diffuse their odor quickly.

Cinnamon comes from the bark of *Cinnamomum zelyanicum,* an evergreen tree in the family Lauraceae. Others in the family are mainly aromatic evergreen trees and shrubs, among them camphor and avocado. Cinnamon is native to Ceylon, where all commercial production of the spice is still carried on; attempts to introduce it elsewhere have never been entirely successful. Cinnamon flavors desserts, cakes, and candy, and it is part of curry powder; its oil, distilled from the leaves, is used medicinally.

Another constituent of curry is ginger, from the underground rhizomes of *Zingiber officinale.* A reedlike plant of wet tropical regions and member of the family Zinigiberceae, ginger is closely related to the banana family, Musaceae. Used extensively as a flavoring for medicinal products, bakery goods, and beverages, ginger is produced in both dried and cured forms. It is prepared by removing the bark from rhizomes of the plant, washing and partly boiling it, and then drying it in the sun. Much of this form of ginger is exported by Jamaica. China supplies preserved green ginger, made by cleaning the rhizomes and boiling them with sugar and water.

Pepper, which has nothing whatever to do with the bell and chili pepper of the family Solanaceae, is one of the oldest and most treasured spices. It is universally incorporated in flavoring, pickling, and preserving food. New trade routes to the "Indies" were sought eagerly in the fifteenth century mostly because of the desire for pepper. Black pepper and white pepper are both obtained from *Piper nigrum,* of the family Piperaceae. The pepper plant is a high-climbing perennial vine that holds on with **adventitious roots,** which grow from leaves or stems. Black pepper is made from the dried ground fruit of the pepper plant; white pepper comes from the dried ground seeds.

Piper nigrum grows in the monsoon areas of Southeast Asia, and today India is its greatest exporter, followed by Indonesia. The United States uses about 40 percent of the pepper produced in the world.

Vanilla, favorite flavoring for ice cream, cakes, and candies, is an orchid, *Vanilla planifolia,* a native of South America's tropical rain forests. A tall, climbing, vinelike plant with fleshy green stems that attach themselves to supports with aerial roots, it bears as fruit a slightly triangular capsule 15 to 23 centimeters (6 to 9 inches) long covering a number of seeds. Freshly harvested vanilla beans have none of the aroma and flavor of the finished product. These come out in the curing, which consists of partially drying the seeds in the sun and then "sweating" them under blankets. The sweating is fermentation, during which the seeds are chemically changed, bringing out the taste.

The first time that cloves were used as a spice or perfume was recorded in China in the third century B.C., and cloves are still among the most important of commercial spices because of their many uses. As a powder, they flavor sweet and savory dishes and are used in pickling and in making sauces and ketchup. Dried cloves are the desiccated, unexpanded flower buds of the evergreen tree *Eugenia aromatica* in the family Myrtaceae (myrtle), which also gives us allspice and the fruit guava. The Moluccas are the original home, but Zanzibar and Indonesia are now the world's greatest producers of cloves, growing 90 percent of the world crop.

Many of the chemicals responsible for the taste and odor of spices have been identified and synthesized. One is the vanilla flavor in your ice cream—it is synthetic and not, alas, the product of a tropical orchid.

COTTON

Of the fiber crops that supply us with clothing, paper, and rope the most useful is cotton *(Gossypium)*, an attractive shrub in the family Malvaceae (mallow). Among its close relatives are the ornamental hibiscus flowers *(Hibiscus rosea-sinensis)* and the edible

5.16 Ripe cotton fruit breaks open and the long hairs of the seed coat are exposed. Some cotton varieties grow as woody shrubs (left). The cotton fibers (right) develop from single cells in the outer layer of the ovule. The mature fibers are the cell walls of the dead cells and are still attached to the seed coat. The cotton gin separates the fibers from the seeds.

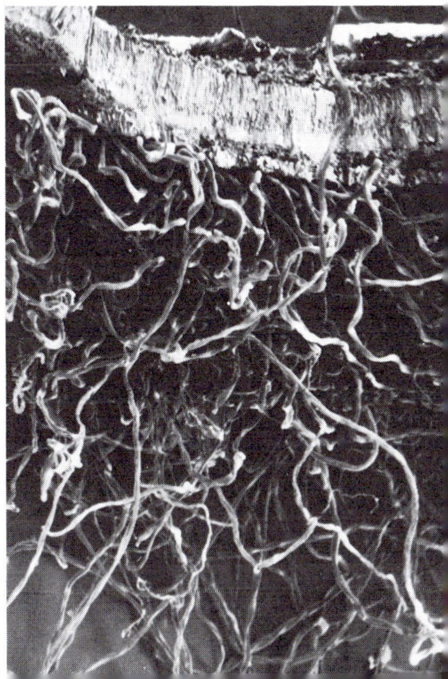

hibiscus okra *(Hibiscus esculentus)*, whose fruits are eaten as a vegetable in some parts of the world. Cotton fruits are not utilized at all; it is the seed hairs or lint on the seed coats of some varieties that are harvested and spun into fibers. These are very long single cells, with cell walls of cellulose.

Only four of the thirty species of cotton currently recognized are cultivated. Two occur in the Old World *(Gossypium arboreum* and *Gossypium barbadense)*. The New World species have a longer lint and are more extensively grown. But each of the four species has varieties, ranging all the way from primitive backdoor perennial cottons, which may grow as high as 5 meters (15 feet) and are made locally into hammocks and clothing, to highly selected annuals used commercially. Most of the cotton in the world's markets grows from these modern varieties on huge plantations, many of which have complex irrigation systems and mechanical harvesting.

A mystery about cotton is how the New World species got started. Both species are very different from the two Old World species. Geneticists have discovered that they combined from wild unlinted New World species and linted African species. No one is sure how, where, and when these two groups came into contact to produce the ancestor of today's New World cottons. Early theories held that people dispersed them, but more recent studies suggest that African cotton reached the New World as seeds that floated across the Atlantic shortly after the continents began to drift apart (see page 460), while the land masses were much closer together. Once these cottons became established and spread, they hybridized with a wild South American species to form a wild linted cotton plant, which at some later time was domesticated. Even in the mid 1950s, no wild linted cottons were known in South America, but more recent explorers have found small populations of the wild plants in extremely arid parts of northeastern Brazil, Peru, and Ecuador.

Cotton is grown throughout the world's tropics and subtropics. A leading producer for more than a hundred years, the United States recently has accounted for half the world's supply. Here cotton growing is highly sophisticated, with mechanical harvesting and even defoliation, which, using chemical herbicide sprays, makes the leaves drop before the bolls, clearing the way for the mechanical pickers to harvest the bolls. The seeds produced by the cotton bolls are the source not only of fiber but also of oil and a high-protein feed for stock.

One variety has the highest quality and longest lint, often 5 centimeters (2 inches) or more, and that is Sea Island cotton. A native of tropical South America, it was taken to the West Indies and from there was introduced to the islands off South Carolina and Georgia during our days as a colony.

Cotton is the most political of all plants grown in the United

States. Early in the nineteenth century, farmers began to substitute cotton for tobacco because demand and price had changed, and the cotton gin, which mechanically separates lint from seed, was invented. The machine could turn over far more bolls, and that of course led planters to demand more slaves to produce greater crops. This cry for growth had much to do with setting the stage for the Civil War. The war, in turn, badly disrupted the world cotton trade, because much of the product came from the southern states where crops had been burned or not planted. As a result, other centers of production opened up for cotton growing, especially in Africa and tropical America. Yet another effect changed life in cities of the American north. Hand picking of cotton was essentially eliminated after World War II, starting the mass migration of agricultural labor from the rural South to the urban North.

5.17 Most of the world's rubber plantations are in Southeast Asia, although the tree is native to South America. Workers make V–shape cuts in the bark; the milky latex oozes from the inner portion of the bark and is then collected.

RUBBER

Rubber, an elastic hydrocarbon, comes from a plant that changed forever the civilized world. It is difficult to realize today that little more than a century ago the world knew no rubber raincoats, rubber-tired wheels, elastic garments, rubber bands, rubber-soled shoes, rubber erasers—the list has no end. Yet all these were derived in a very few years from **latex,** the white sap produced by the tree *Hevea brasiliensis* (in the family Euphorbiaceae), which is the world's primary source of natural rubber today.

Other plant families have members that produce latex, including the Compositae and Moraceae, but the latex from some does not contain rubber. In those with rubber the percentage varies considerably but is generally so low that extracting the rubber is unprofitable. But *Hevea brasiliensis,* Para rubber, contains latex with a high concentration of rubber particles; it has been spread across the globe.

The rubber tree is a native of the Amazon basin in Brazil, where its sap was originally made by Amazonian Indians into watertight containers and balls. Today it is a relatively unimportant crop in its native land, because much of the world's rubber comes from Southeast Asia. The anomaly was brought on by natural pests and diseases. South American leaf blight, a serious disease of rubber trees, occurs throughout the New World tropics but not in the Old World. Plantations in many parts of the New World are impossible because of the blight, but in southeastern Asia the rubber tree thrives.

Plantations in Southeast Asia were started from seeds smuggled out of Brazil by Henry Wickham, an English plant collector. Seedlings were established in 1875 at the Royal Botanical Gardens in Kew and then were shipped to Ceylon and Singapore. Although only seventeen hundred of Wickham's seventy thousand seeds survived, from those came 90 percent of the rubber produced in the world today. Like coffee and chocolate, rubber's center of production thus came to be a continent far from the one in which it originated.

Hevea brasiliensis is a member of the family Euphorbiaceae, which includes several other tropical commercial crops: castor oil, cassava, and the common Christmas plant, poinsettia *(Euphorbia pulcherrima).* The genus *Hevea* contains twenty species of trees, but no other species is a useful source of rubber. Latex is collected by tapping the tree: an incision is made in the bark and the latex flows down the cut into a container tied to the trunk. After collection, the latex is coagulated by mixing it with acid, after which it is pressed between rollers. The resulting strips of rubber are then dried and smoked in curing houses. Industry may then transform the raw latex into something as thin and fine as a surgeon's glove or as tough as the giant tires on heavy construction equipment.

These are samples of the plants that humans take for their food and for the other activities that are part of "making a living." Others could be added. More important than additional details is the understanding that we depend on plants for our survival. Because of this, plants have also supplied much of the basis for the development of our cultural and political institutions.

Summary

About five thousand species of plants are used today for food, fiber, and industrial products. Three members of the grass family are especially important as staple foods: wheat *(Triticum),* maize *(Zea mays),* and rice *(Oryza).*

Wheat was among the first plants to be cultivated systematically and is now the most widely grown. Today's crops are hybrids of cultivated wheat and wild grasses.

Maize is the second most widely grown crop, although half the world's supply comes from the United States. It is a monoecious plant (having male and female flowers on the same plant), and its fruit is the part we eat. It originated in the Western Hemisphere and today's plant is the product of much hybridization.

Half the world's population, however, lives on neither wheat nor maize but on rice, which occurs naturally in swamps. Most is cultivated, and consumed, in Southeast Asia.

Soybeans *(Glycine max)* have long been grown and eaten in Asia; recently they have become important in this country. A short bush that is a member of the legume family, soybeans are useful for both the fat and the protein they produce. Soybeans have recently assumed tremendous economic significance for the United States as an export crop.

For many tropical countries sugarcane *(Saccharum officinarum)* has comparable importance as an export. It is a tall grass with sugar-rich juices in its stems. Molasses and rum as well as much of the world's sugar are made from it.

Sugar is also derived from the sugar beet *(Beta vulgaris)*, which is grown in temperate but not in tropical climates. It is one of the few plants domesticated in modern times.

Potatoes *(Solanum tuberosum)* yield greater volume and value than any of these plants. The potato's tuber (a swollen underground stem) is so rich in calories, vitamins, minerals, and proteins that people can and have depended on it as their sole food. Irish people relied on it to such an extent that when the potato blight ravaged their crops in the mid-nineteenth century, mass starvation and migrations to America followed. It is a staple food throughout much of the temperate zone.

The role of cassava *(Manihot esculenta)* in the tropical diet is comparable to that of the potato in temperate areas. Its root is edible and a good source of calories, though unlike the potato it is otherwise low in nutritional value.

Bananas *(Musa)* when cooked are nutritionally similar to potatoes. Two species from Southeast Asia are the ancestors of all today's edible bananas, which are grown in Asia, Africa, Central America, the West Indies, and the Canary Islands. Besides providing food, banana trees serve as shade for coffee and cacao trees, and their leaves can be made into plates or thatching for a roof.

The coffee plant *(Coffea)* is a shrub native to tropical areas; most is now grown in South America, especially Brazil. Varieties of the plant are few, making it very vulnerable to disease. The fruits are the source of the beverage, but they offer no nourishment.

Seeds of the cacao tree *(Theobroma cacao)* are the source of another popular substance, chocolate; caffeine is also derived from them. A fairly short tree, the cacao has both upright and horizontal branches, grows in flushes, and exhibits cauliflory (the flowers grow directly on the trunk of the tree). Although cacao was first cultivated in Central America, about two-thirds of the world's supply now comes from West Africa.

Tea *(Camellia sinensis)* is probably even more popular around the world than coffee; it too is a drug not a food plant. Leaves and buds are the source of the beverage. The tree probably originated in Burma, but many hybrids have been cultivated. The Chinese grow most of the world's tea, but Ceylon and India export the largest part of the world's supply.

Tobacco *(Nicotiana)* was first brought from the New World to the Old because of its medicinal qualities. It, like the potato, is a member of the family Solanaceae. Its leaves are fermented for smoking. The plant is tolerant of numerous climates, but it is extremely sensitive to soil conditions.

Even before the New World was discovered, explorers were seeking new sources of spices. Most come from plants grown in wetter parts of the tropics, usually islands. Oils from various parts of the plant are the source of the flavor and aroma for which spices are valued. Cinnamon, for example, comes from the bark of a member of the family Lauraceae; ginger from the underground rhizomes of a plant in the family Zingiberaceae; white pepper from the fruit of a plant in the family Piperaceae and black pepper from its seeds; vanilla from the seeds of an orchid; and cloves from unexpanded flower buds of an evergreen in the family Myrtaceae.

Cotton is another plant of the tropics; it also grows in subtropics. The seed hairs of a shrub *(Gossypium)* in the family Malvaceae are harvested and spun to produce this important fiber. New World and Old World species differ significantly, and the origin of the New World species is still debated. Half the world's cotton comes from the United States, where economic and political controversies related to cotton growing have been important historical factors.

Many trees produce the white sap latex, which is the source of rubber, but almost all the world's rubber is made from the latex of *Hevea brasiliensis,* a tree native to Brazil. Most rubber today comes from Southeast Asia.

Review Questions

1. What do the Irish potato and cassava have in common?
2. Genetic engineering is a much discussed and often controversial term and is considered by many to be a new phenomenon. Does this chapter give you any cause to disagree?
3. With a few exceptions, the major crop plants were domesticated early in human history. What are the exceptions and what are probable causes for this phenomenon?
4. Politicians often speak of guns and butter as two incompatible economic objectives. How did the soybean prove or disprove this rule?
5. Contrast the contribution of the New World to the ranks of domesticated plants with that of the Old World.
6. The movement of domesticated plants around the world has provided much of the world's drama for the past 500 years. Discuss some outstanding examples of this drama.

Medicinal Plants

The medical value of plants captured the attention of people millennia ago. As early as 5000 B.C. the Chinese had lists of drugs derived from plants; in 2735 B.C. the emperor Shen Nung is said to have noted more than three hundred medicinal plants. Babylonians, Assyrians, Hebrews, and ancient Greeks and Egyptians also developed such lists. Both Pliny and Galen wrote about drug plants (one root, Pliny claimed, would cure only if one said the patient's and the parents' names while digging it). Around the first century A.D. Dioscorides, a physician for the Roman emperor, wrote *De materia medica;* this compilation of information on medicinal substances was the authoritative reference for fifteen centuries.

After the invention of movable type in the fifteenth century A.D., herbals flourished. Those were illustrated lists of information about plants emphasizing those that were valuable as food or, especially, as medicines. Their authors often adhered to the doctrine of signatures, which said that a plant's shape signified its value in treating a similarly shaped human organ. Much of the herbalists' information was false, much was superstition, but they encouraged a dedication to careful observation of nature.

1 Medicinal plants were grown in such medieval "physic gardens" as the one illustrated. The first botany professors taught plant identification in medical schools. They have been succeeded by pharmacology departments but plants are still important in medicine.

2 Drugs vital to medicine are obtained from plants. For example (on this page) morphine, a pain killer, is derived from *Papaver somniferum;* ipecac, an emetic, from *Cephaelis ipecacuanho;* quinine, used in treating malaria, from *Cinchona;* (on the facing page) reserpine, used in treating high blood pressure, from *Rauwolfia serpentina;* digitalis, for heart disease, from *Digitalis purpurea;* and atropine, for asthma, from *Atropa belladonna.*

Papaver somniferum

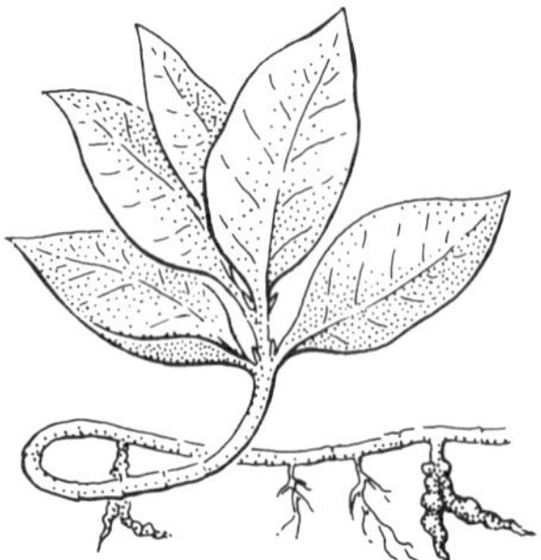

Cephaelis ipecacuanho
(Brazil root)

Cinchona

In 1546 the first pharmacopoeia, a list of medicinal substances and directions for making preparations from them, appeared in Nuremburg. Until then medicines had varied not only in their concentrations but even in their constituents. By the seventeenth century many writers of herbals also directed botanical gardens, which were primarily devoted to medicinal plants. From these medical roots grew the modern science of botany.

Thousands of potions and pastes with presumed or actual therapeutic value have been extracted from herbs, spices, roots, bark, flowers, and, most frequently, leaves. Of the several thousand plants valued today as medicines around the world, most are used by folk practitioners; according to at least one author, Siri von Reis Altschul, there is evidence that 25 to 50 percent of those in non-Western pharmacopoeias are to some degree effective. Some folk remedies long ignored by Western science were later found to be effective. *Rauwolfia serpentina,* for example, was used by Hindus for centuries and recognized by Western medicine in the 1950s as a valuable tranquilizer. Belladonna and digitalis (foxglove) are ancient drugs valuable today.

A number of widely used modern drugs were derived from ancient folk medicines. Characters in the histories of these achievements include quacks and chemists, physicians and botanists, as well as traders, remote tribes, and even governments. The story of opium is one example.

For centuries juice from the pod of the opium poppy *(Papaver somniferum)* was widely used in Mediterranean countries; Theophrastus mentioned it around 300 B.C. An annual plant native to Asia Minor, the opium poppy relieves pain and induces euphoria followed by dreamless sleep. Around 1803 a young German, Sertürner, found and isolated opium's active ingredient, which he christened *morphium* after Morpheus, the god of sleep. Today we know it as morphine. Although it has harmful side effects, including great danger of addiction, morphine is still very widely prescribed to relieve pain. As a result of Sertürner's work, doctors could prescribe measured doses of the ancient pain-killer.

His work had other, more far-reaching consequences, for Sertürner discovered that morphine is an alkaloid (a nitrogenous substance; see Chapter 9). Once morphine was isolated many scientists concentrated on studying alkaloids; from their work came many other important drugs.

Another alkaloid found several years later was the cure for malaria, perhaps the greatest killer throughout history. Malaria can be traced at least as far back as 1000 B.C. In the sixteenth century A.D. part of Italy was abandoned because of it, and waves of malaria plagued Europe during the seventeenth and eighteenth centuries. Malaria was also a serious threat in the United States and made European settlement in Africa close to suicidal.

The cure for malaria was used in Europe for nearly two hundred years before anyone knew its chemical ingredients. It is quinine, which comes from the bark of the *Cinchona* tree native to regions of abundant rainfall on the eastern slopes of the Andes. Linnaeus named the tree in honor of the

Countess of Chinchon, wife of the Spanish viceroy of Peru, who according to legend had been cured of malaria by the bark in 1630 and had introduced it to Spain and Italy. It is more likely, however, that the credit belongs to the Jesuit priests of Lima.

In Europe the cure was known as "Jesuit's bark," but bloodletting was still the preferred treatment for malaria during the first half of the seventeenth century. Then a quack, Robert Talbor, treated and cured Charles II of England by using cinchona bark. Despite continued controversy, cinchona was included in the British pharmacopoeia of 1677 and, by the end of the seventeenth century, it was used around the world to treat malaria.

The only source of the medicine was still cinchona growing wild in South America, and neither the natural history of the tree nor the constituents that made it effective were known. About 1820, however, Joseph Pelletier and Joseph Caventou isolated the ingredient in cinchona that cured malaria: quinine, an alkaloid. Their discovery allowed the effectiveness of particular specimens of cinchona to be measured. Demand for the bark grew, and it became obvious that the South American forests, which had already been exploited for years, could not keep up with the need—and might even be destroyed in the attempt to do so.

The British and the Dutch in particular tried to obtain seeds to cultivate outside the New World. Several expeditions sponsored by these governments failed: plants were killed in transit, or seeds failed to germinate, or the trees that grew from them yielded little quinine. Plantations begun by the Dutch in Java from such seeds seemed likely to preserve the cinchona tree from extinction but unlikely to be a practical source of quinine.

In the 1860s Charles Ledger, an English bark trader living in Peru, succeeded where many others had failed. He sent his Indian servant Manuel to obtain seeds from a stand of cinchona believed to be particularly rich. Manuel did not return for several years, but the seeds he brought became the basis for successful Dutch plantations in Java. (Manuel died as a result of his efforts—according to one story, he was killed by other Indians because he had broken a tribal prohibition against revealing secrets of the cinchona; according to another version, he was jailed by the Bolivian government for selling cinchona seeds for export and died in prison.)

The trees grown from the seeds Ledger sent yielded especially high amounts of quinine, but this fact alone would not have ensured the success of the Javan plantation. The planters soon learned that they faced two special problems: the tree, named *Cinchona ledgeriana* in Ledger's honor, was certain to be cross-pollinated by other inferior varieties unless it was completely isolated, but the tree did not grow unless grafted to roots of more vigorous varieties. Those in charge of the plantation went to great trouble to isolate *Cinchona ledgeriana,* and they grafted seedlings to roots of a more vigorous variety. The results of their labors were successful plantations in Java that gave the world much-needed quinine and the Dutch a virtual monopoly on the drug until synthetic replacements were developed during World War II.

Plant-derived drugs have little effect on major diseases such as pneumonia, typhoid, and venereal disease, which are caused by bacteria. With the introduction of sulfa drugs and antibiotics in the 1930s and 1940s the relative importance of medicinal plants declined. Nevertheless, according to some estimates, more than half of American prescriptions today contain a drug that either comes from a plant or duplicates plant substances.

Rauwolfia serpentina

Digitalis purpurea
(foxglove)

Atropa belladonna
(belladonna)

Structure

PART III

and Function

How is it that plants live, some for hundreds of years? Do they take food from the soil? What are their main parts and what must plants have to survive?

Whether you are looking at a philodendron on a shelf or a towering pine on a hillside, the plant you see developed according to a similar plan—one quite different from the way animals grow. In recent years scientists have learned much about what controls this development, including light, gravity, and substances normally produced by the plant itself. Although much remains to be learned, the knowledge that we now have about plant development (described in Chapters 6 and 7) is vital to agriculture.

Plants require energy to grow and this energy is obtained by photosynthesis and made available to cells by respiration (Chapter 8). Water and minerals too are needed, and these are taken in from the soil by the roots. Thus we see a kind of division of labor: leaves produce energy, "food"; roots make water and minerals available. To bring the food down from the leaves to the rest of the plant, and water and minerals up from the roots, plants have two transport systems through their stems or trunks. These transport systems are described in Chapter 10. In producing energy and moving minerals and water, plants are also playing an important role in cycling water, carbon, nitrogen, and other elements through the biosphere.

Studying how plants develop and survive, we examine the architecure of plants and structure of their cells, emphasizing seed plants and keeping in view throughout how limitations on energy and minerals shape plants, and how plants affect the rest of the living world. The goal: to observe, appreciate, and understand the world more completely.

CHAPTER 6

The Developing Plant

The sequoia huge and enormously tall, the African violet small and delicately flowered, and the pea plant climbing up a trellis as it grows are all plants. However different their shapes and sizes, all plants develop in much the same way. Development consists of three interrelated processes: (1) **growth,** an irreversible increase in size; (2) **morphogenesis,** a change in shape or form; and (3) **differentiation,** increasing specialization of the structure or function of cells or organs. These interrelated processes produce the mature organism—plant or animal. But we will see that although plants and animals have some common features, they do not develop in quite the same way. Knowing how the human leg develops will not tell us how the branch of a plant grows.

THE BASIC PLAN

Plants grow and change shape as their cells divide and enlarge. Just as one might construct a building brick by brick, a plant develops cell by cell. Every plant starts as an elongated axis (see Figure 6.1). New cells are produced in the **meristems,** regions of cell division at each end or **apex** of the axis (we will discuss how cells divide in Chapter 11). Directly behind a meristem is an **area of cell elongation.** Here cells take up more water, lengthening each cell. They elongate primarily in one direction, so the meristem is pushed ahead. Throughout the plant's life the meristems keep producing

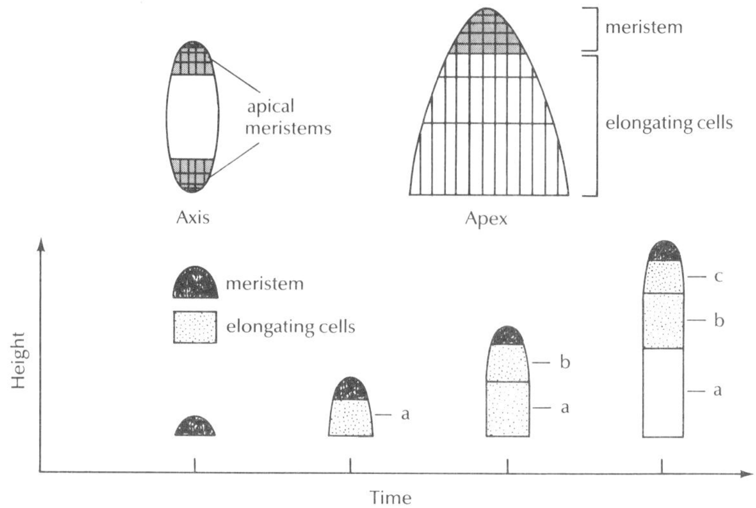

6.1 A plant axis grows as a result of the elongation of cells produced by a region of cell division called a *meristem.*

new crops of cells, which in turn elongate. Mark the tip of a plant at regular intervals of a few millimeters with lines of indelible ink. As the plant grows, the region of cell elongation will be shown by the increased distance between the lines.

Cell division also affects morphogenesis. If plant cells divide in one plane, they produce long strings of cells, called **filaments;** when they divide in two planes, flat sheets like leaves are produced; and when they divide in three planes, cylinders can result (see Figure 6.2).

6.2 Major factors in morphogenesis (the determination of form) are the different planes of cell division and the different rates of division.

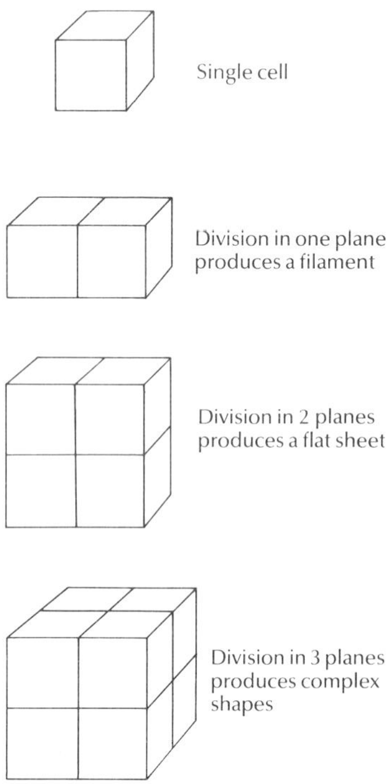

Sequoia, African violet, and pea plant follow this same basic plan. As the plant develops, the cells become different, suited for particular functions; similar cells form **tissues** and tissues form **organs**—roots and stems, leaves and flowers. Seeds seem a logical place to begin a closer look at how these organs form.

FROM EMBRYO TO SHOOT

Seeds A small self-sufficient package, a seed has three main parts: the protective **seed coat;** the **embryo** inside the seed coat; and the **endosperm,** the food that nourishes the embryo (see the seeds shown on page 82). The seed coat is a tough, often waterproof remnant of parental tissues; it develops from the outer cell layers of the ovule. Because the seed's journey is often rough, with drastic changes in environment, the coat of many seeds possesses great resistance to environmental extremes such as moisture and dessication.

The endosperm of gymnosperms is derived from tissue of the ovule, whereas in angiosperms it results from fertilization. It is composed of cells rich in food reserves such as starch, protein, or fats. During the plant's early growth, the endosperm supplies nutrients to the embryo, the part of the seed that interests us most.

The Developing Embryo We have seen in Chapters 2 and 3 that male sperm fertilizes the female egg to produce a one-celled zygote. The zygote then divides, soon producing a ball of cells. This ball becomes **polarized** as cells begin to divide at two points opposite each other (see Figure 6.3). Now the future stem and root of the plant are defined: it is settled (scientists are not sure how) which end will be "up"—becoming the stem—and which "down"—the root. Once the ball is polarized, distinct regions for cell division and elongation develop, transforming the ball into an axis. At each end *(apex)* of the axis is a meristem—known as the **apical meristem**—and behind each meristem is an area of cell elongation. Both ends grow by the same processes of division and elongation, but in opposite directions. At the stem end, cell division produces one or two flaps called cotyledons. Plants with one flap, as we saw in Chapter 3, are called monocotyledons, or monocots; those with two, dicotyledons, or dicots.

In many dicots, the embryo's cotyledons become filled with food and grow to be the largest structures in the seed. Think of a peanut, which splits into two halves when it is shelled; those two halves are the cotyledons. (In peas and beans, too, the cotyledons are what we eat.) Sandwiched between the peanut's two cotyledons is a smaller triangular nubbin at one end; this is the embryo's axis. In monocots there is, obviously, only one cotyledon, and it does not store food. The embryos of some monocots, the grasses, have cylindrical caps covering the two apices: the **coleoptile** on the stem and the **coleorhiza** on the root.

The Germinating Seed The new plant has thus begun to grow inside the seed; it is already a miniature plant. In temperate zone plants the embryo will be **dormant** (that is, its physiological activity will have slowed down greatly) by the time the seed is mature. Many tropical plants have fruits in which the seeds may begin to germinate while still attached to the parent plant.

From plant to plant the conditions necessary for **germination** vary (some of these conditions are discussed in Chapter 15). Say the seed lands in the right location in our temperate zone, with appropriate rainfall, drainage, hours of sunlight, and soil. On a spring day the sunlight heats the soil, the seed takes in water, the cells of the embryo begin to divide and grow again, and the root breaks out of the seed coat. The seed has germinated. The young plant's root will grow faster than the stem, absorbing the mineral nutrients and water without which the plant cannot grow. This rapid initial growth is an adaptation that helps the seedling survive.

When the seeds of grasses germinate, the coleorhiza is quickly

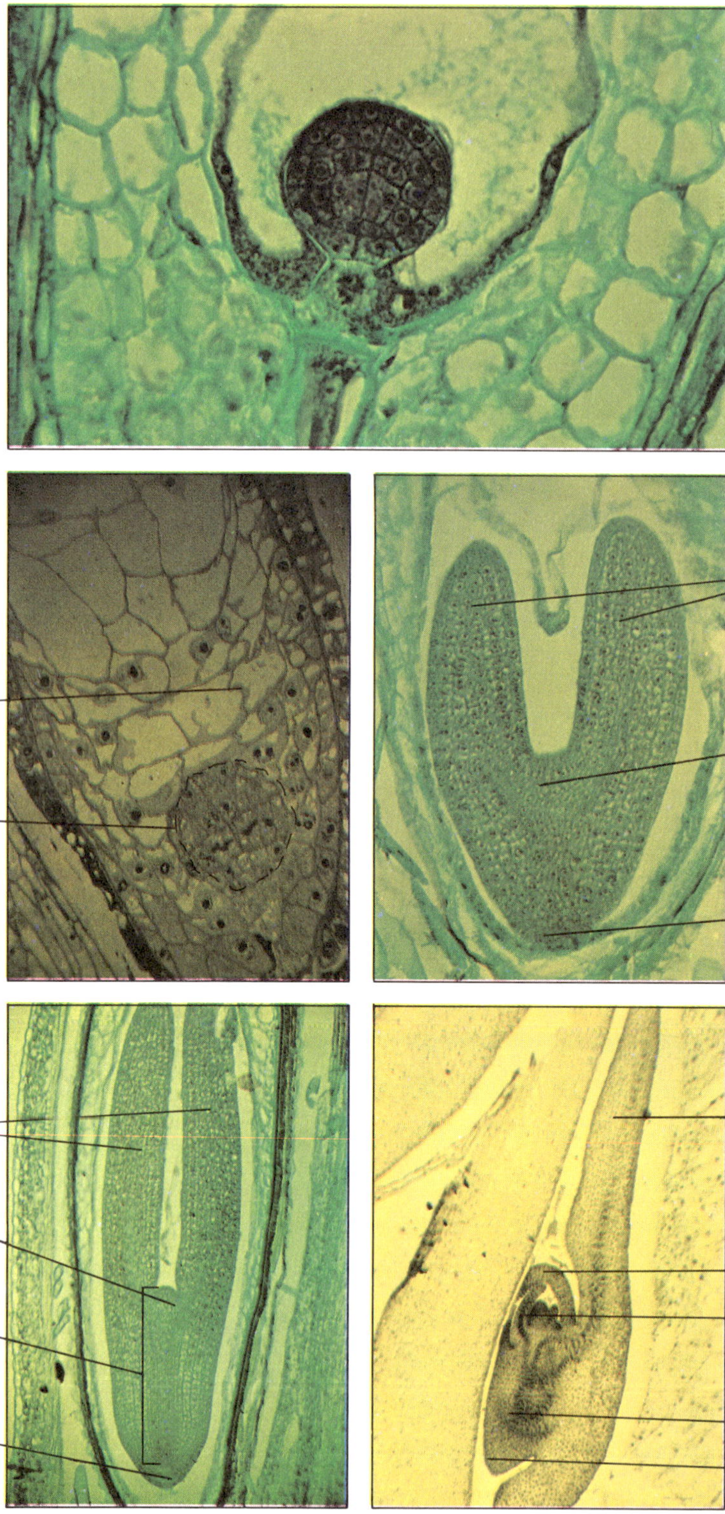
Endosperm
Young embryo beginning to show polarization
Developing cotyledons
Developing shoot meristem
Developing root meristem
Cotyledons
Shoot meristem
Embryonic axis
Root meristem
Cotyledon
Coleoptile
Apical meristem
Root meristem
Coleorhiza

6.3 FROM EMBRYO TO SHOOT

The single-celled zygote undergoes cell division and the young embryo is a ball of cells (facing page, top). The ball begins to elongate and enlarge, nourished by surrounding endosperm cells (facing page, middle left). The *Epilobium lanceolatum* embryo illustrated here develops two young cotyledons as a result of differential cell division (facing page, middle right). The embryo of the mature seed has formed a shoot and root meristem, and vascular tissue connecting the ends of the axis begins to differentiate. Most of the endosperm is gone (facing page, bottom left). Monocot embryos go through similar stages of development, but the embryo of the mature seed has a different morphology. This grass embryo has a single cotyledon, and the shoot and root meristems are covered with a coleoptile and coleorhiza, respectively (facing page, bottom right). When a bean seed germinates, a section of the axis between the cotyledon and root called a hypocotyl elongates and the cotyledons are pushed above the soil. The epicotyl is the part of the shoot above the cotyledons (below). In contrast, a germinating wheat seed (bottom) possesses an elongated coleoptile from which the first leaf emerges.

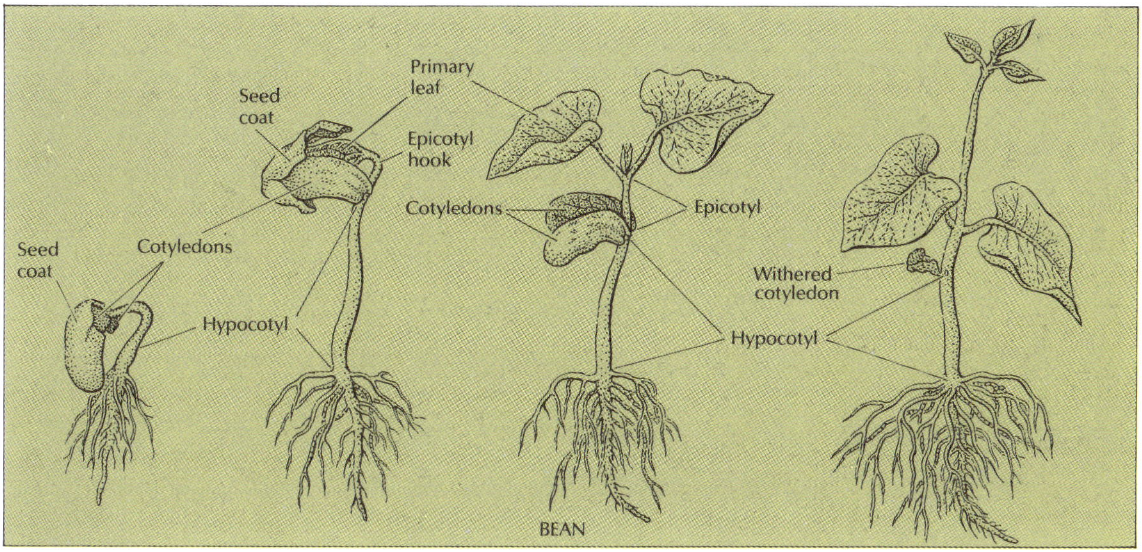

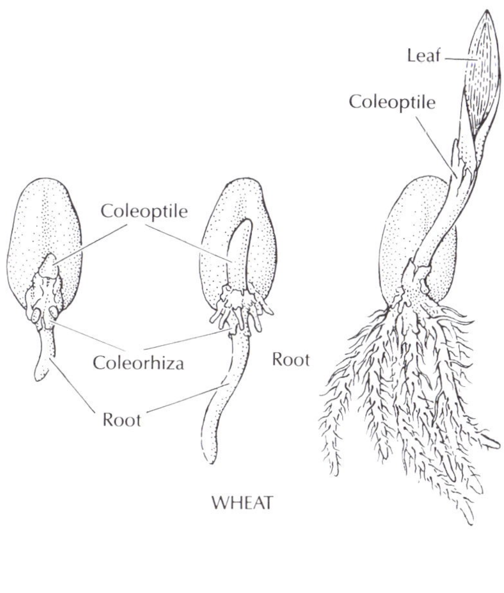

ruptured by the root and disintegrates. The coleoptile on the stem's apex, in contrast, has a meristem and will elongate by continued cell division and enlargement until it emerges from the soil into the air. It forms a kind of tube through which the first tender young leaves grow.

In some dicots, when the seed has germinated the cotyledons enlarge, develop chlorophyll and leaflike shape, and function—briefly—as photosynthetic organs, transforming the sun's energy into food for the plant. Radishes and cucumbers follow this plan. The cotyledons will wither and drop off after leaves develop. Cotyledons of many plants, such as the familiar pea *(Pisum sativum),* remain buried in the ground. Pull the young plant free and you will see cotyledons attached to the base of the stem close to where the root originated.

Some plants, such as species in the genus *Streptocarpus* (the Cape primrose is one), have one cotyledon that becomes as big as a small platter and is the plant's only photosynthetic organ. These plants are often chosen as ornamentals because of their showy flowers. They appear to have only one big leaf, but it is actually a cotyledon. (The name *seed leaves* often given to cotyledons is a misnomer, despite their leaflike shape and the fact that some cotyledons photosynthesize; their origin is unlike that of true leaves, discussed in the next section.)

By the time the plant's **shoot**—the stem and young leaves—has emerged from the soil, the young plant already has a fairly extensive root system. Water and minerals are now needed in increasing amounts for the plants to grow, and the root system is able to supply them. The shoot can grow rapidly. Eventually, the seedling consumes all the food of the endosperm or cotyledon; if it is to continue to grow, it must photosynthesize more food than it consumes. We will look next at the development of the main organ of photosynthesis, the leaf.

LEAVES

So far the plant axis has been described as if it were an unadorned stick that simply gets longer and longer. And some cacti and Euphorbs do have a single axis that is both stem and photosynthetic organ. But by far the greatest number of plants have leaves and side branches. How do these arise?

The Leaf Takes Shape Once again the apical meristem is the source. It is a complex tissue with not one but several centers of cell division. Division in its basal part produces the cells for the stem, which elongates as the plant grows. But on the flanks of the apical meristem are local centers of cell division that grow little

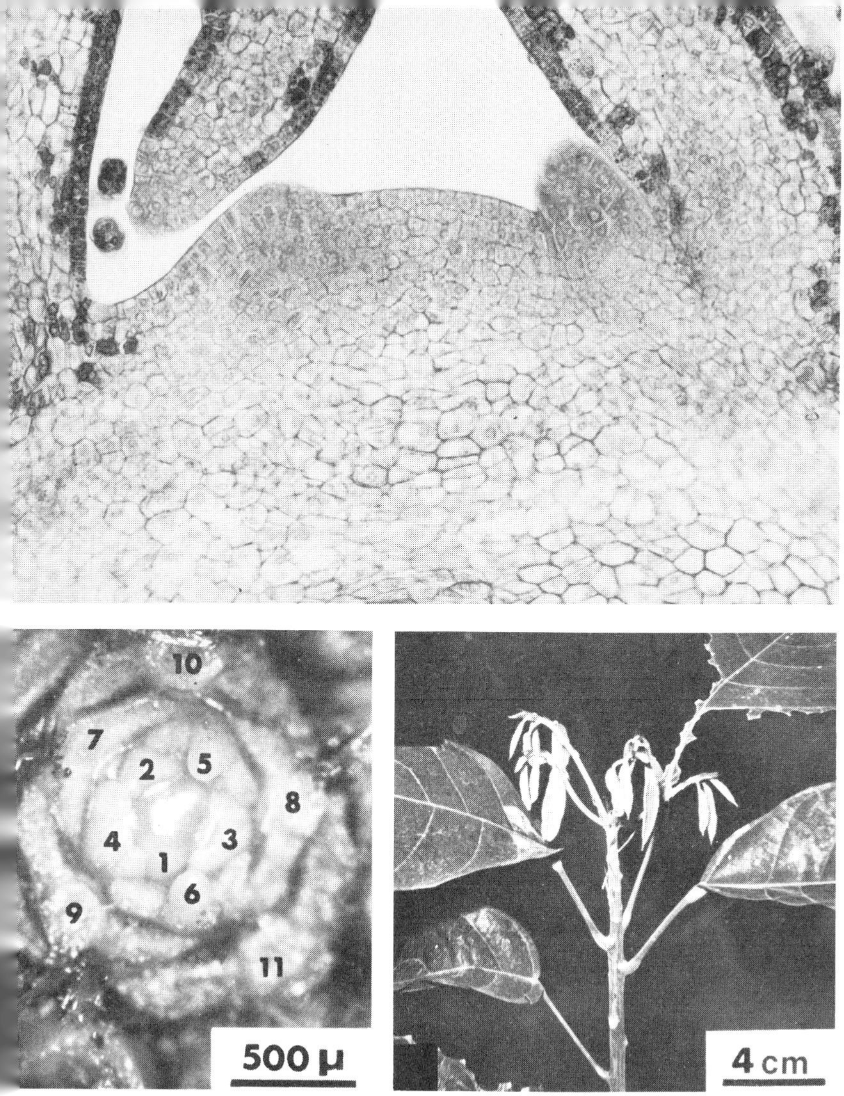

6.4 Cells of the stem as well as leaves and lateral buds arise from the apical meristem. Rhythmic cell division on the flanks of the apical dome produces bumps that become leaf primordia. Lateral bud primordia arise in the axils of the leaf primordia. A longitudinal section of an apical meristem of the chocolate tree, *Theobroma cacao,* is illustrated (top). A view of the meristem after all the young leaves are dissected away is shown (left). The apical dome is in the center and the leaf primordia form a spiral. Young leaves at the tips of young branches of *T. cacao* (right) were produced by the apical meristem in the manner illustrated at left. All the leaves exhibit the spiral pattern of insertion on the stem.

bumps, somewhat like cherries stuck around the top of an ice cream cone (see Figure 6.4).

These bumps, the **primordia** ("first rudiment"), also produce cells by division; they are the beginnings of both leaves and buds. At first they are small copies of the apical meristems. While the plant axis continues to grow along its straightforward route, the primordia develop into leaves and buds. The **apical bud** consists of the meristem and surrounding young leaves and buds.

All dicot leaves are formed in about the same way. At first the leaf grows outward as if to form an axis. As cell division continues, the developing leaf forms more and more of a flattened sheet as marginal meristems along the sides of the sheet produce new cells. Figure 6.5 shows how a shape like a maple or oak leaf can be formed from such a tonguelike structure: cells divide at different rates in

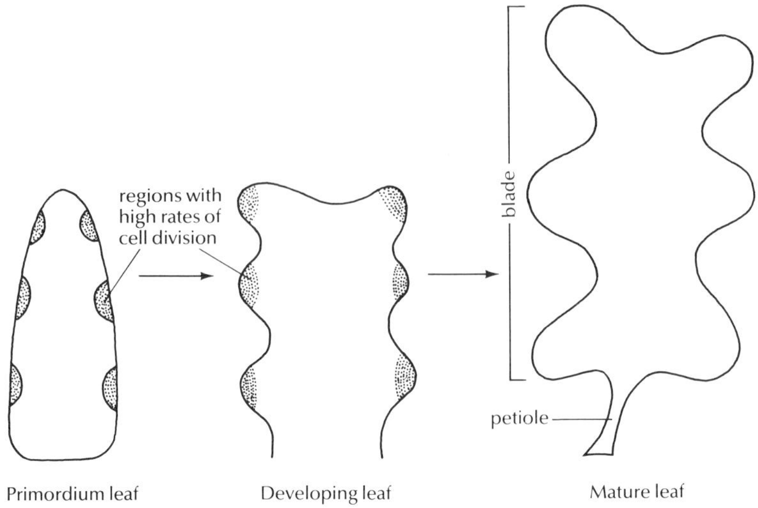

6.5 Morphogenesis of leaves. Differential rates of cell division are important in establishing organ morphology. Shape is also influenced by differential cell enlargement.

6.6 A grass leaf elongates much like a stem or root, except that the meristem is at the base and new cells elongate and push older regions ahead.

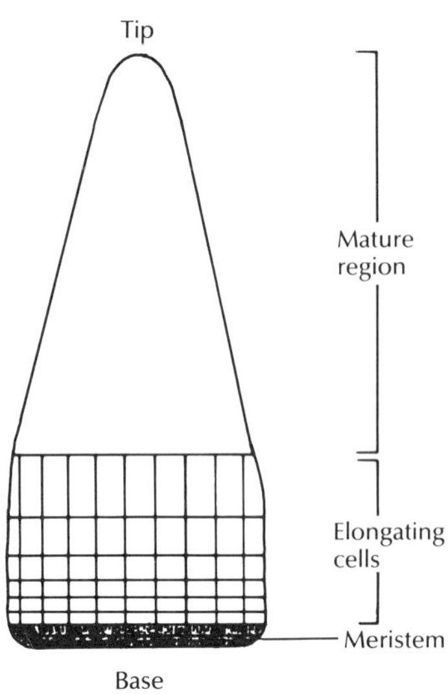

different regions. Young leaves clearly visible at the end of a stem usually have completed most of their cell division. From now on the leaf will grow mostly as the result of cell enlargement, just as the stem does.

The mature shape of most dicot leaves includes a leaf stalk—**petiole**—and blade—**lamina.** In some plants the blade has several segments called **leaflets**, and the leaf is then called **compound.** We have all seen leaves with toothed edges (elms) or with lobes, leaves whose blades are linear or circular. We do not know why leaves have their many shapes and differences, except that some variations appear to be adaptations to the plants' habitats, as we will see in Chapter 15.

Although leaf primordia are usually formed in the same way in monocots as in dicots, the further growth of their leaves is often different. The strap-shaped leaves of grasses, for example, have a meristem at their base, a **basal meristem** (see Figure 6.6). It produces cells that elongate and push the older parts of the leaf upward. Gently pull a blade of grass off its stem. The part closest to the stem is soft, tender, and light green; the blade's upper parts are old and tough, usually a dark grayish green. A leaf of this type, actually a leaf axis, grows very much like our hair. Both the hair and the grass continue to grow at the base. Mow the lawn and the younger grass leaves continue to grow at the base even though the tips are cut off. Consider the lawnmower plus compost pile to be the equivalent of a grazing animal such as a cow and you will see the adaptive significance of the grass's basal meristems (see also page 408).

The Arrangement of Leaves A plant may look at first to be a confusion of leaves growing any old way, but close examination shows that the leaves have a regular pattern, distinctive for

6.7 The pattern of attachment of leaves on the stem is constant for each species. Many plants, such as the rose (left), produce leaves in a spiral. This reduces shading of older leaves by those younger. The spirals can be described numerically. In the diagram (above), count the spirals from one leaf to the next leaf in the same position. You should count three. Now count from one leaf up to but not including the next leaf in the same position. You should count eight. The fraction $3/8$ is that part of the circumference of the stem between leaves in the same position. Some plants, like *Coleus* (center), do not have the spiral arrangement. They are opposite each other and each pair is at right angles to the adjacent pairs. This results in four ranks of leaves at right angles to each other.

each kind of plant. The pattern is so consistent that leaf arrangement helps in classifying plants. It results because each kind of plant produces primordia regularly and predictably.

Thus *Coleus,* a common houseplant shown in Figure 6.7, will have each leaf opposite the other, but each pair is on a plane perpendicular to the next pair of leaves below, making four vertical ranks of leaves, with every other leaf in its own rank. For contrast, look at how leaves are arranged on many grasses. Inserted in the stem either opposite each other or alternately, they form only two ranks.

Among other leaf arrangements the commonest is the spiral, like that of the roses. The form of the spiral differs from one kind of plant to another, but it is consistent for all plants of one kind. This arrangement results in the greatest exposure of leaves to sunlight. If you have a mathematical bent, you may enjoy knowing that the spiral arrangement of leaves on stems, and before that their production by an apical meristem, can be described by a formula. In leaves that grow in spirals, simply count the spirals, then the number of leaves up to but not including the one in a duplicate position. Now you have two figures, such as a 3 (for three spirals) and an 8 (for eight leaves). Expressed as a fraction (3/8), the figures give the fraction of the stem's circumference between successive leaves. The numerator and denominator express the chance that one leaf will be exactly above another.

The Mature Leaf Leaves are the primary organs of photosynthesis, absorbing the sun's light and converting it to energy, food for the plant (Chapter 8). In Chapter 10 we will examine how the leaf is involved in moving fluids all the way from root to topmost branches. The form enabling the leaf to perform these functions is complex; for now, we will mention just its principal parts.

As the cells divide and enlarge, giving the leaf its shape, they also become more specialized. Three main parts of the leaf can be distinguished with a microscope. The **epidermis** forms the leaf's skin and some of its cells, the **guard cells,** are specialized for forming pores called **stomata** (singular is *stoma*), important in movement and evaporation of water. Internal cell layers, the **mesophyll,** contain numerous chloroplasts where photosynthesis takes place. Finally, each leaf has conducting tissue, the **veins,** through which water and nutrients are transported. We can see these passing through the blade; the pattern of veins varies from one kind of plant to another.

BUDS AND BRANCHES

The apical meristem of a shoot is enclosed in a bundle of leaf primordia and young leaves. You can see it without a microscope but with difficulty, because it is usually less than 0.5 millimeter (0.02 inch) across. The young leaves clustered at the end of a stem surrounding the meristem, however, are clearly visible. Looking down the stem into the angles formed by the bases of the leaves and the stem itself, you will see bumps that look like small apical buds. This region is the **axil** of the leaf; it is a lot like an upside-down armpit (your armpit is an *axilla*).

Lateral Branches In seed plants every leaf has a bud in its axil. The leaf and bud are attached to the stem at the **node,** shown in Figure 6.8. The areas between nodes are called **internodes.**

An **axillary bud,** like apical buds, has an apical meristem and leaf primordia. Its meristem is produced by the apical meristem of the stem much as leaf primordia are. But the axillary bud meristem is usually formed later than the leaf primordium that subtends it.

Once it is formed, the axillary bud is equipped to begin cell division and to form both leaf primordia and axillary buds of its own. As it grows it forms a **lateral branch.** If more than one axillary bud is formed, more than one lateral branch might grow, but not all at once. Some of the axillary buds can remain dormant, like the embryo in a seed.

Forms of Growth Look at the form of a tree or shrub. It is determined largely by the position and number of lateral branches and their orientation. By observing its shape we already know something about the plant's history, because the position of side branches coming off a main branch is identical to the placement of leaves on the main branch, determined by the position of the primordia. Remember how leaves are positioned after a regular pattern followed by every plant in a species. Lateral branches grow from axillary buds, so that they too have a specific, constant pattern.

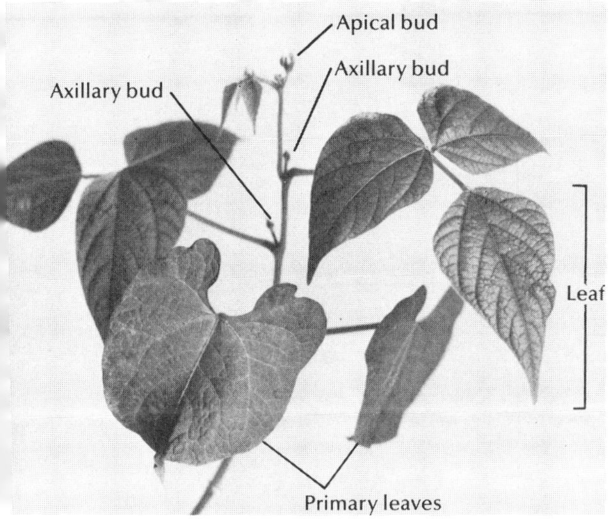

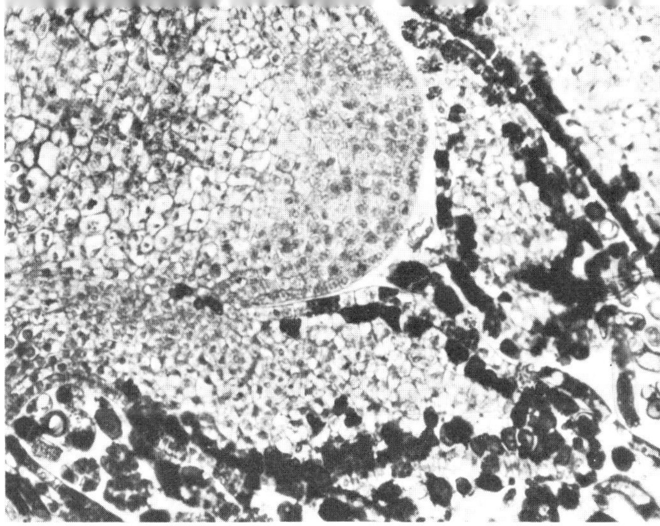

6.8 Leaves of flowering plants, such as this bean plant (above, left), have buds at their bases and these buds develop into lateral branches. Consider the adaptive significance of this growth pattern for an immobile organism. Axillary buds consist of an apical meristem and a variable number of leaf primordia (above, right). They are carbon copies of the shoot apical meristem from which they were produced. The insertion of leaves on a shoot is a node (left), and axillary buds are on top of the base of the leaf petiole when it meets the stem (the axil). The region of the stem between nodes is an internode.

Trees have only one stem or at most a few, but shrubs may have many. Pines and redwoods usually have one main stem with lateral branches regularly arranged around it, composing from bottom to top an almost perfect cone. Some oak trees have a spread-out shape, crooked branches making a complex network. Most palm trees, in contrast, have one trunk and no lateral branches at all. The palm also differs from pine or oak by the shape of its trunk: from near the top to the bottom, where a tangle of finger-thick roots widen the base, the trunk's diameter is almost identical.

The growth of axillary buds determines the shape that shrubs, pine, oak, or palm will have. On the young shoot of a shrub, axillary buds produce many stems. Pine trees send out lots of lateral branches from these buds, but they are subordinate to the central axis, the trunk. A palm tree is an extreme example of apical domi-

6.9 A plant's growth pattern is largely determined by the amount of lateral branching resulting from the growth of axillary buds. The tangled pattern illustrated by the oak (left) is due to frequent termination of apical bud growth and formation of lateral branches. Apical dominance is almost absent. Complete apical dominance is illustrated by the coconut palm (center). These palms have a solitary shoot and a tuft of large leaves. Conifers, such as these pines and hemlocks (right), typically have one vertical main shoot (tree trunk) and a regular series of horizontal lateral branches. These lateral branches have a much shorter life span than the main shoot.

nance: it is usually just a straight stem with no branches, the opposite of shrubs, which have many lateral branches. **Apical dominance** means just what it appears to say: the apical bud is dominant and growth of axillary buds is inhibited. In contrast, the oak gets its many-branched, spreading, scraggly look because apical buds abort and many axillary buds grow, starting side branches. In some species you can observe this phenomenon by pinching off the apical bud (a common technique in horticulture to control the form of plants or quantity of flowers); nearby axillary buds will then begin to grow.

Juvenile Foliage A curious thing happens in the young plants of some species: the leaves have a different shape or arrangement than older plants of the same species. **Eucalyptus** trees provide a good example of this "juvenile" and "adult" foliage (see Figure 6.10). *Theobroma cacao,* the tree producing seeds from which chocolate is made, has different arrangements for leaves on vertical and horizontal branches. Each vertical branch grows for a time and then its apical bud dies, releasing from dormancy several axillary buds that produce horizontal lateral branches. An axillary bud then produces a vertical branch to repeat the sequence. From one side the mature tree shows tiers of branches.

On a bean plant the first pair of juvenile leaves are simple and are opposite each other on the stem; all subsequent leaves have three leaflets and alternate on the stem. Common ivy is mostly the juvenile form, with horizontal stems and no blossoms; mature ivy pro-

6.10 Many plants have juvenile foliage, which differs in form from the leaves of the older plant. A striking example is this *Eucalyptus*, where the young trees have gray, tongue-shaped leaves and each node has two opposite leaves. As the tree matures, it produces dark green, scimitar-shaped leaves, which have an alternate arrangement on the stem.

duces vertical stems bearing flowers. The water fern *Marsilea* has several forms of juvenile leaves, from straplike shapes to two- to four-paired leaves to the final shape, which looks like a four-leaf clover.

We know little about what controls juvenile and adult growth, but we assume that the plant's form and manner of growth help adapt the plant to its environment, though costs and benefits are not always clear to us. Large-leaved plants in the tropics, leafless cacti, and miniscule duckweed—all reflect the effects of natural selection on form and its constant companion, function.

REPRODUCTIVE ORGANS

Leaves and buds and the apical meristems from which they originate are **vegetative** organs whose functions are nutrition and growth. From apical meristems **reproductive** organs such as

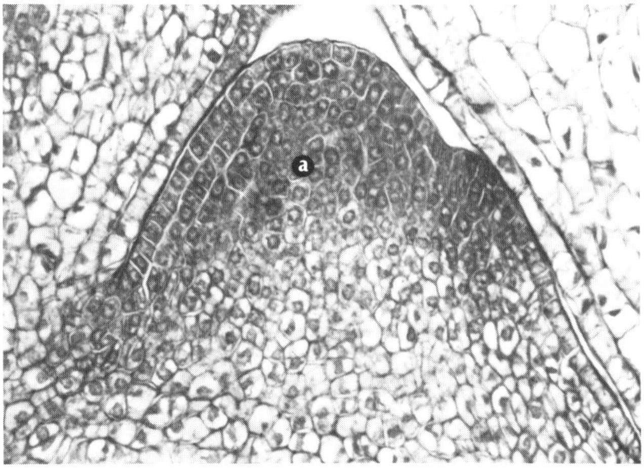

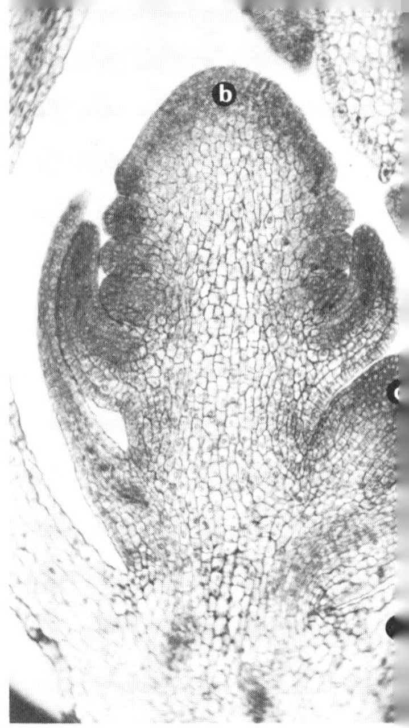

6.11 Flowers arise in the same fashion as leaves—from the apical meristem. When the Columbine apical meristem is triggered to produce a flower, the apical dome begins to elongate and floral bud primordia begin to form (right, (a) vegetative apex at flower initiation). The transformation of a vegetative to a floral apex results in the formation of a flower-bearing shoot or *inflorescence* (far right, (b) developing flower bud, (c) lateral flower bud, (d) bract, (e) leaf base). The terminal floral meristem and two lateral floral meristems develop floral organ primordia (bottom, right (f) floral apical meristem, (g) flower part primordia, (h) lateral flower bud). The apical flower is mature and the axillary flower bud will soon open. The formation of a complex flower is perhaps the most dramatic example of plant morphogenesis. Its beauty is the product of the same processes of differential cell division and elongation that produce leaves, stems, and roots.

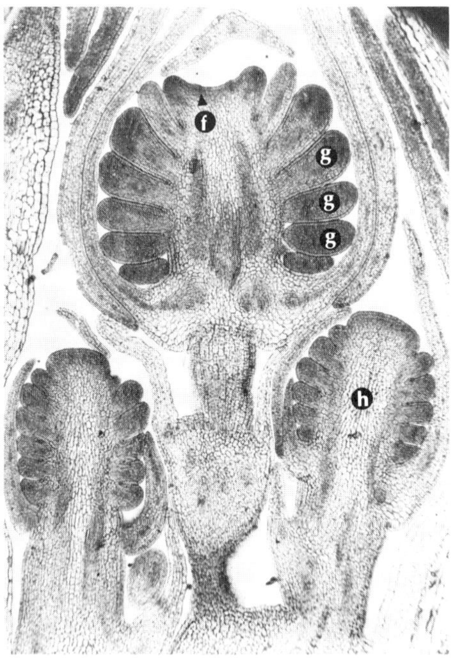

flowers and cones also form. Because they have the same source, these reproductive organs are closely related to leaves, as you can see in many flowers that produce a leaf in place of a petal or stamen.

The Flower Why are flowers produced instead of leaves? What changes a vegetative meristem to a floral meristem? We can identify many of the factors that trigger flowering (see pages 210–217), but what happens between the release of the trigger and

6.12 The pineapple and apple plants present contrasting behavior of apical meristems. A pineapple fruit (above, left) is formed from a complete *infiorescence*. The plant's apical meristem produces leaves, flowers, and leaves once again. The leafy shoot on top of the fruit can form a new plant. Specialized branches on apple trees called *short shoots* have axillary buds that form flowers (above, right). The apical meristem of these branches produces leaves and axillary buds that form flowers the following year.

the onset of flowering is unknown. We can, however, describe the morphogenic events, the physical details seen once flowering is begun.

Flowers are formed by the floral meristem much as leaves are made by the vegetative meristem; they result from variable rates of cell division in the surface layers of the apex. Just as axillary buds are produced by the vegetative apex, axillary flower buds can be produced by a floral meristem (see Figure 6.11). In this fashion a whole cluster of flowers may be formed on a stem, or they may appear in particular sequences with axillary vegetative buds.

There is enormous variety in the patterns of vegetative growth and flower development, as Figure 6.12 illustrates. On fruit trees the apical buds usually remain vegetative and produce only leaves, but some of the axillary buds, or buds on some specialized lateral branches, will produce flowers. Plants as diverse as tulips and rhododendrons do the opposite: their apical buds make flowers and axillary buds provide for further vegetative growth. Early in life the tomato plant begins to produce floral buds at regular intervals and continues to do so until frost; the apical buds remain vegetative but produce floral buds as well as leaves and axillary buds. The pineapple fruit *(Ananas comosus)* is most interesting: it is derived from a stalk of flowers, but the tuft of leaves on top of the fruit is made by a vegetative apex. You can cut off this top and the apical meristem will produce a whole new pineapple plant.

The Fruit What is a fruit? Most of us would not even hesitate before answering. It is a thing like an apple, or a pear, or a cherry, or all the kinds of fruit in the bins at the supermarket. The biological description of a fruit, however, includes many of the things that at the supermarket we call vegetables: string beans, pumpkins, peppers, cucumbers, walnuts (in the shell), even oat seeds. To the biologist these are as much fruits as peaches and plums. Why? Because they, like all fruits, are derived at least in part from flowers.

In its simplest form a fruit consists of a seed and the enlarged ovary of the pistil. A string bean is a good example.

As the petals and stamens of the bean's flowers wilt, the ovary swells. This growth, like the growth of other organs, results from the division and enlargement of cells. As the ovary enlarges the ovules inside are developing into bean seeds. The bean pod is the wall of the ovary and within it the seeds have the same position as the ovules had in the ovary. When the bean fruit is ripe, the top of the pistil often still shows at the end opposite the stem. Sometimes you can also see dried-up sepals at the base of the pod.

Some fruits, like apples, are more complicated (see Figure 6.14). Here the ovary wall is only part of what we eat. The rest of the fruit is made from bases of stamens, petals, sepals, and the top of the flower stalk. Opposite the stem end of the apple, you can usually see the withered dark remains of the floral organ tips. Inside the core, which is the ovary, the seeds are buried.

The pineapple, too, is complicated. The fruit is made up of the thickened flower parts of all the flowers on the stalk. By contrast, the seeds of grasses, which are really fruits, are deceptively simple in structure, with a thin hard wall around the ovary to which the walls of the ovules adhere closely. Each kernel of corn is a fruit.

Most fruits do have seeds, but some that we commonly see do not, such as seedless grapes and bananas. The latter have little black specks—imperfect seeds. Seedless oranges and other fruits are cultivated with no seeds. All these fruits, however, still develop from flowers; how they can be propagated despite their lack of seeds will be discussed in Chapter 12.

Cones Cones are also products of apical meristems. Their structure differs from flowers in important ways (see page 102), but cones develop from meristems in much the same way as flowers do. Although cones and flowers seem so different now, it would be very difficult to tell them apart if all the extinct conifers and angiosperms were still present.

THE STEM

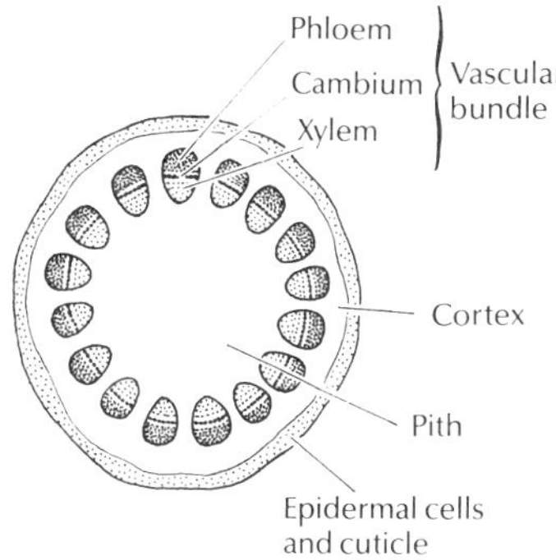

6.13 This cross section of a nonwoody stem has an anatomy characterized by concentrically arranged vascular bundles.

As the shoot emerges from the soil and the leaves form, the apical meristem continues to produce new cells. The tissues eventually produced by apical meristems are called **primary tissues** and their growth **primary growth.** All vascular plants elongate as the result of primary growth. Cells in the area of elongation behind the apical meristem do not only elongate, however; clumps of cells also differentiate. The results in the stem are structures that transport minerals and water and help to support the seedling as it grows taller.

Primary Growth As the cells in the region of cell elongation continuously elongate and differentiate, long cablelike strands of cells are created, running through the plant axis vertically. These strands, the **vascular bundles,** have two principal parts: the **xylem** (from the Greek word for "wood") and **phloem** (from the Greek word for "bark"). All vascular plants have xylem and phloem. The xylem carries water and dissolved minerals; phloem moves sugar and other organic materials produced and used by the plant.

In the center of the axis, other primary tissues differentiate from cells produced by the apical meristem; these are thin-walled cells known as the **pith.** Similar cells of the outer, more peripheral regions of the axis are called the **cortex.** Cortical cells frequently contain chloroplasts and can also serve as storage cells. And a layer of cells covers the axis—the **epidermis,** somewhat like our skin.

Still other primary tissues with highly specialized cell types are **fibers.** Of great interest to the plant and ourselves, these strings of thick-walled cells beneath the epidermis may be 50 centimeters (20 inches) long. For the plant they are tough flexible rods beneath the surface, somewhat like the steel plies in a radial tire. Often they occur in multiple layers, providing greater strength for the stem. Sugarcane stems have so many hard fibers that only a hard whack with a machete will sever them.

Fibers in the stems of flax plants are used to make linen cloth. Hemp fibers, which can be twisted into rope, are produced by various species of the genus *Cannabis,* which some of you will recognize as the source of a material providing a common but still controversial form of recreation. *Cannabis sativa* (marijuana) was legally cultivated for many years over large areas of the Midwest because of its fibers. Its seeds are highly valued for bird feed, and its leaves and flowers are valued for their hallucinogenic qualities; from the viewpoint of human society, it is a uniquely endowed plant.

Many dicots and most monocots go through their lives with only primary tissues (see Figure 6.15). They grow in width only by enlargement of the cells produced by the apical meristems. Because they produce little xylem or wood, the interior of their stems re-

6.14 DEVELOPMENT OF FRUIT FROM FLOWER

All fruits are derived completely or in part from flowers. The development of the bean (photos, facing page, top) illustrates how the growth of the ovary produces the fruit. The table (facing page) lists the many kinds of fruits and a few are illustrated (below). The pea, like the bean, is a legume—a type of dry fruit—while a tomato is a berry—a type of simple fleshy fruit. A tomato is an enlarged ovary. What we usually call berries, such as the strawberry, are in reality aggregate fruits. The fleshy part is derived from the flower stalk, and all the "seeds" are really fruits. Pineapples are multiple fruits, formed from the whole flower stalk.

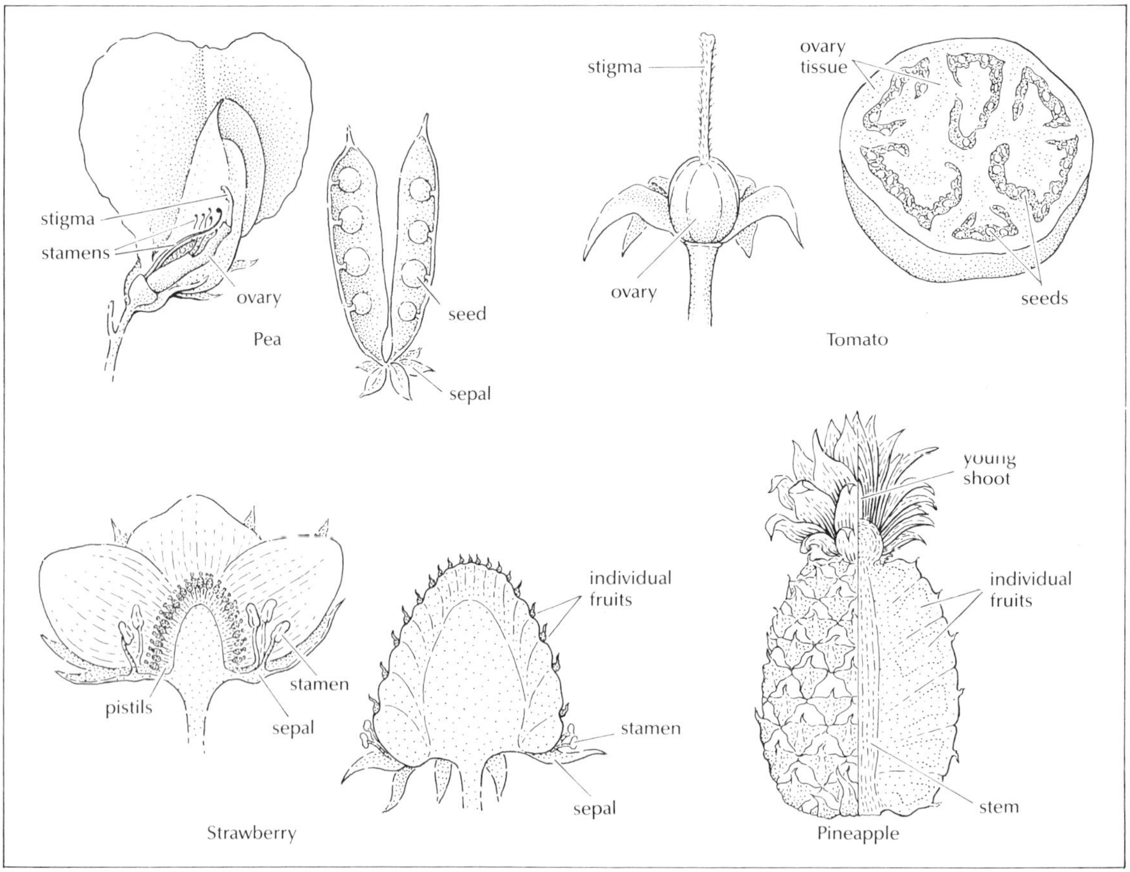

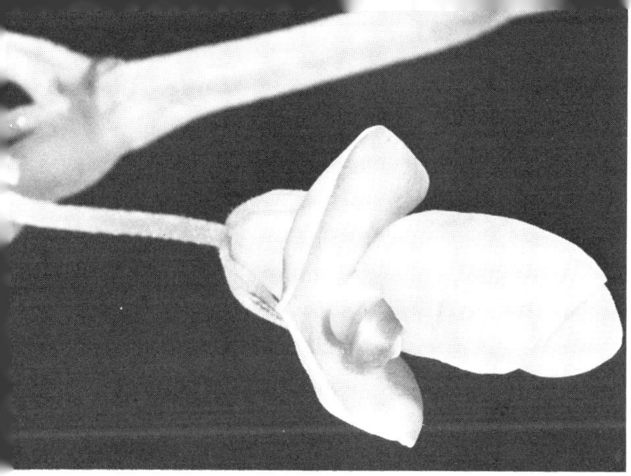

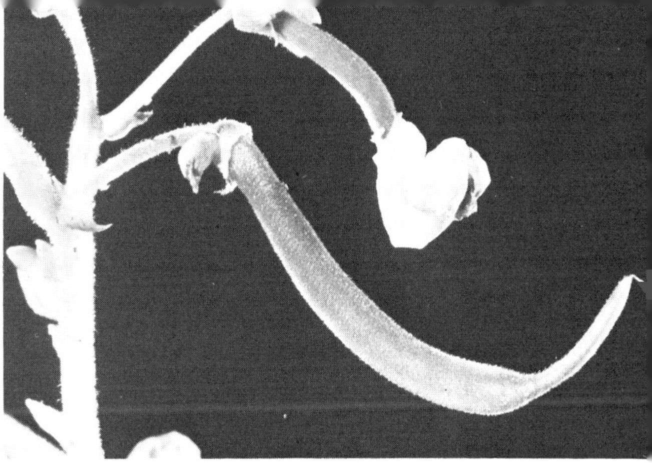

Classification of Fruits

Type	Description	Example
SIMPLE FRUITS	The fruit is derived from a single pistil.	
Simple dry	The fruit wall becomes papery, leathery, or woody when the fruit is mature. Kinds include:	
	Dehiscent (splits along definite lines or at definite points when mature, releasing several to many seeds):	
	Legumes: derived from a simple pistil and opening along two sides	Pea
	Capsule: derived from a compound pistil and opening in various ways	Poppy
	Indehiscent (does not open at maturity and has only one or two seeds):	
	Achene: having a seed coat or a single seed not fused with the fruit wall	Buttercup
	Grain: having a seed coat of a single seed fused with the ovary wall	Corn
	Samara: a winged fruit	Maple
	Nut: large, one-seeded, with a woody or stony wall	Hazelnut
Simple fleshy	The fruit wall, or a portion of it, becomes fleshy. Kinds include:	
	Berry: the entire ovary wall ripens into fleshy fruit tissue.	Tomato
	Drupe: the fruit wall is divided into outer fleshy layers and stony or woody inner layers. The fruit has one seed.	Peach
	Pome: the entire fruit wall is fleshy, but the inner layer is harder. The fruit has multiple seeds.	Apple
AGGREGATE FRUITS	Fruits derived from a number of simple pistils of a single flower. Some aggregate fruits also are composed mostly of other parts of the flower.	Raspberry Strawberry
MULTIPLE FRUITS	Composed of a number of closely associated fruits derived from different flowers, these fruits form one body at maturity.	Pineapple

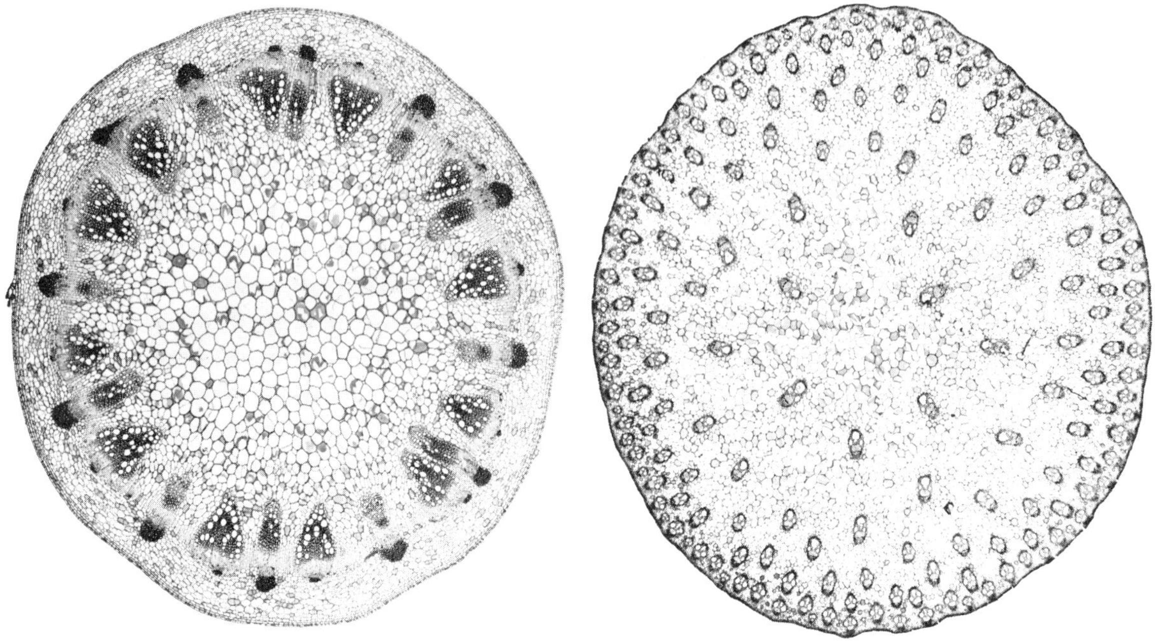

6.15 This cross section of a young sunflower *(Helianthus)* stem (left) illustrates common patterns of dicot stem anatomy. A circle of separate strands of xylem and phloem tissue (vascular bundle) is bounded on the outside by cortex and on the inside by pith. The epidermis is the outside layer of cells covering the stem. Cambium cells between the xylem and the phloem will produce new xylem and phloem cells, and in time a solid ring of vascular tissue will be formed. This cross section of a corn *(Zea)* stem (right) illustrates anatomical patterns of monocot stems. Vascular bundles are randomly distributed throughout the stem, and a cortex and pith are not separated. Vascular bundles do not produce a cambium, so woody secondary tissue is absent.

mains soft; they are called **herbaceous** plants. Most such plants are annuals. Among these plants are beans, corn, petunias, and daisies. Other dicots and the gymnosperms have yet another kind of growth, known as secondary growth.

Secondary Growth The center of a tree is solid dead wood. The tree cannot grow thicker from the inside out. The bark on the outside also looks dead. How then does it grow thicker?

Between the wood and the bark is a thin layer of dividing cells, the **vascular cambium** (see Figure 6.17). It is a meristem derived from primary tissue produced by the apical meristem. Only a few cells thick, the vascular cambium is responsible for all the tree's lateral growth, whereas the apical meristem initiates vertical growth. All cells produced by the vascular cambium form **secondary tissue.**

The long, thin, bricklike cells of the cambium differentiate in the vascular bundles from cells that lie between the future xylem and phloem, forming a ring between them throughout the whole stem except at its tips. The cambial cells' long axis is parallel to the surface of the plant axis. As cambial cells divide in a plane parallel to this axis, they produce secondary phloem and xylem cells in an alternate fashion, resulting in a ring of xylem on one side of the cambium and phloem on the other side.

As the cambium produces new xylem and phloem cells in the vascular bundle, cells in areas between the vascular bundles are also induced to divide, and the result is a continuous ring of cambial cells. Consequently, there is a ring of secondary xylem and secondary phloem, like the bread of a sandwich, enclosing the cambial layer, as shown in Figure 6.17. In woody plants this growth soon results in a large thick-walled core of xylem cells surrounded by a very thin cylinder of cambium and a larger cylinder of thin-walled phloem cells. Now we have a central cylinder of wood, then a thin ring of vascular cambium, and then an outer layer of bark.

The vascular cambium does not produce only xylem and phloem, however. Generations of cambial cells in specific regions produce files of thin-walled living cells that traverse both xylem and phloem. These **vascular rays** (shown in Figure 6.16) help transport materials. Rays are easy to spot in wood, especially if the wood is dry—it will split along the rays. And you can tell the plane in which a piece of wood has been sawed (cross section, longitudinal section, or tangential section) by examining the rays in different planes.

Even the girth of the mighty redwoods, which are big enough to have a tunnel through them that can accommodate an automobile, is the result of this pattern of growth—a ring of single cells adding their microscopic products year after year. Every division produces another cambial cell and either a xylem cell or a phloem cell, both of which will eventually die, while each new generation of cambial cells carries on the work. Thus it is in all **woody plants:** their increase in thickness, year by year, is the result of a thin, ever-advancing, ever-changing tide of living cells.

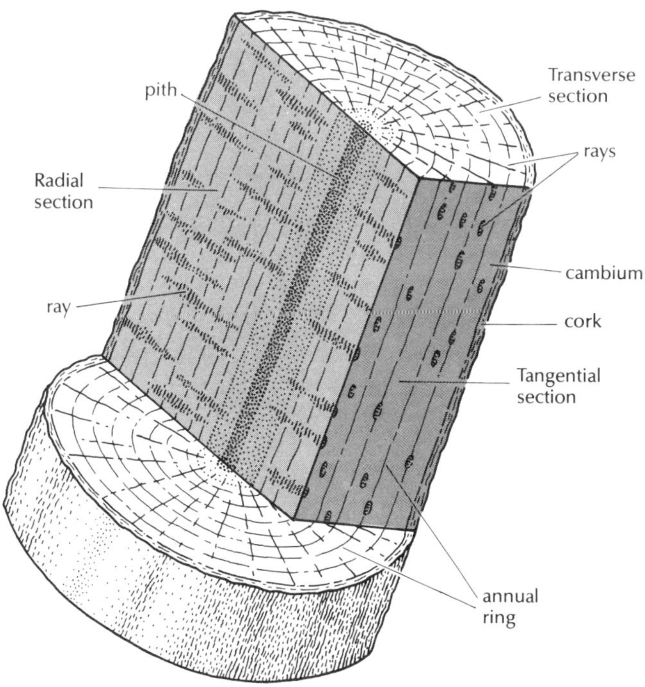

6.16 A cylinder, such as a plant stem, has three basic planes of section. A cross section is obvious. A radial section is a longitudinal section on any radius. A tangential section is a longitudinal section that is not on a radius. The plane of section of lumber can be easily determined by the appearance of the rays.

6.17 SECONDARY GROWTH: STEMS

The sequence of cross sections of successively older regions of an alfalfa stem (below, left) illustrates the initiation of secondary growth. The young stem (top) has only primary tissue, i.e., tissue derived from the apical meristem. The vascular bundles consist of xylem on the inside and phloem on the outside with vascular cambium in between. Further down the stem (middle photo), divisions of the vascular cambium have produced new xylem and phloem cells. These are secondary tissues. In addition, parenchyma cells between the vascular bundles have differentiated into cambium to complete a circle of cambium around the stem. The third section, further down the stem (bottom photo), has a solid ring of secondary xylem and phloem. This is why the circumference of woody stems increases with age. See the illustrations (facing page, top) of the cross sections and longitudinal sections of a stem as it increases in age and girth. A woody stem is mostly secondary xylem. It is surrounded by a layer of cambium cells which also produce phloem. The ring of phloem is narrow because a cork cambium forms in its outer part each year and produces cork cells which progressively wear away. The pattern of growth of rings (facing page, bottom right) results from the rate and type of xylem cell production during different parts of the growing season. Primary growth leads to elongation of an axis and secondary growth produces an increase in width. A woody stem, such as a tree, is a constantly elongating and expanding cone (below, right).

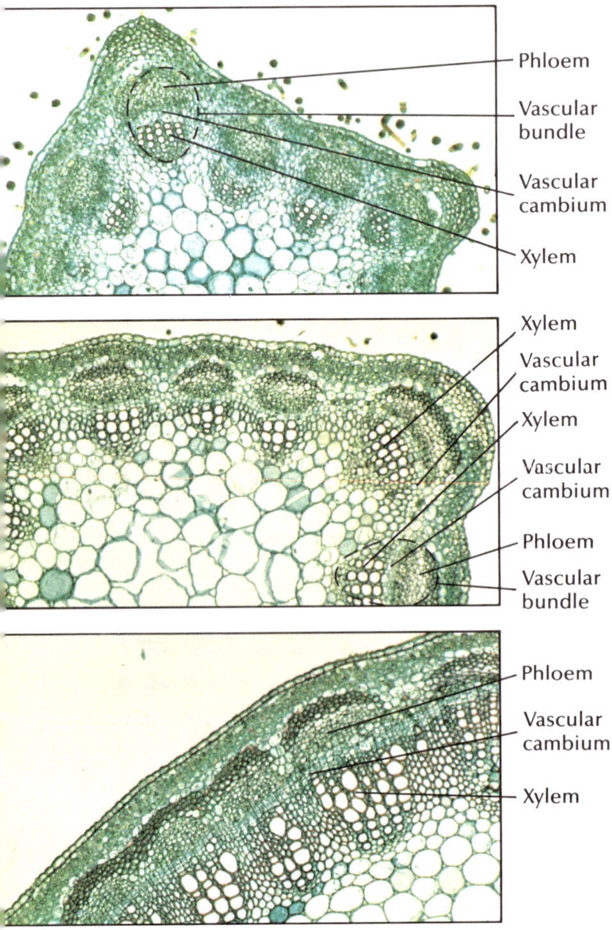

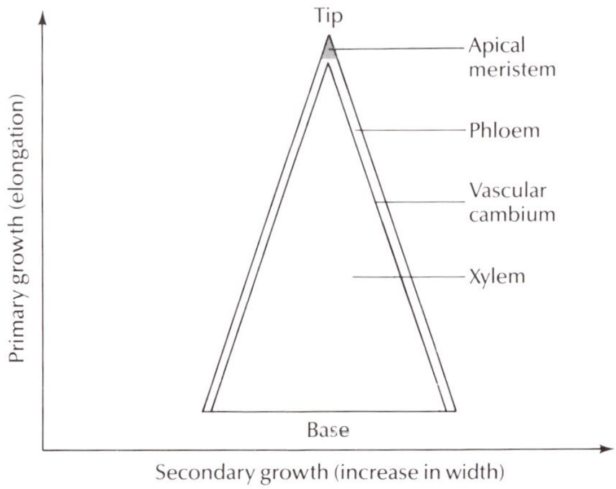

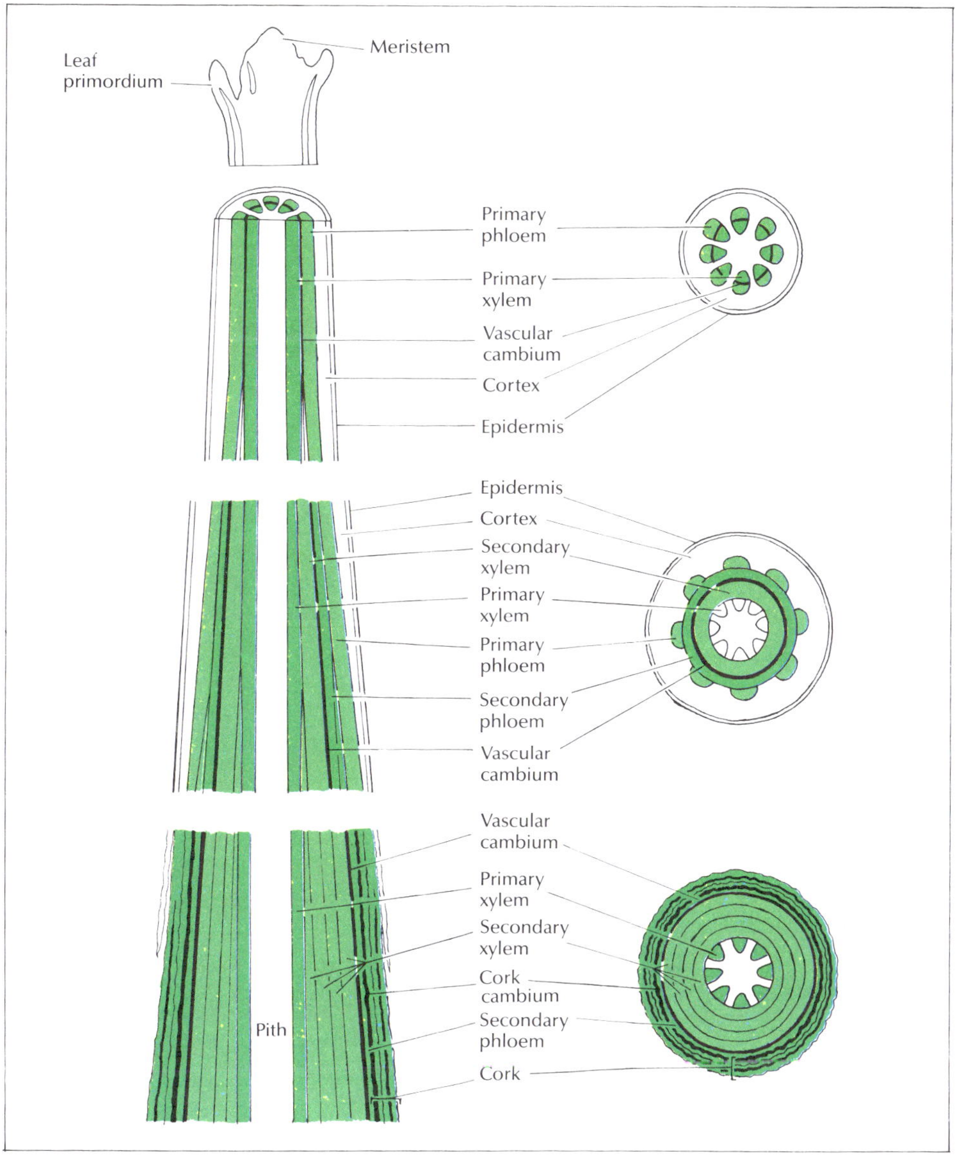
Meristem
Leaf primordium
Primary phloem
Primary xylem
Vascular cambium
Cortex
Epidermis
Epidermis
Cortex
Secondary xylem
Primary xylem
Primary phloem
Secondary phloem
Vascular cambium
Vascular cambium
Primary xylem
Secondary xylem
Cork cambium
Secondary phloem
Cork
Pith

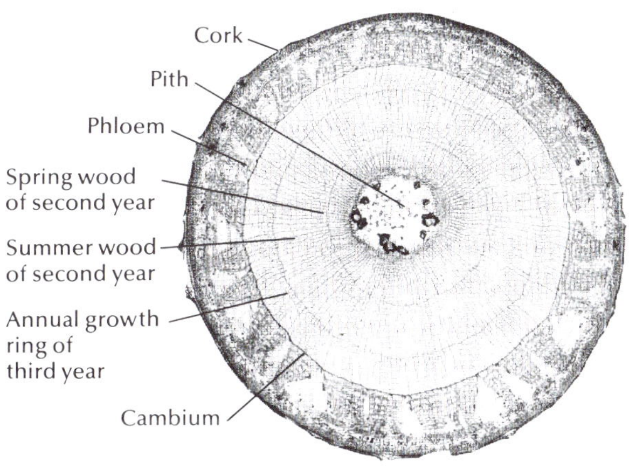
Cork
Pith
Phloem
Spring wood of second year
Summer wood of second year
Annual growth ring of third year
Cambium

Visualize the trunk of a tree as a constantly expanding, upward-growing cone, with the expansion resulting from the simultaneous activities of primary (apical meristem) and secondary (cambial) growth. Trees grow in width so slowly that growth is imperceptible unless we let a long interval go by. The effect of this constant increase in width is demonstrated by the following experiment. Drive a nail into a tree trunk, or cut out a small chunk with an axe, or tie a strong rope tightly around the trunk. After a couple of years, the nail will have disappeared. The axe wound will be at least partly covered over. The rope will probably have been broken and part of it may be buried by the bark, visible only as a swelling underneath. The bark, like a semisolid plastic, flows over and obliterates objects.

Young trees, you may have noticed, usually have good-sized branches near the base, but older trees may have bare trunks for the first 4 to 6 meters (15 to 20 feet) up, or even higher. What happened to those branches? Perhaps they died and broke off, but what happened to the stumps? You can find the answer in any room paneled with real wood. Each of those decorative knots in the wood is part of the stump left by a branch buried by the growing tree.

A more remarkable example of this growth in width comes from some French and German forests where, after the battles of two world wars, timber survived that local sawmill owners now refuse to cut. Why? Because secondary growth has buried so many bullets, bits of shrapnel, and fragments of bombs and grenades that cutting into the wood is to risk ruining a blade.

If by constantly depositing cells the cambium gradually produces wood that will surround and bury any foreign object, then cambial cells must sometimes divide longitudinally, perpendicular to the plant's vertical axis. Without some divisions in this plane, the cambium would not increase its circumference. This constant expansion of the trunk will completely swallow the base of a branch broken off close to the trunk.

Annual Rings Trees like the giant redwoods *(Sequoiadendron giganteum)* of California and the bristlecone pine *(Pinus aristata)* of California and Nevada are said to be the oldest living things, and in a way they are. Most of their mass, however, is a big pile of wood, dead as a two-by-four stud in a house. The oldest wood, at the center of the tree, was produced by the remote ancestors of the present cambial cells. The living part of the tree is no older than the grass growing at its base; it is created by the continuously dividing cambial cells. The tree is like a coral reef: a thin layer of life residing on a vast heap of ancient skeletons.

A tree can be read like a living record of the years through which it grew; the record is in the **annual rings** of its secondary xylem. Like some ancient city—Athens, Ephesus, Rome—the tree has been built up, layer by layer, through millennia. By digging down through the layers of civilization at the site of an ancient city, archaeologists can reconstruct ancient lives: how people lived, what they ate, the tools and weapons they used. Similarly, by studying the annual rings of a tree, we can reach far back in time to learn something about how it grew.

In times of drought, annual rings are very thin because of a short growing season. We know that about seven hundred years ago, in 1290, a drought that struck the American Southwest was so severe that it might have forced the pueblo dwellers of Mesa Verde in Colorado to leave their cave cities and wander elsewhere in search of food. We know of the drought from the wooden ridgepoles with which they roofed their houses; no other record hints at a reason for these abandoned cities and their vanished dwellers. All the ridgepoles were cut before 1290, showing that from that year on they built no new houses there; no one was left to build. The very thin annual rings in trees from that year show that it was a year of exceptional drought.

Even forest fires leave their blackened traces in the annual rings, while years of plenty, with good rainfall, show up as thick rings. **Dendrochronology,** or tree time, is one of the most sophisticated techniques for studying the recent past. A tree does not have to be felled to tell its story. Dendrochronologists can learn much by simply cutting a plug from its trunk; the hole quickly heals over.

Cork Cork is another secondary tissue. All woody trees have cork (most people call it "bark"). How does it develop? Why is it rough, cracked, or scaly on many kinds of trees?

Look at the cortex in Figure 6.18. A ring of cambial cells called the **cork cambium** develops in this tissue, and the inner walls of outer cells produced by its divisions become impregnated with a waxy waterproof substance, **suberin,** and then the ring of cells dies. Generations of such cells produce the cork. (Bottle corks are cut from the bark of the cork oak, which lives in Spain and Portugal.) The cork cambium produces new cells, and the old sheets of dead cells are worn off the outside. That cork can regenerate itself is evident in the cork oak *(Quercus suber),* where the cork is harvested every few years from the same trees.

What about the ridges and cracks that we see in almost every kind of tree bark? The tree is growing in thickness from cell divisions in the cambium beneath the cork. As the cells divide and the tree trunk swells, the outer part has to give, and so the cork splits. Sometimes you can even see sap oozing through in pine trees, whose cork is scaly and can have wide splits from new growth pressing beneath its surface. In trees like the sycamore and eucalyptus, the cork cambium develops unevenly in strips and plates, leaving long strips and pieces of bark hanging so loosely that wind at last blows them off.

Stems and branches of woody plants also have openings in the

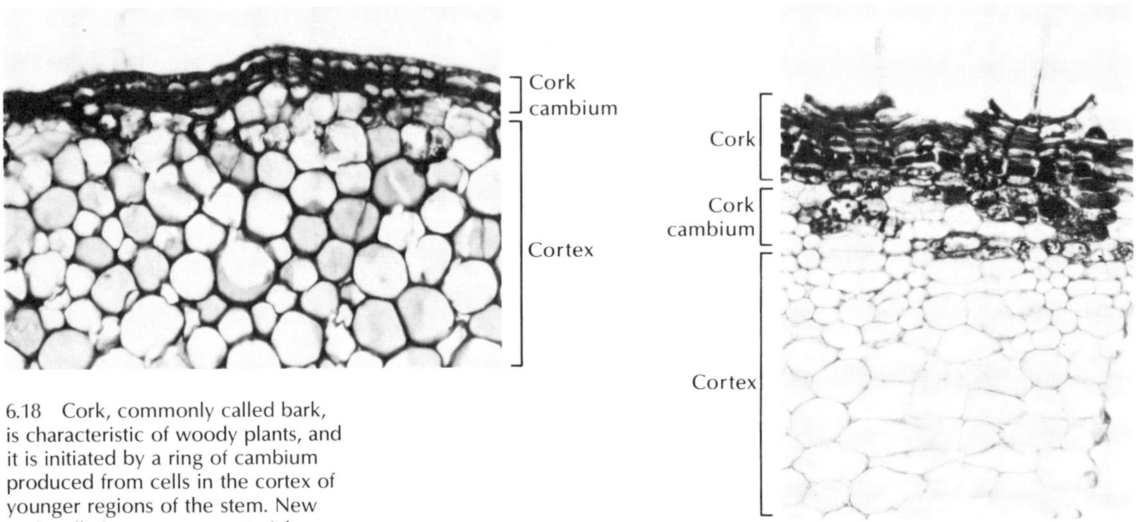

6.18 Cork, commonly called bark, is characteristic of woody plants, and it is initiated by a ring of cambium produced from cells in the cortex of younger regions of the stem. New cork cells form a waxy material called *suberin* and soon die. As new layers of cork are produced, older cells on the outside slough off. Cork production is usually seasonal, and new cork cambium arises within the cortex. Combined with formation of new vascular tissues, this leads to the disappearance of cortex. Subsequent layers of cork cambium arise within older phloem tissue, so that while xylem is added to the inside of the stem, phloem is continuously removed from the outside.

bark called **lenticels.** These openings permit the exchange of gases between the atmosphere and living cells in the stem, which require oxygen to live.

ROOT GROWTH

There are significant differences in the anatomy of young roots and young stems (see Figure 6.19). Roots, too, have apical meristems, but in a root this meristem also produces a **root cap,** formed from cells on the surface of the apical meristem. Because the cap is continuously worn down by pushing through earth, new root cap cells must constantly be produced. In addition, the patterns of primary tissues of roots are somewhat different from those of stems, and the origin of lateral roots is completely different from the origin of lateral branches.

Two distinct single-celled rings grow outside the phloem in the young regions of roots. The outer ring of these cells is the **endodermis;** the inner ring is the **pericycle,** an internal, secondary meristem. It produces secondary root primordia, from which lateral roots arise. Thus lateral roots develop not from the external layers of the meristem, as branches do, but from inside the parent root.

Root primordia, like bud primordia, develop apical meristems, and the root elongates in the same way as any plant axis—by the elongation of cells behind the root meristem. But during early growth the lateral roots first have to get to the outside: they grow between the cells of the cortex and split the parent root open while they do so. Once outside, each lateral root grows other lateral roots, which in turn branch out, so that fairly soon they have formed a wide network. As the root elongates and absorbs water and minerals, very delicate extensions of the epidermal cells are formed in the root's younger regions. These are the **root hairs,** which are usually so numerous they cover the ends of the roots like fuzz.

Many plants, including carrots and dandelions, have one large, thick primary root, called a **taproot,** and some smaller secondary roots (Figure 6.19). Other plants, such as grasses, have a spreading tangle of thin secondary roots, a **fibrous root system.** As a result of these variations in their root systems, plants obtain water from different soil depths.

There is as much and often more of a plant below ground as there is above it, and just as branches of adjacent trees and shrubs intermingle, so do their roots (Figure 6.19). Branches are free to be moved about by wind, rain, and snow, but roots of neighboring plants can entangle themselves with astonishing thoroughness. Roots of the same or related species will even fuse, just like two branches grafted together (see page 192). Root grafts are common in forest trees. Like the mycelial network connecting a group of mushrooms, which stand apart aboveground, the apparently separate trees in a forest can be connected below ground into one large colony. Vulnerability to disease spread through these roots is a great hazard when many trees of the same species are planted close together.

PLANT AND ANIMAL DEVELOPMENT: A COMPARISON

The sequence of development summarized here is obviously different from that of animals. One underlying reason is that plant and animal cells differ. Animal cells do not have internal pools of water or fairly rigid walls, and they often move. In fact, animal cells move quite a lot, often migrating in groups during an animal's development. But where plant cells are initiated, there they stay. The plant is built up cell by cell.

From the time it is an embryo, an animal grows in rather straight-line fashion. The individual may grow for a long time, as some fish do, but growth after birth occurs within organs already established. Specialized cells make additional specialized cells, but no organs are added.

In plants growth occurs by production of new cells and by their elongation. But plants also add organs because some structures act as **permanent embryos.** The apical bud, for example, acts in a way like the plant embryo in the seed. At each apical meristem the development of mature tissues is constantly repeated. Each embryolike apical meristem makes many copies of itself, and each copy

6.19 ROOTS

Lateral roots are produced by a ring of cells called the *pericycle* just inside the endodermis layer. This cambium forms root primordia which grow through the cortex by means of elongation of cells produced by apical meristems. First seen from the outside as bumps (photo, facing page, left), they soon rupture the root epidermis (facing page, bottom photo). These new roots will in turn produce lateral roots and so on, ad infinitum. The large amount of the plant underground and the extent of entanglement of roots and trees are illustrated by these partially exposed root systems of *Casuarina* trees near a beach on Kauai, one of the Hawaiian Islands (bottom left photo, this page). The roots were exposed by waves of high tides. Roots of adjacent trees often fuse together as illustrated by these *Ficus* tree roots on an exposed cliff (below, right). Because of these connections, a forest of the same species of trees is in some respects like a single large organism. A longitudinal section of a root tip (facing page, bottom drawing) reveals an important difference between roots and shoots. Root apical meristems produce a covering, a root cap, as well as cells which elongate and induce growth. The root cap serves as a battering ram as the root grows through soil, and older cells are constantly being removed. At the same time, new root cap cells are produced by the apical meristem. The dark region in the center consists of differentiating vascular tissue. A cross section of this region of the root (facing page, top) reveals a central core of xylem with a variable number of radial extensions. Phloem differentiates between these xylem ridges. This pattern of vascular tissue contrasts with the vascular bundles of herbaceous stems. Root vascular tissue is surrounded by a ring of meristematic cells, the pericycle, which produces root primordia. The endodermis is outside the pericycle. This cell layer is important in maintaining an osmotic gradient between the xylem sap and the soil water (Chapter 10). An older region of the root illustrates the pattern of secondary growth. A cambium forms between the xylem and the pockets of phloem and soon surrounds the xylem. The production of secondary xylem and phloem by this cambium and the development of a cork cambium in the cortex produce a tissue pattern very similar to woody stems. When all roots are essentially the same size and there is no single main root, the system is termed *fibrous*. Some plants have large main roots which produce smaller lateral roots, which produce smaller lateral roots, and so on. This is termed a *tap root* system. See root systems illustrated (below, left).

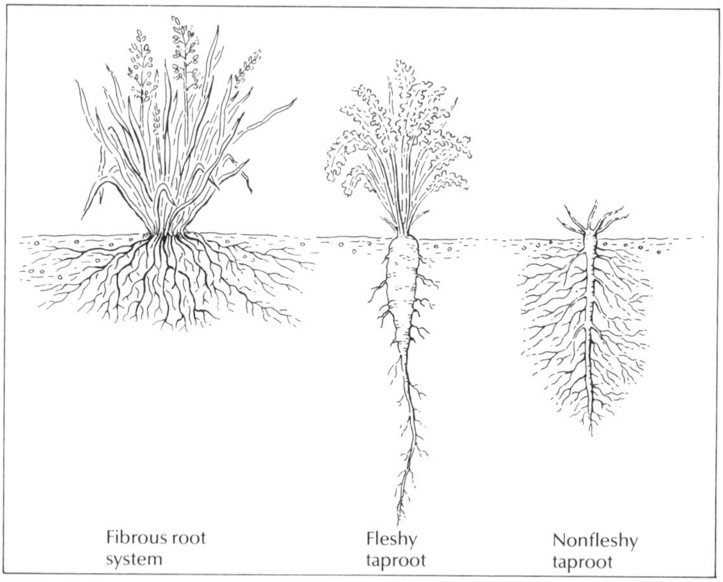

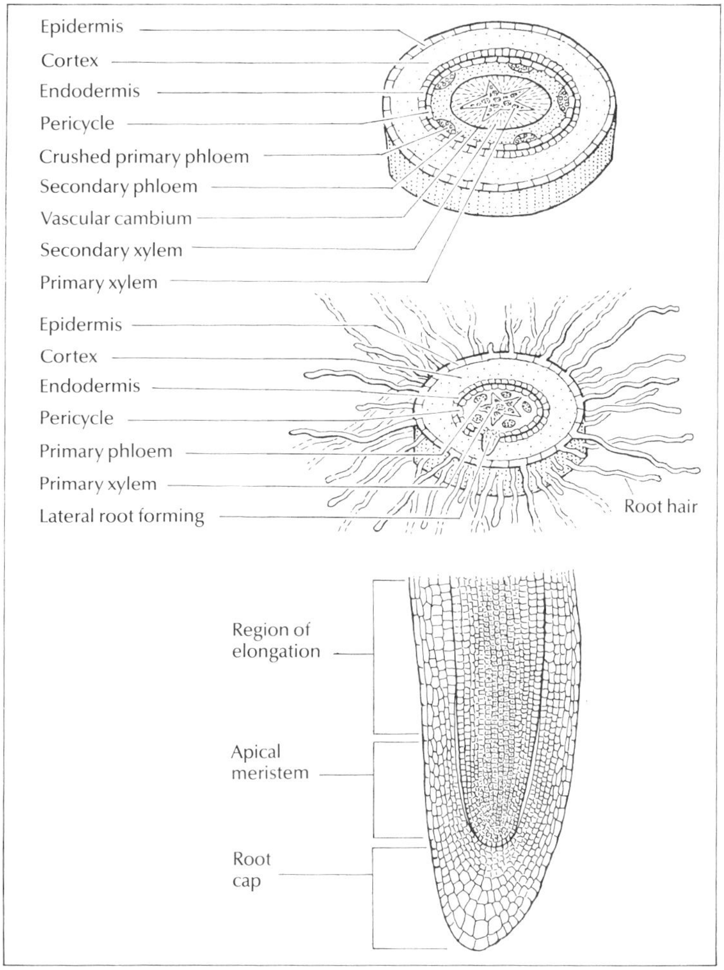
Epidermis
Cortex
Endodermis
Pericycle
Crushed primary phloem
Secondary phloem
Vascular cambium
Secondary xylem
Primary xylem
Epidermis
Cortex
Endodermis
Pericycle
Primary phloem
Primary xylem
Lateral root forming
Root hair
Region of elongation
Apical meristem
Root cap

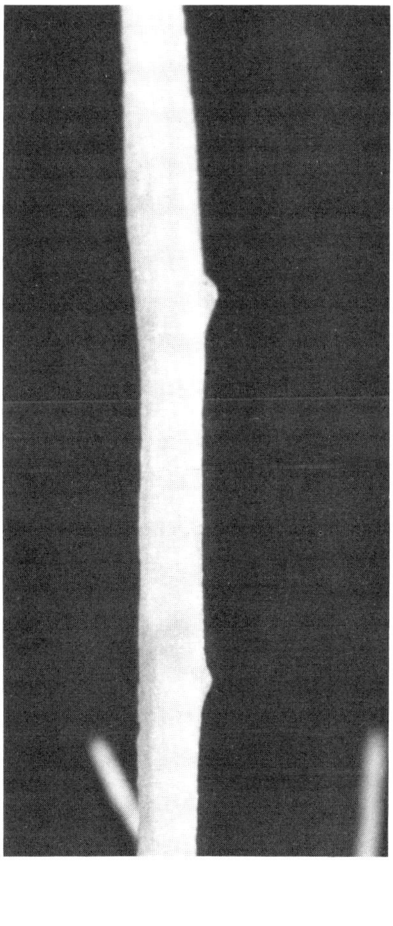

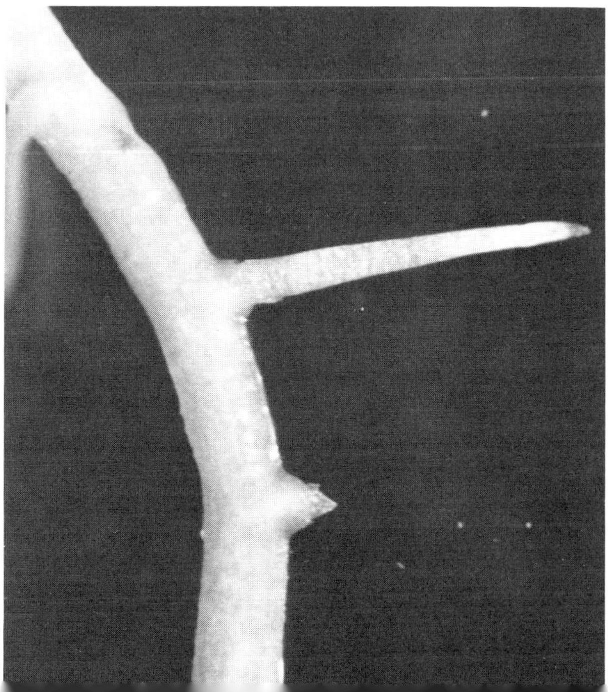

can be separated from the parent stem and initiate new axes and new apical meristems. Animals develop by enlarging on an established form; plants develop by adding on.

The plant biologist and the animal biologist therefore ask their questions differently as they initiate research on problems of development. Among the more fascinating questions in botany is what controls the plant's development. Why does a plant flower or not flower? What controls the ripening of fruit or the rate of cell elongation? Twentieth-century scientists have made some progress in answering these questions—the topic of the next chapter.

Summary

All plants grow according to a similar plan, one very different from the pattern that animals follow. Plants are built up cell by cell; these cells do not move. They have fairly rigid cell walls and grow when they take up water. This growth occurs in regions known as areas of cell elongation; new cells are produced by cell division in other areas called meristems. During their development plants, unlike animals, also add new organs—leaves, branches, flowers.

After a zygote is produced by fertilization, it divides and develops into a ball of cells, which becomes polarized, forming an axis with meristems at each end. Behind each meristem is an area of elongation. The axis grows by cell division and elongation, producing an embryo. At the stem end of the axis, one or two cotyledons also form. These, along with the embryo and the endosperm, which provides food, and the seed coat, which gives protection, are the main parts of the seed.

The seed may be dormant for some time. Then at germination the embryo's cells again divide and grow, and the root breaks through the seed coat. In some dicots the cotyledons provide stored food for the seedling or even photosynthesize for a short time; in other plants all the seedling's food comes from the endosperm. The roots develop rapidly, bringing water and minerals to the rest of the plant, and eventually the shoot emerges through the soil.

At the flanks of the shoot's apical meristem are the primordia; these cells divide and develop into leaves and buds. The apical meristem produces primordia in a regular and predictable manner that differs from species to species.

The leaf of a dicot will first grow outward and then into a flattened sheet; its final shape is determined by the rates of cell division in meristems along the sides of the sheet. Leaves may be simple or compound and will usually have a petiole (stalk) and lamina (blade). In grasses, a meristem is at the base of the leaf and the oldest cells are at the tip; this results in a strap-shaped mature leaf.

Every leaf has a bud in its axil, the angle formed by the juncture of stem and leaf. This axillary bud has a meristem and leaf primordia and will form a lateral branch. The growth of axillary buds thus determines the shape of a shrub or tree. If the apical bud is dominant and axillary buds inhibited, branches will be few.

In some species leaves have one shape or arrangement when the plant is young and another when it is mature. Beans, ivy, and eucalyptus are familiar examples.

Leaves, buds, and apical meristems are the plant's vegetative organs; the reproductive organs—flowers, cones, and fruits—grow in a similar way. Varying rates of cell division in the layers of an apical meristem will determine a flower's shape, and the meristem may also produce axillary flower buds. Why a flower rather than a leaf is produced by a meristem, however, is unknown. The cones of gymnosperms are produced by meristems in much the same way as are the flowers of angiosperms.

Fruit develops from the flower, or part of it. The ovary in the flower's pistil grows by division and enlargement of its cells, and the ovules develop into seeds. At its simplest a fruit is a seed and an enlarged ovary, but not all fruits have fully developed seeds.

All the tissues produced by the apical meristem are known as primary tissues. In the plant axis these tissues include primary xylem and primary phloem, which make up the vascular bundles; thin-walled cells called the pith at the center of the axis; the cortex in the outer regions of the axis; fibers, which are thick-walled cells that strengthen the stem; and, covering the axis, the epidermis.

Some plants grow only by enlargements of the cells in these tissues; other plants also have secondary growth, which produces secondary tissues and increases in diameter. In these plants the tissues produced by the apical meristem form another meristem in the vascular bundles—the vascular cambium. As cells of the cambium divide, secondary xylem is formed on the inner side of the cambium and secondary phloem on the outer. Traversing both xylem and phloem are the vascular rays. The cambium also divides longitudinally, increasing in circumference.

In woody plants the cambium produces xylem and phloem year after year. The dead secondary xylem are the annual rings, the width of which reflects the length of the growing season that year.

A ring of cambial cells also develops initially in the cortex; generations of these cells form the cork of the tree. As the cork wears off a new cork cambium is formed in the phloem. Lenticels, openings in the bark, allow gases to pass between the air and the stem's living cells.

The anatomy of the mature root looks much like that of the stem, but there are differences in the younger portions. The apical meristem at the root end, moreover, produces a root cap. Lateral roots, unlike lateral branches, develop from inside the parent root—from secondary primordia produced by the pericycle. The lateral roots grow other lateral roots, producing a wide network. In some plants there is one dominant root, the taproot; other plants have a network of thin secondary roots, a fibrous root system. Extensions of the epidermis known as root hairs also grow on younger regions of the root. The roots of adjacent trees and shrubs often intermingle and may even fuse.

Review Questions

1. Most flowering plants and skyscrapers are examples of apical growth. Compare how these two structures increase in size.
2. The seed is a nice example of "packaging." Explain. What does the fruit have to do with the way a plant makes its living?
3. Outline the basic features of leaf development.
4. Discuss the significance of apical dominance to plant form.
5. Compare the development of leaves, buds, flowers, fruits, and cones.
6. What is the difference between primary and secondary tissues?
7. How are the vascular cambium and the cork cambium similar? How are they different?
8. How do derivatives of apical meristems of shoots and roots differ?
9. Describe the origins of lateral branches and lateral roots.
10. Compare the way you develop with the way an apple tree develops.

CHAPTER 7

Regulation of Development

We have sketched in Chapter 6 a picture of the formation of plant meristems, the growth of seedlings, and the development of leaves and flowers. But what causes some cells to form a meristem, other cells to elongate, others to form flowers?

One important hint came in 1880 when Charles Darwin, two years before he died, published *The Power of Movement in Plants* with his son Francis. In the experiments that supplied material for their book, carried on in Darwin's village garden and house in Down, Kent, they observed that the tips of grass seedlings bend toward light. Today we call this movement **phototropism,** movement toward or away from light. Darwin proved that only the very tip of the coleoptile (the closed cylinder covering the tip of the grass as it grows upward through the soil) was sensitive to light. When Darwin covered the tip so that it could not be illuminated, no bending occurred. He concluded that some substance in the coleoptile tips diffused downward to the region of cell growth, where the bending occurred. His elegant experiments have been repeated many times by schoolchildren, university students, and scientists.

AUXIN AND ITS EFFECTS

It took forty-six years to prove experimentally the existence of the substance suggested by the Darwins' work. In 1926 the Dutch-American botanist Frits W. Went demonstrated that there

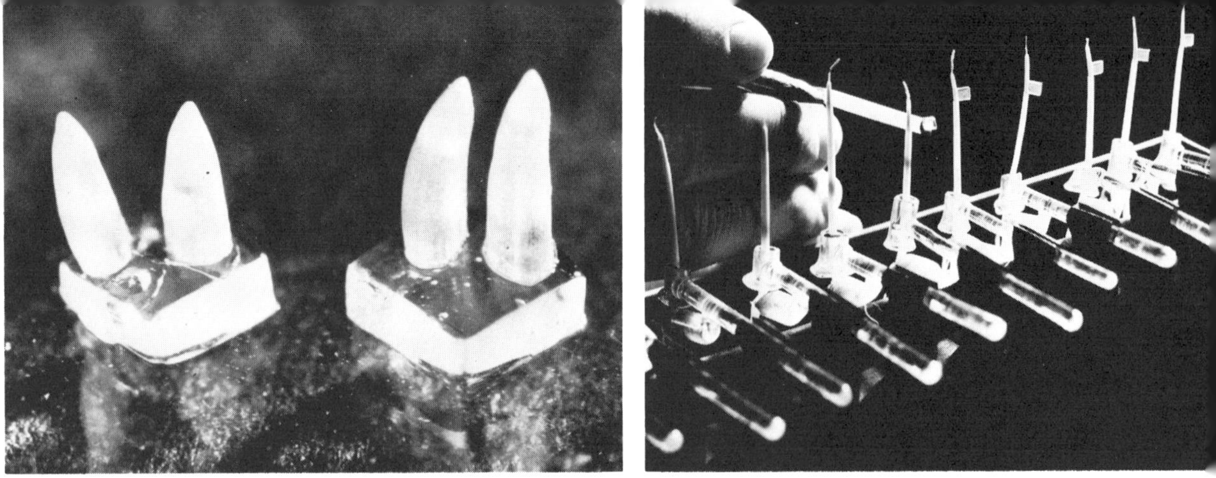

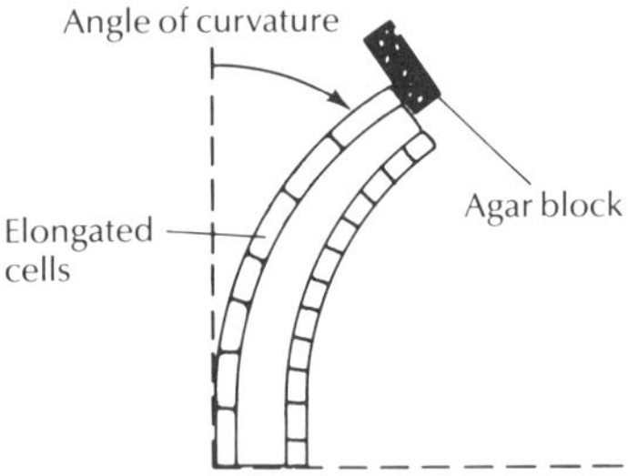

7.1 The oat coleoptile curvature test is the classic and most sensitive bioassay for the auxin, indole acetic acid (IAA). Coleoptile tips are placed on agar blocks and the IAA drains into the agar (above, left). The agar is subdivided and blocks are placed on one side of decapitated oat coleoptiles (above, right). The primary leaf is detached from the base and serves as a support for the agar block. The coleoptile will curve away from the agar (left), and the amount of curvature is proportional to the amount of auxin.

was such a growth-promoting substance. Went placed blocks of agar (the absorbent gelatinlike material obtained from red algae) under coleoptile tips, resting each tip on a block for a few hours so that any substance in the tip would diffuse downward into the agar (see Figure 7.1). Then he transferred the agar blocks to one side of the top of oat coleoptiles whose tips had been removed, keeping the decapitated coleoptiles in darkness. Sure enough, the coleoptiles began to curve away from the side of the agar blocks. Went called the substance **auxin,** from the Greek word meaning to increase. Measuring the coleoptiles' curvature, he found that the amount of auxin that had diffused onto the agar was proportionate to the amount of curvature induced in the decapitated coleoptiles.

The consequences were enormous. Scientists were now certain that they had found at least one substance that causes plants to grow, and they had a technique for measuring it. Auxin causes plant cells to elongate and it leads to the bending Darwin observed because the cells on the side away from the light elongate more than those on the side facing the light. Still to be discovered was where auxin occurs in a plant, its chemical composition, and how it works. Some answers came quicker now.

Meristematic tissues are the main source of auxin. Not until 1934 was its chemical composition unraveled. Isolating auxin from fungi, botanists and chemists working together in Holland and at the California Institute of Technology provided good evidence that it is indole-3-acetic acid (IAA). Only in 1946 was IAA conclusively proven to occur in higher plants.

IAA can be called a **hormone:** it is an organic compound (that is, one containing carbon) produced in small amounts in one part of the organism (in this case, the meristems) and transported to another area where it has an effect. A relatively simple compound, IAA has significant effects. Its primary one is to induce elongation of cells, the basis of plant growth. It does this by inducing elasticity of the cell wall. In a growing cell the wall is plastic, stretching as the cell takes up more water, so that it becomes larger. The walls then become rigid, and the cell grows no more. Multiply this one cell by thousands, all elongating, and obviously the plant will grow. Can we then induce plants to grow ever larger by adding auxin? Let us take a closer look at plant cells.

Plant Cells: Taking in Water Thin-walled plant cells usually have about fourteen sides. Isolated, each would form a sphere, but press many spheres together and you get fourteen-sided shapes. Take a bottle half-filled with soapy water and give it a good shaking. If the bubbles that form are large enough, you will see some with all fourteen sides.

The wall surrounding the cell's living material, its **protoplasm,** lets water, dissolved salts, sugars, and other compounds into the cell. But two membranes can be barriers to the substances. One, the **plasmalemma,** surrounds the cell's living material; the other, the **tonoplast,** surrounds the water-filled **vacuole,** the largest part of the mature plant cell. Water will enter or leave the cell depending on the concentration of dissolved salt particles in the vacuole compared with their concentration in the water surrounding the cell. This movement is **osmosis** (from the Greek word for push), of which you may have heard; both its discovery and its way of working are useful to know.

A century and a half ago, René Joachim Henri Dutrochet, a French physician, was investigating the manner in which cells take up water. He set up an experiment where he stretched a pig's bladder across the wide mouth of a glass funnel, which he had filled with a salt solution. He turned the funnel upside down and put its wide end in a pan filled with water (see Figure 7.2). The salt solution begin to rise in the funnel: it was taking up water from the pan. The cells in the pig's bladder let water in, but at the same time kept salt from going out.

Water will always move to a solution that has a higher concentration of solutes. Put a cucumber into a concentrated salt solution and water in the vacuoles of the cucumber cells will flow out, into the salt solution, because it has more solute than the cell vacuole. After a while the cucumber becomes a pickle. Thus the principle of pickle making is osmosis.

Cells grow because of osmosis, and all plants' growth depends on it. In growing cells, the vacuole is filled with a solution of salts and

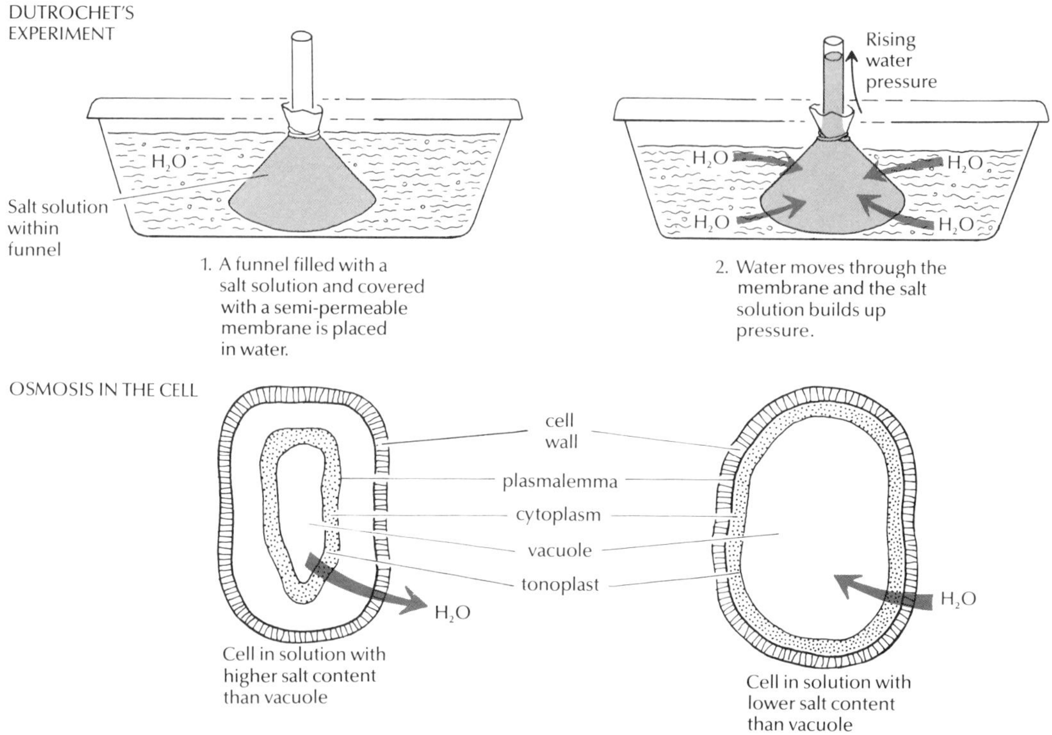

7.2 When two salt solutions are divided by a barrier that will only allow passage of water molecules (a semipermeable membrane), water will diffuse from the solution with the lowest salt concentration. When water enters cells, a pressure is created because of the cell walls. When water leaves cells in leaves and nonwoody stems, wilting results.

other compounds that is more concentrated than the salt solution outside it. While the membranes inhibit salt from flowing out, the vacuole gains water. As a result, water will keep coming into the cell until the solute-to-water ratio in both solutions is equal or until the cell wall reaches the limits of its stretchability.

Cell Walls The relatively rigid cell walls are made of **cellulose,** which is itself composed of long strings of molecules of glucose (a six-carbon sugar). This composition makes cellulose a **polymer**—a compound consisting of long linked chains of repeated compounds. Nylon and rubber are polymers, the former an artificial and the latter a natural compound. Starch is also a polymer of glucose. (Cellulose and starch differ in the way their glucose molecules are attached to each other.) Cotton fibers are nothing more than long strings of cellulose that were once the walls of long hair cells developed from the epidermis of seeds.

Cellulose molecules are arranged in small **fibrils,** or threads, layered in the cell wall somewhat like cords in an automobile tire. New cellulose fibrils are produced by the cell and incorporated into the cell wall randomly, but the outer layers of the wall have an ordered arrangement of fibrils, all fibrils in each layer lying parallel. The fibrils are cemented together in a matrix of **pectin,** the same substance a cook adds to jellies to make them stiff. But like a tire, the cell wall has some stretch built in.

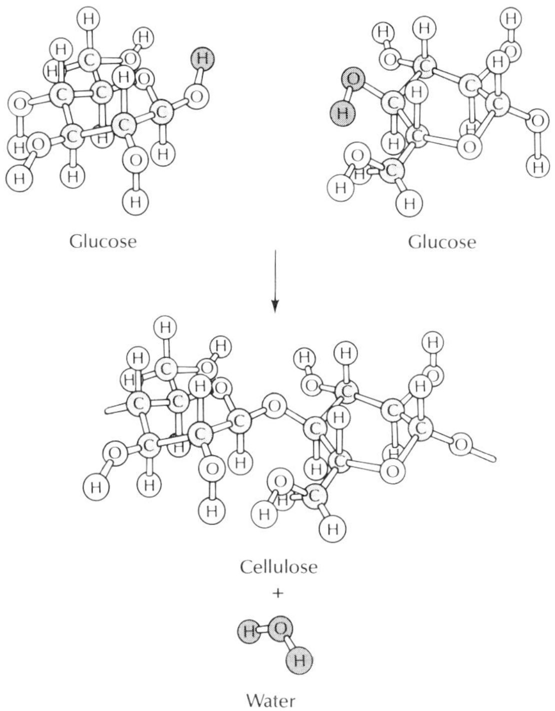

7.3 Cellulose is a polymer, a chain of glucose molecules. When glucose molecules bond together, a molecule of water is lost. Starch is also a polymer of glucose.

IAA apparently loosens the binding between these cellulose fibrils in young cell walls, increasing their stretchability. When the limits of the wall's ability to be stretched are reached, the cell wall will exert a back pressure, countering the osmotic pressure, and the cell will take up no more water—it will cease to grow.

Limits to Growth A plump water-filled plant cell is described by a word that sounds like the condition it names: **turgid,** or swollen; the noun is **turgor.** Because their cell walls are stretched tight, turgid cells provide much of the structural support for nonwoody tissues like leaves. They are like paper bags full to bursting with water-filled ballons: they stand straight because the balloons give them rigidity (see Figure 7.4).

We have seen that IAA can make the walls of cells that have growth potential more flexible. As these walls stretch they exert less back pressure and the cells can take up more water, but the walls are not weakened: more cellulose fibrils are woven into them. The cell does not end like a balloon, its wall made large and thin by stretching. It is bigger in all ways, with greater volume, with bigger walls that are just as strong as when it began taking up water and eventually holding more cytoplasm. Cell growth, then, is an irreversible increase in water capacity and a consequent growth of the cell as a whole.

Why does a cell stop growing? Knowing that auxin makes the cell

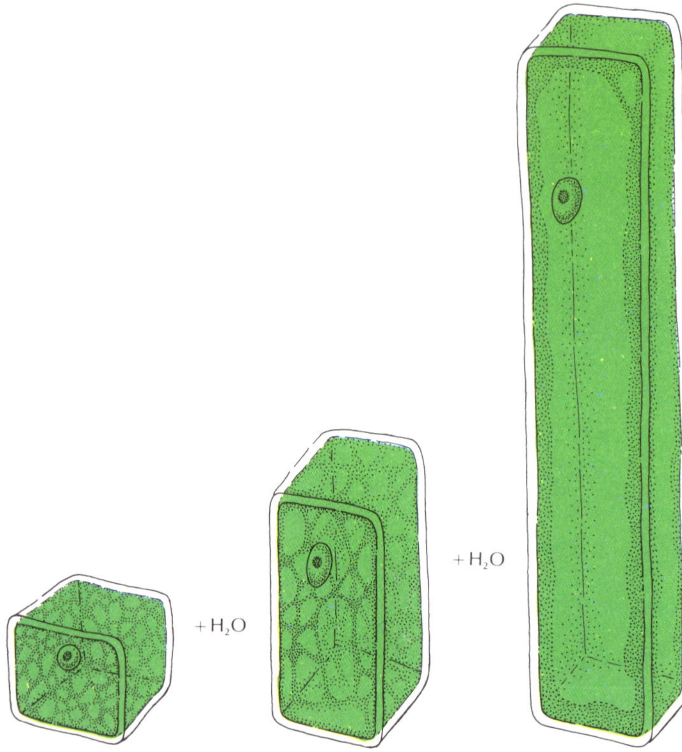

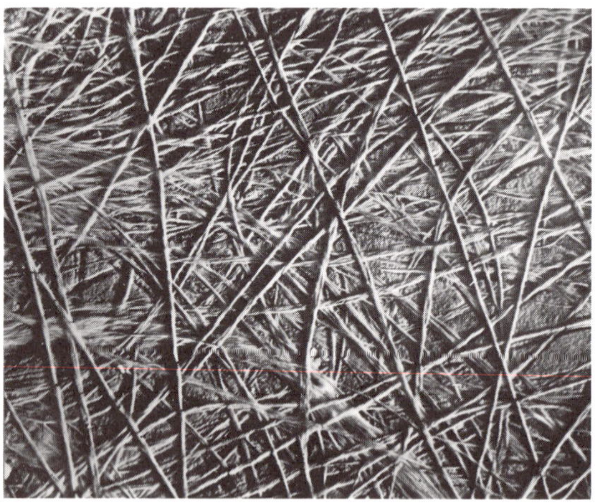

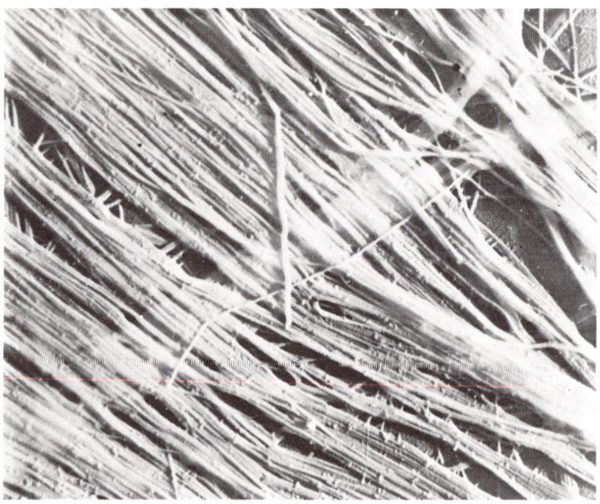

7.4 Electron micrograph shows cellulose microfibrils deposited on the inside of the wall of growing cells in a random pattern (below, left). As the cell expands, the fibrils are deposited in an ordered arrangement (below, right). Plants grow by means of cell enlargement, which occurs through water uptake accompanied by cell wall synthesis (right).

wall plastic, and that the cell can get larger and larger, can we assume that it might simply keep growing ad infinitum? Obviously not. Cell walls grow more rigid as they enlarge, in some cells because lignin is added to the walls. **Lignin** is a complex compound that makes wood tough and durable. As the xylem cells in vascular bundles mature, increasing amounts of lignin are deposited in their cellulose walls. At first it is laid down in a spiral, so that the cells can still stretch; the lignin deposits simply expand like strings. As the cells mature, however, lignin deposition goes on increasing, and finally the cellulose fibrils are cemented together and the cells lose their stretchability. The cells eventually die, leaving the thick lignified walls to serve as water pipes and support pillars (see Chapter 10).

Probing Deeper How does IAA work? Finding the answer is like peeling the layers off an onion (the layers are really modified leaves). It is known that IAA induces plasticity in cell walls, but there is still another layer: how this is done is not known. Chemical reactions are involved, but what kind and how many? Is the effect on the walls direct, or is it a secondary one, resulting from other effects elsewhere in the cell? Work continues on such questions.

Researchers have found that synthetic chemicals related to IAA can also produce growth in some tissues that have a capacity for growth but are deprived of their natural IAA. But neither IAA nor any of its chemical cousins can promote growth when applied to intact plants, even to dwarf varieties. Other substances, it follows, must also affect growth.

OTHER CHEMICAL GROWTH REGULATORS

Plant cells cannot grow without IAA, but other hormones also influence the plant's growth. In addition to the auxins, they include cytokinins, gibberellins, and ethylene (a gas). The classifications are based on the effects with which the hormones were first associated. In fact all these hormones usually have multiple effects. Auxins, for example, also alter **abscission** of fruit and leaf—the natural separation from the plant that causes leaves and fruits to fall in season.

Cytokinins The first cytokinin discovered was a substance influencing the division of cells (hence the name; *cyto* is "cell," *kinin* is "divide"). Evidence has also been found that these hormones influence the differentiation of tissue to form meristems and that they can inhibit the aging of plant tissues such as occurs when a leaf turns yellow. More will be said about cytokinins in Chapter 12.

Gibberellins A class of compounds with a structure more complex than that of IAA, the gibberellins are among the best-known plant hormones. Their memorable name comes from a fungus, *Gibberella fujikoroi,* long known in Japan as the cause of "foolish rice" disease, which causes plants to grow unnaturally tall and thin. Japanese scientists succeeded in isolating from the fungus a compound that also induced this "foolish" behavior. Western scientists, ignorant of the research until after World War II, finally isolated and purified gibberellic acid (GA) and other gibberellins and determined their structures in 1959. The existence of GA in angiosperms was confirmed by B. O. Phinney and coworkers at UCLA in 1959. (See Figure 7.5 on page 206.)

Phinney was also able to determine a good deal about how GA works by using dwarf corn plants. He found that gibberellin overcomes the growth deficiency, and he found out why: plants will not grow normally without gibberellin, and the genes of dwarf plants have no provision for making it. When GA is applied to dwarf plants, they grow tall. The dwarf corn test thus provided a **bioassay** for this compound (or rather, these compounds, because more than forty versions of gibberellin have already turned up). (A bioassay or biological assay tests the response of living organisms to determine the quantity of a substance too minute to be identified chemically.) Phinney and others went on to discover that gibberellins occur in all normal plants, where, like IAA, they are produced in the meristematic regions.

Gibberellins do strange things beyond making dwarf plants grow tall and normal plants grow taller than normal. They cause rosette plants such as African violets and leaf lettuce, whose leaves usually grow on short stems close to the ground, to produce long stalks with flowers, which is referred to as **bolting.** (An example appears in Figure 7.5.) And they have commercial value in that they increase the size of some commercial fruits like the Thompson seedless grape. The grapes reaching consumers' tables suddenly were larger some years ago, the effect of gibberellins. This manipulation is physiological, not genetic. Gibberellins are also important in stimulating germination of some seeds.

Like other naturally occurring plant hormones, some things about the gibberellins are well known, including where they are manufactured, what they can do, and how they can be synthesized. But how they do their work is still a mystery.

Ethylene: The Ripening Gas It may seem odd to classify a gas as a hormone, but plants do produce a gas that helps regulate growth, apical dominance, flowering, and aging. It is **ethylene** (C_2H_4), and it too is produced in meristems. Even at very low concentrations, it, like other hormones, affects other tissues. But plant scientists have to be endlessly careful about jumping to conclusions. Experiments had seemed to place on ethylene the responsibility for effects on growth previously attributed to auxin. Further experiments indicated that auxin in fact stimulated the synthesis of ethylene, which then affects growth. Auxin was the instigator but not the instrument of the effect.

One example comes from Hawaii. For many years synthetic auxin was sprayed on pineapple fields there to induce the plants to flower at a preset time. Growers had assumed that auxin caused the flowering, but finally they found their error. The IAA induced formation of ethylene in the plants, and the ethylene caused the plants to flower. Growers could save money by omitting the auxin and spraying the plants directly with ethylene-producing compounds.

You may recognize the name *ethylene;* it is a common by-product of petroleum refining and is also the building block for synthesizing many plastics. And the word *plastic* brings to mind just how fuzzy is the distinction between "natural" and "unnatural." A member of Congress recently raised public protest about the gassing of tomatoes to make them turn red. The innocent public, she said, was being subjected to unnatural and possibly toxic materials by food producers. Of course the gas was ethylene, and far from being "unnatural," it occurs in the plants themselves. It was simply being applied sooner than the ripening tomatoes would produce it themselves.

Many changes occur as a fruit ripens. Starches are transformed into sugars, which make fruits sweet, and cell walls soften and break down, making the produce easy to bite into and eat. During the ripening, flavors and odors we associate with ripe fruits appear, set off by, among other things, the plant's synthesis of ethylene, which stimulates ripening. Ethylene literally pours out of ripe fruit, detectable and measurable by special instruments. Conduct an experiment with ethylene, putting some ripe and some green bananas into a closed container. Then see how quickly the green bananas ripen by comparison with other green ones placed in a container by themselves.

A fruit ripens as its tissues age or **senesce.** Senescence leads to death, which is as much a part of the plant's development as growth. All these growth regulators, as well as some others we have not examined, affect senescence, just as all are involved in development. But specific events are regulated not by a specific chemical but by a "cocktail" of chemicals. When you consider that each chemical can have different effects on different cells and at different concentrations (for example, a hormone can stimulate growth at one concentration and inhibit growth at another), and when you think of the many forms of growth regulators (the gibberellin molecule has about a hundred possible variations), the complexity of combinations is overwhelming.

APPLYING HORMONES

This swarm of possibilities is one reason why it has been so difficult to learn how these chemicals work. Many parts of the plant's development, however, can be controlled because at least

7.5 HORMONES AND DEVELOPMENT

The effects of plant hormones are summarized in the diagram (below, left). Gibberellins affect plant germination (see facing page, middle), and they are involved in the production of an elongated flower stalk, as in rosette plants. The application of gibberellin can exaggerate this phenomenon, as illustrated here with the cabbage plants (photo, below, right). Auxins help to control the formation of abscission areas in the petioles of leaves and fruits. In the photo of the snap bean plant (facing page, bottom), the abscission areas are the short, slightly thickened portions of the petioles, just beyond the point of branching. When the cells in this layer no longer receive the proper amount of auxin as a result of leaf aging, the cells begin to self-destruct and the leaf soon falls off. Cytokinins stimulate cell division and appear necessary to induce cell division in many tissues. These cultured soybean cells (facing page, top) will divide and produce new cells only when cytokinin is included in the culture medium.

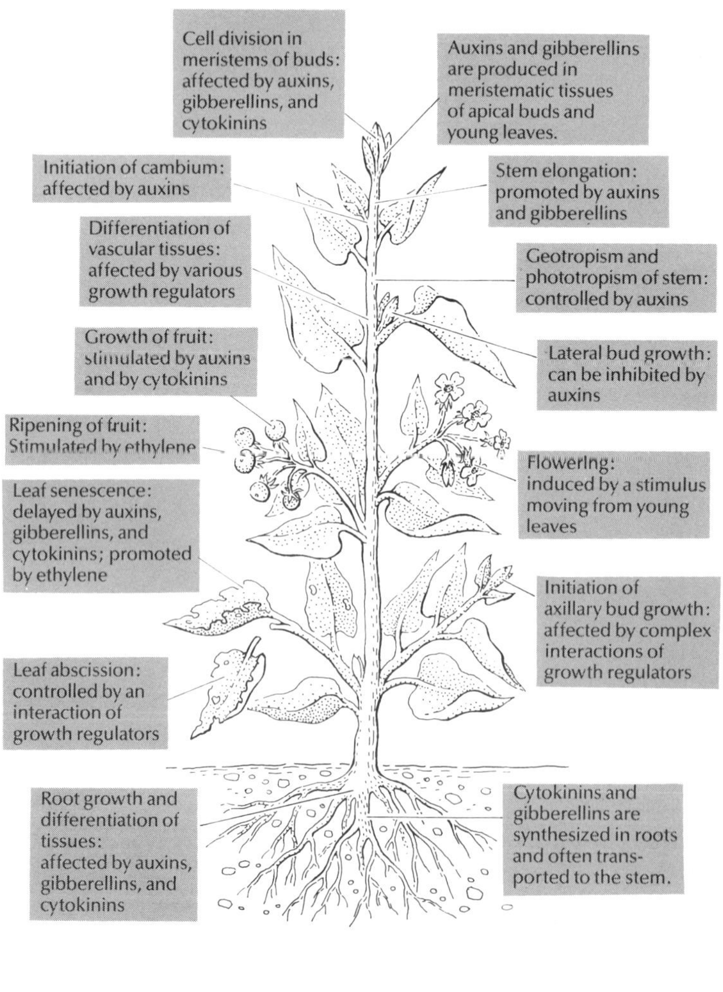

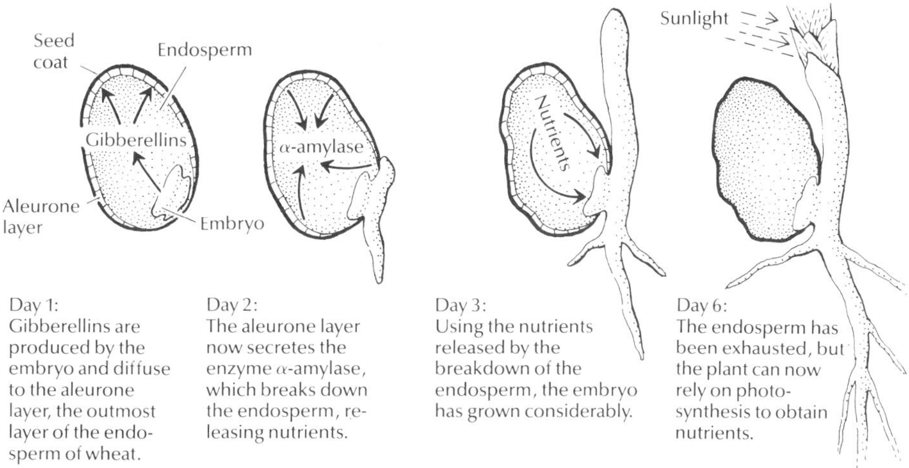

Day 1:
Gibberellins are produced by the embryo and diffuse to the aleurone layer, the outmost layer of the endosperm of wheat.

Day 2:
The aleurone layer now secretes the enzyme α-amylase, which breaks down the endosperm, releasing nutrients.

Day 3:
Using the nutrients released by the breakdown of the endosperm, the embryo has grown considerably.

Day 6:
The endosperm has been exhausted, but the plant can now rely on photosynthesis to obtain nutrients.

some things about how the plant works are understood; but the increased production of food and fiber that we must achieve if all humanity is to prosper can be assured only if much more is learned.

Controlling Fruit Ripening Manufactured ethylene is important to growers of some fruits and vegetables today. Tomatoes, peaches, bananas, and pineapples are picked green for shipment to market. They are harvested at the moment when ripening has been triggered but fruit growers can control the ripening. When they want uniform and rapid ripening, they can apply ethylene and the fruit quickly ripens, all at the desired time. That means growers have some control over when they send their crops to market. With further development this technology could mean that farmers would be less at the mercy of bad weather, which can delay ripening and prevent their selling when prices are high; they might even control the price by ripening and shipping fruits when demand is high.

Ripening can also be delayed when that is desirable. Peaches, pears, and bananas can turn to mush so fast that the fruit grower desperately needs to be able to slow their ripening. Fruits kept in an atmosphere high in nitrogen and low in oxygen will not ripen, because production of ethylene depends on oxygen. Just as refrigerated freight cars and trucks have made it possible to supply a year-round variety of fresh fruits and vegetables (by preventing spoilage), so shipping fruits in a nitrogen atmosphere could provide us with even more freshly ripened, better-tasting fruits and vegetables at any time of year. The fact that fresh apples are available the year round is an example of how we take the benefits of controlled-environment fruit storage for granted.

Many urban-bred people may not know it, but any fruit that is left to ripen on the plant will taste better than another picked green and then ripened, simply because the ripening process is more complete. Being able to inhibit ripening, on the other hand, means that fruit can be left to ripen on the plant until a chosen date, then the ripening can be stopped during shipment and restarted after delivery. Ripening goes further while the fruit is on the tree, a fact known to anyone who has tasted a field-ripened peach, pineapple, banana, or mango.

Fruit growers can also control fruit drop by controlled applications of hormones. Leaves and fruits drop off when a layer of cells near the end of their petioles dies, a process known as abscission. Until then, hormones have flowed from both stem and fruit or leaf, keeping the cells alive. Too much hormone can cause leaf fall just as well as lack of it. The leaves of the cotton plant can be made to fall before the cotton is harvested, an advantage in harvesting. The farmer just supplies the plant with a certain amount of synthetic hormones. Synthetic auxins can also selectively kill plants, which makes them important as herbicides.

Weed Killers The word *herbicide* has a distasteful past because of the thousands of tons applied during the war in Vietnam. They were sprayed from aircraft to strip leaves from trees in the Vietnamese forests, laying bare the jungle trails over which the North Vietnamese supplied their armies in the south. No one knew if wholesale defoliation of forests might leave lasting devastation, but fortunately this does not appear to have happened.

A great deal must yet be learned about herbicides, even those most frequently used. The most common weed killer, 2,4-D (2,4-dichlorophenoxyacetic acid), is used to kill only specific dicots (see Figure 7.6). In the low doses that kill dicot weeds, it has little effect on monocots because they have an enzyme that inactivates the 2,4-D molecules. Suburban homeowners can spray their lawns without hurting the grass, and farmers can apply defoliants on their cornfields and kill only dicot weeds. Hoeing a cornfield is exercise that few people do willingly; instead, weed killers permit low-tillage agriculture (see Chapter 4). And 2,4-D has no harmful effects on the soil because it is quickly broken down by soil organisms. Other common herbicides do not affect growth directly but specifically inhibit photosynthesis.

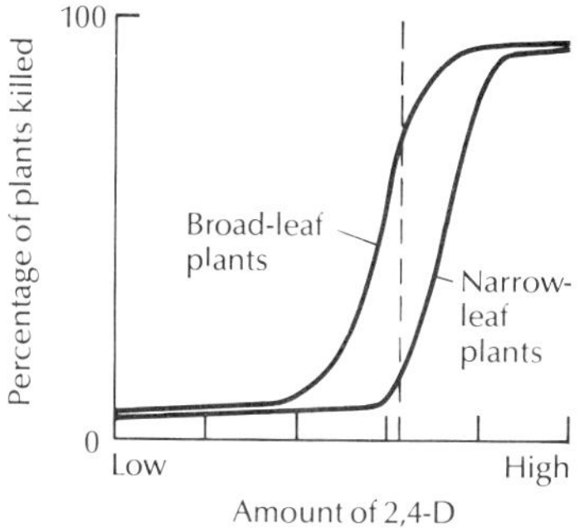

7.6 The importance of selective herbicides, such as 2,4-D, in growing cereal crops is shown by these curves. The concentration indicated by the dashed line will kill most broad-leaf weeds but will not affect cereals, which are grasses.

Applying precise dosages of a herbicide is a necessity, because low concentrations can produce one effect whereas the opposite might occur at higher concentrations. The use of herbicides has increased greatly since the early 1960s and the demand continues to grow. They have helped make agricultural food and fiber production more efficient. Specificity and safety of herbicides, however, are a matter of great concern.

By all these uses, and more, we continue to reap practical benefits from Darwin's country garden experiment. It did not take long to find that many plant hormones will induce stem cuttings to form roots, an important practice in horticulture that will be discussed in Chapter 12. Rooting compounds usually are a mixture of an inert powder with a synthetic hormone like indolebutyric acid. The grower dusts one end of a stem cutting, as from a rose, or the base of an African violet petiole, with the powder and sets it in moist sand or soil. The chemical induces cortical cells near a vascular bundle to divide and form a root primordium, which grows out of the stem or petiole just like a secondary root. This is an **adventitious** ("unusually placed") root, and when the cutting is planted it will take hold and grow into a whole plant. The stems of plants like willows, beans, and corn form adventitious roots spontaneously, but being able to induce them artificially is a clear advantage and the method is much applied in horticulture.

LIGHT AND DEVELOPMENT

Light controls much more than just bending in plants. Its effects start with photosynthesis, which influences everything about a plant. But can all the ramifications of photosynthesis be separated from other factors of development affected by light, such as color, shape, and size? To find out we must break down the subtle effects of light shining on a plant to see what parts of the plant are affected by it.

Analyzing Light and Pigments Sunlight is more than an absence of darkness; more lies in its "golden" rays than the poet suspects. The physicist analyzes sunlight as a mixture of wavelengths of light, of colors. Plants have substances that absorb specific colors, or wavelengths, of light (see Figure 1.3). These substances are **pigments,** compounds that absorb some colors in the visible spectrum of light and reflect others. A blue pigment reflects blue light—it does not absorb it—and the object appears to us as blue. Chlorophyll, the pigment that absorbs the wavelengths of light effective in photosynthesis, reflects green light, which is why we see most plants as green.

Is green light, then, important in photosynthesis? No, because the green light is reflected. Visible light can affect the chemistry of plant cells only when it is absorbed by a pigment. A pigment can be matched with a physiological event in a plant by measuring both the wavelengths absorbed by an isolated pigment and the effects produced by the same wavelength.

To conduct such an experiment, white light can be separated into its component parts with a spectrophotometer. A prism in the instrument produces a spectrum of visible light—a "rainbow" with narrow bands of all colors from red on one end to blue on the other. Different colors can be screened out so that perhaps only a band of one color shows. Now an isolated pigment in a small glass container—chlorophyll, for example—is exposed to that band of light; if the pigment does not absorb that color, it will register on a photoelectric cell behind the glass container. An **absorption spectrum** can then be plotted (see Figure 7.7). From that we determine that chlorophyll absorbs red light (it does not register on the photoelectric cell) and blue light while most of the green light is reflected.

We can then compare this absorption spectrum with an **action spectrum,** which plots the physiological effect of specific wavelengths. A leaf, for example, can be exposed to various wavelengths and the strength of the response (say, the amount of photosynthesis) to them plotted, as in Figure 7.7. The absorption spectrum of chlorophyll matches the action spectrum of photosynthesis: chlorophyll absorbs red and blue light but very little green light, and red and blue are the wavelengths effective in photosynthesis.

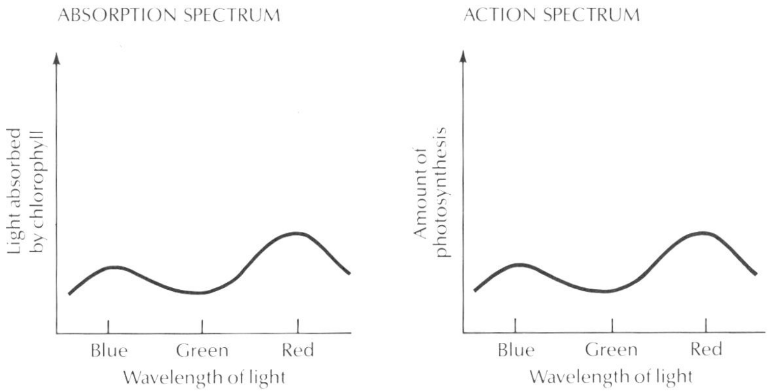

7.7 An absorption spectrum describes the kind of light a pigment absorbs, and an action spectrum reflects the relationship between the kind of light and a physiological process. In this case, blue and red light are absorbed by chlorophyll, and they are the most effective wavelengths in stimulating photosynthesis.

The duration and intensity of light—that is, the light energy required to induce a plant to respond in a specific way—are also criteria for identifying a pigment as the intermediary in a light-induced reaction (see Table 7.1). Photosynthesis, for example, does not take place when a green leaf is exposed to low-intensity red light.

Phototropism People often speak of how houseplants and seedlings "seek the light," but plants, with no nervous or muscular system, cannot seek anything. The reaction is **phototropism** (*photo*, "light"; *trophy*, "turn")—the plant's "movement" (in fact, it grows instead of moving) toward or away from light. Light striking a plant on one side makes auxin from the apical meristem collect on the opposite side in the region of cell elongation, stimulating cell elongation on that side and creating a bend in the plant stem. But what causes the unilateral ("one-sided") transport of auxin from one side to the other? What is the relationship between a pigment that has absorbed light and the transport of auxin down one side of the

Table 7.1 Effects of Duration and Intensity of Light on Light-Induced Reactions

	Photomorphogenesis (Blue pigment)	Phototropism (Yellow pigment)	Photosynthesis (Green pigment)
BLUE LIGHT			
High energy	Yes	Yes	Yes
Low energy	No	Yes	No
RED LIGHT			
High energy	Yes	No	Yes
Low energy	Yes	No	No

stem? Like so many important questions, very little is known about this one.

Another problem is determining the pigment responsible for phototropism: the action spectrum shows that blue light is the effective wavelength, and it is also known that yellow pigments absorb blue light. One yellow plant pigment is immediately recognizable; the **carotenoids** are the family of yellow to red pigments responsible for the color of carrots, daffodils, and tomatoes and for much of the color in autumn leaves. Another group of yellow plant pigments are the **flavins,** which have important functions in cell metabolism. The action spectrum of phototropism is similar to the absorption spectra of these two pigments. But action spectra are always relatively crude compared to absorption spectra, making it difficult to trace the relationship with these two pigments precisely, and the identity of the phototropic pigment is still a matter of controversy.

One thing is known with certainty, however: phototropism is an adaptation that affects a plant's survival. Plants must be favorably oriented to sunlight in order to photosynthesize, and they must photosynthesize to develop and reproduce.

Many roots grow away from light, a reaction called **negative phototropism.** Here auxin transport, its movement in the tissues, and differential growth (the bending away from light) are similar to that of stems in phototropism. There is one obvious difference, however: auxin must be moved in an opposite direction. Why auxin is transported differently in stems and roots is another unknown.

Photomorphogenesis A plant seedling growing in total darkness looks completely different from one grown normally in light. This is one profound difference between plants and animals. An animal kept long in darkness may be unhappy and pale, but it still looks much like an animal living normally in the light. A bean plant grown in the dark has long internodes, its leaves are narrow and unexpanded, its stems are thin with an apical hook, and it has no chlorophyll. What happens if this plant is put out into the sunlight?

The change is striking. The narrow leaves expand and turn green as their chloroplasts develop chlorophyll. The apical hook straightens. The internodes remain long, a reminder that a plant's stems tell much of its history. Newly developed internodes, however, will be short.

A dark-grown plant like the one described above and the example in Figure 7.8 is said to be **etiolated.** Natural etiolation displays itself in plants that are adapted to growing in full sunlight but are growing in the shade by chance. Like tomato plants in a shady home garden, or like many plants grown indoors, these will grow up tall and spindly. With too little light they become at least partially etiolated,

7.8 Light- and dark-grown plants of the same age would seem to be different organisms. The dark-grown plant (left) has long internodes, small leaves, and lacks chlorophyll. It is said to be *etiolated.*

possibly beyond help even by artificial lights. A bank of fluorescent lights above a plant provides only one-tenth or less of the light intensity from sunlight on an average day.

Etiolation is an example of **photomorphogenesis,** the light-regulated development of plant form. The plant pigment responsible for photomorphogenesis has been found, as we will see in the next section.

Photoperiodism In 1928 botanists W. W. Garner and H. A. Allard, on a project in the Department of Agriculture Experimental Station in Beltsville, Maryland, were trying to develop improved varieties of tobacco. One day they found an exceptionally large plant growing in the fields, and naturally they tried to obtain seeds from it, hoping to breed a significantly larger plant for commercial tobacco growers.

Finding no way of inducing the plant to produce flowers, they put it into the greenhouse. There at last it flowered in December, long after the normal flowering time. They now had seeds and ran tests to learn which factors induced the plant to flower. The length of day, they found, was crucial, and with this discovery Garner and Allard opened a new field in biology: **biological clocks,** or measurement of time in organisms.

Garner and Allard went on to examine the flowering behavior of an array of plants, establishing that there was a critical time for flowering when a plant was exposed to light. Some plants flowered before the critical period, some afterward. These were named **short-day plants** and **long-day plants,** and the phenomenon itself was called **photoperiodism.** They found too that many plants

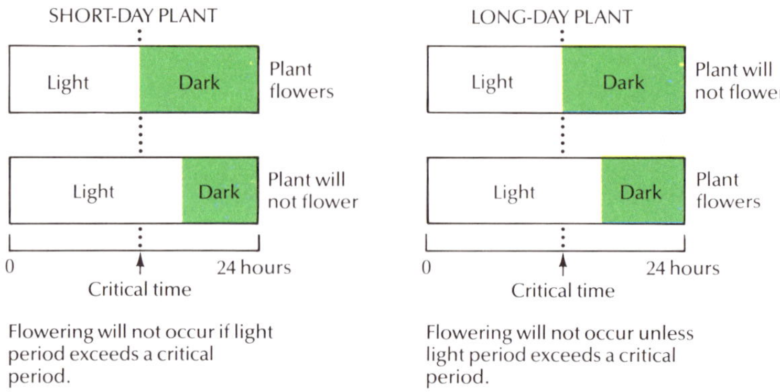

7.9 Photoperiodism refers to a critical time for the dark period. The absolute length of the light or dark period is not the determining factor.

flowered independently of photoperiodism, and they called these **day-neutral plants.**

Subsequent research unearthed a surprise: it is the relative length of the night, not the day, that is critical in photoperiodism. Short-day plants must have a *specific* uninterrupted dark period to flower, whereas long-day plants will flower only if the dark period does not exceed a critical length of time (see Figure 7.9). It is not the absolute length of each day or night that is important but the time up to or after the critical moment.

Garner and Allard's discoveries were revolutionary. Knowledge of photoperiodism is indispensable to the flower industry, for one. Easter's date changes to suit human convenience, but no matter when the holiday occurs, Easter lilies are always ready because they are photoperiodic and the light period (or dark period, if you prefer) can be controlled in greenhouses, so that they flower when we want them to. Poinsettias are ready at Christmas, or any other time we may want them, because they, too, are photoperiodic. They are short-day plants, like those giant chrysanthemums that are in bloom just when we need them to greet the football season.

Photoperiodic phenomena occur in many kinds of organisms: reproductive behavior in birds, hibernation in mammals, and release of eggs in worms are examples. The study of biological rhythms grew out of these experiments, and everyone who has experienced jet lag after traveling long distances by air has some feeling for the subject. Photoperiodism is one of the most active subjects of pure and applied biological research, and the whole thing came from the search for a better tobacco plant.

Red Light and Blue Pigments But are we any closer to knowing which plant pigment is responsible for photomorphogenesis? First a look at the results of some experiments on germination of seeds.

In the middle 1930s, botanists at the Smithsonian Institution in

7.10 These Chrysanthemum plants are short-day plants and will only flower if the dark period exceeds a critical length. If this critical dark period is interrupted with a short period of red light, the plants will not bloom (left). If the red light is followed by far-red illumination, the plants will act as if they did not receive the red light and they will flower (right). Far-red light reverses the effect of red light.

Washington, D.C., were investigating the action spectrum for germination of lettuce seed. Many seeds will germinate only in light, among them the seeds of many weed plants. The Smithsonian's scientists found that red light was the most effective in stimulating germination, and they also found that long wavelength red, or **far-red light,** inhibited germination of seeds. By the early 1950s, research workers at the Beltsville laboratory realized that the stimulating effect of red light could be reversed by exposing the seeds to an additional dose of far-red light. This light, in turn, could be negated by another dose of red light; the seeds would then respond to the kind of light in the last dose.

To see what these light experiments proved, we will need a further look at photoperiodic flowering. The action spectrum turns out to be similar to that of seed germination: red light promotes flowering, far-red reverses its effect. This pattern brought to mind etiolation, which happens when a plant is deprived of light. It was found that red light can turn a dark-grown or etiolated plant into one that grows normally in light, and far-red light inhibits the effect of red.

Could photoperiodic flowering, germination, and etiolation be mediated by the same pigment? The answer was found in the Beltsville Experimental Station, where scientists isolated a blue pigment that absorbed both red light and far-red light. They found this combination with a very sensitive spectrophotometer developed to study ripening in apples. Called **phytochrome** (*phyto,* "plant"; *chrome,* "pigment"), its concentration in plants is so low that it cannot be detected by the eye, even in etiolated plants devoid of chlorophyll. It is universally found in plants, and it is called the **photomorphogenic pigment.**

Quite a bit has since been discovered about the chemistry of phytochrome and the processes it controls. The most interesting and perplexing questions are still unanswered, though. How does this light-activated pigment react or interact with other components of cells to transform a vegetative into a floral apex, and how does it inhibit growth of internodes or initiate seed germination?

How do we know that photosynthesis is not involved in these reactions (for chlorophyll also absorbs red light)? At least this question has an answer. Chlorophyll cannot be the effective pigment system because photomorphogenic effects are produced in etiolated plants by a very small amount of red light. Etiolated plants do not have chlorophyll. Even in plants with chlorophyll, effects are obtained at an intensity of red light too low for effective photosynthesis. Also, the photomorphogenic effects are photo-reversible: one kind of red light will inhibit the effects of another, which does not happen in photosynthesis. And photosynthesis can be inhibited by chemicals that have no effect on events mediated by phytochrome.

"Florigen" Photoperiodic flowering is especially intriguing because a chemical "message" is certainly transmitted from young leaves to the apex of the stem. This fact was discovered by subjecting one leaf of a short-day plant to a light regime designed to induce flowering, while exposing the rest of the plant to light that would not induce flowering (see Figure 7.11). The plant flowered. The flowering stimulus had to come from the leaf that had been subjected to flower-inducing light. It even proved possible to measure how fast the flowering stimulus was transmitted from leaf to apex by removing parts of leaf and petiole at different periods following the flower-inducing light period. The rate of movement proved to be consistent with the idea that chemicals were being transported.

Once the existence of chemical transmission was verified, it was reasonable to suspect that a specific flowering hormone, or **"florigen,"** was the agent. Despite many experiments designed to prove or disprove this theory, the best efforts of large numbers of scientists have failed to isolate and identify such a hormone. The most likely solution is that a number of growth regulators or other metabolites are involved; small differences in their concentration in the apical meristem's cells may well trigger the flowering.

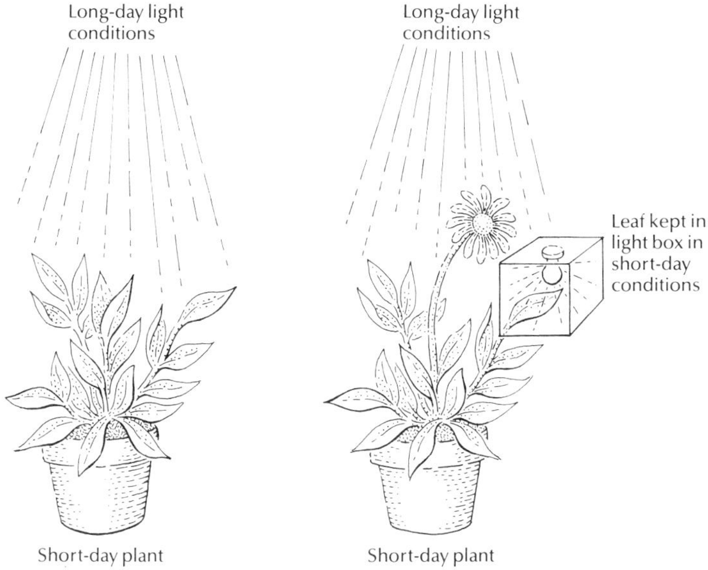

7.11 The transmission of a floral stimulus can be demonstrated easily. A young leaf kept in a flower-inducing light period while the rest of the plant is in a non-inducing light period will send a signal to the vegetative apex and a flower will form. The nature of this chemical message is unknown.

One day the material that controls flowering in plants will be known, and a large part of the national economy will be affected. Knowing how to pull the trigger that sets the apical meristem to producing flowers, not leaves, will affect not only the floral industry but also agriculture. Farmers might delay flowering and thus prevent damage to crops by early cold weather; or they might speed up flowering to shorten the growing season or make it possible to harvest ripe grain and fruit at a more favorable time.

GRAVITY

Most plant stems grow up and most roots grow down, but how do plants tell up from down?

This behavior is called geotropism (*geo,* "earth"; *tropism,* "turning"). Roots respond positively to gravity, stems usually respond negatively, and the general up-and-down growing is **geotropism.**

Some of gravity's effects on plants are remarkably like those of light on phototropism. When roots or shoots are placed at an angle to the force of gravity, their reactions show this parallel. A unilateral distribution of auxin causes differential cell elongation and a turning up or down of the stem or root, as shown in Figure 7.12.

But how do the cells know where they are with respect to gravity? Present theories are not particularly convincing. Another mystery is why the shoot apex reacts differently to the distribution of auxin than the root apex does. Why is auxin transported in a different direction in the two organs? Another complication in root growth comes from evidence that much of the auxin effective in growth is

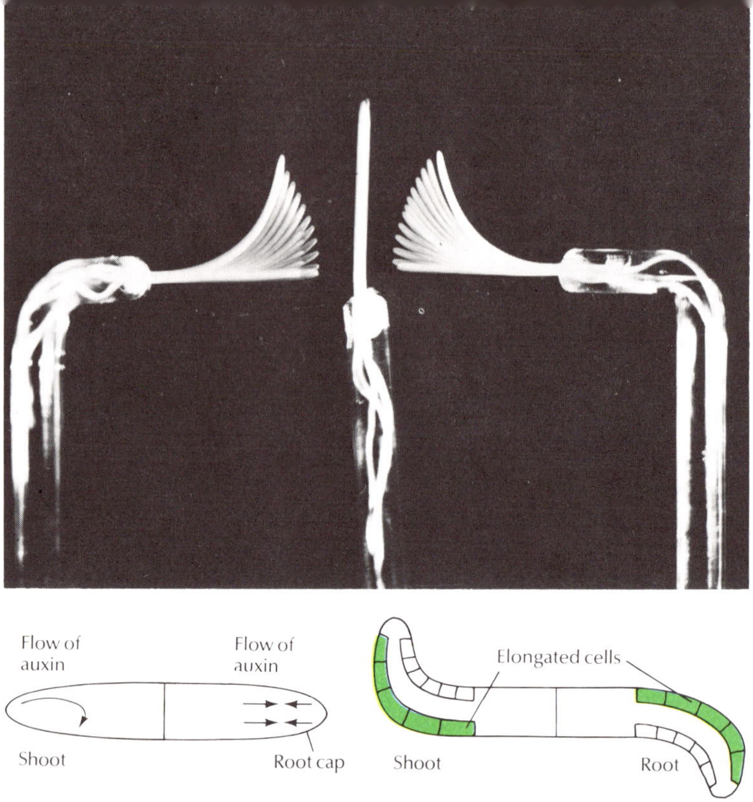

7.12 Positive geotropism is exhibited by these oat seedlings when placed at right angles to the force of gravity (photo, above). Time-lapse photographs at ten-minute intervals illustrate the reaction of the seedlings as auxin is redistributed and the cells in the growing region of the coleoptile undergo differential elongation. Most shoots respond negatively to gravity, and most roots respond positively (right). The geotropic growth response in each organ results from differential cell elongation. In the shoot tip, cells on the bottom grow more than those on the top and the shoot grows up. This response is caused by a uinilateral distribution of auxin produced at the tip. The root cap is the site of geotropic sensitivity in roots, and it produces growth substances that are transported to the region of cell elongation. Growth substances from more basal regions of the root are also transported to this region, and the interaction of growth substances from root cap and root results in the inhibition of cell growth in the bottom half of the root. The root, then, grows down.

transported to the root tip, not from it. Root growth is also influenced by an inhibitor from the root cap.

Shoots generally grow upward, but many plants have shoots called runners or rhizomes that grow horizontally to the ground, at right angles to the force of gravity. Try to fight crabgrass or Bermuda grass in the lawn, or try to raise strawberries, and you will soon be familiar with these. Runners snake along the surface and rhizomes grow underground.

The lateral branches of many plants, particularly the conifers, also grow horizontally; they are like aerial runners. If the top is removed from a conifer with a single trunk, a new vertical shoot, not a horizontal one, will be produced near the top. How does the tree signal such a change? Strange too that some plants with horizontal shoots, kept in the dark or dim light, will start producing vertical shoots. This reaction has been shown to be controlled by phytochrome.

MYTHS ABOUT PLANT GROWTH

Life is full of immensely popular beliefs that cannot be substantiated by scientific observation by trained observers working independently. One of these is that plants respond to music. Lots of people are convinced that they do, and even that plants grow especially well in response to a specific instrument or composition. You

can buy a record that is supposed to make plants grow better if you play it to them often. But none of the claims has ever been confirmed in carefully controlled experiments.

Plants, like other organisms, can be damaged or even killed by high-frequency sound, but in no other way can sounds we hear make any difference to plants. They have no specialized nervous systems for receiving and interpreting sound; in fact, they have no nervous systems at all. Just as light must have receptors (pigments in plants and animals) to have an effect, so sound must have receptors, and in plants there are none. And why should we think that plants would respond to our recently created noises any more than they respond to the noises of their natural environment? Artificial sounds, after all, are very recent beside the plant's long evolutionary history. Do shore plants grow better when they are constantly soothed by the sound of the surf, or mountain plants under the rush of the wind?

A variation of this theme is that plants respond to human emotions. Many people like to believe that their plants share their depressions or elations; a best-selling book collects such tales, and a lie detector operator won notoriety by claiming that his instrument can record the reactions of plants to human emotions. Alas! It is not so. A lie detector works by measuring the amount of electrical conductance caused by the secretions of salts by the skin. People under stress usually perspire more than people who are not. Leaves also

7.13 Both strawberries (left) and ivy (right) have creeping stems called *runners* and *rhizomes*. Runners are generally on the surface of the ground and rhizomes beneath, but the distinction is fuzzy. The ivy stem can climb vertical surfaces because it produces modified roots called *haustoria*, which cling to surfaces. The response of these stems to gravity is intermediate between that of most stems and roots.

secrete salts on their surface, but the two do not compare; cause and effect are entirely different and in plants the salt secretions have nothing to do with emotions, which plants are not equipped to have. Plant scientists have tried to replicate the lie detector experiments but without success.

One of the most absurd claims in this cult about plant responses is that plants will respond to good and bad thoughts. Loving thoughts will make them prosper, say the believers, but a person who thinks ill of them can cause plants to wither and die. Any weekend gardner can testify that torrents of blasphemy and evil thoughts will not cause the crabgrass in the lawn to turn up its "toes" or slow the flourishing growth of weeds in the flower and vegetable beds.

As a botanist, I am all for people feeling kindly toward their plants, for there is no denying that they respond to care and suffer from neglect. But this response reflects all factors in their growth, and if you tend them well while cursing them roundly for the trouble they give you, they will grow just the same.

Summary

Hormones, light, and gravity help regulate the plant's development. At the most basic level, growth occurs because the plant's cells take up water. In a growing cell the solution inside the vacuole will have a higher concentration of salt than does the solution outside. As a result of this osmotic gradient, the tonoplast, a semipermeable membrane surrounding the vacuole, will inhibit the passage of salts out of the vacuole while water passes into it.

As the vacuole takes up water, more fibrils of cellulose (a compound composed of long strings of glucose molecules) are woven into the cell wall, increasing the cell's volume. The cells grow more rigid as they enlarge because of the addition of lignin to the walls. Eventually the deposits of lignin will cement the cellulose fibrils together, and the wall can stretch no further.

The stretchability of the wall, and hence the cell's ability to grow, can be increased by IAA (indole-3-acetic acid), an auxin. It is but one of the substances important in controlling the plant's growth; others are the cytokinins, gibberellins, and ethylene. All these are classes of hormones—substances produced in one part of the plant that are transported to and affect another part.

Charles Darwin had concluded that there was some substance in plants that was transported from one part to another and stimulated growth. Frits Went later demonstrated that such a substance existed, and in the 1930s and 1940s the composition of this substance, IAA, was discovered and its occurrence in higher plants shown.

One effect of IAA is seen in phototropism, the bending of plants toward light. Somehow IAA, which is produced in meristems, is transported to the side of the plant away from the light, increasing the elasticity of cell walls there and hence inducing cell elongation. Auxin also affects abscission, the separation of leaves and fruits from the stem.

The class of hormones called cytokinins influences cell division, formation of meristems, and the aging of plant tissues. Gibberellins affect a plant's size and, in some plants, stimulate the germination of seeds. Gibberellin can make lettuce and many other plants bolt—the production of a long flower stalk; its absence can produce dwarf plants. Ethylene, a gas, helps regulate growth, apical dominance, flowering, and aging; it stimulates the ripening of fruit.

All these hormones have multiple effects that vary with the concentration of the hormone. Despite their complexity and the fact that scientists still do not know the precise mechanisms by which they regulate development, hormones are already widely used in agriculture and horticulture. Ethylene is sprayed on plants to induce flowering or ripening of fruits; hormones may be applied to induce abscission so that a crop can be harvested; synthetic auxins are used as herbicides.

Light affects growth in multiple ways: by photosynthesis, phototropism (bending toward or away from light), photomorphogenesis (light-regulated changes in form), and photoperiodism (responses regulated by the relative lengths of day and night). Specific wavelengths of light are absorbed by particular pigments in the plant; the green pigment chlorophyll, for example, absorbs red and blue light while reflecting most green light. To study how light affects a plant's physiology, scientists first separate it into its wavelengths and then plot an absorption spectrum for a specific pigment, showing which wavelengths the pigment absorbed. This absorption spectrum can then be compared with an action spectrum, which plots the effect of specific wavelengths on physiological reactions.

Phototropism is an obvious effect of light on plants. Roots often show negative phototropism, bending away from light. Carotenoids (yellow-to-red pigments) and flavins (yellow pigments) are candidates for mediating phototropic reactions.

If a plant that normally grows in sunlight is grown in the dark, it will be etiolated; tomato plants grown in the shade, for example, are tall and spindly. Red light can reverse the effects of etiolation; it can also stimulate germination and flowering. Far-red light will reverse the effects of red. The pigment phytochrome mediates these effects. Universally found in plants, it occurs in very low concentrations.

To flower, some plants—short-day plants—must have a specific amount of uninterrupted time in the dark. Long-day plants, on the other hand, must not be in the dark for more than a specific time. The critical periods vary from species to species. Such photoperiodism is found in many other organisms as well.

Some chemical message, however, must be transmitted from leaves to stem to induce flowering. Although some assumed that a specific hormone, which was called florigen, must therefore exist, it has not been found. More likely, several hormones act to trigger flowering.

Plants also respond to gravity; roots usually grow down and stems up. The distribution of auxin accounts for this geotropism just as it accounts for phototropism, but why it is distributed differently in stems and roots is unknown. Some shoots and branches grow at right angles to the force of gravity, rather than vertically.

Plants thus respond to many elements in their internal and external environments. There is no good evidence, however, that they react to music or other sounds, much less to human emotions.

Review Questions

1. Auxin, water, and the cell wall are all involved in cell growth. Define their specific roles and how they relate to cell growth.
2. Research has not provided very good answers to questions about the basic mechanism of action of growth regulators. Why has this been such a difficult task? The gibberellins probably offer the best hope at present for such an understanding. Defend this statement.
3. The pigments regulating photosynthesis, phototropism, and photomorphogenesis absorb blue light. Assuming you had a source of blue light and red light devise an experiment to distinguish between the different pigments regulating these processes.
4. What is meant by photoperiodism?
5. Plant *A* flowered when subjected to 12-hour "days" and plant *B* flowered when subjected to 10-hour "days." Plant *A* was designated a short-day plant and plant *B* was designated a long-day plant. Interpret this apparent contradiction.
6. How are phototropism and geotropism similar? How are they different?
7. Every few years, claims are made that plants respond to music and other sounds. How can these claims be tested?

CHAPTER 8

How Plants Obtain Energy

To grow, plants need water, carbon dioxide, minerals, and light, and the temperature around them must not exceed certain limits. Leave one of these factors out and growth stops. But once given these materials, the plant's organic and cellular machinery will begin the intricate work of producing food for the plant, and incidentally for all other living things. The thousands of plant species living today are the products of millions of years of evolution shaped by selective pressures that in many cases resulted from limitations on one or more of these elements.

The complex production going on in the plant acts rather like a chemical factory. "Chemical factory" might bring to mind dozens of buildings, large and small, connected by pipes and tubing, with steam hissing and smoke pouring out of stacks. But the things such a factory does can be and are done in an area generally so small that no one had seen it before the microscope was invented: this area is the cell. Multiplied millions of times, the plant cell provides the energy and compounds for the structure and development described in Chapters 6 and 7. As in the human cell, chemical compounds must be changed to release energy (a process known as *respiration*), but plant cells also accomplish a productive miracle that neither human cells nor a factory can duplicate: they transform light energy into chemical energy, a process known as **photosynthesis.** This ability, possessed only by green plants and some microbes like blue green algae and some bacteria, is the prime cause for the unique features of plants.

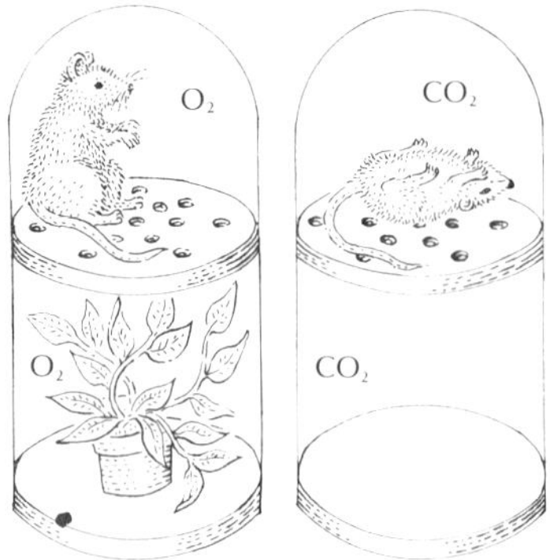

8.1 Joseph Priestley's famous eighteenth-century experiment demonstrated that plants produce oxygen.

PHOTOSYNTHESIS

Joseph Priestley, an English chemist, discovered the element oxygen in 1774 and with it the answer to a puzzle: why do animals sealed in airtight chambers invariably die? He found that animals use up all the oxygen. Priestley went on to discover that plants somehow reverse this process: they produce oxygen.

Then in 1779 Dutch physician and chemist Jan Ingenhousz discovered that only the green parts of plants produced oxygen. Furthermore, he found that plants without light *consume* oxygen just as animals do. More discoveries followed until finally in 1804, a Swiss, Theodore de Saussure, summarized the process of photosynthesis in one chemical expression:

$$\underset{\text{(6 molecules of carbon dioxide)}}{6CO_2} + \underset{\text{(6 molecules of water)}}{6H_2O} + \text{light energy} \longrightarrow \underset{\text{(a sugar)}}{C_6H_{12}O_6} + \underset{\text{(6 molecules of oxygen)}}{6O_2}$$

This expression states that carbon dioxide plus water plus light energy produces sugar and oxygen. The sugar becomes food for the plant and some of the oxygen is released to the atmosphere; that is how plants have supplied oxygen to the earth's atmosphere.

How does the plant accomplish this feat? We saw in Chapter 1 that sunlight is used to split water molecules into hydrogen and oxygen: oxygen is released; hydrogen is attached to other compounds, forming sugar. This is an accurate but very simplified account of photosynthesis, which involves many chemical reactions. The first series of reactions are known as **light reactions,** because they cannot begin without light.

Light Reactions From the light energy beamed from the sun, a plant usually absorbs only a small percentage, and only a portion of this can be used in photosynthesis because chlorophyll absorbs only visible light, particularly red and blue. Chlorophyll is found in membranes within the **chloroplasts,** the structures in plant cells where photosynthesis takes place (see Figure 8.2 and the Supplement on page 242).

What happens when chlorophyll absorbs light? With the additional energy that this green pigment absorbs from sunlight, it splits the water molecules (H_2O) within the chloroplasts, producing oxygen molecules and hydrogen. Exactly how it does this we do not know, but we do know many of the results (see Figure 8.3). As stated earlier, some of the oxygen diffuses into the atmosphere. The hydrogen by itself is no good to the plant. It is attached to a complex organic compound, which acts as a short-term hydrogen **carrier.** This carrier holds the hydrogen until it is attached to carbon in the next phase of photosynthesis.

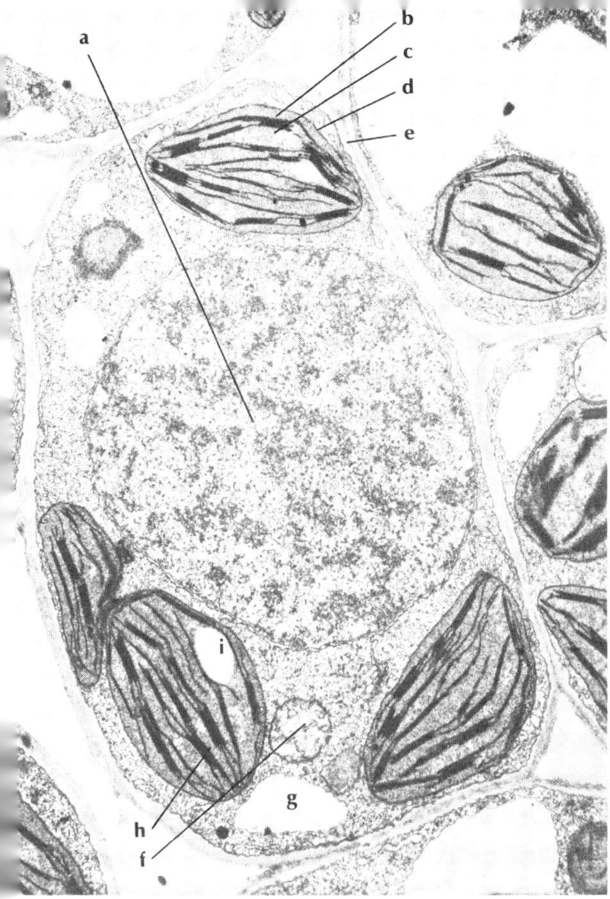

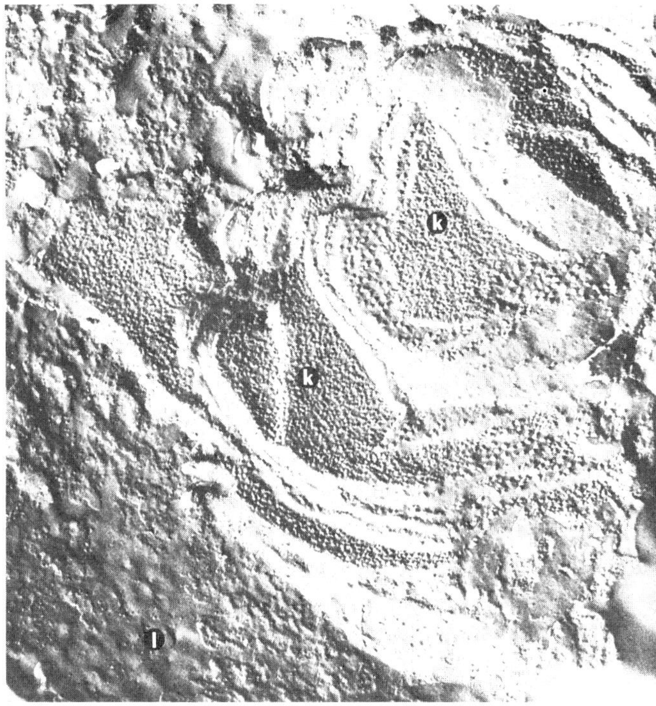

8.2 An electron microscope photo (left) of a section through the cell of a tobacco leaf illustrates the relative sizes of chloroplasts and other cell organelles, (a) nucleus, (b) chlorophyll-containing internal membrane system, (c) stroma, (d) chloroplast envelope membranes, (e) cell wall, (f) mitochondrion, (g) vacuole, (h) chloroplast, (i) starch, (j) intercellular space. The spinach chloroplast (right), (k) photosynthetic membranes inside chloroplast, (l) chloroplast envelope membranes) was frozen and split apart in much the same fashion as a block of ice is fractured when hit with an axe. Different membranes are characterized by the placement and size of particles. These are proteins embedded in layers of lipid. The chloroplast membranes can be broken and separated and the types of chemical reactions they perform correlated with the structure of the membranes.

The most important effect of this light-induced splitting of water, however, is that it produces energy that is used to bind chemicals together; when the chemical bonds of these compounds are later broken, energy will be released and be available to do further work. (For a review of why this is so, see the Supplement on page 25.) In photosynthesis the energy derived from splitting H_2O with light energy (the light reaction) is used to make adenosine tri-phosphate (ATP) from adenosine di-phosphate (ADP) found in the chloroplasts. ATP acts as a middleman, capturing and transferring energy; it is the common energy "broker" in plant and animal cells. There are other energy brokers, but ATP is the most important.

Dark Reactions The light reaction is the unique reaction of the chemical factory represented by the plant and is a chemical reaction that animal cells cannot do. We now know that plants need light so that chlorophyll can split H_2O and produce energy and hydrogen as fuel. The chemistry of life, however, is essentially based on the chemistry of carbon, which is the backbone of all organic compounds. A plant has many compounds based on carbon, among them cellulose, starch, and sugar. Such compounds also contain hydrogen, which is bonded to the carbon. The bonding, or

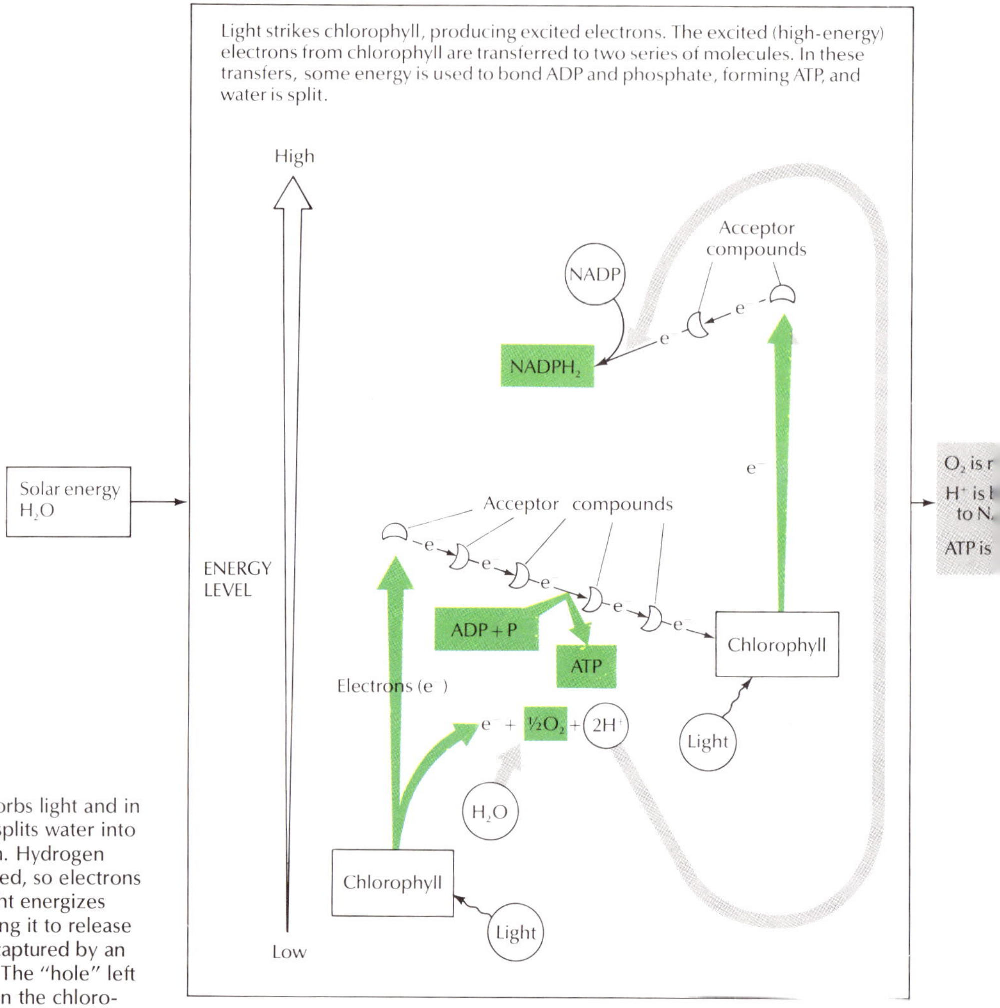

8.3 Chlorophyll absorbs light and in some unknown way splits water into hydrogen and oxygen. Hydrogen ions (H^+) are produced, so electrons (e^-) are available. Light energizes chlorophyll by inducing it to release electrons, which are captured by an acceptor compound. The "hole" left by the lost electrons in the chlorophyll molecule is filled by the electrons produced by splitting water. The electrons captured from chlorophyll are passed on by a series of acceptor compounds. In the process, energy is lost (Chapter 1, Supplement), but some is captured and used to form ATP. Before the electrons hit the bottom of the energy gradient, chlorophyll energized by light sends them back up to a higher energy level. Other acceptor compounds pass them on to another complex organic compound (NADP). They meet the hydrogen ions released from water and NADP is reduced by the addition of two hydrogen atoms. The hydrogen carried by NADPH will be used to reduce the CO_2 captured in the dark reaction of photosynthesis.

tying-together, of carbon and hydrogen molecules is the second part of photosynthesis. This is how it works:

Carbon dioxide dissolved in water diffuses into chloroplasts, which contain chlorophyll. Here an enzyme comes into play. **Enzymes** are organic compounds (proteins) found in all living cells; they have the peculiar property of being able to provoke chemical reactions without themselves being changed (see Figure 8.4). In short, they are organic catalysts.

The enzyme immediately accelerates the attachment of CO_2 to a sugar with five carbon atoms, producing a six-carbon sugar, which immediately breaks up into two three-carbon sugars. This primary product is called 3-phosphoglyceric acid (PGA). These simple sugars (PGA) then react with the hydrogen originally released by the splitting of water during the light reaction. One hydrogen atom is attached to a carbon in each PGA. Thus in the end, each carbon dioxide molecule absorbed by the leaf from the air has a hydrogen atom obtained from water attached to it.

Combining hydrogen with carbon uses energy. During photosynthesis the energy needed to make this combination is obtained from the energy broker adenosine tri-phosphate (ATP). Phosphate is broken off at the end of the ATP molecule by an enzyme, and some of the energy from breaking this bond is used to transfer hydrogen from the carrier compound, which captured it from water, to PGA containing the newly captured CO_2. Just as it takes energy to bond hydrogen with carbon, so too energy will be released if hydrogen later is removed from a carbohydrate, and that is why carbohydrates are fuels.

This **fixing** of CO_2 and adding hydrogen to it is called the **dark reaction** of photosynthesis, because it does not require light. All it needs is a source of transferable hydrogen and of energy in the form of ATP, both of which were supplied by the light reaction. Products of the light reaction are quickly depleted, so the total photosynthetic reaction is light-dependent.

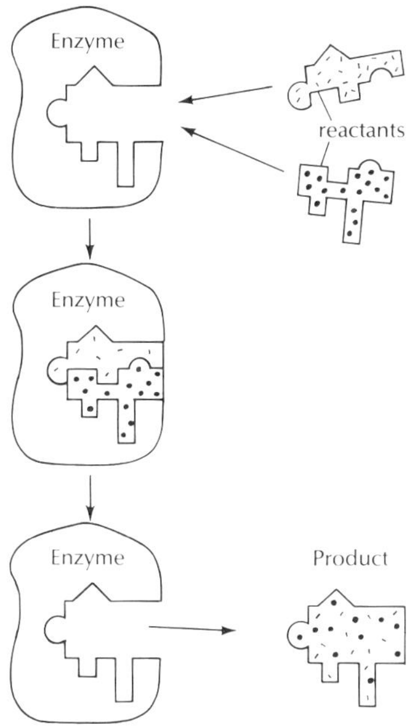

8.4 Enzymes are proteins that act as organic catalysts. This diagram illustrates the "lock and key" model of enzyme action.

Products of Photosynthesis An outline of the light reaction and these two phases of the dark reaction may help to review the steps in photosynthesis:

1. Light Reaction

$$\text{water} + \text{ADP} + \text{phosphate} + \text{X (hydrogen acceptor)} + \text{light} \longrightarrow O_2 + \text{ATP} + \text{XH}$$

2. Dark Reaction, part one

$$CO_2 + C_5\text{-compound} \longrightarrow 2C_3\text{-compounds (PGA)}$$

3. Dark Reaction, part two

$$2\text{PGA} + 2\text{XH} + 2\text{ATP} \longrightarrow 2C_3\text{-compounds-H} + 2\text{X} + 2\text{ADP} + 2\text{ phosphate}$$

Then the three-carbon compounds are used both to regenerate the five-carbon CO_2 acceptor and to make first glucose, a six-carbon compound, and subsequently starch. Thus, part of the carbon incorporated into the plant as a result of photosynthesis keeps the carbon-fixation cycle going and part is new fuel for the plant to use in growth.

Plants can rapidly convert the sugar synthesized during photosynthesis to other compounds involved in growth, but carbohydrates, especially sugars and starches, are usually the products that the plant stores. These, along with fats, are the reservoirs of energy in a plant; they are small "batteries" that received their charge from solar energy. When the chemical bonds of these compounds are broken, energy will be released and available to do further work. Related compounds like cellulose, though not utilized by the plant

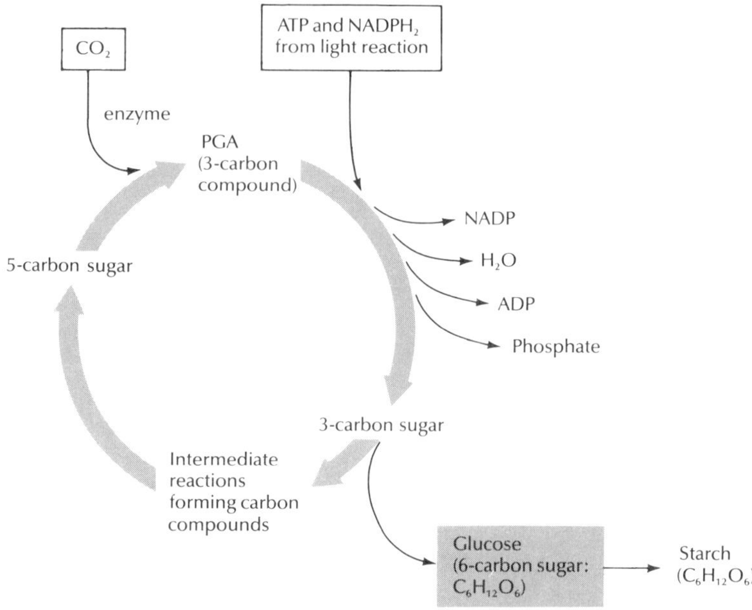

8.5 The dark reaction uses the ATP and hydrogen produced in the light reaction of photosynthesis to reduce the CO_2 captured from the atmosphere into a 3-carbon compound (PGA). This is how hydrogen is initially inserted into the food chain. As its name implies, light is not required for the dark reaction, but it would soon stop if ATP and hydrogen were not supplied by the light reaction.

as a source of energy, make up much of its mass and provide much of the biosphere's fuel because they are consumed by animals (Chapter 1). Our oil, coals, and natural gas are derived from these plant-produced compounds that were made by photosynthesis 300 million years ago and stored away in the earth.

Light and Plant Adaptations A single-celled green alga is immersed in water, in which gases and minerals are dissolved. Osmosis controls the cell's uptake of water; gases such as CO_2 diffuse in and out of the cell, and the plasmalemma regulates its uptake of minerals (see Figure 8.6). Since gases, water, and minerals are in the cell's immediate environment, it needs no special structures to acquire or transport them. Water is abundant in this environment, but CO_2, minerals, and light can be limiting. As will be seen in Chapter 9, minerals required for growth are generally limiting in the open oceans, but light is the major requirement for growth that is limiting in deep bodies of water.

Water is fully exposed to sunlight, but light can penetrate the water just so far due to absorption and scattering. The deeper the water, the less light there is, until finally it fades out entirely. That is why aquatic plants grow only in shallow water near the shore or on the ocean's surface. Deeper waters have no plants at all because light is completely absorbed by surface water. Most of the ocean, therefore, has little plant life. Even the giant kelps, which grow almost as tall as the biggest trees on land, are restricted to the shallower waters of the continental shelves.

Single-celled algae must have all the machinery to carry out photosynthesis as well as to utilize the products of photosynthesis for growth. But what about land plants? A major theme in their evolution has been their exploitation of light. The first land plants were really algalike, and like the algae, they lived where water was abundant. As a result of their adaptations to land, their photosynthetic organs are no longer in water; these organs occur in the plants' aerial parts, those aboveground, and so are in direct contact with the light. They are concerned primarily with light absorption and with the exchange of gases, particularly CO_2 and O_2. The parts below ground live in an environment much like the one-celled alga, which is probably similar to the first green plants. Roots take up both water and minerals in the soil and transport them to the aerial parts, which send back to the roots the food they need to grow.

This division of labor has been a major factor in the development of the great diversity of land plants, and one that seems to have some disadvantages. Since absorption of light and exchange of gases are required for photosynthesis, and since air in most environments has relatively little water in it, the aerial parts of plants have a special problem. How do they absorb gases without having their cells dry up? The way the plant conducts this balancing act is discussed in Chapters 10 and 15.

A Tool for Biochemical Investigations We have seen that the light reaction splits water molecules and provides the hydrogen and energy for the subsequent synthesis of carbohydrates. It is the unique part of photosynthesis, and it is still very much of a mystery. In spite of many years of intensive effort, we still do not know how light and chlorophyll combine to break down water, though a lot is known about what happens afterward. Once this great challenge of biology is overcome, our understanding of nature and our chances of meeting the world's need for energy will improve greatly.

Much more is known about the dark reaction because of a breakthrough permitting investigation of the path carbon follows in photosynthesis. This investigation was made possible as a result of atomic research previous to and during World War II, which made available radioactive isotopes of elements. In the late 1940s, Melvin Calvin and coworkers at the University of California at Berkeley used a radioactive isotope of carbon, carbon 14, to trace CO_2 during photosynthesis. They put plants in an atmosphere containing CO_2 gas made with carbon 14, and they killed leaves or green algae cells at selected times after exposure to $^{14}CO_2$. They separated compounds extracted from the plant cells by using another new technique, paper chromatography. When a small amount of plant extract is placed on filter paper, and one end of the filter paper is placed in a solvent, compounds will be distributed at different

8.6 LIGHT AND PLANT ADAPTATIONS

Major features in the evolution of land plants have been adaptations for the absorption of light energy and for the protection of photosynthetic organs from dessication. Single-celled algae (below) are not limited by water but they can only obtain adequate light near the surface of water. Liverworts (below) live on damp soil and only the surface exposed to air has a waxy cutin layer. Vascular plants (facing page) have aerial parts, adaptations for capturing light energy, but waterproofing, as seen in the waxy cutin layer of the liverwort, is the price paid. Photosynthesis can only take place when they can transport enough water to their leaves. The graph (facing page, bottom right) illustrates that photosynthesis for all green plants is regulated by light intensity and concentration of carbon dioxide.

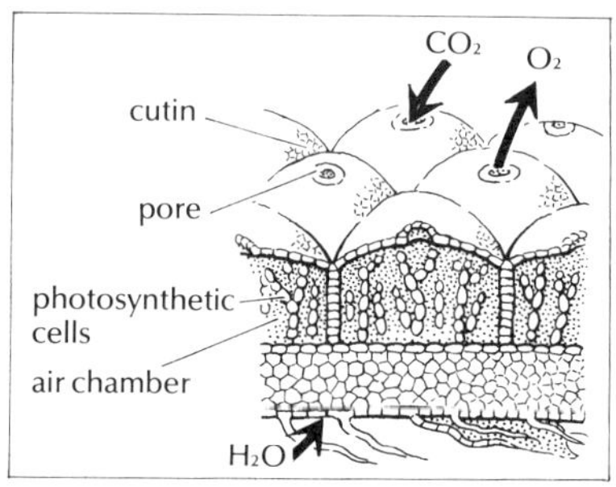

Liverwort

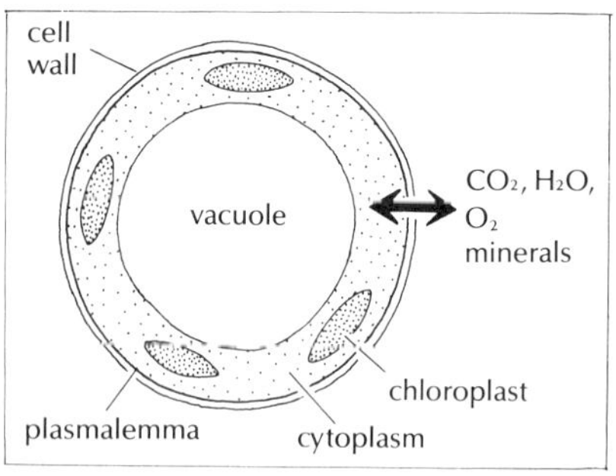

Single-celled alga

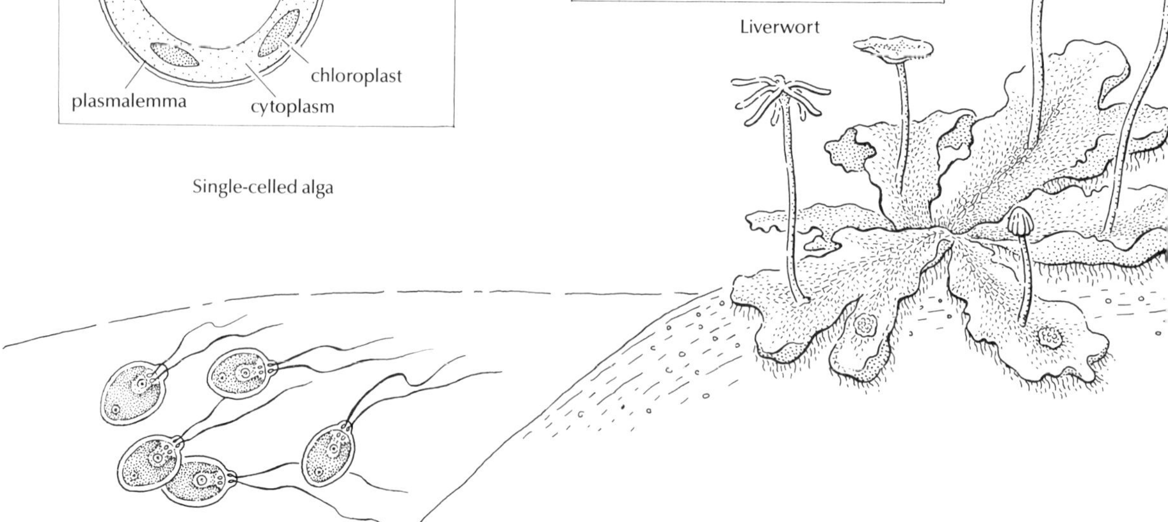

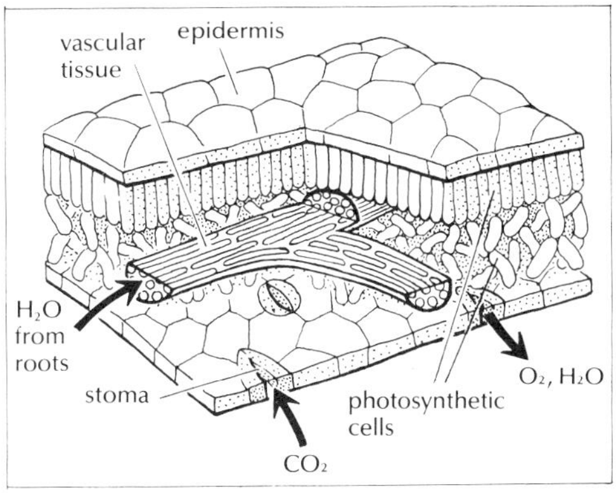
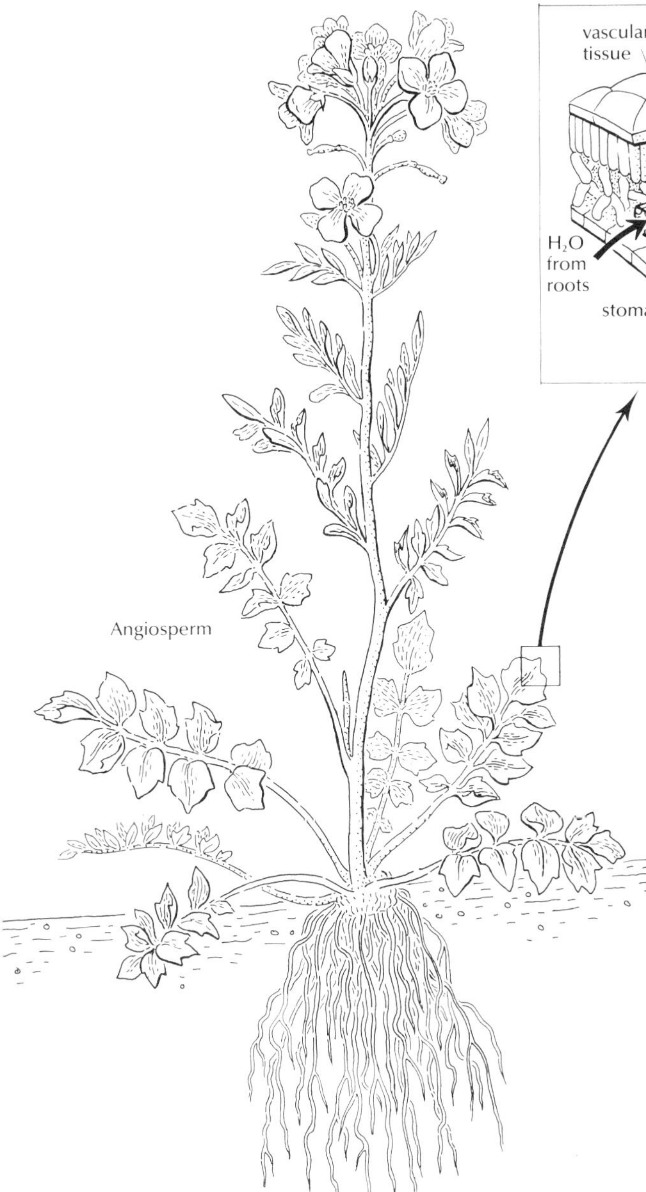
vascular tissue
epidermis
H_2O from roots
stoma
CO_2
photosynthetic cells
O_2, H_2O
Angiosperm

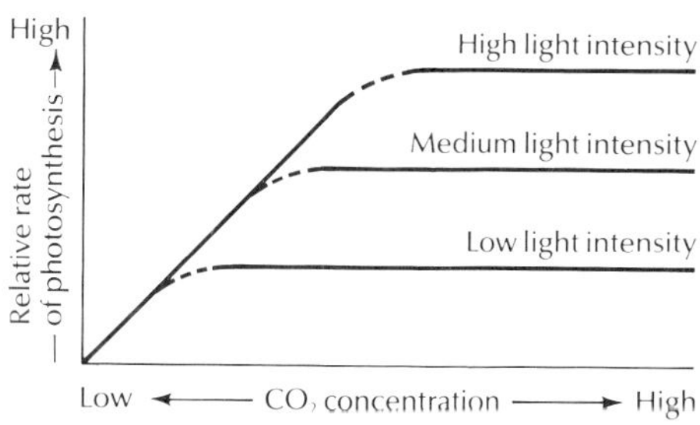
High
Relative rate of photosynthesis →
High light intensity
Medium light intensity
Low light intensity
Low ← CO_2 concentration → High

places on the paper depending on their solubility in that solvent. About the same thing happens to a drop of ink on a blotter; the ink pigments separate as the solvent spreads.

Compounds containing carbon 14 can be located by exposing the dried filter paper to sheets of X-ray film. The radioactive isotope exposes the film, by giving off electrons that bombard the emulsion, and the dark spots on the developed film are used to determine where the radioactive compounds are on the paper. The separated compounds can be removed from the paper and identified by chemical techniques. In this way, Calvin's laboratory had traced the main steps in the path of carbon during photosynthesis by the middle 1950s, and for his work he was rewarded with the Nobel Prize.

Radioactive isotopes are now essential in biochemical research. The pathway of carbon during cell respiration, discussed in the next section, was found by using glucose containing carbon 14. After feeding cells with ^{14}C-labeled glucose, the compounds in cell extracts that contain carbon 14 can be identified, and the rate at which carbon is transferred from glucose to CO_2 can be determined. When physicists first began to break atoms apart, they had some idea of the destructive potential of their research, but they did not foresee that their work would also become essential for the study of life, and would contribute so much to its understanding.

RESPIRATION

The sugars formed in photosynthesis are storehouses of energy. The plant must release the chemical energy stored in these compounds bit by bit to do the work of living. Plants (and every consumer in the food chain) do this by "burning" the hydrogen in these compounds, releasing CO_2 and heat to the atmosphere and making energy available for work. (See the Supplement on page 25 for a review of how hydrogen burning releases chemical energy.)

The process must begin, we have seen, with photosynthesis, in which carbon is **reduced** (combined with hydrogen) when the sugar glucose is formed. Remember the general expression*

$$CO_2 + H_2O + \text{energy} \longrightarrow [CH_2O] + O_2$$

When we "burn up" energy or "fuel" our bodies, we are **oxidizing** (removing hydrogen from) the compounds based on carbon and hydrogen. This burning is essentially the reverse of photosynthesis:

$$[CH_2O] + O_2 \longrightarrow CO_2 + H_2O + \text{energy}$$

This process, known as **cellular respiration,** occurs in all cells.

* $[CH_2O]$ is a generalized formula for a carbohydrate—carbon + water.

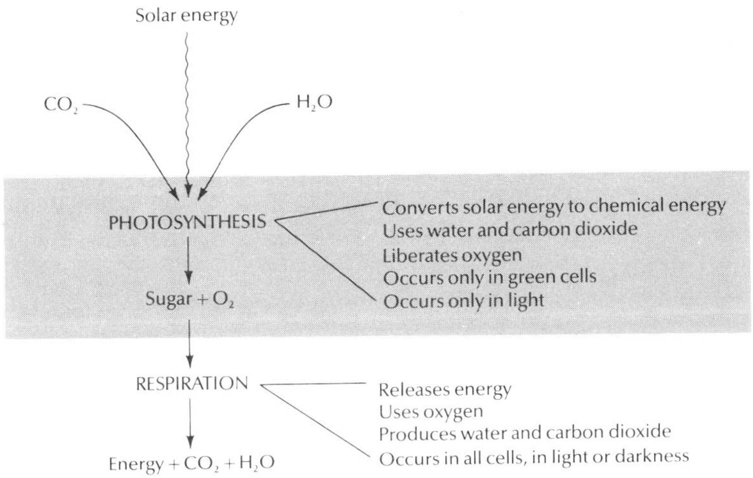

8.7 Both photosynthesis and respiration provide energy for cells. Solar energy is converted to chemical energy during photosynthesis. This chemical energy is made available to cells during respiration and some is transformed to heat. The carbon and hydrogen joined together during photosynthesis are split apart during respiration, and CO_2 and H_2O are available as raw materials for photosynthesis.

The energy in the expression is both heat and chemical energy. The chemical energy is ATP, which was produced in the light reaction of photosynthesis and used in the dark reaction to reduce carbon—to bond the hydrogen and carbon (from CO_2 in the air) and form sugar. Now hydrogen is removed from the sugar, oxidizing the carbon and releasing energy, which is used to make ATP.

Heterotrophic (other-feeding) organisms obtain all their energy by burning organic compounds to produce ATP; then ATP supplies energy to make carbohydrates, proteins, and fats; to fuel cell division; to transport materials through membranes; and to power many other processes in cells. Plants, too, carry on all these processes, plus photosynthesis.

Fermentation Respiration is a complex process, involving about thirty steps. It shares many of its first steps with **fermentation,** an energy-creating process that does not require oxygen; it is also known as **anaerobic respiration.**

Fermentation takes place in the cell's **cytoplasm** (all the cell's protoplasm outside the nucleus). In this process, glucose is broken down into two molecules of a three-carbon compound called **pyruvic acid,** and some of the energy obtained during this partial burning of glucose is used to synthesize ATP. Depending on the organism, further reactions occur: in animal muscle cells, lactic acid ($C_3H_6O_3$) is produced from pyruvic acid, ending fermentation; in yeast cells the end comes when ethyl alcohol (C_2H_6O) and carbon dioxide are made from pyruvic acid.

Both these products include carbon and hydrogen, meaning that the burning of glucose is incomplete. (And there lies the explanation of why alcohol is not the weight-watcher's friend.) When pyruvic acid is formed, only 10 percent of the energy in glucose has been

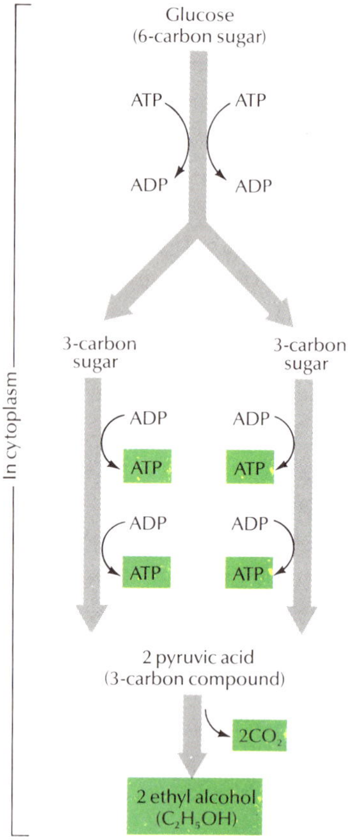

8.8 The breakdown of sugar to pyruvic acid takes place in the cytoplasm and does not require oxygen. Very little energy is obtained in the process. The end product of respiration can be alcohol. In such cases, anaerobic respiration is called *fermentation.*

released, and only 2 ATPs per molecule of glucose are produced. This is in contrast to 38 ATPs formed during the complete process of respiration where CO_2 and H_2O are the end products.

Many bacteria and some fungi such as yeast (discussed in Chapter 4) can live without oxygen or with very little, and for this reason they are called **anaerobic organisms.** The wine, beer, chocolate, soya, and bakery industries all depend on fermentation. In making wine, yeast cells (which are present on grapes or sometimes added) break down sugar in the grapes' juice to alcohol and carbon dioxide and obtain energy for their own growth. Fossil fuels too are partly a product of fermentation: they were produced from plants long covered by water or mud where oxygen could not get to them. Larger molecules in this buried vegetation were broken down by anaerobic organisms, but the basic hydrocarbons remained. Though later modified by heat and pressure, the energy left in them by this incomplete combustion is still available to supply us with heat, electricity, and power for transportation.

Aerobic Respiration The fossil fuels like petroleum probably will someday be replaced in part by ethyl alcohol, which is produced by fermentation. It might, for example, replace gasoline in internal combustion engines. Burning ethyl alcohol is similar to what happens in **aerobic respiration.** In both, the products obtained from fermentation are burned further, producing CO_2, H_2O, and energy. They are very different, however, in that burning alcohol releases energy all at once as heat; aerobic respiration in cells releases energy in small increments. These small amounts of energy are employed in the synthesis of ATP.

Aerobic respiration takes place in **mitochondria** (except in bacteria and blue green algae, which do not have mitochondria), sausage-shaped organs much smaller and more numerous than chloroplasts. The mitochondria, floating freely in the cytoplasm, are truly the powerhouses of the cell (see Figure 8.9). Here pyruvic acid loses a molecule of CO_2 and forms a two-carbon derivative of acetic acid ($C_2H_4O_2$). This two-carbon compound is the endpoint for all the sugars, fats, and proteins in the cell.

The two-carbon compound is picked up by a four-carbon carrier compound called oxalacetic acid and a six-carbon compound results. In the course of a complex metabolic cycle, two CO_2 molecules are released, and the four-carbon oxalacetic acid is regenerated. The cycle is frequently called the Krebs cycle after Sir Hans Krebs, the British biochemist who received a Nobel Prize in 1953 for discovering it. Now all the carbon from pyruvic acid has gone and all the CO_2 initially incorporated from the atmosphere during photosynthesis has been released and the carbon cycle is complete. Pyruvic acid is thus the fuel remaining from fermenting glucose, and it is thrown into the cellular furnace by the oxalacetic acid, which is

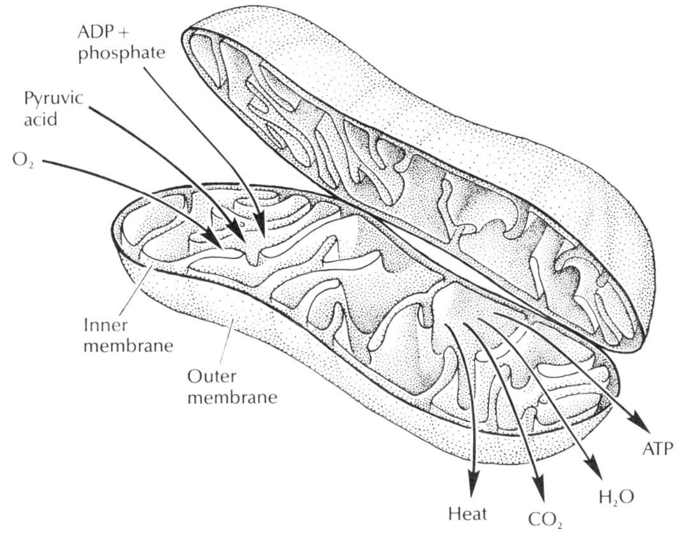

8.9 Mitochondria are filled with an enzyme soup. The reactions of the Krebs cycle take place in this soup in spaces between the internal membranes. The series of oxidation-reduction reactions, during which most of the energy released in respiration takes place, occurs on the surfaces of these convoluted internal membranes. In all organisms except bacteria and blue green algae, carbon is returned to the atmosphere by the mitochondria.

constantly being regenerated. The latter compound, oxalacetic acid, may be thought of as the fireman feeding the cellular furnace with remnants of glucose.

During the Krebs cycle, in which CO_2 is released, hydrogen atoms are stripped from the participating compounds and acceptor compounds are reduced. A series of oxidation-reduction reactions takes place on the surfaces of the internal membrane of the mitochondria. Energy is released and some is used to make ATP and some is lost as heat. The final oxidizing reaction combines molecular oxygen with hydrogen to form water. The cycle that began when water was split during photosynthesis is thus completed when water is re-formed as the last reaction in the step-by-step burning of sugar. (For a summary of the process, see Figure 8.10).

Oxygen fires this cellular furnace. If the oxygen supply is cut off, there is no terminal receptor for hydrogen, and the chain of oxidation-reduction reactions producing energy for most of the ATP generated during the burning of glucose will come to a halt. Aerobic organisms soon die of suffocation, because fermentation provides too little energy to sustain them.

The effect of oxygen on energy production can be seen every time you watch a race like the 400-meter dash. This race is usually run with such all-out effort that near the end it is common for the leg muscles to be completely out of oxygen. When the oxygen is gone, the cells begin to carry out fermentation and produce lactic acid, which inhibits the muscles' action and causes crippling cramps. In order to give their cells extra oxygen, athletes often used to breathe pure oxygen before a race, a practice now forbidden in competition. The oxygen they breathe combines with hydrogen to form H_2O and

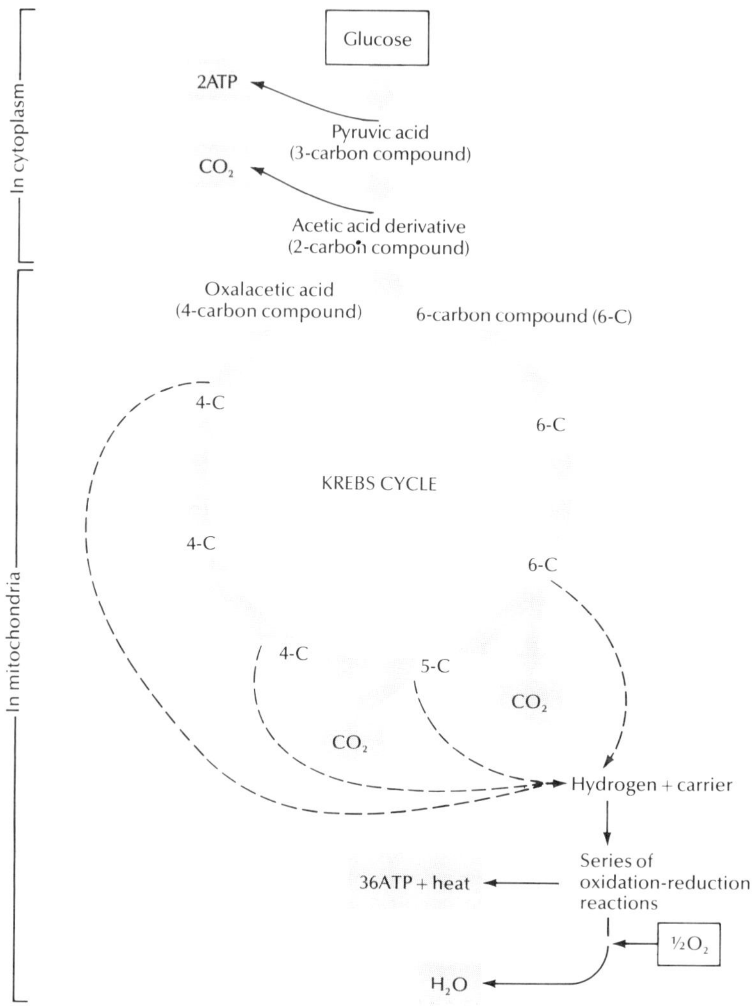

8.10 Aerobic respiration takes place in the mitochondria. In this cellular furnace, all the carbon remaining from glucose is released as carbon dioxide. During the metabolic cycle that accomplished the release of CO_2, hydrogen is knocked off the carbon compounds and, in the form of hydrogen ions (H^+) and electrons (e^-), is passed along a complex series of oxidation and reduction reactions. ATP is formed during this burning of hydrogen and the reactions end with the formation of water. ATP formed in this manner supplies most of the energy for the activities of all heterotrophic organisms and tissues.

ATP: the ATP powers the contraction of muscle fibers that provide the leg motion in running.

Each mole of H_2O formed in the mitochondria yields 52 kilocalories of energy and three moles of ATP are produced. Formation of a mole of ATP requires about 8 kilocalories so 24 kilocalories have to be captured by the cell for every mole of H_2O that is formed:

$$\text{Formation } H_2O \longrightarrow 52 \text{ kcal} + 3\text{ATP}$$
$$\text{Formation of 3ATP requires 24 kcal}$$

The other 28 kilocalories are released as heat. Respiration then is almost 50 percent efficient:

$$24 \text{ kcal} \div 52 \text{ kcal} = 0.46 \times 100 = 46\%$$

Half the energy released by respiration is lost as heat. The heat given off by our own bodies is produced by mitochondrial respiration just as in plants. Yet plants do not feel hot. We pictured the plant cell's structure as a bag of water wrapped in a film of cytoplasm and bathed in a thin stream of water. Heat is produced very diffusely because all that water acts as a heat sink (it absorbs heat) and the thin leaves, like vanes of a radiator, radiate heat to the air. Heat, however, can be measured quite easily in plants. Germinating seeds, for example, respire very rapidly and the heat they produce is easy to measure and feel if we put a large number of seeds in an insulated container. Skunk cabbage *(Symplocarpus foetidus),* a common large-leafed monocot of marshes and wet places, is well known for the heat it produces. The large lilylike flower has a central column bearing the sexual organs, and for a short time when the flower is ready for pollination, respiration is so intense that this structure is warm to the touch. When you take a picture of this or any plant with film sensitive to infrared light, you will obtain an image of the plant just as you would from any heat-generating body, such as a poorly insulated building or an animal.

THE CARBON CYCLE

The mitochondria, we have seen, are the powerhouses of the cell; carbohydrates are the fuel for the powerhouses, which produce ATP; and ATP provides the energy for much of the work that has to be done in cells. As the site where CO_2 is released in green plants, fungi, and animals, mitochondria are also important as indispensable components in the **carbon cycle** which is summarized in Figure 8.11. At each step in the food chain, respiration releases CO_2, making it available for recycling.

In a constant exchange, carbon is carried from air or water into green plants and into organisms that ingest the plants, then back into the atmosphere following respiration and back to plants again. Most carbon enters the food chain through the stomata ("pores") in

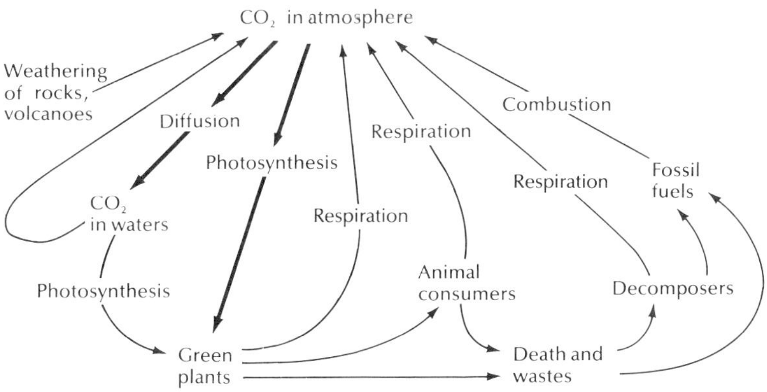

8.11 The food webs of the biosphere trace the pathway of carbon. As carbon compounds pass through every cell, some is returned to the atmosphere and some is passed on to fuel other organisms and cells. This rapid turnover of carbon maintains a relatively constant but small pool of CO_2 in the atmosphere. The rapid return to the atmosphere of carbon bound in fossil fuels is an event of our time.

the leaves of plants. Most CO_2 is released by soil microbes that decompose plant and animal litter. If all decomposition were to stop right now, we would run out of CO_2 in the atmosphere in twenty years. And with no CO_2, photosynthesis would end.

The rate at which CO_2 is released varies over the world. Because high temperatures and abundant water speed up decomposition, the breaking-down of organic matter by microbes is most rapid in the tropics and much slower in temperate forests—a fact that can lead us into some speculation. If all the world were cool, temperate, coniferous forest, we might soon find that not enough CO_2 was available for recycling. For example, 1 acre of Douglas fir forest fixes 6 tons of CO_2 and one large tree contains as much CO_2 as 20 acres of air. Where cool winters, dry summers, and acid soil inhibit decomposition by soil organisms, all this CO_2 can be tied up in the trees for a long time; a large dead tree in the California mountains may not decompose for a hundred years or more. As a result, a lot of fuel—dead trees—accumulates in temperate forests, feeding the devastating fires that are their characteristic. In contrast, the turnover of organic matter in wet tropical forests is very rapid.

Carbon atoms are cycling and recycling without cease between the atmosphere and the organic compounds produced by living cells, but the CO_2 in the atmosphere does not fluctuate as much as we might expect. It remains fairly stable at 0.032 percent everywhere, because of mixing by wind and because of the oceans, which cover 70 percent of the earth's surface. Carbon dioxide is dissolved in oceanic waters, which help regulate the quantity in the atmosphere. The amount of gas dissolved in water depends on temperature: the warmer the water, the less gas dissolved. A rise of one degree in ocean temperature would increase the CO_2 level in the air everywhere. Assigning causes to changes in atmospheric gases is difficult because of these relationships, but evidence is building that there are significant increases in CO_2 levels in the atmosphere, presumably as a result of the burning of fossil fuels. There is considerable debate on the implications of this phenomenon.

Ocean temperature may well have been higher in the late Paleozoic era, some 300 million years ago. More CO_2 would therefore have been available to be fixed by terrestrial plants, perhaps generating the luxuriant tropical forests that then covered much of the world. Atmospheric concentrations of CO_2 are generally limiting to photosynthesis. Plants photosynthesize much better if the CO_2 concentration is increased as much as ten times that of the atmosphere. Increasing concentrations of CO_2 is common practice in greenhouses, producing plants and flowers bigger than those you can grow in your garden. Unfortunately, this technique is not practical for crops grown on a large scale.

Whatever their cause, the Paleozoic forests provided the raw ma-

8.12 The coal forests of the Carboniferous period of the Paleozoic era formed our fossil fuels. The prolific primary production of that time might have been due to a higher CO_2 level in the atmosphere. Plants grow better at elevated CO_2 concentrations, and it is possible that the addition of CO_2 to the atmosphere as a result of burning fossil fuels will stimulate photosynthesis.

terial for the fossil fuel we burn today. We are now completing the carbon cycle for this organic material, in which decomposition and release of CO_2 was begun but not completed long ago.

Summary

Plant and animal cells perform many of the same chemical reactions, but only green plant cells can photosynthesize, that is, transform solar energy into chemical energy. The distinctive phase of photosynthesis is the light reaction: chlorophyll in the chloroplasts absorbs sunlight; with the energy absorbed, water is broken down into hydrogen ions and oxygen. Oxygen is then released into the atmosphere; the hydrogen ions are picked up by an organic compound and subsequently passed along to carbon compounds during the dark reaction. Some energy from the light-induced splitting of water is used to form ATP (adenosine tri-phosphate) from ADP (adenosine di-phosphate).

The succeeding steps are known as the dark reactions because light is not needed. Carbon dioxide entering leaf cells from the atmosphere is attached to a five-carbon sugar and the resulting six-carbon compound breaks up into two three-carbon compounds (PGA). ATP supplies the energy to transfer hydrogen from the carrier compound to PGA, which serves as the building block for sugars and starch. These dark reactions have been determined by using radioactive isotopes.

The availability of light can be an important limiting factor for plant growth because light is necessary for photosynthesis. Oceans, for example, have plant life only on the surface. The evolution of land plants resulted in photosynthetic structures that are in the air, and hence in direct contact with light, but as a result they are in danger of drying out. Balancing the needs for light and for water are constant problems for land plants.

The energy taken in, transformed, and stored by photosynthesis must be released in small steps to be used by the plant. Respiration is the process releasing this energy. Most simply, in respiration hydrogen in the sugar produced by photosynthesis is burned; the carbon combines with oxygen to form water. CO_2 and heat are released; and energy in the form of ATP is made available to the organism. In fermentation or anaerobic respiration, no oxygen is used and the burning is incomplete, so less energy is released by these processes than by aerobic respiration. Lactic acid, pyruvic acid, or ethyl alcohol may be by-products of fermentation. Aerobic respiration, carried on in mitochondria in plants, fungi, and animals, involves the final burning of the products of fermentation. In a complex process called the Krebs cycle, CO_2 is released and hydrogen is removed from carbon compounds. In the course of a series of oxidation-reduction reactions ending in the combining of hydrogen with water, energy is released. It is either lost as heat or is used to make ATP.

Aerobic respiration produces almost twenty times more ATP than fermentation. Aerobic respiration, however, is only about 50 percent efficient, for approximately half the energy released goes not into the formation of ATP but is lost as heat.

Respiration completes the carbon cycle begun when CO_2 was absorbed by green plant cells. Most CO_2, however, is returned to the atmosphere not by respiration of plants and animals, but by that of microbes. This release of CO_2 by microbial decomposition is most rapid in the tropics as a result of their high temperatures and abundant water. Without CO_2 photosynthesis ends. The amount of CO_2 in the atmosphere remains fairly stable because of the effects of wind and ocean, but it is thought that more was available in the Paleozoic era, generating luxuriant tropical forests. When the concentration of CO_2 is increased, the rate of photosynthesis usually increases as well. Carbon dioxide appears to be increasing in our atmosphere but its causes and effects are a matter of debate.

Review Questions

1. Plant and animal cells have very similar chemical reactions, but the light reaction is unique to those cells with chlorophyll. Why?
2. The basic chemical apparatus for the dark reaction of photosynthesis is

possessed by most heterotrophes. Why is it that only green plants are autotrophic? How do they link the light reaction with the dark reaction?

3. A forest of giant marine kelps and a forest of Douglas fir trees are similar in height. There are plants at different levels beneath the Douglas fir, but this does not happen in the kelp forest. Why?
4. Why is carbon-14 a basic tool of biochemists?
5. Carbohydrates are the main products of photosynthesis. Where do plants obtain the elements that constitute carbohydrates?
6. Why is it impossible for a large organism to exist in an anaerobic environment?
7. A newly discovered organism is examined with an electron microscope and found to possess mitochondria in its cells. What will the discoverers be able to say about the organism's environment? Can this organism carry out fermentation?
8. Some organisms, such as kangaroo rats, obtain almost all their water from so-called metabolic water. Their diet is dry seeds. Where does this water come from?
9. The National Association for the Protection of the Carbon Cycle enthusiastically reports all forest fires in its newsletter. Can you suggest a rationale for this unconventional enthusiasm?

SUPPLEMENT

Plastids

Plastids are self-replicating organelles found in all living plant cells. Plastid species vary greatly in form and function, but all are descendants of proplastids found in meristematic cells. Proplastids, which also come in many shapes, are small, often with a diameter of a micrometer (μm) or less, so they, like plastids, must be studied with an electron microscope. All plastids and proplastids are enclosed by two membranes. Mitochondria are also surrounded by two membranes, and they are close to proplastids in size, but the structure of proplastids differs from that of mitochondria in several ways. Proplastids lack the invaginations, or intrusions of the inner surrounding membrane into the internal matrix, characteristic of mitochondria, and proplastids frequently have starch grains. In fact, all starch, whether in a leaf or in a potato tuber, is formed in plastids. When meristematic cells divide and enlarge, proplastids also divide, by pinching into two parts in a manner similar to the fission of bacterial cells.

Seedlings grown in the dark are etiolated (Chapter 6), and their white or pale yellow leaves contain a species of plastid called an **etioplast.** As leaf cells enlarge, their proplastids enlarge; older leaves of an etiolated plant will have etioplasts 3–6 μm in diameter. As etioplasts enlarge, they syn-

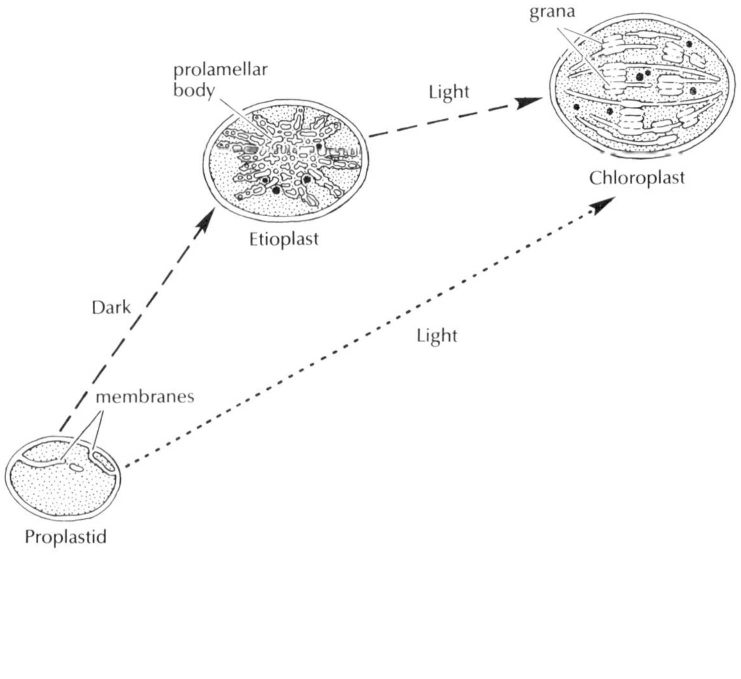

1 All plastids originate from small, self-replicating organelles in meristematic cells called *proplastids.* If the plant is grown in the light, proplastids grow in size, form complex internal membranes with chlorophyll, and become chloroplasts. If tissues or plants are kept in the dark, proplastids develop into etioplasts, which do not have chlorophyll. Light transforms etioplasts to chloroplasts.

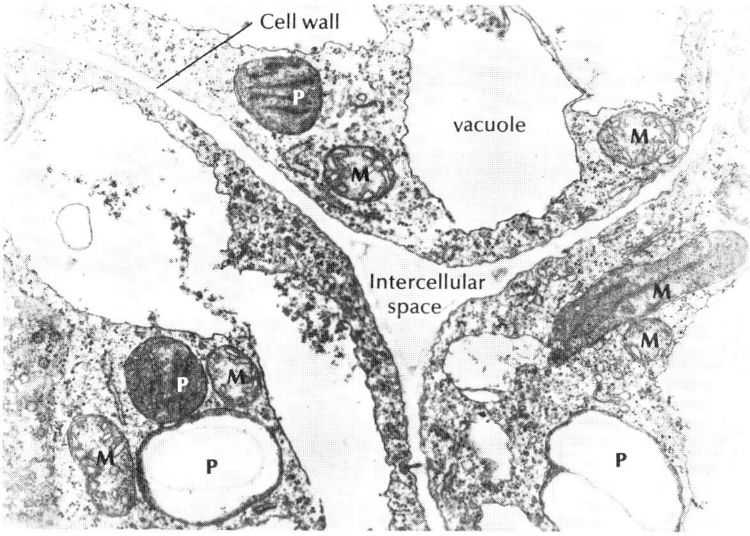

2 Sections of proplastids (P) in plant meristem cells can be identified by membrane sheets or starch grains. Sections of mitochondria (M) are also visible. This section of cells in the basal meristem of sugar cane leaves was photographed with an electron microscope.

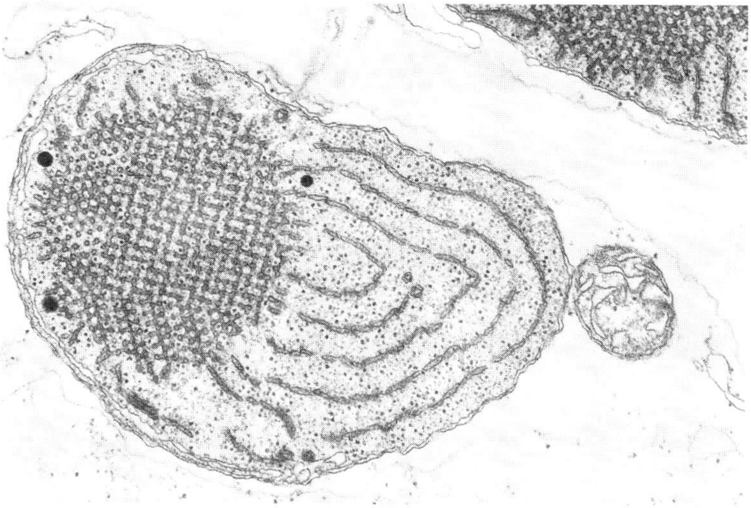

3 Etioplast. Developing from proplastids when leaves develop in the dark, etioplasts are much larger than proplastids and feature a complex arrangement of internal membranes called the *prolamellar body*. When exposed to light, this body disappears as the etioplast is transformed into a chloroplast. This is a section of an etioplast from a dark-grown sugar cane leaf.

thesize internal membranes that look like flattened sacs. The characteristic internal structure of etioplasts in fully expanded leaf cells is a crystalline arrangement of internal membranes called the **prolamellar body.** When etiolated plants are exposed to light, etioplasts will begin to form chlorophyll immediately and within twenty-four hours the plant will look like a normal green plant (except for its spindly appearance described in Chapter 6). During this greening, the leaf cells will grow, the immature chloroplasts will replicate, membranes will be rearranged, and new membranes will be synthesized. As the etioplast is transformed into a chloroplast, it increases in size and proteins are synthesized in the stroma of the chloroplast. These proteins are the enzymes involved in the dark reaction of photosynthesis. Angiosperms do not form chlorophyll in the dark, but all other green plants do. The reason for this difference is unknown.

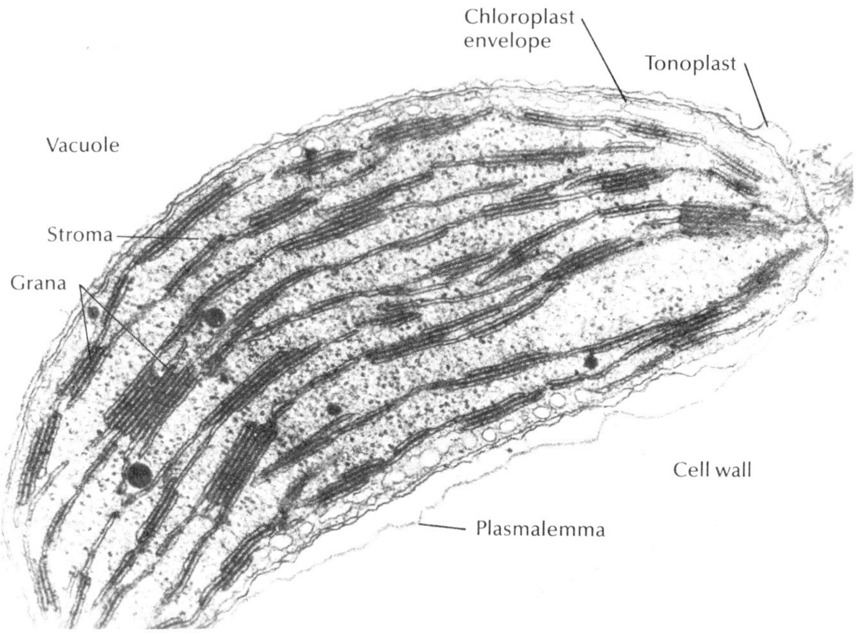

4 Fully intact and functional chloroplasts can be isolated from leaves. Much of our knowledge about photosynthesis has resulted from work with isolated chloroplasts. This electron microscope photo of a section through a sugar cane chloroplast clearly shows the internal membranes where the light reaction takes place. The grana are stacks of flattened sacs, and all the grana are interconnected. The dark globs are lipid droplets. The dark spots in the stroma are ribosomes (see Chapter 11), important in the synthesis of proteins.

All living cells of plants have plastids. Root cells have small plastids with few internal membranes. These plastids usually turn green when the root is exposed to light. Even a carrot will turn green when placed in the light, and the same is true of other underground organs, like tubers, which contain large starch-storing plastids. Cut a potato in half and leave it in the light and its interior will turn green. Cells in plant stems are often green, due to chloroplasts in the cortical cells. Strip the bark off a woody plant, and a green layer will be exposed. This is the cambial zone whose cells have chloroplasts. Whether the cells are from the pistil, ovary, seed endosperm, or leaf epidermis, plastids are present. Given the right environment, most plastids can become chloroplasts, plastids containing chlorophyll.

Chloroplasts vary greatly in size and shape. They range from disk-shaped objects 5–10 μm in diameter, found in many land plants, to long, folded, strap-shaped objects, found in the cells of some filamentous green algae. While many chloroplasts store starch, some store fat globules and other proteins in the form of crystals. Leaf cells of angiosperms typically have fifty to a hundred chloroplasts that are about 10 μm in diameter. They move within the streaming cytoplasm, and are usually arranged along the cell walls like plates next to a wall. While leaf chloroplasts are frequently disk-shaped, they do change shape and form protrusions and knobs.

The life of plastids in a developing leaf exposed to light is quite different from that of plastids in an etiolated plant. The leaf primordium has proplastids that replicate as the cells of the leaf meristem divide. As the leaf cells enlarge, proplastids develop into chloroplasts and continue to replicate until cell growth is completed. Development into chloroplasts consists of enlargement and the formation of the internal membrane structures characteristic of chloroplasts. Every plastid is surrounded by two membranes, called envelopes; the internal membranes are derived from the inner envelope. These inner membranes form flattened sacs (think of deflated air mattresses) that extend across the plastid. These are the same sacs formed in etioplasts, but in the next step chloroplasts differ. In places, these flattened sacs are folded in a complex manner to form stacks of closely adhering membranes resembling a stack of coins. These stacks of membranes are called **grana** (singular, granum). All internal membranes of chloroplasts have chlorophyll and carry out the light reaction of photosynthesis. Most land plants have chloroplasts with grana, but some algae, photosynthetic bacteria, and blue green algae do not. It is obvious, therefore, that grana are not necessary for photosynthesis. Their adaptive significance is a matter of speculation.

Most work on chloroplast membranes has been devoted to the internal, chlorophyll-containing membranes, but in the past few years techniques have been developed to isolate the envelope membranes and to study their structure and composition. It has been found that the inner envelope is the one that is selectively permeable: it regulates what goes in and out of the plastid. This function is reflected in its structure and protein composition, which are different from those of all other membranes of the chloroplast.

There are many other variations in the structure and function of chloroplasts, but space only permits a brief discussion. One other species of plastids, however, is very much a part of your world. This is the **chromoplast** (colored plastid). Every time you eat a tomato, peel an orange, or gaze upon a daffodil, you are involved with chromoplasts. The fate of all chloroplasts (and of all plastids except for those in apical meristems) is death. In dying, however, many chloroplasts exhibit a flash of yellow, orange, or red. The young, green tomato ages and ripens (Chapter 7), and as its cells senesce, their chloroplasts lose chlorophyll and protein and synthesize carotenoids, which have a yellow to red color, depending on the tomato. The internal membranes of the chloroplast become greatly reduced in size, and the chromoplast is filled with carotenoid in the form of platelike crystals. A similar sequence takes place in the rind as an orange ripens (even those without added color), except that the carotenoid is orange and is dissolved in large globules of fat within the chromoplast. The color of many flowers and autumn leaves is also the result of a transformation from chloroplast to chromoplast characteristic of senescence in these plants. Like the life of many organisms, that of plastids is a fascinating and still little understood series of events.

CHAPTER 9

Plants and Minerals

Through photosynthesis and respiration, the plant fulfills its energy requirements. Some of the sugar it made from solar energy is transported to the roots, where it is burned to produce energy. The plant uses some of this energy to obtain minerals—another necessity for its survival—moving them from the soil solution across the cell membranes of root hairs into root cells. The scope of plants' activities in extracting and moving minerals is incredible; by comparison, human activities in mining look small.

MINERAL REQUIREMENTS

All the complex organic compounds making up a plant are built from chemical elements. To determine which elements plants require and in what amounts, they have been grown in distilled water to which known elements are added. From such experiments botanists have compiled the following list of required elements:

- *Macroelements:* 99 percent of the plant's dry weight (its weight if all water is removed) consists of carbon (C), hydrogen (H), and oxygen (O). Significant quantities are also required of nitrogen (N), phosphorus (P), potassium (K), sulfur (S), calcium (Ca), and magnesium (Mg).
- *Microelements* or *trace elements:* Iron (Fe), copper (Cu), manganese (Mn), zinc (Zn), molybdenum (Mo), boron (B), cobalt (Co), and

chlorine (Cl) are needed only in small amounts. Some plants also require sodium (Na) as a microelement.

After carbon, hydrogen, and oxygen, which are obtained from the atmosphere and make up most of the matter in plant cells, nitrogen is the most common element in plants, making up 1.5 percent of dry plant tissue. Nitrogen too is obtained ultimately from the atmosphere; though it exists for short times in combined form in the soil. Plants obtain the other essential elements from the soil. (We will see in Chapter 10 just how they mine, concentrate, and transport these.) The major elements comprise 3.3 percent of dry plant tissue; trace elements appear in very low concentrations. Molybdenum, for example, is only 0.00001 percent, or one-hundred thousandth, of dry plant tissue.

From the law of the minimum (discussed on page 43), we know that the yield of a crop or an individual plant is limited by the availability of the least plentiful essential mineral. If a single mineral is missing, the plant will not complete its life cycle. Some of these elements appear in just a few of the cell's chemicals; others are widely distributed. If a plant is deficient in a mineral, it will first show characteristic symptoms (see Table 9.1).

Phosphorus, for example, is used to make ATP and chloroplast membranes; deficiencies are often shown by yellowing, loss of leaves, or retarded growth. Plants lacking magnesium, a component of chlorophyll, often change color. Some plants become pale; others develop purplish red leaves with green veins. Some amino acids include sulfur, which is also used in making lipids, the fatty or oily substances that are part of membranes and are also reservoirs of energy. With insufficient sulfur, many plants will turn yellowish and be stunted.

Some plants need more of the trace elements like molybdenum than others, but these elements usually are not limiting, though some soils do lack them. Most chemical reactions in cells are mediated by enzymes, which in turn are often activated by particular microelements, although only vanishingly small amounts are needed. Iron, for example, is needed for synthesizing chlorophyll and is part of the compounds used in the light reactions of photosynthesis; often more iron is required than is readily available in the soil. Yellow leaves will often but not always signal that sufficient iron is lacking.

Because it is so abundant in salt (NaCl) and other compounds, we might think sodium would be important to plants, but most do not need it at all, and those which do need very little.

How easily an element can be transported from one part of the plant to another is important in determining the effects of a mineral deficiency. Nitrogen, for example, needed for protein and other compounds, is very mobile; if it is limiting, it will be carried from older leaves to rapidly growing regions of the plant. The mature

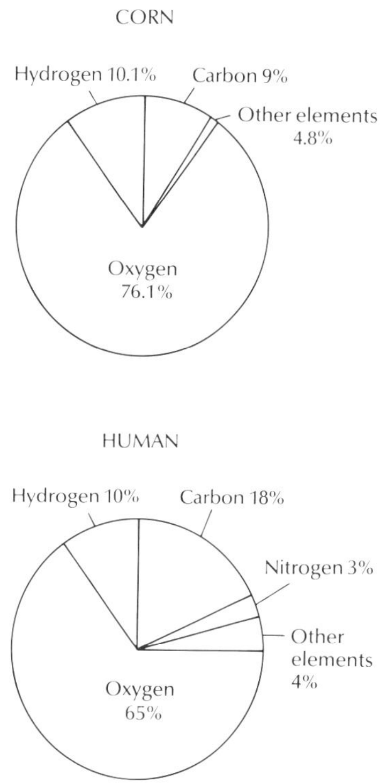

9.1 The major elements in both plants and animals are oxygen, carbon, and hydrogen. The dilute plant protein is concentrated in the muscles of animals, so nitrogen is generally a larger component of animal tissues than of plant tissue.

leaves will then show nitrogen deficiency, often by yellowing leaves (chlorosis). Calcium, on the other hand, stays put; if it becomes limiting to the plant during its growing season, the young leaves will show a calcium deficiency, often indicated by growth abnormalities.

THE NITROGEN CYCLE

The Role of Nitrogen More nitrogen is needed than any other element absorbed by roots from the soil. Lack of it often limits growth. First, it is needed for protein, an extremely important

Table 9.1 Roles of Selected Mineral Elements and Symptoms of Deficiencies

Mineral	Roles	Symptoms of Deficiencies
Nitrogen	Part of amino acids, proteins, nucleic acids, coenzymes, ATP, chlorophyll	Stunted growth, chlorosis (yellowing), leaf fall, purplish coloration; older leaves are most affected
Potassium	Enzyme activator; required for protein synthesis and opening and closing of stomata	Necrosis (dead, dry spots), mottled chlorosis, weak stems; older leaves are most affected
Calcium	Part of cell wall; affects membrane structure and permeability	Death of root and shoot tips, stunted roots; younger leaves are most affected
Magnesium	Part of chlorophyll; enzyme activator; maintains ribosome structure	Mottled leaf chlorosis, leaf tips turn upward; older leaves are most affected
Phosphorus	Part of nucleic acids, nucleotides, sugar phosphates, ATP, enzymes	Stunted growth, dark green leaves, maturity delayed; older leaves are most affected
Sulfur	Part of proteins, thiamine, biotin, enzymes	Chlorosis of young leaves between the veins
Iron	Part of cytochrome; enzyme activator; used in chlorophyll synthesis	Chlorosis of young leaves between the veins; stems are short and slender
Boron	Carbohydrate movement	Death of stem and root tips; roots are swollen; leaves are twisted, pale at bases
Manganese	Enzyme activator; catalyst	Chlorosis, necrosis between veins; younger leaves are most affected
Zinc	Used in synthesis of chlorophyll and IAA; part of some enzymes	Chlorosis, leaves and internodes are stunted, leaf margins are distorted; older leaves are most affected
Molybdenum	Electron carrier; used in nitrogen fixation	Chlorosis and death of young leaves

component of cells. Proteins are chains made of links called **amino acids,** and all amino acids contain nitrogen (see page 61). The nitrogen is reduced (that is, combined with hydrogen) to form an amino group (NH_4), and these amino groups are the building blocks of protein molecules. Protein can make up as much as 10 percent of the plant's weight, except in seeds like soybeans, where it is stored for later nourishment of the embryo and can form 40 to 50 percent of the seed's weight. This means 6 kilograms of nitrogen per 100 kilograms of seed, and world production of soybeans contains about 10 million tons of nitrogen. This does not include the nitrogen in the rest of the plant.

Other compounds in plants also contain nitrogen, among them the **alkaloids,** which are produced in large amounts by many plants. They are used in synthesizing many of our medicines, including quinine, belladonna, antibiotics, ephedrine, cortisone, and reserpine (see the Supplement on page 159). They are also drugs (opium, cocaine), hallucinogens (mescaline, LSD, marijuana), poisons (strychnine), and stimulants (caffeine).

One of the most widely used alkaloids is nicotine, produced by the roots of the tobacco plant and transported to the leaves; it can account for up to 3 percent of the weight of tobacco prepared from dried leaves. Nicotine and other alkaloids have been used as medicines, poisons, and stimulants. Like many of the other plant alkaloids such as pyrethrum, obtained from a *Chrysanthemum,* nicotine is an effective insecticide. Their effectiveness as insect killers probably explains why plant alkaloids are made in great abundance and variety by so many plants. Insects, after all, are the leading herbivores. A mouthful of alkaloid-rich leaves or stem tissue has an unpleasant and sometimes fatal effect on the chewer, so alkaloid production can be a significant adaptation. Compounds containing nitrogen, such as alkaloids, must contribute more to the plant's survival than they cost in using up scarce nitrogen.

Nitrogen is also a component of ATP and other compounds that help transfer energy in cells. The nucleic acids, chemicals that are part of the cell's genetic machinery (discussed in Chapter 11), also contain nitrogen. Considering the cellular demand for nitrogen, we should not wonder that nitrogen counts so much in plant nutrition.

Obtaining Nitrogen Like carbon, nitrogen is ultimately obtained from the atmosphere, of which 78 percent is N_2. Atmospheric nitrogen is very stable, however, and does not easily combine with other elements because a triple bond joins the two nitrogen atoms ($N{\equiv}N$); a lot of energy is required to break this bond. Higher plants cannot use atmospheric nitrogen. Before they can take up nitrogen, it must be either in oxidized form such as nitrate (NO_3) or nitrite (NO_2), or in reduced form such as ammonia (NH_3). These compounds enter plant roots from the soil.

Proteins and amino acids can occur in the soil, but plants do not

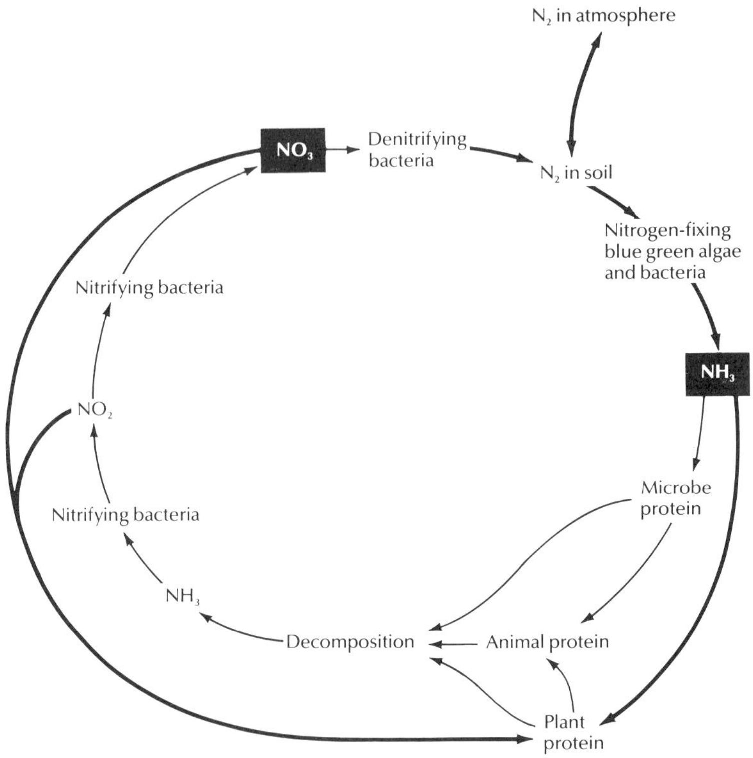

9.2 Depicted here is the cycle of nitrogen fixed by bacteria and blue green algae.

get their supply of nitrogen from these either. One reason is that the soil is teeming with animal life, which gets to the amino acids and proteins first. Instead, the roots of plants take up their nitrogen as nitrate (NO_3), nitrite (NO_2), or ammonia (NH_3). Ammonia and nitrite in the soil are converted by microorganisms into nitrate, however, so the roots mostly take up nitrate. Nitrate must be reduced to NH_3 before the plant can incorporate it into amino acids. The process is similar to the one by which a plant fixes CO_2 (Chapter 8). Essentially, N_2 is reduced by adding hydrogen, a process that requires energy.

But where does the nitrate come from? How does N_2, from the air in the soil, become NH_3 or NO_3? One important source is humanity. Each year 10 million tons of nitrogen are added to the soil by chemical fertilizers. Most is produced by the Haber-Bosch process, which Germany developed during World War I to be sure of having enough combined nitrogen, the basic constituent of explosives. Under heat and pressure, N_2 and $3H_2$ are combined with the aid of an inorganic catalyst. The result is $2NH_3$, which can be oxidized to NO_2 and NO_3. The source of the hydrogen is methane—natural gas (CH_4), a fossil fuel that ultimately comes from photosynthesis. Because this method of fixing atmospheric nitrogen requires a large amount of natural gas, any rise in the price of already expensive fossil fuel is almost immediately reflected in the price of fertilizers and of foods.

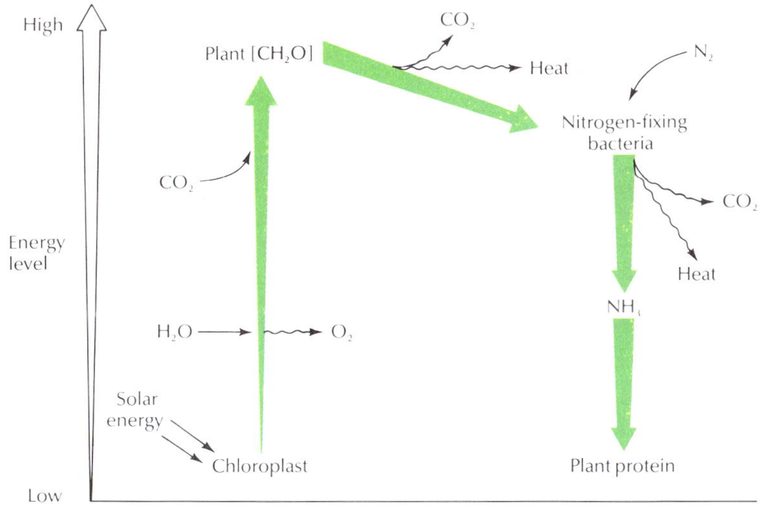

9.3 The reduction of nitrogen requires energy and this energy is provided by the products of photosynthesis. Reduction of nitrogen, like reduction of carbon, is powered by solar energy.

The only ways we have found to duplicate nature's reduction of nitrogen are by this or similar methods: using high temperatures, pressure, inorganic catalysts, and expensive fuels. Nitrogen fertilizers are badly needed to grow food for a ceaselessly growing population. If an easier, cheaper way to make them could be found, the world's supply of food would be under far less pressure.

Nitrogen Fixation: Biological Sources Nature provides several sources of reduced nitrogen: bacteria, both some that live free in the soil and others that form symbiotic relations with plants, and blue green algae, as well as nonbiological sources. In the amount of nitrogen fixed, **symbiotic bacteria** are by far the most important source. These bacteria live in the **root nodules** of some plants (see Figure 9.3), combining N_2 and hydrogen by means of an enzyme to form ammonia, some of which the bacteria uses, while some is secreted into the plant's roots. In these symbiotic associations the bacteria are supplying the plant with nitrogen while the plant makes the sugar that the bacteria use for energy. Because the sugar is produced by photosynthesis, this assocation depends, in the end, on light (see Figure 9.4).

The most common nitrogen-fixing symbiotic genus of bacteria is *Rhizobium*, which lives in root nodules of members of the legume family, such as clover, peas, beans, vetches, and alfalfa. Observe these root nodules by pulling up a clover plant from a lawn. The little bumps on the roots, often pinkish, are nodules. The color is from a chemical very closely related to hemoglobin, the red pigment in our blood. In both plants and humans, the pigment combines with oxygen. That is how our own cells obtain oxygen; in root nodules when the pigment is combined with oxygen it protects the nitrogen-fixing enzyme, which does not function well if much free oxygen is around.

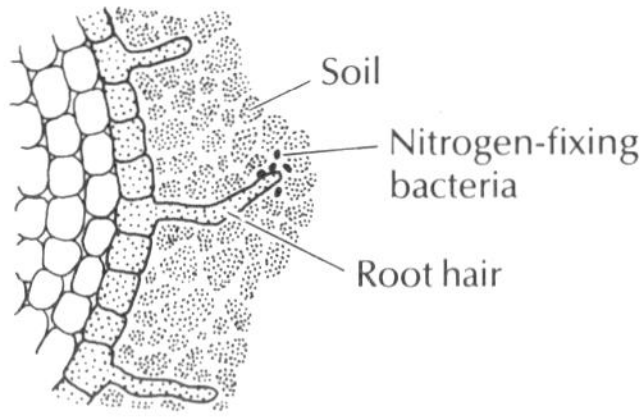

1. Nitrogen-fixing bacteria that come into contact with the root hairs of a legume multiply.

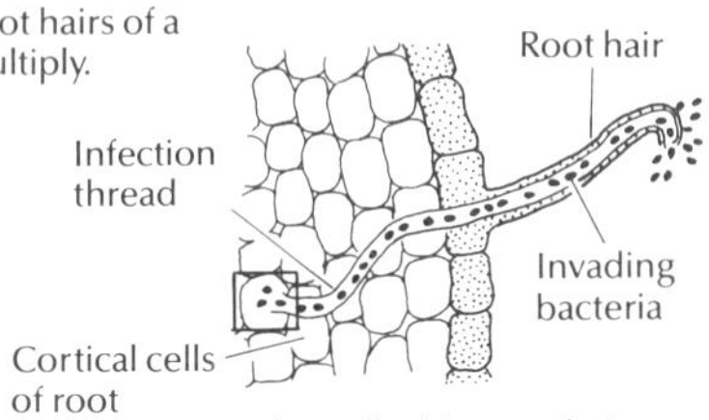

2. The wall of the root hair is softened and the bacteria enter it, penetrating as far as the cortical cells of the root.

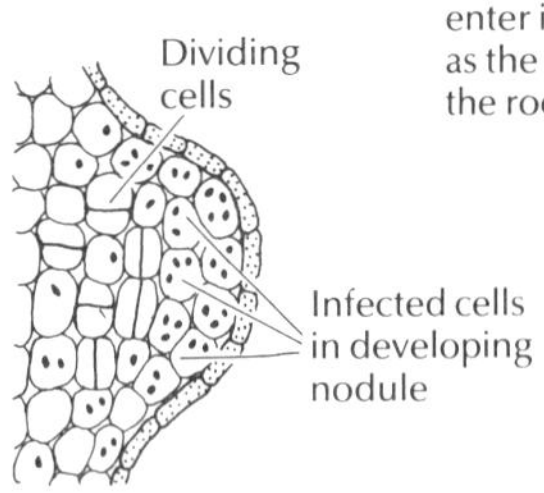

3. Cortical cells in the infected region divide; those containing bacteria are enlarged. Thus a mass of new tissue—the nodule—is formed.

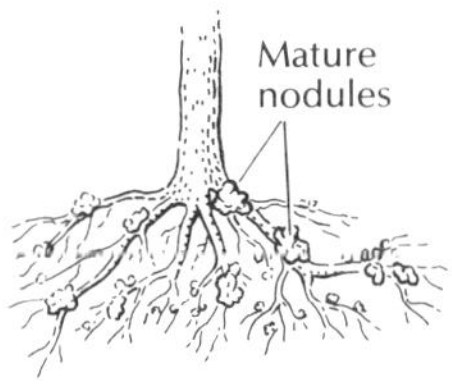

4. The mature nodules are large enough to be seen with the naked eye.

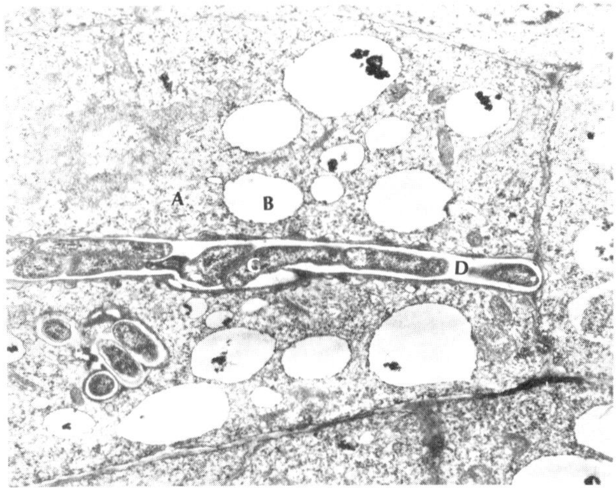

Detail of infection thread

A. Cytoplasm of root B. Vacuole C. *Rhizobium* cell D. Infection thread

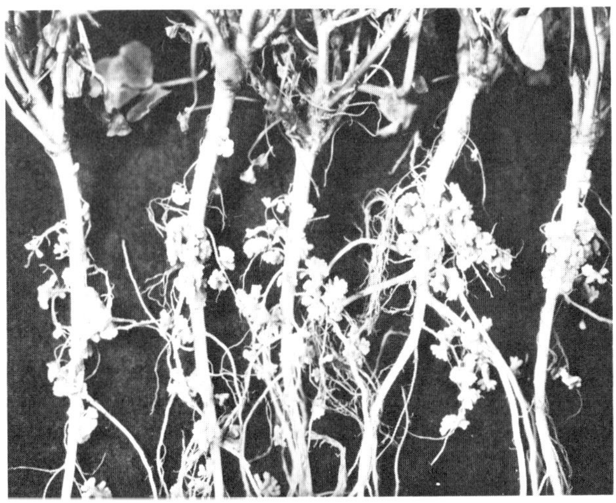

Root nodules

9.4 (Left) Nitrogen-fixing bacteria live in the soil and penetrate root hair cells. They invade cortical cells and induce cell divisions in the cortex, which develop into nodules. Nitrogen-fixing bacteria invading cells (top, right) are separated from the cell cytoplasm by a membrane derived from the cell's plasmalemma. The bacteria cells divide and fill infected cells. Legumes (bottom, right) can have numerous nitrogen-fixing root nodules easily visible to the eye.

About eight plant families other than the legumes have species with nitrogen-fixing root nodules. Like the wild legumes, these are found on nitrogen-poor soils, giving them an important ecological role. In northern forests where the alder tree (shown in Figure 9.5) is

9.5 Alder *(Alnus)* root nodules have a type of bacteria different from the root nodules of legumes but, like legumes, they can fix a great deal of nitrogen. The ability of alders to fix nitrogen is increasingly appreciated, and they are being employed to increase forest production.

a common species, it fixes nitrogen in great amounts. It promotes fertility of the soil so effectively that the timber industry is learning to appreciate it. In fact, the alder is so good at fixing nitrogen that high levels of nitrates appear in streams whose banks are covered with alders. Alder could do as much for commercial forests as clover has for pastureland.

The genus *Ceanothus,* or California lilac (Figure 9.6) is prominent in California flora, and it too fixes nitrogen. *Ceanothus* grows on dry shallow soils with little organic content, making its ability to fix nitrogen important to other organisms in the environment.

Mutualistic relationships also connect nitrogen-fixing bacteria and animals. The termite can live on cellulose because protozoans in its gut break down the cellulose, and the termite lives on the protozoans. Cellulose, however, does not have nitrogen and the protozoans, like all animals, require a source of amino acids to make

9.6 *Ceanothus* (top, left) is very noticeable in the California flora because of its showy blossoms. Some species have abundant root nodules and play a vital role in producing nitrogen in their plant communities. The root nodules (top, right) are really modified roots, with a greatly thickened cortex containing cells with nitrogen-fixing bacteria (see electron micrograph, right).

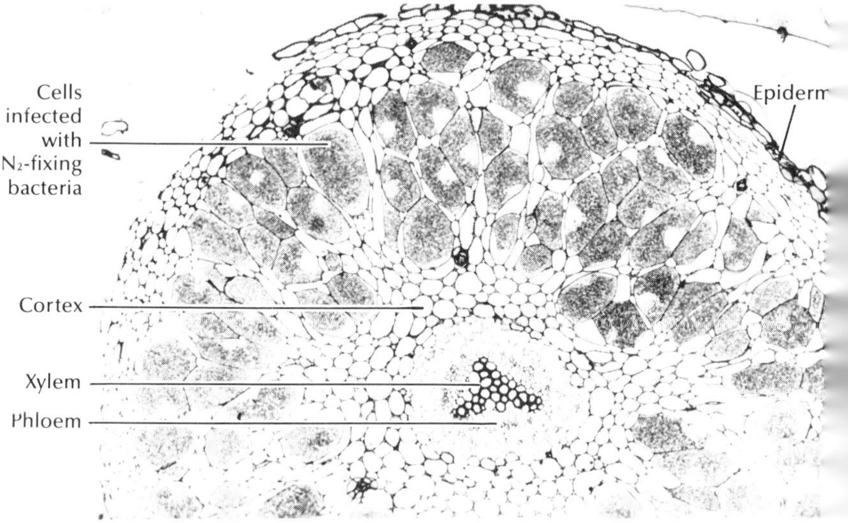

protein. But bacteria living in the termite gut can fix nitrogen, and they provide food for the protozoa. Thus the termite is an ecosystem with three trophic levels. Cows and other ruminants also have symbiotic microorganisms that break down cellulose, but these require reduced nitrogen. A cow can live on old brooms and newspapers so long as you also provide some source of reduced nitrogen such as urea.

9.7 Insectivorous plants use their leaves to capture nitrogen. The fly caught in the sticky secretion of leaf hairs of the sundew will eventually be digested by enzymes in the secretions and its nitrogen-containing substances will be absorbed by the leaves. The second photo is only four hours after the capture. The pitcher plant (top, right) and the Venus's-flytrap (right) have different but equally effective methods to obtain additional nitrogen.

Free-living microorganisms such as some blue green algae and bacteria fix nitrogen, but they contribute only slightly to the nitrogen supply, because they utilize most of what they fix for their own growth. Some blue green algae also have mutualistic symbiotic associations with plants. They are found in the root cortex of some cycads. A small aquatic fern, *Azolla,* floating in great numbers on the surface of ponds and lakes, holds nitrogen-fixing blue greens in pouchlike structures on the underside of its leaves.

Nitrogen and Insectivorous Plants Commercial exploitation and unplantlike behavior have made most of us aware of the Venus's-flytrap *(Dionaea muscipula)*, the plant with traplike leaves that capture insects, although most people have not considered the significance of this and other meat traps. Pools of water formed in the bases of modified leaves of pitcher plants *(Sarracenia)* drown insects and other small animals; the sundew's *(Drosera)* leaves are covered with hairs exuding a sticky goo that entraps and digests small animals. Bladderworts *(Utricularia)* are aquatic plants that also have an elaborate trapping mechanism.

All these plants with meat traps have something else in common. Venus's-flytraps, sundews, and some pitcher plants are bog plants.

Bogs have very little nitrogen available for plants, because their lack of oxygen and their high acidity inhibit decomposition of animal and plant protein. Large tropical pitcher plants are epiphytes (that is, they grow on other plants or in places that give them support but not food), and nitrogen is often limiting in their soilless environment. In all these plants the proteins of the captured insects are broken down by enzymes to amino acids, which are absorbed by the leaves. By trapping meat the plants survive in environments that otherwise supply little nitrogen.

Nonbiological Fixation of Nitrogen Chemical fertilizers are not the only sources of nonbiological nitrogen fixation. Volcanic eruptions and lightning create extremely high temperatures, and both produce nitric oxide (NO). **Photochemical reactions** (light energy producing chemical changes) then transform NO to nitrite (NO_2), and rain washes the NO_2 into the soil where bacteria oxidize it to NO_3. The NO_3 is then taken up by plants, reduced to NH_3, and incorporated into protein.

Similar reactions are set off by the high temperatures at the electrodes of automobile spark plugs. By one estimate, the nitrogen produced in this manner equals half the world's annual supply of fertilizer. Nitrogen fixation can even be provided by the high temperature at the tip of a burning cigarette.

Loss of Nitrogen When plants die and when animals eat them, leave droppings on the ground, and themselves die, nitrogen-containing compounds are returned to the soil by decomposition, continuing the nitrogen cycle. (Figure 9.8 is a summary of this cycle.) Microorganisms in the soil then reprocess these compounds, transforming proteins into ammonia, which is taken up again by plants in the form of nitrates dissolved in soil water and then converted once again into organic compounds.

But as nitrogen is cycled, some is inevitably lost. In any field that is under cultivation, part will be lost when topsoil is washed away by rain and melting snow and when water filtering through the topsoil leaches away nitrates, washing them out of the fertile layer and into streams. Bacteria in the soil break down nitrates and release nitrogen into the atmosphere, a process known as **denitrification.** Thus bacteria and blue green algae start the cycle by reducing nitrogen, and bacteria finish it by completely oxidizing the combined nitrogen and returning N_2 to the atmosphere.

But nitrogen fixation and denitrification do not always balance each other very well. Since the agricultural and industrial revolutions and resulting population explosions, we have been removing large amounts of nitrogen from the soil. In these times of intense agriculture, much nitrogen loss occurs because we harvest the plant for use as food, which is to say, we remove nitrogen compounds

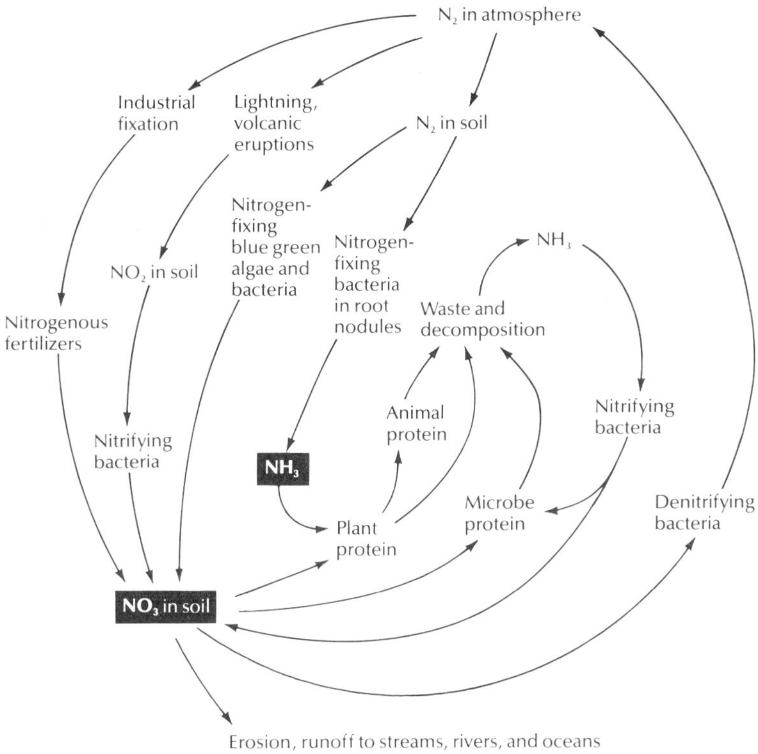

9.8 Nitrogen enters the terrestrial ecosystem in different ways and the cyles are complex. Nitrogen cycles in a similar way in aquatic ecosystems.

from the soil and transport them long distances. Nitrogen leaves the soil for good with the harvest and is not recycled back to the land. Free-living, nitrogen-fixing bacteria actually add little to the soil, using the biggest share of the nitrogen they fix for their own growth. Most of the crops we grow for food and fiber do not have symbiotic bacteria to fix nitrogen, and it is not always possible to rotate cash crops with nitrogen-fixing legumes. As a result, fertilizers of all kind are indispensable, making nitrogen fixation one of the top priorities in plant science.

OTHER MINERAL CYCLES

Each element essential to plants, like nitrogen, has a specific cycle, and some cycles, like the sulphur cycle, are quite complex (see Figure 9.11 on page 260). The only gases plants must have are carbon, nitrogen, and oxygen; of these only carbon and nitrogen have to be reduced before plants can use them. We saw in Chapter 8 how carbon is cycled through the biosphere. Other elements can come from the weathering of rocks as soil is formed, or they can be imported by floods. River floodplains are fertile because minerals, obtained from the weathering of mountains, are washed

9.9 Oligotrophic lakes (left) evoke visions of rainbow trout, but the trout, if any, are periodically placed there by the Fish and Game Department. This lake has too few nutrients to support the growth of sufficient algae to feed large numbers of trout. Its crystal water indicates the lack of life. In contrast, the eutrophic pond (right) has an overabundance of nutrients that can support much primary production and thus feed a great many fish.

downstream and onto the plains during annual floods. Without such natural fertilization, soils that are heavily cropped quickly become depleted of minerals. This is the reason for the historic fertility of Egypt's Nile Valley (Chapter 2).

Fire is another natural way of fertilizing soil with minerals. The ash that covers the ground after a fire is concentrated minerals from the burned plants and trees. Tropical "slash and burn" agriculture is a way of temporarily supplying minerals to depleted soils.

Eutrophication Mineral cycles are constantly moving in terrestrial environments and in water. They are sometimes strikingly visible in bodies of water, but people grow confused about the causes. It is easy to reach false conclusions when we compare a mountain lake, in which we can see clear to the bottom in 3 meters (10 feet) of water, with a pond green with scum on a dairy farm. The lake is clean and beautiful; the farm pond looks strangled with pollution. This judgment is not necessarily valid.

Some mountain lakes are clear because little or no life clouds their waters. They have few fish because almost nothing is there for fish to eat. Minerals are scarce, and living is hard for the very few plants that grow there. To use the lake for fishing, we have to stock it artificially. The lake is **oligotrophic** (poorly nourished). The open oceans are essentially oligotrophic, too, their surface waters low in minerals. This, of course, is why few fish are caught there; the big sea fisheries are all in the shallower areas of the continental shelf where runoff from land or upwellings from deep water are rich with minerals. Extending national boundaries 320 kilometers (200 miles) into the ocean is a violent issue in international politics because that

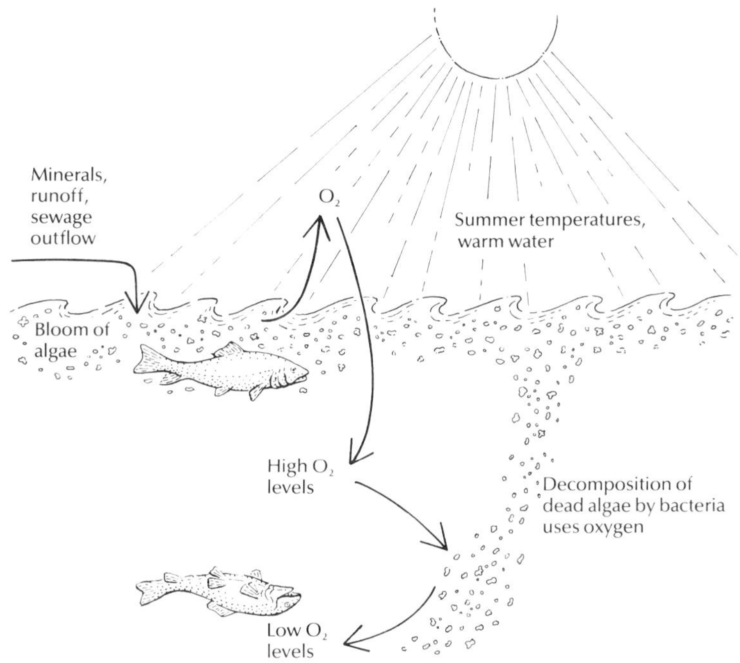

9.10 Advanced eutrophication of bodies of water can result in massive deaths or even total elimination of organisms intolerant of low oxygen levels in the water. Bacterial decomposition of organic matter demands oxygen. The more organic matter in the water, the more respiration, and the greater the demand for oxygen. Warm weather worsens the situation because less oxygen is dissolved in warm water.

is about where the shelf and good fishing end and oceanic depths begin.

The farm pond that looks so polluted compared to the clear lake is green from its cover of algae, which have (and require) minerals in abundance. The algae in turn are food for fish. The pond is **eutrophic** (well-nourished). Depending on how well it is nourished with minerals, it may be full of life, including fish; but this eutrophication can also go too far. If minerals are too plentiful, the algae may increase explosively, and decomposition of dead plants will use up all oxygen in the water. At lower depths many plants and fish will suffocate and die. Hot weather worsens the problem because algae and bacteria grow rapidly and less oxygen is dissolved in the water.

In many places today eutrophication is artificially made. One of the best known examples is Lake Erie. Pollution from raw sewage as well as fertilizers draining from surrounding farmlands has so advanced eutrophication that many consider it a "dead" lake. Though Lake Erie has changed, it is not dead. It has a very large **biomass** (amount of living matter) and is still very productive of life forms, but the forms are different. Blue green algae have increased at the expense of green algae, and lake trout have all but disappeared, replaced by large numbers of carp and other fish that tolerate water with little oxygen.

Eutrophication is reversible. Lake Erie flushes out every three

9.11 MINERAL CYCLES

The biosphere is a closed system for matter. All organisms are composed of elements that cycle around the biosphere among the living and between living and nonliving. Biogeochemical cycles describe the circulation, concentration, deposit, and release of the elements of which organisms are made. Oxygen gas makes up 21 percent of the atmosphere, but it was produced and is maintained by the splitting of water by green plant photosynthesis (facing page, left). Oxygen is a part of organic compounds, which are cycled through food webs. Oxygen is combined with hydrogen during cell respiration to form water, and oxygen is continuously released from water through photosynthesis. Phosphorus (facing page, middle right) is used in large amounts by organisms and enters the food chain through plants which mine it from soil and water. It recycles as a result of decomposition but much is lost as sedimentation in deep oceans. The cycle is maintained by weathering of phosphate-containing rocks and production of phosphate fertilizer from such rocks. The dissolved phosphates produced by weathering are taken up by plants. As in the nitrogen cycle, microorganisms play a major role in the sulfur cycle (facing page, bottom). Weathering (below), resulting from wind, water, ultraviolet radiation, and temperature changes, is essential to operations of biogeochemical cycles. Weathering over eons wears away continents and all life feeds on the resultant debris. This gnawing away of continents feeds deep ocean sediments and buried nutrients are lost to the biosphere. In time, ocean floors are elevated to form continents, the weathering cycle continues, and long buried elements rejoin the biosphere.

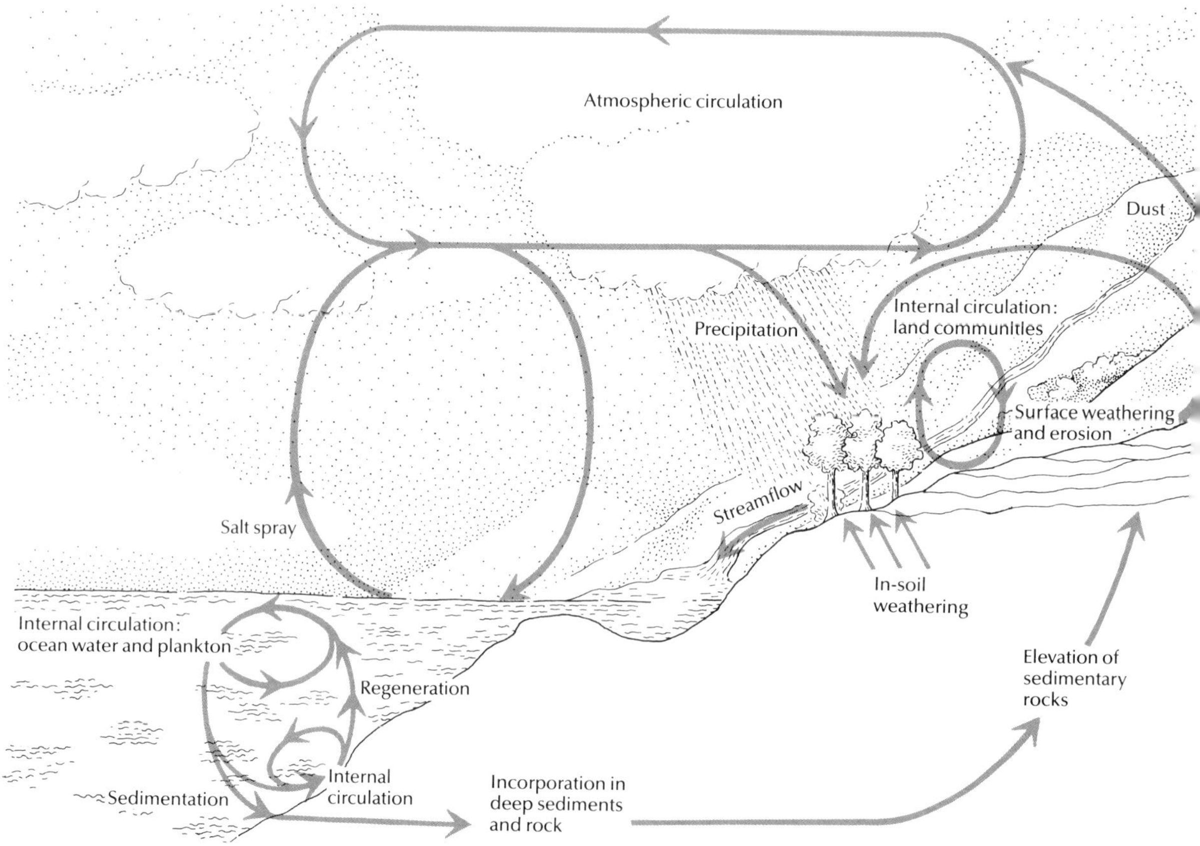

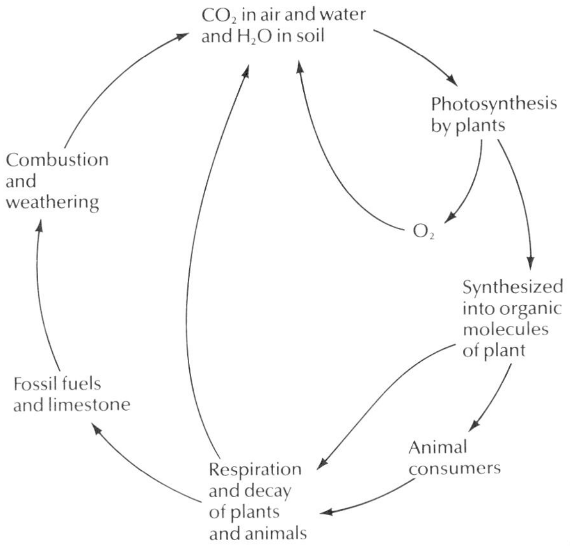
CO_2 in air and water
and H_2O in soil
Photosynthesis
by plants
O_2
Synthesized
into organic
molecules
of plant
Animal
consumers
Respiration
and decay
of plants
and animals
Fossil fuels
and limestone
Combustion
and
weathering

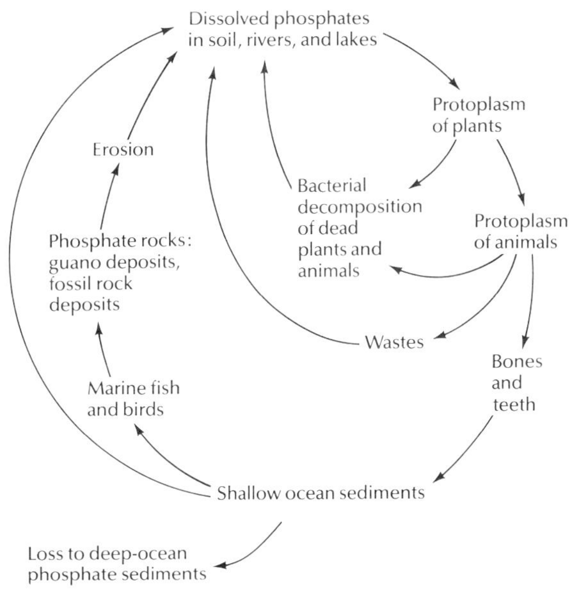
Dissolved phosphates
in soil, rivers, and lakes
Protoplasm
of plants
Bacterial
decomposition
of dead
plants and
animals
Protoplasm
of animals
Wastes
Bones
and
teeth
Shallow ocean sediments
Loss to deep-ocean
phosphate sediments
Marine fish
and birds
Phosphate rocks:
guano deposits,
fossil rock
deposits
Erosion

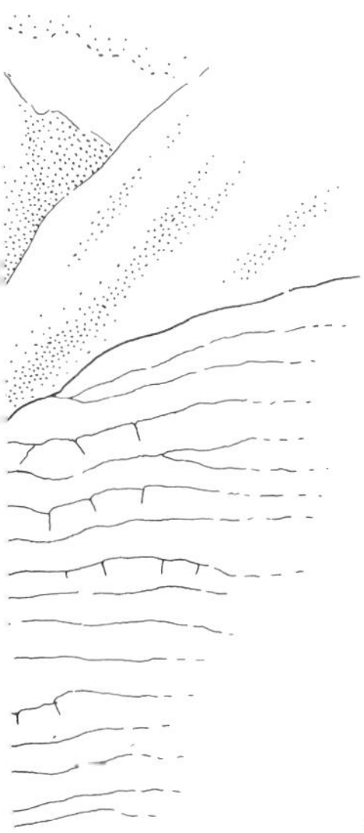

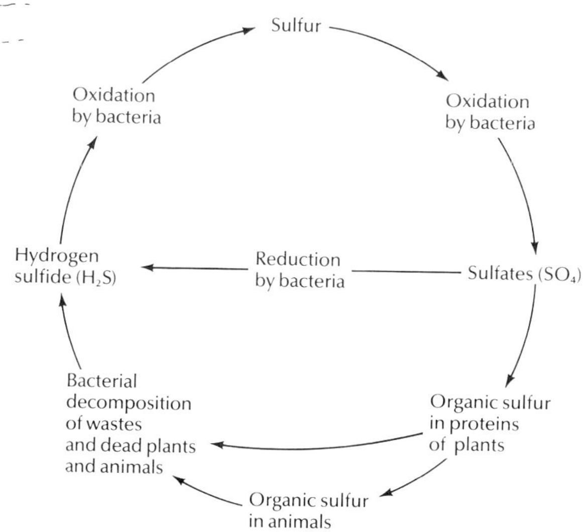
Sulfur
Oxidation
by bacteria
Oxidation
by bacteria
Sulfates (SO_4)
Reduction
by bacteria
Hydrogen
sulfide (H_2S)
Organic sulfur
in proteins
of plants
Organic sulfur
in animals
Bacterial
decomposition
of wastes
and dead plants
and animals

years, so that if the flow of minerals were lessened, eutrophication could soon be reversed. It is not necessary to remove all minerals. The law of limiting factors states that only one essential mineral needs to be limiting and growth will be inhibited. In Lake Washington in Seattle, eutrophication was decreased by controlling minerals entering the lake.

We have shown great talent at speeding up euthrophication, making it seem that all cases of eutrophication are due to human intervention, but natural eutrophication is fairly common. Normal drainage from a watershed or drainage area can provide plenty of phosphorus and other elements from weathered rocks to encourage plant growth, and nitrogen-fixing blue green algae provide the missing nitrogen.

Even regions in the oceans are eutrophic. Estuaries, the ocean ends of rivers, are always highly productive because nutrients carried downstream by the rivers are deposited there. Deep offshore waters may also have eutrophic regions, as do the waters off the coast of Peru, where the upwelling of the Humboldt Current carries minerals and cold water from the depths to the surface. Algae are abundant in this well-fertilized and oxygen-rich water, supplying food to enormous numbers of fish and making this one of the world's great fishing grounds.

Pollution Chemical pollution can add different substances, with results very unlike eutrophication. Many of the chemicals heedlessly flushed into lakes and rivers in thickly populated cities and industrial areas poison plant and animal life.

Some of the worst chemical offenders are not poisons, however, but chemical compounds that stimulate eutrophication. These are the chemical detergents from dishwashers and washing machines. Rich in phosphorus and other minerals, detergents pour into lakes, rivers, and ocean harbors from sewage outfalls, and explosive growth of algae all too often results. Most sewage treatment plants eliminate only solids, leaving the effluent rich in minerals (see Figure 9.13). It would be nice to think that biodegradable detergents would remove the problem, but despite advertisers' claims, these serve up minerals even faster than other detergents.

PLANTS AND SOIL

That which looks like plain dirt to most of us is in fact a richly diverse ecosystem. Much of the world's biomass, after all, is underground, and much of what is underground is directly connected with what goes on above ground. Plant roots, for example, are an important part of the underground biomass. All the teeming life in the soil is supported by products of the photosynthesis of

9.12 The toxic soils in this strip-mined area of West Virginia provide a blue image when infrared-sensitive color film is used. The deeper the blue, the greater the toxicity. Very little life remains in this area, which has been polluted by sulfuric acid formed from sulfur-bearing clays exposed during coal mining. Vegetation reflects much of infrared radiation and provides a red image in infrared-sensitive film.

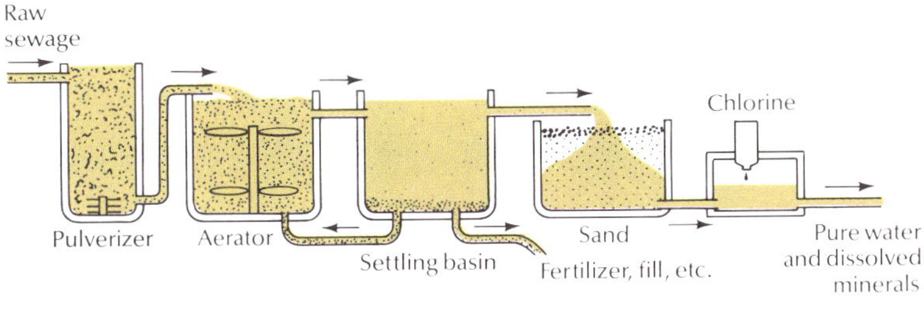

9.13 There are different types of sewage disposal systems but in all cases the raw sewage is decomposed by bacterial action, after which remaining solids (sludge) are collected. The water is filtered and purified and discharged into bodies of water. This can lead to eutrophication because the effluent contains minerals. Very few sewage plants remove these minerals because it is an expensive process.

9.14 Eggplant, zucchini squash, and marigolds grow prolifically in this hydroponic system. The containers have small rocks, which support the roots, and the roots are continuously bathed by a nutrient solution.

plants. Most of the CO_2 exchange so important in the carbon cycle takes place in the soil. Soil is much needed by life in general, yet it is not really necessary to plants.

A paradox? We are so used to the idea of planting seeds in the ground that soil and plants seem indivisible. They are not. If you microscopically examined plant roots growing in the soil, it would be clear that the roots live not in dry soil but in the watery spaces between soil particles (see the Supplement on page 270). Root hairs are very much like algae, immersed in water, and that is the secret of living without soil.

Soil does support the plant, but that is a purely mechanical function. It was proven many years ago that plants do not "feed" on soil; they obtain nothing from the soil but water and dissolved minerals. In fact, the idea of feeding plants is an anthropomorphic misinterpretation. There is no such thing as plant food; supply a plant with water, minerals, carbon dioxide, and light, keep it at a proper temperature, and it will grow.

Hydroponics That plants can grow without soil is proven by hydroponics, a method of growing plants in water enriched with dissolved minerals and aerated to give the roots enough oxygen. (Remember that the roots do not photosynthesize and so do not produce O_2.) The plant stems are supported by a grid, their roots are immersed in the mineral solution, and they grow better than they would in soil.

Vegetables and fruits are grown hydroponically in many parts of the world. Hydroponic gardens take up little space, do not have to be weeded because only the desired seeds are planted, and are less subject to soil diseases. In addition, plant pests are more easily controlled. The supply of minerals and temperature can also be carefully monitored. It is, in fact, in many ways an ideal technique, except that it is still more expensive than gardening in earth, and cannot yet be done on a large scale.

Mycorrhizae The roots of many, perhaps most, plants form close symbiotic associations with fungi. In these **mycorrhizal** ("fungus root") associations, fungal hyphae ("threads") form a layer over the root's surface and penetrate between the cells of the root cortex (see Figure 9.15). Many types of mycorrhizal fungi also penetrate the cortical cells themselves. The fungi obtain organic nutrients from the roots. Studies show that mycorrhizae, in turn, ease absorption of dissolved minerals and that many plants do not grow well without the associated fungus. The fungi are a direct link between decomposing organic matter and root cells, a particularly important connection in soils around conifers, where bacterial decomposition of organic matter is low. In conifer forests fungal hyphae weave multitudinous links with plant roots, a vast underground network joining the roots of all the trees. On this symbiotic association depends the cycling of matter in many segments of the biosphere.

Attempts to get plants to grow on land strip-mined for coal prove that mycorrhizae are indispensable. Plants would not grow because mycorrhizal fungi were missing from the soil; inoculation with the fungus made it possible to reestablish plant cover. The practice is now standard in reclaiming strip-mined land.

The mycorrhizae illustrate how very complex are organismal interactions in the soil. The **rhizosphere** (root zone) is intensely studied, because it is here that inorganic compounds begin their trip through the living elements of the biosphere.

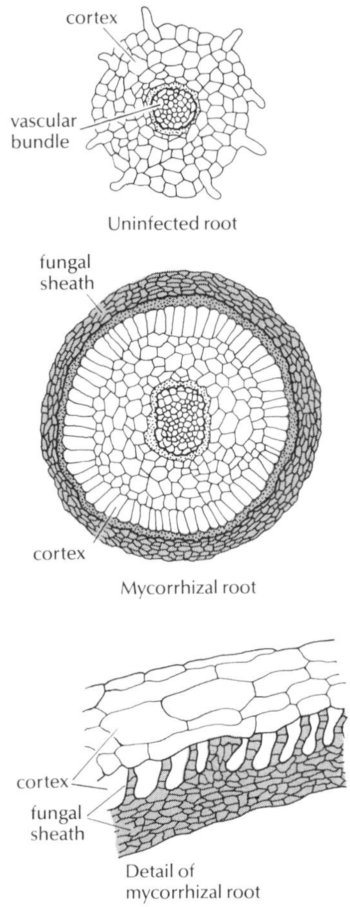

9.15 Mycorrhizal fungi can form a sheath around roots and penetrate between the outer cortical cells. This common symbiotic relationship helps the root to obtain minerals, and the fungus receives organic material.

MYTHS ABOUT PLANT NUTRITION

History is full of tales about what is good for plants, a sign of how important they are to our survival. Fact and fantasy have long been associated with organic gardening. A great many well-meaning people feel that this method is ecologically desirable as a way of avoiding waste and counteracting the massive quantities of synthetic or "artificial" fertilizers, herbicides, and pesticides used in agriculture.

Let us consider organic gardening factually. Plants concentrate minerals from the soil. Cultivate soil or remove crops at harvest time and inevitably the soil will be depleted of minerals unless they are somehow returned to it. Intensive cultivation and removal of crops is necessary to feed a constantly and rapidly growing urban world population, meaning that vast quantities of minerals must be returned to the soil. Before commercial fertilizers were applied on a large scale, much of the soil in the United States was seriously depleted of minerals.

Most minerals are returned to the soil as inorganic fertilizers. Phosphate, for example, is taken from rock with a high phosphorus content. Nitrogen is applied as synthesized ammonia, urea, or nitrates.

"Organic fertilizer" usually means animal manure, and of course manure works, though with some disadvantages. It is a very dilute form of fertilizer, it is bulky, and it is expensive to transport. Human sewage might be used, as it has always been in many parts of the world; it probably will be put to more and more use in the United States as surer ways of purifying it are found. But it also has problems. Transport costs make it expensive to ship this bulky and dilute fertilizer from population centers to Kansas wheat fields. Unless it can be cheaply concentrated, it is in a dilute form and hence must be applied often, elevating the cost of fuel for tractors. Manure also takes time to break down to inorganic compounds; it must be applied early in the season. In many parts of the country, however, the ground is not completely thawed then, and spring rains wash the manure into streams, causing eutrophication. Some evidence warns us that toxic heavy metals accumulate in human sewage and plants might also concentrate them, threatening human and animal health.

In short, manure in all its forms can help us in agricultural production, but it cannot supply all our fertilizer. It is useful in individual gardens and small farms, but is unlikely to supply all the mineral replacements needed for increasingly intensive agriculture.

Plants with nitrogen-fixing root nodules, working with such symbiotic bacteria as *Rhizobium,* can also help supply nitrogen. Legumes are the best of these so far, but even crops of soybeans still need additional nitrogen fertilizer for optimum growth. And even under optimum conditions, yields of soybeans are small relative to those of corn. Unfortunately, most of our food crops do not have root nodules.

How realistic then is organic gardening as the sole form of supplying nutrients to crops? Organic gardening has customarily meant using nothing but organic or natural fertilizers such as compost and manure. But a misunderstanding may be confusing things here. Minerals from manure and other derivatives of living organisms can be used by plants only after microbes have decomposed them to their elements. Fresh cow manure is not only useless as a fertilizer, it is harmful to plants; microbes must first do their work, which takes time. The nitrogen cycle makes this sequence quite clear. Nitrates released from decomposing cow manure are exactly the same as those produced in a fertilizer plant. The same is true of other essential elements, irrespective of source.

Plants take in their essential minerals not in organic but in inorganic form, regardless of the fertilizer used. The difference between the "natural" and the commercial products is not their chemistry but how they were produced. Grow a plant organically or with commercial fertilizers and its chemical content will be exactly the same, because it is determined by the plant's genes and not by its environment. There is no evidence that the same kind of grains, vegetables, or fruits grown in similar conditions will differ in taste or texture whether they are grown organically or with commercial fertilizers.

On the other hand, organic gardening practices are very often beneficial for the soil, improving its physical characteristics by breaking up some soils to let oxygen circulate and binding others for better retention of water. Organic material helps to balance soil flora and fauna, both important for plant growth. The slow release of minerals from organic matter can also benefit growth. Efficient farmers are now and have always been conscious of the benefits of recycling organic matter.

The important question about organic versus commercial fertilizers is not which is best but what the world needs and what it can get to improve agricultural production. If the nation's farmers practiced only pure organic gardening, the corn and wheat fields would soon be depleted of minerals; production of food would decline disastrously.

For more than half a century the demand for commercial fertilizer has been loud all over the world. Even in countries like China, where much animal and human manure is used, manufactured fertilizer is wanted. China imports large quantities and has a crash program to build nitrogen fertilizer plants of its own. They will continue to need supplemental chemicals as they rush to feed more and more people from a finite area of cultivable land.

As the need for fertilizer grows, it must continue to clash with the climbing cost of mined natural gas. The discovery of a cost-effective way to fix nitrogen that does not employ a scarce and expensive fossil fuel will have a tremendous impact on the economy and hence the politics of the world.

Summary

Carbon, hydrogen, and oxygen make up almost all of a plant's dry weight. Nitrogen is the next most abundant element, and phosphorus, potassium, sulfur, calcium, and magnesium are also present. Exceedingly small amounts of other elements, trace elements, frequently activate enzymes mediating essential chemical reactions. As the law of the minimum states, a plant's growth will be limited by the availability of the least plentiful of these essential minerals.

The availability of nitrogen frequently limits growth. Nitrogen is a basic constituent of amino acids, which make up protein, of ATP, and of nucleic acids. It is also an ingredient of alkaloids that many plants make in abundance. Atmospheric nitrogen is very stable, however, and plants obtain nitrogen not from the air but through their roots.

Nitrogen is fixed in the soil by several natural sources. Biological agents include both free-living and symbiotic bacteria and blue green algae; nonbiologic sources include volcanic eruptions and lightning. These result in the formation of nitric oxide (NO), which is changed by photochemical reactions to NO_2, which is then washed into the soil by precipitation. Great quantities of reduced nitrogen are also added to the soil by fertilizers, which are usually made by the Haber-Bosch method. These are costly, however, for by this method the hydrogen necessary to reduce nitrogen is obtained from natural gas.

Free-living bacteria and blue green algae oxidize NO_2 in the soil to NO_3, which plants take up, reduce to NH_3, and incorporate in proteins. These bacteria use most of the nitrogen themselves, however. Symbiotic bacteria living in root nodules will combine N_2 from air in the soil and hydrogen, forming NH_3, some of which is secreted to the plant. Most of these symbiotic bacteria are *Rhizobium* living in root nodules of the legume family; eight other plant families, however, are known to have species with nitrogen-fixing nodules.

Some plants have adapted to nitrogen-poor environments by trapping insects, from which they obtain nitrogen. Epiphytes, which live in soilless environments, and insectivorous plants living in bogs are examples.

The nitrogen that plants take up is returned to the soil when they die and their protein is transformed by decomposers to ammonia, nitrites, and nitrate. Denitrifying bacteria return nitrogen from the soil to the atmosphere by breaking down NO_2, releasing nitrogen. Nitrogen is also lost from the soil when topsoil is washed away, when water leaches nitrates into streams, and when crops are harvested and then transported. As a result of these processes and the fact that most of our crops do not have nitrogen-fixing bacteria, nitrogenous fertilizers are very often necessary to keep land fertile.

The action of water is important in making minerals available to plants; and minerals, such as phosphorous and sulfur, also cycle through ecosystems. Floods bring minerals to the soil, as does the weathering of rocks. Fire is another natural source of minerals, one which tropical slash-and-burn agriculture makes use of. The ashes of burned plants are a source of concentrated minerals.

A body of water that is eutrophic—well-stocked with minerals—may appear polluted because of its cover of algae, which provide food for fish. Eutrophication may go too far, when an excess of minerals encourages such explosive growth of algae that subsequent decomposition decreases the oxygen to such an extent that other organisms cannot survive. Eutrophication, however, can be reversed. Human actions have increased eutrophication (detergents are often a cause) but natural eutrophication is also common.

Minerals that plants obtain through their roots are essential, but the soil in which they grow is not. It does provide support, but plants can be grown hydroponically—without soil, in aerated and mineral enriched water.

Roots and fungi form important symbiotic associations in the soil, mychorrizal associations. The fungi, whose hyphae form a layer over the root's surface, penetrating between and sometimes into cells, get organic nutrients from the roots and aid the roots' absorption of dissolved minerals.

Whether a plant obtains a mineral from an organic source such as manure or from a synthetic fertilizer, it takes up the same inorganic compound. Manure is dilute, bulky, expensive to transport, and it takes time to break down into the compounds that plants use. Organic gardening does have some advantages: improving soil structure, balancing soil flora and fauna, and releasing minerals slowly. On the whole, however, it seems highly unlikely that enough land could be kept sufficiently fertile solely by organic methods to match the world's present food production, much less meet the needs of the future.

Review Questions

1. What is an essential mineral?
2. Why is nitrogen in such demand as a plant nutrient?
3. What is meant by "biological nitrogen fixation"?
4. Compare carbon fixation with nitrogen fixation, including the relationship between them.
5. Even if all parts of most crop plants were returned to the soil, nitrogen would still have to be added to the soil to maintain productivity. Why?
6. Up to a point, the discharge of raw sewage into a body of water can improve fishing. Why?
7. Acid soils inhibit bacterial growth and mycorrhizal associations are often required for plants to grow in such soil. Explain this phenomenon.
8. "Feeding" plants and "plant food" are common terms. What is wrong with this terminology?

SUPPLEMENT

Soil

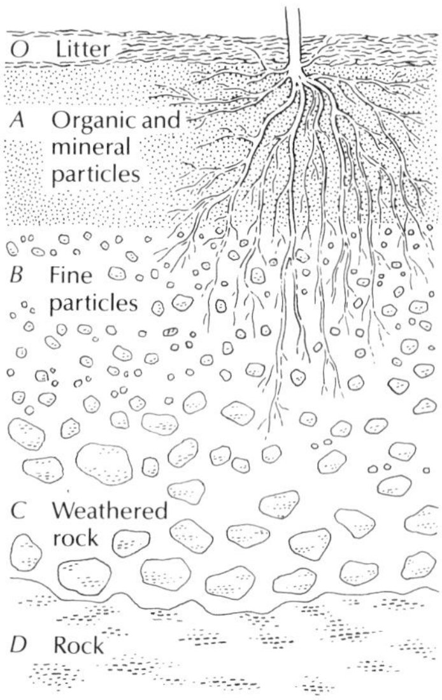

1 Soil profiles are divided into horizons. Horizon *A* corresponds to topsoil. Topsoil is characterized by more organic matter and smaller particles of inorganic matter than deeper horizons. Soils differ in the relative amount of sand and organic matter.

Soil is composed of small inorganic mineral particles, decayed organic matter, living microorganisms, a host of invertebrate animals, and air and water. The small, sometimes microscopic, mineral particles are produced by the action of weather on the large rocks that make up the earth's crust. Chemical processes and the abrasive action of heat, cold, wind, rain, and ice (called **weathering**) disintegrate the rocks on the surface of the earth into small particles of their constituent minerals, mostly quartz, feldspar, and hornblende. Soils may also be transported from one place to another by wind, rain, or glaciers. Furthermore, organisms within the soil alter its composition.

All these changes are continuous. If a vertical section of a particular soil is made (say, a highway embankment), a series of distinct layers, known as **horizons,** is usually found. This **soil profile** (see Figure 1) can reveal the history of the soil's formation.

The mineral particles in the soil are composed principally of the atomic elements magnesium, sodium, and potassium and are classified according to size. The very smallest particles (less than 0.002 millimeter in diameter) are called **clay;** coarser particles (0.002 to 0.05 millimeter) are called **silt;** and the most coarse (0.05 to 2 millimeters) are called **sand.** The varying amounts of clay, silt, and sand give a soil its **texture,** which greatly affects the soil's ability to hold water, its exposure to air, and how well it can support plant growth.

The arrangement of these particles determines the soil's **structure.** The particles form aggregates known as **peds.** Atmosphere and water penetrate the spaces between the peds; here too live the soil's microorganisms. Moisture that is held in these spaces or adheres to the particles is the main source of water for plants and is known as **capillary water.** The soil's other components are dissolved by this water, forming the soil solution from which plant roots take up nutrients. Some of the water penetrating the soil, known as **gravitational water,** seeps downward to underground streams or the water table and is not available to plants. For most plants it is best if one-fourth of the volume of soil is water and one-fourth is air. If the soil is saturated with water, air will be driven out and plant roots will be deprived of needed oxygen.

Most of the soil's organic material is decaying plant material, with lesser amounts of decaying animal matter and living microorganisms. The microorganisms, especially bacteria, fungi, and protozoans, degrade dead

2 Soil is the home of countless organisms and earthworms are among the best known and most important. They pass soil through their guts and improve the soil's fertility and structure in the process. Their microplowing turns over the soil and their ceaseless activity can in time bury large boulders.

plant and animal matter, using much of it as nutrients for their own growth. What they do not use remains in the soil as part of a substance called **humus,** which has great capacity for holding water. Organic material in the soil is also a source of nitrogen.

A dynamic relationship exists between the soil's organic and inorganic components. Animals living in the soil, such as earthworms, slugs, snails, beetles, and rodents, continually churn and mix its constituents. The roots of plants and trees also mix soil elements and break down rock, which helps to stabilize the soil's structure and retard soil loss. Organic matter also granulates the inorganic matter in the soil, creating a loose structure that can be easily tilled.

We can see the relationship between the organic and inorganic matter and understand the effects of texture and structure better by comparing clay and sandy soils. The small particles characterizing clay soils absorb water readily, but these soils can easily become saturated, cutting off the air supply needed by the plant roots and by microorganisms. In sandy soils, in contrast, water tends to sink so rapidly that little moisture may be available to plants, but between the large particles that characterize these soils, air spaces are plentiful. Organic material, however, promotes good soil structure; it encourages the aggregation of particles, thus increasing the amount of space for air (see Figure 3). If a clay soil has good structure, it will thus have the advantages of both clay and sandy soils.

The amount of rainfall, the decomposition of organic matter by microorganisms, and cultivation make soil relatively acidic or alkaline. The degree of acidity in turn affects the solubility and availability of nutrients and can markedly affect the kinds of plants that can grow in a soil. Some crop plants, such as alfalfa, barley, and sugar beets, tend to grow better in slightly alkaline soils; berries and many flowering shrubs like rhododendruns grow well in acidic soils.

3 Plant roots require water and oxygen, so good soil must meet both requirements. Many clay soils have few air spaces and sandy soils have little water-retention capacity.

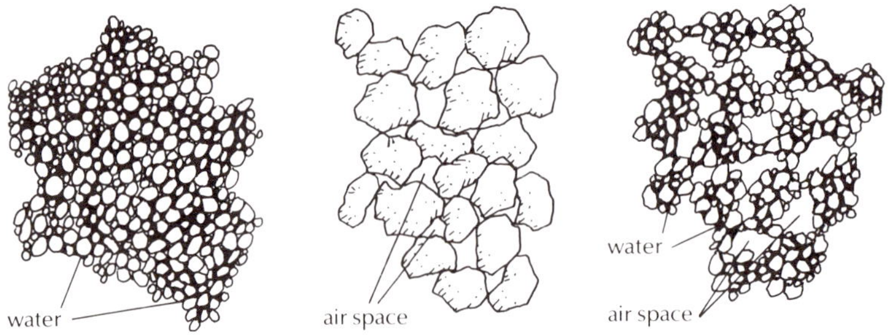

Clay (water-holding capacity but little air)

Sand (air spaces but poor water-holding capacity)

Clay soil with good structure (water and air both available)

4 The world's soils are classified on the basis of physical characteristics, and mineral and organic content. The best grain-producing soils are chernozems.

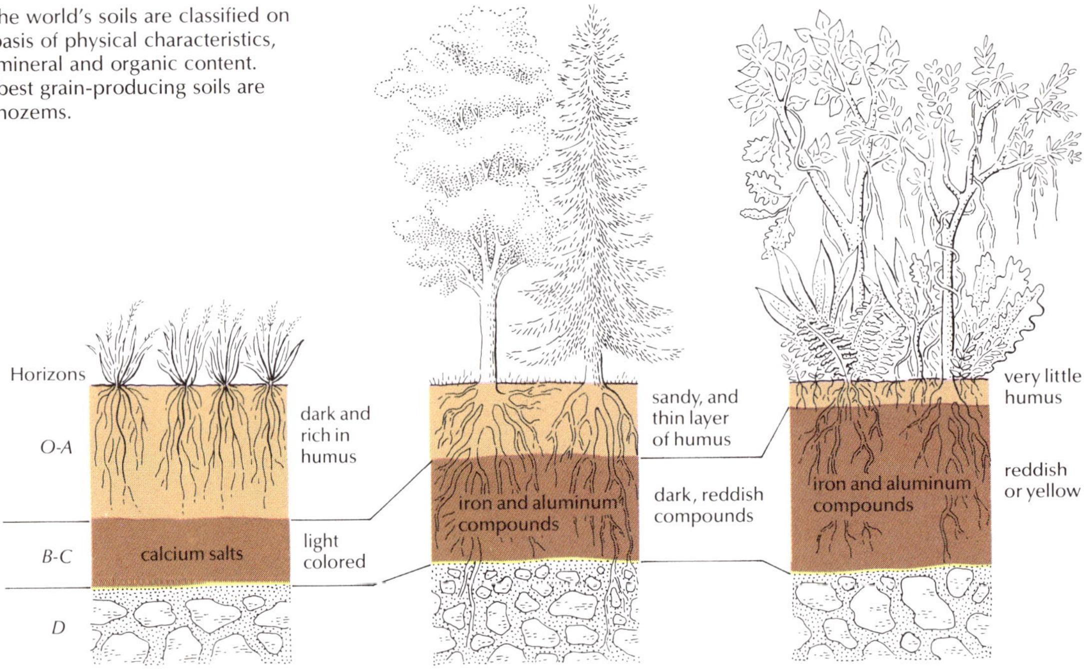

Chernozem (temperate grasslands; temperate climate)

Podzol (temperate, coniferous forest soil; wet, cool climate)

Laterite (tropical rain forest soil; wet, warm climate)

Climate, mineral composition, and soil structure create several major soil types: **podzols** are reasonably fertile soils occurring in temperate climates; **latosols** are relatively infertile soils in fairly tropical climates; **chernozems** are highly fertile soils in temperate climates; and **sierozems** are infertile soils of arid regions, including the great deserts of Africa and Asia (see Figure 4). Obviously the varying fertility of these soils has implications for world agriculture. Because chernozems are so fertile, over 90 percent of the world's grain is produced in them. For those who live in

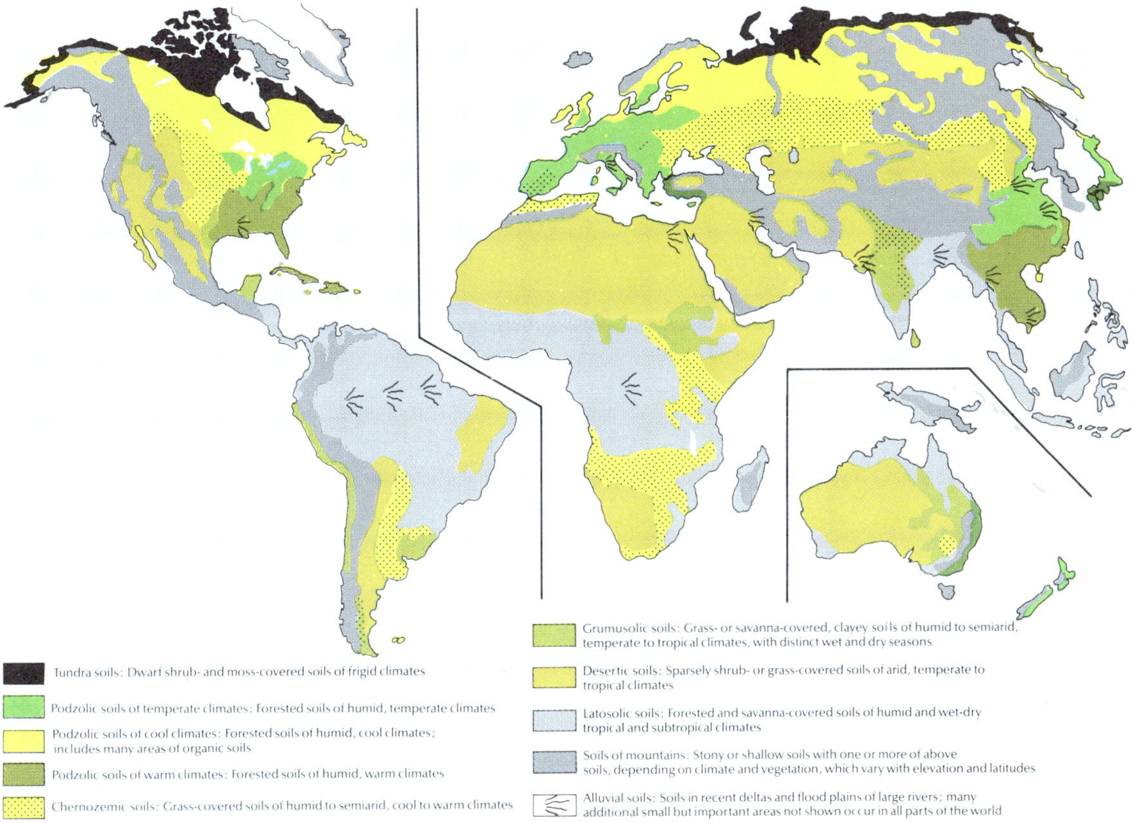

5 Ability to feed the world's population is related to the amount of quality soil. The affluence of the United States is in large measure due to its rich soil and favorable climate.

tropical climates in which latosols are the predominant soil type, producing enough food becomes a greater problem.

Soil can be lost through the abrasive action of running water, wind, and ice upon its surface, a process called **erosion.** Sloping terrain or land that is impermeable to water is vulnerable to water erosion, and arid soil can be literally blown away by wind, creating dust bowls. Overcultivation can weaken the soil structure, making the soil vulnerable to erosion (for further discussion, see Chapter 15). Vegetation and use of particular tilling and planting methods can protect the soil from erosion.

CHAPTER 10

Moving Water, Minerals, and Food

The work plants do to extract minerals from the soil outclasses the best efforts of human technology by far. Our worldwide production of copper, lead, zinc, and iron each year is estimated to be close to a billion metric tons. The worldwide accumulation of minerals by plants, however, is estimated to be about five times the amount that humans are able to get out of the earth with all the latest and most efficient machinery.

Mining the minerals, getting them from soil into roots, will do little for the plant; they must be transported through roots and made available to leaves and meristems. In plants both small and large, the principle is always the same: the nutrients are dissolved in water that is carried through the xylem's dead cells to the leaves, from which some of the water escapes to the air. But mining, concentrating, and transporting the minerals requires energy. Food—sugars produced by photosynthesis in the leaves—is carried downward through living cells of the phloem to the roots, giving energy to the cells that take up water and minerals from the soil. Each plant thus has two transport systems, one for minerals and water, another for food.

Water plays an essential part in all this. It carries the dissolved minerals; it is the source during photosynthesis of hydrogen used in synthesizing carbohydrates, the storehouses of energy; it gives leaves and other tissues support by keeping the cells turgid. And in plants, as in all living things, water is necessary for the cells to work, because it is the solvent for the cell's chemistry. Without water most of the chemical reactions in cells will not take place; one reason is that many of these biochemical reactions involve the addition or subtraction of water.

THE TRANSPORT OF WATER

A plant's need for water can be easily observed. Houseplants left unwatered show an unmistakable symptom: they wilt. First the leaves and then the petioles attaching them to the stem and branches droop, and finally a nonwoody plant's stem becomes limp, giving the whole plant a collapsed look. Why does this happen?

Water pressure usually keeps the cell walls stretched and helps give leaves and nonwoody stems form and structural rigidity, just as a rubber balloon assumes its shape and rigidity when it is filled with air. If a leaf or herbaceous plant loses water, its rigidity goes too, and it starts to wilt. A nonwoody plant growing in well-watered soil or resting in a vase filled with water has cells full of water; it has, in effect, a water-filled skeleton (see Figure 10.2). But even if the plant cannot take up more water, why doesn't the water in its cells still keep it upright? Why does it wilt?

The reason lies in the same phenomenon by which water reaches the leaves—a process known as **transpiration.** It is based on **evaporation,** loss of water, to the surrounding air. This evaporation is due to heat provided by solar radiation. But in transpiration, water is lost not from a large water surface but through tiny openings in the leaves, the **stomata** (from the Greek word for mouth). Whether the roots are taking up more water or not, the plant is almost always losing some water through the stomata.

A plant wilts because it loses more water by transpiration than it takes in. The leaves are affected first, because their cells are first to lose water by transpiration. Photosynthesis then comes to a stop, meaning that no more energy is being produced to do work. If the plant continues to lose water, cells will cease growing, and root cells that take up water-dissolved minerals from the soil will starve and cease their activity. As more water is lost, the whole chemistry of the plant's cells is altered, and soon the plant is dead.

Transpiration and the Leaf's Anatomy It takes little experience in gardening or caring for houseplants to know that, if caught in time, a wilting plant can be revived. The sorry-looking thing will take up newly available water, transporting it to its leaves, and in a surprisingly short time will look healthy and vigorous again. This uptake and movement of water in plants involves the world's greatest water transport system. Take the flow of the Nile, the Mississippi, all the earth's rivers; add waters from aqueducts

10.1 Many leaves have specialized cells at the ends of veins which transport water to the leaf surface when the plant is well watered.

and irrigation canals, modern or maintained since Roman times, carrying water for drinking, washing, industry. Plants move far more water than all these together.

The force for moving this water depends first on loss of water; the stomata control the plant's water pump. These are in the lower and, frequently, upper surfaces of leaves. These surfaces are formed in all leaves by a layer of flat irregular-shaped cells, which seem to fit together much like pieces of a jigsaw puzzle, and form a tissue called the **epidermis** (see Figure 10.5 on page 280). A common type of leaf anatomy has a layer or two of columnar cells beneath the upper epidermis. With their many chloroplasts, these cells are the most active in photosynthesis. They are **palisade cells** (a *palisade* is a defensive enclosure, a kind of stockade made of vertical stakes) and belong to the most common kind of cells in plants, called **parenchyma.** They are thin-walled unspecialized cells, and they make up

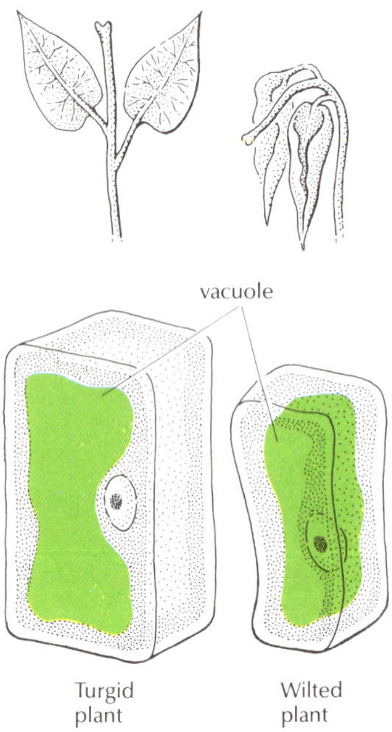

10.2 Turgid cells are full of water, and they provide the skeleton for the soft tissues of plants. Flaccid cells have lost much of their water, and the lack of a skeleton results in a wilted plant.

the cortex and pith, as well as cells of apices, cambium, and many other parts of plants.

Beneath the palisade parenchyma are a few irregularly arranged layers of loosely attached and asymmetrical parenchyma cells, the spongy parenchyma. They also have chloroplasts and conduct photosynthesis. Palisade and spongy parenchyma are called the **mesophyll** ("middle of the leaf"). Bounding the lower side of the mesophyll is the lower epidermis. The veins generally are about in the middle of the mesophyll.

We can think of the leaf as many water-filled bags that must be arranged for efficient exchange of gases but, at the same time, protected from drying out. Plants have many different arrangements of cells to satisfy these requirements, but all have many **intercellular spaces.** The cells are attached to each other, but they have irregular shapes, leaving a good percentage of the surface free to the internal space of the leaf. When the stomata are open, this space is continuous with the external air. The leaf's internal structure has one functional characteristic vital in the exchange of gases between leaf and atmosphere: every cell has surface exposed to this intercellular space. Carbon dioxide and oxygen can therefore freely diffuse into and out of each photosynthetic cell.

A film of water coats each cell, and from here evaporation takes place. The intercellular space is saturated with water vapor, so the leaf parenchyma lives in a pond much as the algal ancestors did. The epidermal layers are an almost transparent waterproofed layer covering this cell-filled pond, but, through the stomata, contact is kept intermittently with the external environment. As long as the roots bring in enough water to keep up with the water vapor lost at the leaf's surface, the intercellular space remains saturated with water vapor. When water begins to be limiting for the leaf's cells, the stomata close.

The opening and closing of the stomata are controlled by the **guard cells,** which are specialized epidermal cells; they are usually sausage shaped. Unlike other epidermal cells, the guard cells have one edge that is not attached to the walls of adjacent cells. When a leaf holds a lot of water, its guard cells are turgid, their adjacent walls separate, and a pore (stoma) is formed, opening a direct connection between atmosphere and the leaf's intercellular space. To become turgid the vacuoles of the guard cells must acquire solute, creating an **osmotic gradient** (an unequal concentration of salt in solutions on either side of a semipermeable membrane), so that water will then diffuse into the vacuole.

It is not known what triggers the transport of salts (such as potassium) in and out of these vacuoles. We do know, however, that transport of salts takes energy, and guard cells also differ from most other epidermal cells in having numerous chloroplasts. (Other epidermal cells frequently do have chloroplasts, but these are usually few, small, and poorly developed.) Photosynthesis undoubtedly is involved with the guard cell's function, but we do not know quite how. Although most carbon enters the food chain through stomata, remarkably little is known about these valves that regulate its passage.

When the leaf has lost too much water, guard cells lose water too, and the pore closes. Even with plenty of light, CO_2, and minerals, plants cannot photosynthesize when the stomata close. Midway through a hot summer day, stomata of leaves exposed to the sun are closed, yet even then some water vapor escapes through cracks between the guard cells.

Protection from Desiccation Plants lose a startling amount of water. One corn plant may throw off several gallons a day. A palm tree in the tropics releases hundreds of gallons daily. Air above the canopy of trees in a tropical forest has so much water evaporated from the leaves that a haze hangs over the forest. This massive release of water by dense vegetation continually affects a region's meteorology. Air over a desert area, with few plants releasing water, nearly always remains extremely dry.

From their very beginnings, land plants have had to contend with a major problem: desiccation, drying out, of their aerial parts. Many adaptations have evolved. Both leaf surfaces are waterproofed by a waxy layer, the **cuticle** (it makes leaves shine), which developed early in the history of land plants; without it, the aerial parts would soon wither. The stomata are holes in this protective coating. Even with this waterproofing, however, water is lost by evaporation.

Water loss in some plants is curtailed even when the stomata are open because the guard cells are completely or mostly on the underside of the leaf, which is cooler and has a slower rate of evaporation. Many **xerophytes** (*xero,* "dry"; *phytes,* "plants") have further protection; their stomata are sunken and the depressions are full of hairs (see Figure 10.3). Like brushy hollows in a dried-out meadow, these damp hairy depressions help prevent evaporation. (They act like sponges absorbing dew, and the hairs restrict air movement and hence evaporation.) Other plants have waxy plates over the stomata; these act as barriers, decreasing evaporation.

Some plants have adaptations that seem more extreme. Members of the club moss genus, *Selaginella* (Chapter 14), are popularly known as "resurrection plants" because they shrink and wither when the environment is extremely dry, only to expand to full size when they get rain. Some are sold dried out and packaged; they revive speedily in a bowl of water.

The leaves of many plants, such as grasses, roll up when desiccated, reducing their surface-to-volume ratio. As any structure grows in volume, its surface area increases more slowly, a basic relationship that accounts for many structural adaptations in plants

10.3 Sunken stomata decrease the diffusion of water vapor from the leaf's interior. In this case, a cycad leaf's epidermal cells, covered by a thick layer of wax, form a volcano-like mound around each stoma. Cycads, with their large palmlike leaves, often grow in hot, dry areas.

(see Figure 10.4). A thin leaf has an extremely high ratio of surface to volume, ideal for exchanging gases with the air around it. Because water is evaporated as water vapor or gas, the thin leaf also gives the most efficient geometry for water loss. On the damp floor of a forest, an environment in which light is limiting, a large thin leaf is the most efficient structure for photosynthesis. But in the desert, light is not limiting, water is, and the plant must be able to photosynthesize without drying out. Reducing the amount of surface compared to volume is the solution. Most cacti are shaped like cylinders or spheres, having little surface in relation to their volume. The thick fleshy **(succulent)** leaves common to many xerophytes reflect this same adaptation.

Many desert plants, and even unwatered houseplants, show another adaptation that cuts water loss by decreasing surface area. The ocotillo *(Fouquieria splendens)*, which grows in deserts of our South-

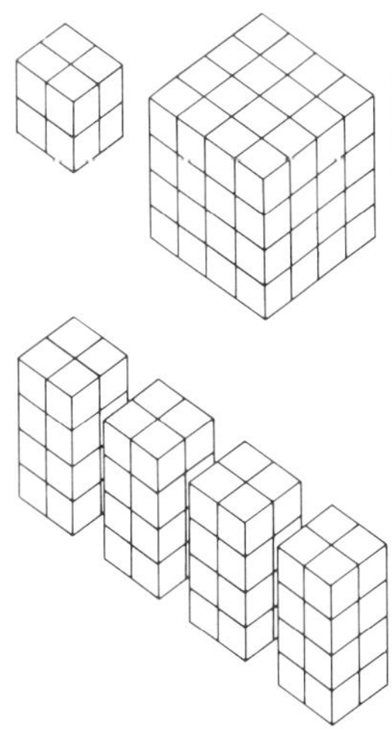

10.4 If an object increases in size so that the surface area is squared, the original volume will be cubed. Thus large, bulky objects, such as tree trunks, football linemen, and supertankers, will have a small surface-to-volume ratio; whereas thin, small volume objects, such as grass leaves and solar water heaters, have a high surface-to-volume ratio.

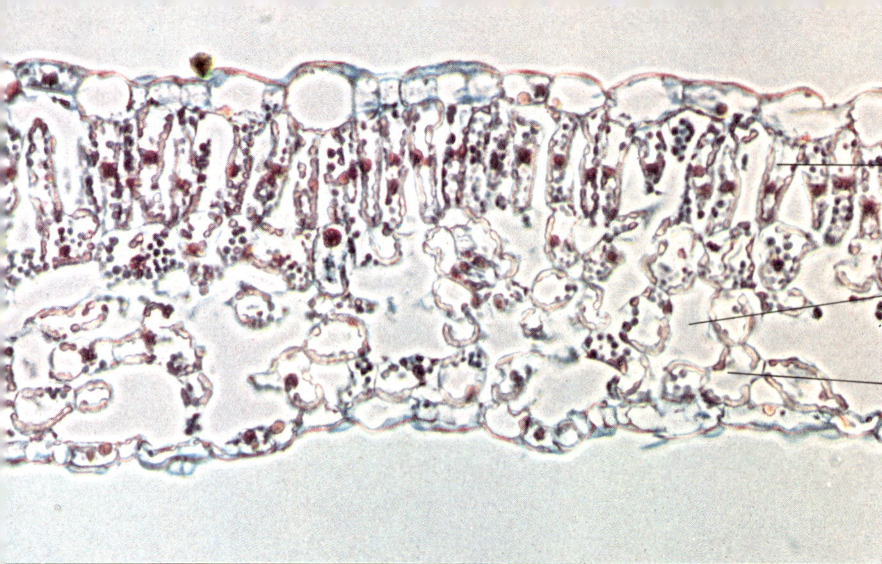
Palisade cell
Intercellular space
Cell of spongy parenchyma

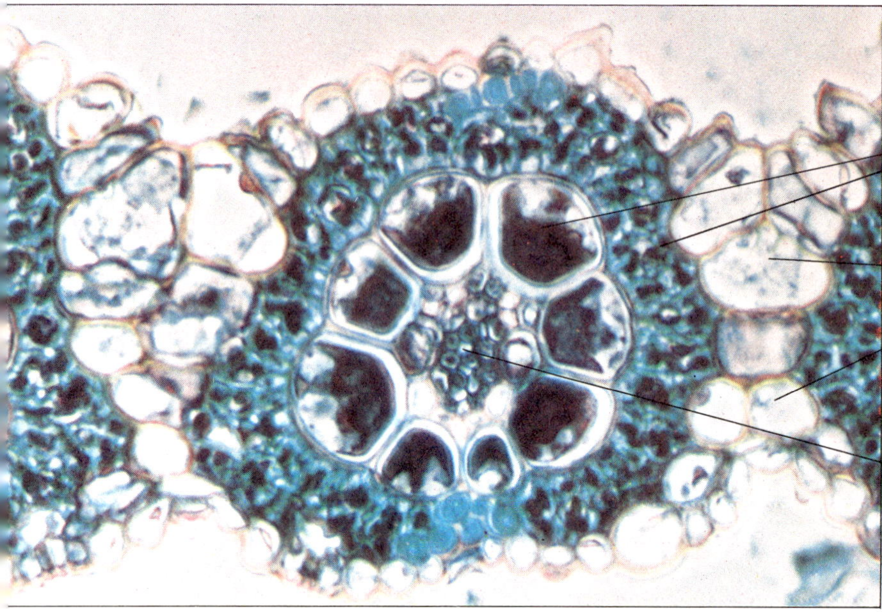
Chloroplast-containing cells
Colorless cells (when they lose water the leaf rolls up)
Vascular bundle

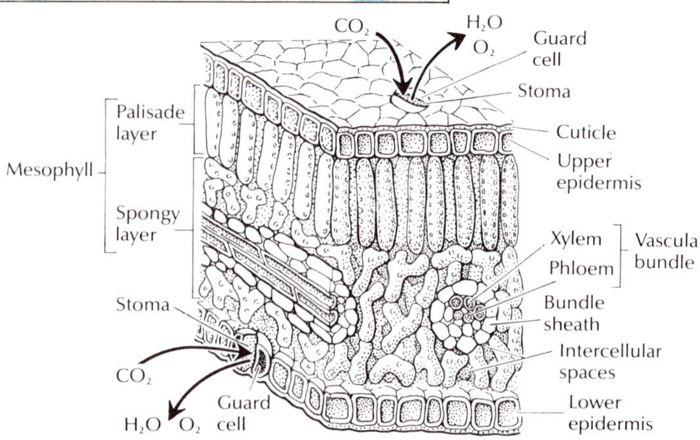
CO2
H2O
O2
Guard cell
Stoma
Cuticle
Upper epidermis
Palisade layer
Mesophyll
Spongy layer
Xylem
Phloem
Vascular bundle
Bundle sheath
Stoma
Intercellular spaces
CO2
Lower epidermis
H2O O2
Guard cell

10.5 LEAVES: STRUCTURE AND FUNCTION

A cross section of a tobacco leaf (facing page, top) reveals an anatomy similar to that of many dicot leaves. Two types of mesophyll cells, palisade parenchyma and spongy parenchyma beneath it, are sandwiched between upper and lower epidermis. Another type of leaf anatomy, one common in grasses, is illustrated by a stained section of buffalo grass *(Buchloe dactyloides)* (facing page, middle). Vascular bundles are surrounded by a ring of large chloroplast-containing cells called a *bundle sheath,* and this in turn is surrounded by a layer or two of mesophyll cells. The large cells without chloroplasts beneath the upper epidermis control the rolling of the grass leaf. When they lose water, the leaf rolls up and when they are turgid, they force the leaf to unroll. The irregular shape of mesophyll cells in a grass leaf and the large intercellular spaces are illustrated in the highly magnified section of a grass leaf (below, left). The spongy parenchyma of dicot leaves has a similar appearance.

Transfer of gases and evaporation of water take place at the boundary of intercellular space and cell wall surface. The surface of a lily leaf (below, right) shows how a leaf can be like a sieve when the stomata are open. It is through the stomata formed between two guard cells that oxygen and water leave the interior of the leaf and carbon dioxide enters. All the carbon in your body once entered stomata similar to these. The drawing (facing page) depicts the three-dimensional structure of the leaf. The guard cells, so vital in regulating transpiration and diffusion of carbon dioxide are characterized by unattached and unevenly thickened cell walls on one side of the cell. When water enters the vacuoles, force is exerted on the thickened walls, which act like springs which always bend outward, and an opening, a stoma, is formed. When the leaf experiences water stress, water leaves guard cell vacuoles, the walls straighten out, and the stoma is closed (below).

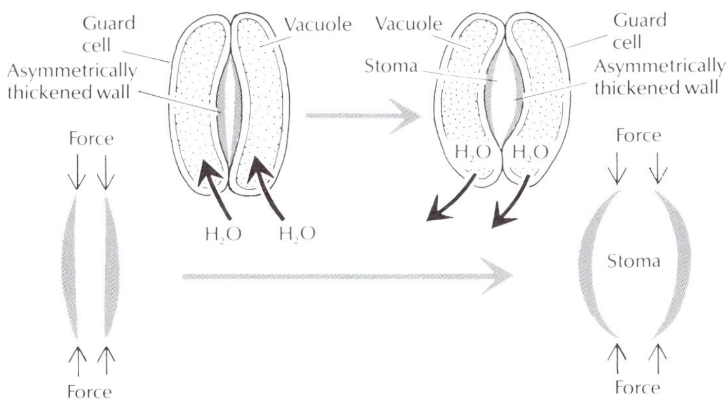

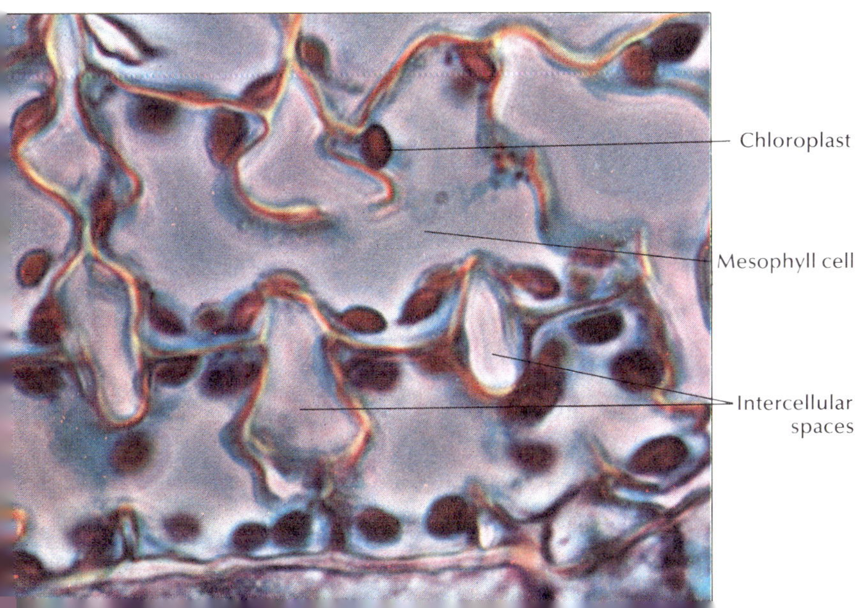

10.6 The ocotillo loses its leaves when little water is available. This reduces the surface-to-volume ratio and conserves water. When water again becomes available, it will produce leaves (right). This can happen several times each year. When moisture is abundant, the plants will also flower.

west and of northern Mexico, is a good example. During an extended drought the leaves fall off; the plant remains leafless, conserving water until rain comes or until it is watered, when it produces a new batch of leaves. While it had all its leaves, the surface area the plant exposed to the relatively dry air was large; losing leaves eliminates a large surface for evaporation.

Arid or semiarid lands, too dry for most crops, occupy a significant percentage of earth's terrestrial surface and are found in many heavily populated parts of Africa and Asia. Recurrent drought adds many hectares each year. About the only hopes for making such land bear crops to feed more and more millions are plants adapted to grow in extreme dryness. For people in dry countries, the difference between varieties of grain that are drought resistant and others that are likely to die under arid conditions can be the difference between life or famine.

A small leaf will have hundreds of stomata on its surfaces, and they present a paradox. Water evaporates most rapidly around the outer edge, the perimeter, of a pore. Evaporation therefore is most rapid when there is a large perimeter area compared to the pore's diameter. Water evaporates more rapidly from many small pores than from a few large ones, because the size of the perimeters added together is much greater. Paradoxically, the leaf has evolved a most efficient system for evaporation of water, even though this system contributes to the plant's greatest enemy, desiccation. Land plants have evolved in a kind of constant juggling act: How can the plant procure the CO_2 it requires without doing itself in? Water, CO_2, and O_2 are passed through open stomata, even though the plant risks simultaneous dehydration. But transpiration also provides the force pulling the water, and the nutrients dissolved in it, from roots to leaves.

Moving Water Many plants lift water a few centimeters, but in trees water rises to the height of a thirty-six story building. Pipes leak and pumps fail and dams crumble, but the plant's "machinery" continues throughout its life, fueled by solar energy. How can trees like sequoias raise water to such heights?

At one end of the network, roots take up water; at the other, leaves lose it to the outside air. Roots and leaves must therefore be connected, which is the function served by the vascular system discussed in Chapter 6. The roots grow every which way among the soil's particles and take in water and dissolved minerals. Root hairs absorb this mineral-enriched water, which then goes into the root cortex and xylem. From there it is drawn into the plant stem. But the water-conducting xylem is made up of dead cells; it has no pumping mechanism. What, then, causes water to move upward in the stem, against gravity?

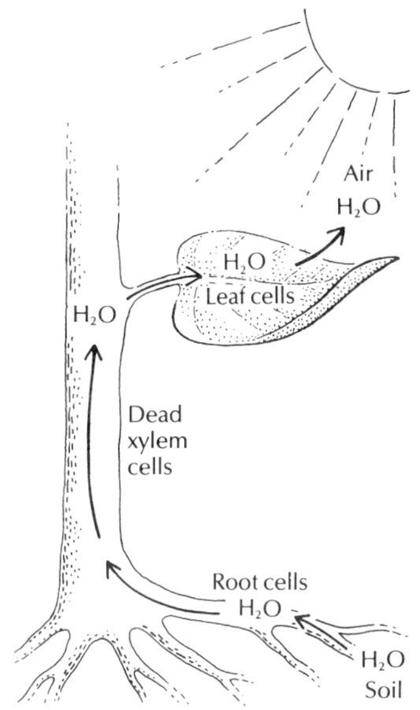

10.7 The water conduction system of plants consists of dead xylem cells. Water enters this system through living root cells and leaves it by evaporation through leaf stomata. Solar energy provides the carbohydrates which root cells burn for energy to take up water. Solar energy also pulls water through the plant by evaporating water at leaf surfaces.

We have seen that transpiration is the force; as water vapor is lost through the stomata, water is pulled upward through the leaf. But explaining the strength of this pull requires other forces, too. We will need a new image of how water flows, not rushing from pool to pool, or running in streams, or gushing from spring or faucet, and certainly not in large bodies. In plants, water moves in pipes as fine as a silk thread. Water in threads acts very different from the water we see anywhere else. Picture the cables on a bicycle, running inside thin plastic-covered steel tubes from handlebars to brakes and shifting mechanism. Confined in the plant's water-bearing channels, the water column acts much like the cable. It hangs together, molecule by molecule, in a thread that can pull and be pulled.

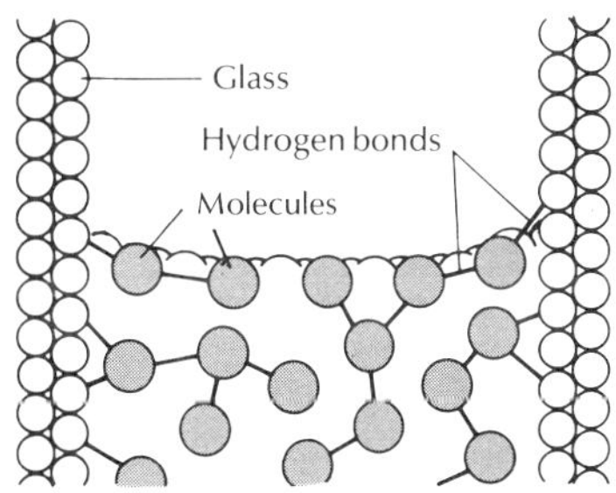

10.8 Water molecules, attracted to each other and also to the oxygen atoms in glass, adhere to the sides of a glass tube, forming the crescent-shaped surface. *Cohesion* is the attraction between the water molecules. *Adhesion* is the attraction between molecules which makes water adhere to the sides of a glass, or to dead xylem cells.

In any kind of vascular plant, the forces that move water against the pull of gravity are always the same. Transpiration is the force, and the "cables" of water are made possible by **adhesion** of water to the walls of xylem elements and **cohesion** between one water molecule and another.

You can see adhesion in a glass of water on a table. Look closely: the water's surface curls upward where it touches the glass. To see it even better, dip a transparent soda straw into a soft drink or cup of coffee. Lift the straw out and most of the coffee will run out the bottom, but some of it stays in (adheres to) the straw. This is called **capillary action.** This force, multiplied many times, helps transpiration pull water from cell to cell through a tree up into branches, twigs, and finally leaves.

Cohesion, the attraction between water molecules, keeps the

threads of water intact within the conduits. A similar attraction works in the chemical reaction that produces water from hydrogen and oxygen. The hydrogen is attracted by the oxygen. Likewise, in the plant's fine threads of water, this attraction between hydrogen atoms and oxygen atoms causes cohesion between molecules of water and keeps the threads unbroken as they are pulled upward. Attraction between molecules also makes water adhere to the xylem cell walls.

Water-Conducting Cells Dead xylem cells transport water. A nineteenth-century German botanist, Edouard Strasburger, demonstrated this by a famous experiment. He boiled a 12-meter (40-foot) section of a wisteria vine in a large pot. All cells in the stem had to be dead, yet the leaves at the top did not wilt; clearly, the wisteria vine was still pulling water through all 12 meters of its length.

In dying, many xylem cells in angiosperms form channels because holes are formed in their end walls. The cell walls remain, however, and these strings of open-ended cells, one on top of the other, make a tube.

But why don't the cell walls collapse, now that they are no longer turgid? As they develop, the primary walls of these cells change. The cellulose of which the walls are composed does not disappear, but something is added along with new cellulose: lignin (see page 202). The tube now has much harder walls than those of the primary cell. Thick cellulose walls impregnated with lignin are the tree's woody structure, and with the xylem cells dead already, the main structure does not change even if the whole tree dies. The newly formed plumbing system is also its supporting skeleton.

The xylem's conducting cells vary somewhat in different kinds of vascular plants. In conifers all the conducting cells are **tracheids** (shown in Figure 10.9), tapered like spindles at one end, overlapping each other at the thin end like a long row of spindles partially pushed alongside one another. In this type of cell, the upward-moving water goes through thin-walled areas, the **pits,** along the tapered ends of each cell. Angiosperm xylem has both tracheids and varied open-ended cells of miscellaneous lengths called **vessel elements.** The long pipes or vessels formed by these elements are a more efficient system for conducting fluids than the conifers', and the gradual change from tracheids to specialized vessel elements is an interesting step in the evolution of vascular plants (see Figure 10.9).

Vascular bundles extend into the leaves, where they form a system of connected veins. Each vein consists of xylem and phloem cells. Small veins include only a few cells of each, and in many dicotyledons the ends of veins are single tracheids. The gridwork of veins keeps each leaf cell near a water pipe and the system for transporting food.

Consider how these forces work together. Plants lose water by evaporation through their stomata. When the stomata are open, this loss is constant: molecule by molecule, the gas diffuses to the open air through the stomata, and as it escapes, more water molecules are pulled upward, replacing the losses. Inside the plant is a constantly moving supply of water, beginning at the roots, with adhesion holding it along the walls of the conduits, cohesion holding the threads together, and evaporation from the leaf surface pulling the whole column upward.

Cohesion is a negligible force in a pipe as big as those that carry water in our homes (half an inch or so in diameter) because the molecules of water are too diffuse for the molecular attraction between them to take effect. But in a thread of water as narrow as that within the plant's xylem elements, molecular attraction is very powerful; and many measurements have shown that the tensile strength of this cable of water is enough to lift water to the tops of our tallest trees—over 100 meters (300 feet).

This pulling force shows in the thickness of trees while they are actively transpiring: the tree's diameter actually shrinks. Every small pipe in the xylem is a bit thinner, because of the pull exerted by the thin threads of water and the force of adhesion drawing the walls closer together. These microscopic changes in millions of cells cause small but easily measured changes in the trunk's diameter.

WATER UPTAKE

What happens at the other end of the plant's pumping system: the roots? How do the roots take up water from the soil? And once they have taken it up, how do they start it on its way?

A partial answer at least is osmosis (see page 199). It starts with an osmotic gradient. Water diffuses through the semipermeable membrane and equalizes the salt solutions by diluting the stronger one. Root cells take up water by osmosis because the salt solution in their vacuoles is more concentrated than the soil solution.

If the root hair cells cannot maintain a concentration of salt higher than that of the soil solution, water will move *out* of the root cells and the plant will die. For that reason salt flats have very few plants, sometimes none; seedlings cannot survive because the roots become dehydrated. Land recaptured from the sea by diking must be rinsed often with fresh water before crops can be planted. When the Romans destroyed the city-state of Carthage in 146 B.C., they supposedly spread salt over the ruined land to prevent resettlement. The Romans, like most people throughout history, had a practical knowledge of osmosis.

The osmotic problems of a salty environment should make us wonder about plants and animals living in or on seawater. They do take up a lot of salt, but they have ways of dealing with excessive

10.9 XYLEM

The different appearance of wood is reflected in the form of xylem cells composing it. Pine wood (photo, below left) consists of narrow, spindle-shaped cells called *tracheids*. Water passes through pits in the side walls of tracheids. Birch wood (below center) has vessel elements with openings in end walls. Oak wood (below right) has long conduits called *vessels,* consisting of short cells without end walls. The interior of vessel elements is revealed in a photograph made with the scanning electron microscope (facing page, bottom left). Veins in leaves are like rivers, carrying water and minerals to cells and receiving photosynthetic products from them (top photo, facing page). Every cell in a leaf is close to a vein as seen in the cross section of a corn leaf (middle photo, facing page). Most plant secondary tissue is xylem (facing page, bottom right). It consists of dead, thick-walled conducting cells, called *vessel elements,* and tracheids, fiber cells, and parenchyma cells.

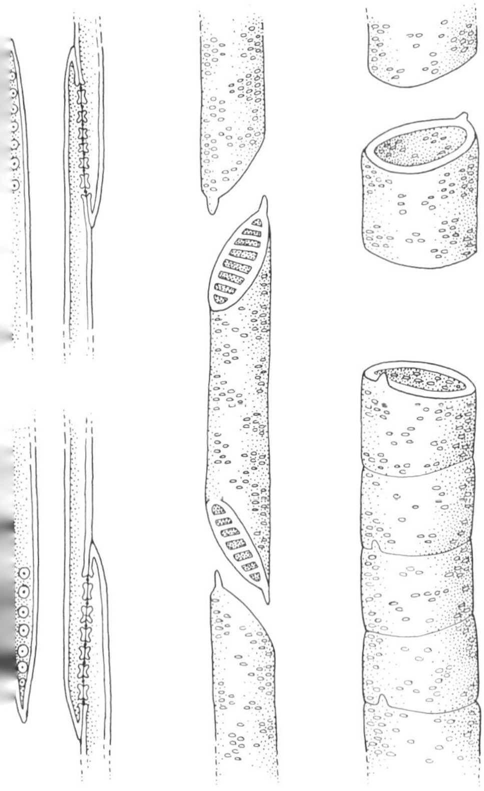

Pine Birch Oak

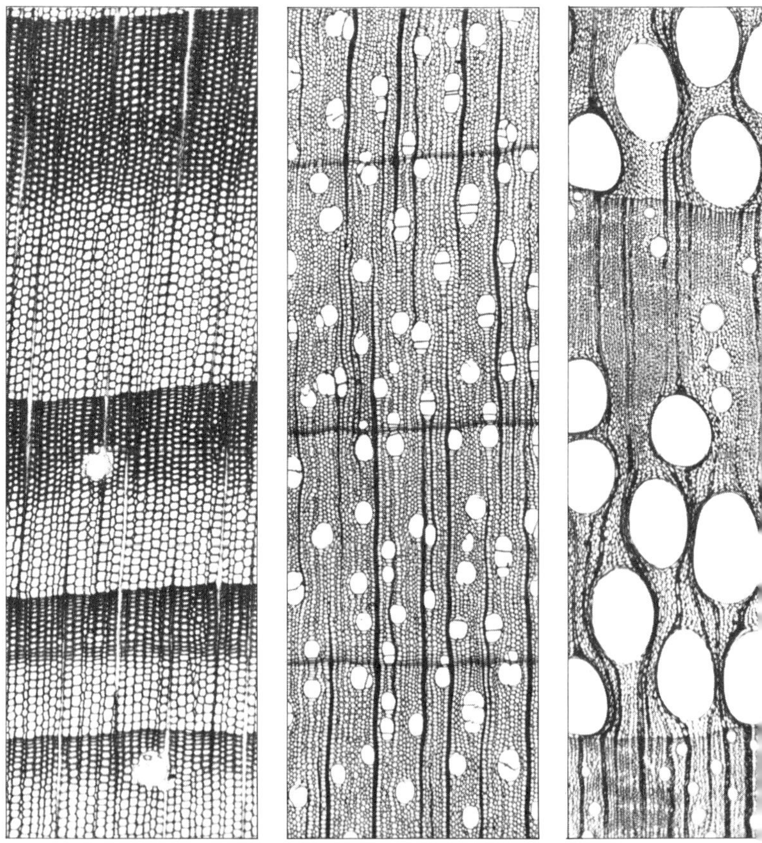

Pine Birch Oak

Vein

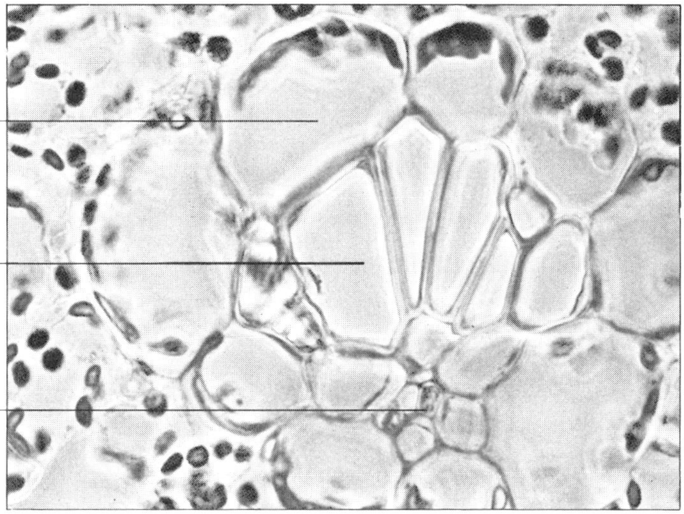
Chloroplast-containing cell
Vessel element
Phloem

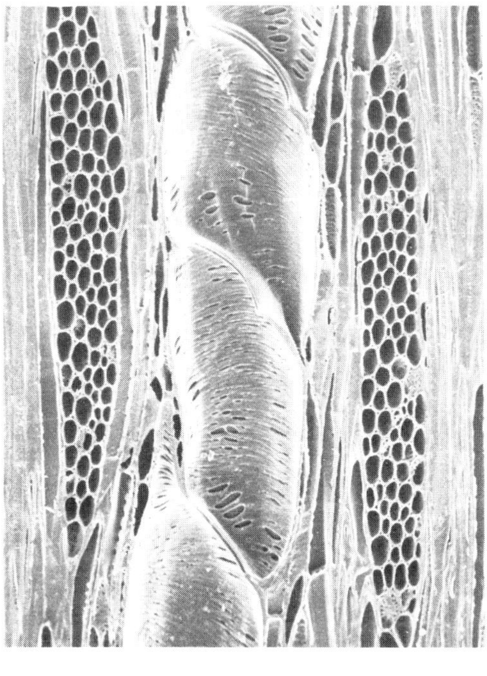

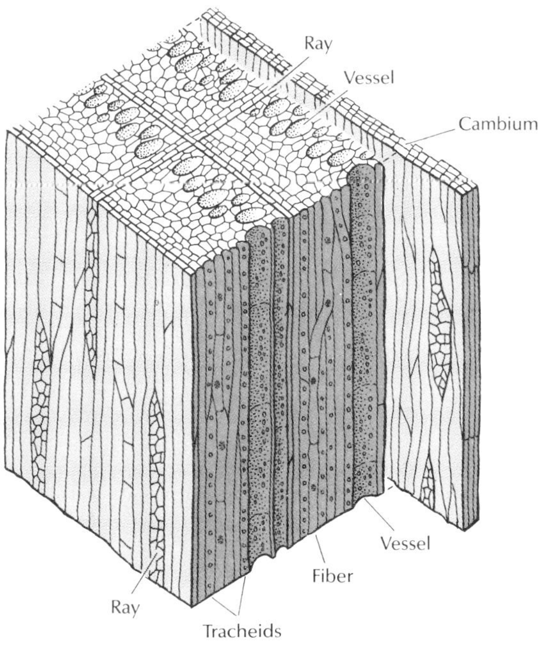
Ray
Vessel
Cambium
Vessel
Fiber
Ray
Tracheids

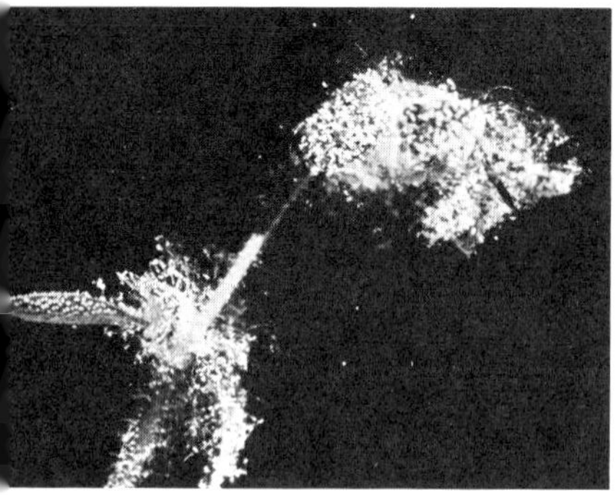

10.10 The large amount of salt many halophytes can secrete is demonstrated by the salt crystals and salt whiskers formed on the stems and leaves of this plant of alkalai heath *(Frankenia grandifolia)*. Salt glands secrete a salt solution and rapid evaporation forms the salt whiskers.

salt, such as specialized organs for excreting it (see Figure 10.10). For example, cord grass *(Spartina)* has specialized cells of the leaves that collect and excrete salt. Similar difficulties are solved by **halophytes** (*halo*, "salt"; *phytes* ,"plants"), land plants growing in habitats where the soil water is so salty its concentration can exceed that of seawater.

How do roots get water in the first place? Each rootlet among particles of soil grows a dense crop of root hairs outward from its surface (see Figure 10.11). These hairs are in fact elongated epidermal cells, and they take up water directly from the soil. The size of the absorptive surface of roots is often unappreciated. A grass plant, for example, can have 300 to 400 miles of roots, providing an absorptive surface one-half the size of a football field.

For a clear idea of how water moves around, visualize the plant's interior as having two parts. All the intercellular spaces and cell walls, the plant's nonliving space, is like a diffuse sponge. Water and minerals can diffuse throughout this open area by capillary action (remember adhesion and cohesion). The sponge is kept in continuous contact with the solution in the soil by root hairs. Isn't the soil solution therefore continuous with the transpiration stream in the xylem of the root, making an osmotic gradient impossible? It would be continuous except for the **endodermis** (*endo*, "inner"; *dermis*, "skin"), a ring of differentiated cells outside the vascular cells (Figure 10.12). These outer cells have a ring of waxy material, the **Cas-**

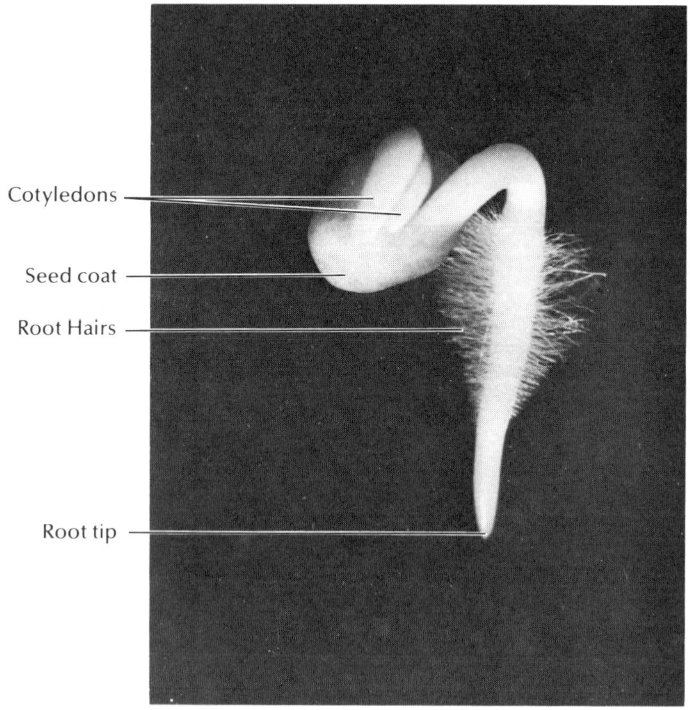

10.11 Water and minerals enter plants through root hairs, elongated cells of the root epidermis. Root hairs arise where xylem is differentiating, behind much of the growing region. Like most plants, the radish root is quite long before the shoot begins to grow.

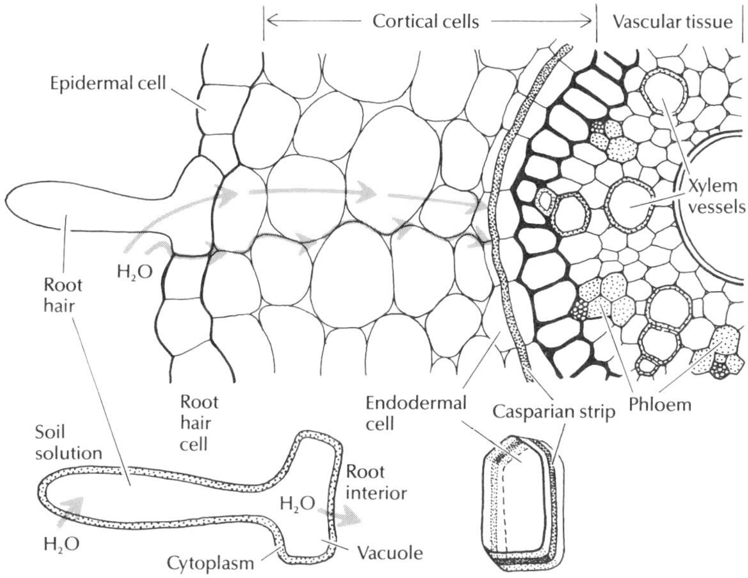

10.12 Water enters cells in two ways. One is by means of osmosis, with water entering first into the vacuoles of root hair cells and then into cortical and other root cells. Water also diffuses through the porous cell walls but only to the endodermal cells, where passage is inhibited by a waxy deposit in the cell walls. All water and salts in the transpiration stream in the xylem must pass through the plasmalemma of endodermal cells, so an osmotic gradient is maintained.

parian strip, around the middle of the walls in a plane parallel to the inner and outer cell walls. Solution from either the vascular tissue or cortex must move through the endodermis. All water and salts from the soil must pass through the **plasmalemma** of the endodermis. This selectively permeable membrane maintains the osmotic gradient from root hairs to xylem.

In the plant's water transport system, all water taken up by the roots moves through the dead xylem cells on its upward way. Dissolved in it, mineral salts also taken up by the roots move in the transpiration stream to leaf cells, where they accumulate in the vacuoles. Because the leaf cells lose water in transpiration, their salt concentrations are relatively high. This creates an osmotic gradient between leaf cells and the transpiration stream in xylem and the veins, which helps keep the stream of water moving upward through the xylem. But remember that the cells through which the water moves are dead; they do not pump water along.

MINERAL UPTAKE

Mineral salts are more highly concentrated inside the root hair cells than in the soil solution, so root cells take up water by osmosis. But where do the minerals come from, and how do they get through the semipermeable cell membrane?

No one has found a completely satisfactory explanation, but somehow dissolved minerals do get into the roots, as proven with radioactive isotopes. Submerge a plant's roots in a solution of potassium 41 and chlorine 36 and the vascular tissue of the roots will show radioactivity. Thoroughly wash the roots, rinse, and wash again to make sure that no radioactive particles cling to the outside parts, and the vascular tissue still shows radioactivity. Potassium 42 and chlorine 36 somehow have passed through the outer layer of root cells.

Mineral Carriers The hypothesis proposed to explain this phenomenon states that **ions** (charged elements or molecules) are carried into the interior of the root system actively aided by chemical compounds called **carriers,** which combine with the mineral ions outside the root cells and release them once they are inside. For the ions this is a one-way trip. They cannot get outside again, but the carriers can and do. After releasing ions they go back through the cell membranes, pick up more minerals, and do it again.

That something like this has to happen is now reasonably certain. Take barley roots that have no chlorine in them and dip them into a solution containing 0.2 part per million (ppm) of radioactive chlorine. Within an hour the roots will hold 40 ppm, a 200-fold increase over the chlorine in the solution outside the roots. With a difference like that, you might conclude that the chlorine inside the roots would diffuse to the solution outside so as to equal the osmotic balance. But it does not, proving that the root cells are able to concentrate the minerals they absorb.

Further experiments show that when a mineral reaches a specific concentration, the rate of absorption levels off. Then a second transport system takes over. This system apparently takes care of selectivity. It seems to screen out minerals that the plant does not require, or requires only in small quantities. Suppose you set up an experiment, immersing roots of a barley plant in a potassium solution. As we saw in Chapter 9, potassium is essential to plants, and the roots will take up a great deal of the potassium solution. But what will happen if we add sodium to the solution? Sodium is a very close relative of potassium in the *periodic table of elements* (a classification by atomic number of the elements), but it is not essential to most plants. It will be absorbed along with the potassium, but only up to a certain level.

What is the practical value of this selectivity to the plant? Imagine a plant growing in one of the many desert or semidesert areas, where sodium is often more highly concentrated than potassium. Yet it is potassium that the plant requires, and the discrimination of the secondary transport system keeps it supplied with potassium, while preventing it from being saturated with sodium. Many details of the way this ion transport system operates have not yet been worked out.

Work is going on in the roots of plants all the time. The energy to fuel the extraction of minerals from the soil does not, however, come

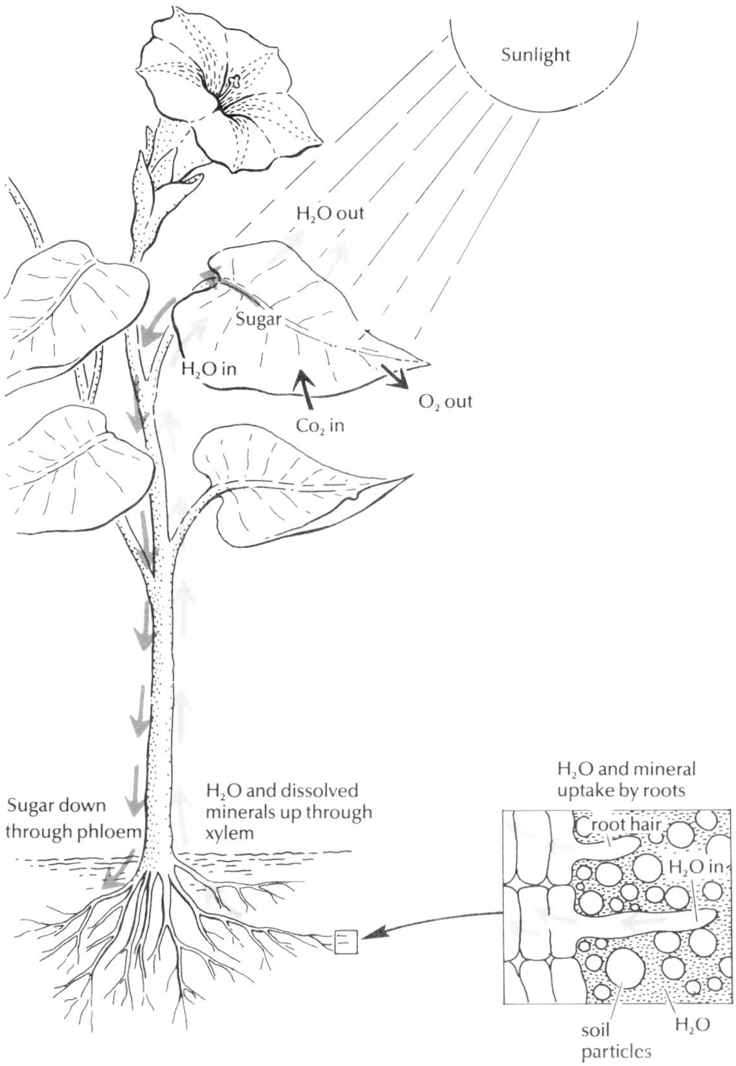

10.13 Solar energy is responsible for the synthesis of food, which is conducted downward through living phloem cells to the roots. Roots use energy from this food to maintain an osmotic gradient, which will move water into roots. Water and salts are pulled up through the dead xylem as a result of solar-powered evaporation of water from the leaf surface.

from the large amounts of mineral substances that the plants mine. It comes down from the leaves, in the form of sugars and other organic compounds produced by photosynthesis. The energy broker for transporting ions in root cells is, as you might suspect, ATP produced from sugars (see page 225).

Transpiration and Mineral Transport Is transpiration a waste? Vast amounts of water are moved through the plant, but only a small part is used. All the minerals required by the plant move in the transpiration stream, so that the loss of water, which appears excessive, is the price paid for a steady supply of minerals.

Remember that plants mine a very dilute ore and concentrate it in their tissues, a fact demonstrated by the ashes of burned plants, which are a very concentrated source of minerals. It takes a steady stream of dilute mineral solution to achieve this concentration. The energy to evaporate water from the leaf surface, which maintains this transportation system, is solar, so if too much water is not lost by evaporation, transpiration costs the plant very little. Just as carbon is made available to organisms in the biosphere by means of solar energy, minerals are brought up from deep underground in the solar-powered transpiration stream to begin their cycling through the biosphere.

Plants have another problem handled by transpiration: the great heat load on leaves from solar radiation. Water vapor evaporating from the surfaces of leaves keeps them cool. (Think of how rapidly evaporating water cools your skin.) Without this water-cooling system, they could heat to lethal temperatures.

TRANSPORT OF FOOD

The roots must get all their energy ultimately from the leaves. Food comes to them as **sucrose,** a sugar that is highly soluble in water and does not readily react with other substances. These are ideal properties for a material that must be moved a long distance. Produced in the leaves by photosynthesis, the plant's food must be carried to other parts of the plant where food is needed to provide energy for growing and doing work. This food transport takes place in the phloem. Though upward transpiration by xylem and downward translocation by phloem cells have some similarities, there are major differences.

Movement in the Phloem Cells For one thing, although both food and water travel through long strings of cells, the phloem cells are not dead but very much alive. Living phloem cells form "pipes" with perforated end walls that look like sieves, the **sieve tubes** (see Figure 10.15 on page 294). Movement of the sugar solution downward in the phloem is thought to be driven by **mass flow,** a source-sink movement. Something at the bottom (roots) of the transport system keeps using up material that the top parts (the leaves) put into it. It is a little like the upward-moving system in reverse. Water carried by the xylem is pulled up by transpiration; the sugar-rich solution, commonly called **sap,** is pulled down by the roots.

Sometimes the solution in the phloem cells appears to move in both directions at once, but how it can do so is not yet known. Studying the movements of solutions in the phloem cells is quite difficult. But something of their movement can be learned with radioactive isotopes. Carbon 14 can be incorporated into photosynthetic products (see page 229) and its direction and rate of movement can be easily determined.

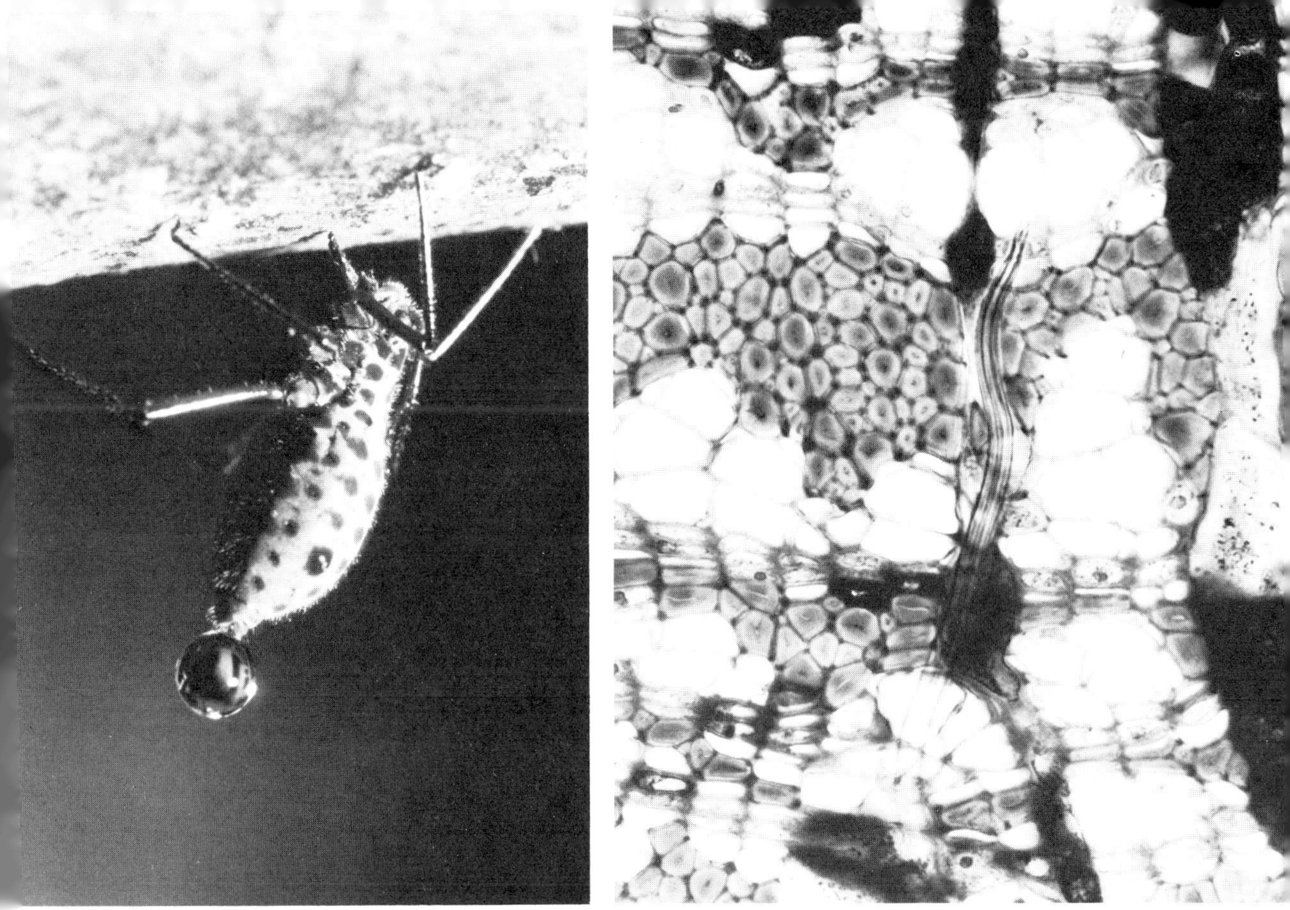

10.14 The aphid (left) has proven useful for studies of phloem transport. Its mouthparts penetrate phloem cells and materials being conducted are forced, under pressure, into the aphid. The sweet sap from the phloem is excreted by the aphid as "honeydew", which is collected by ants. The aphid may be removed from its still-inserted stylets (see microscopic view of stylets, right); and the phloem sap, pumped through the stylets, may be collected and analyzed.

Nature provides another most useful tool for studying movement in the phloem, the aphid (see Figure 10.14). This tiny insect, loathed by gardeners, is equipped to tap the sugar-rich solutions carried in the phloem much as the mosquito taps human blood. The aphid's mouth parts are precisely adapted to transport the sap of plants. Instead of a mouth it has a **stylet,** a thin flexible tube that can be pushed right through the outer cell layers of a plant leaf or young stem until it reaches a living phloem cell rich in sugary sap. The sap is under pressure and streams up the stylet into the aphid. The insect can extract so much sap that copious waste exudes from its afterparts in a clear drop of "honeydew," which in turn is often collected by ants, giving aphids the nickname "ant cow."

The gardener dislikes aphids because they can remove so much sap that a plant starves to death. On the other hand, some scientists love them because their stylets can get down to live cells of the phloem; no scientific tool yet invented can do so with such accuracy. The researcher lets the aphid start feeding, then knocks it out with ether, after which the body can be cut away leaving only the stylet, through which sap continues to flow. Because the sap may keep coming for days, enough can be collected to analyze its contents and calculate its rate of flow in phloem.

10.15 PHLOEM

Phloem as seen in the light microscope is observed in cross section (facing page, top left) and longitudinal section (facing page, bottom right). The cross section also shows fiber cells. Sieve plates, the end walls of sieve tube elements, are observed in longitudinal sections (facing page, bottom left). The dark bodies in the sieve tube elements are proteinaceous material called *slime bodies.* The phloem is bounded by ray cells. The drawing of phloem tissue (below, left) is a three-dimensional view. During their differentiation, sieve tube elements lose their nucleus, but each sieve element has a closely associated cell called a *companion cell* (facing page, bottom left) which does have a nucleus. They are connected by numerous plasmodesmata, and the companion cells function as the control center for synthesizing new proteins in both cells. Sugars, unlike water, are transported by living cells. Leaf cells assimilate carbon dioxide into sugars which are transported to the roots. Root cells respire the sugar and give off carbon dioxide in the process (below, right).

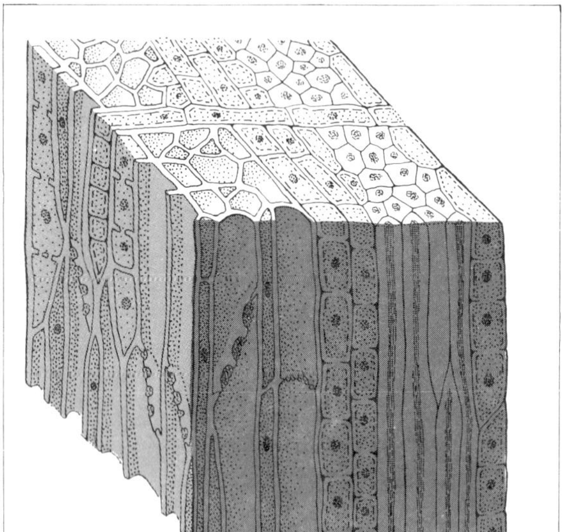

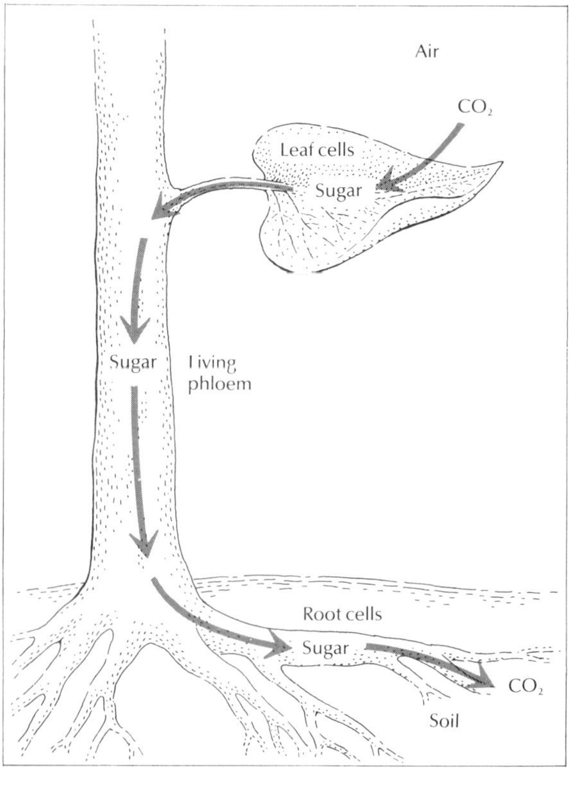

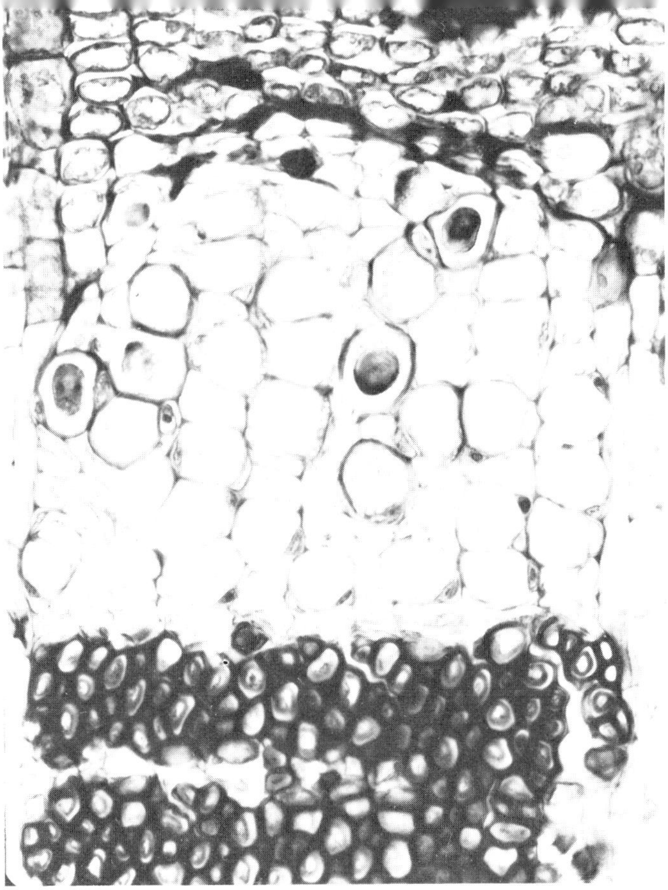

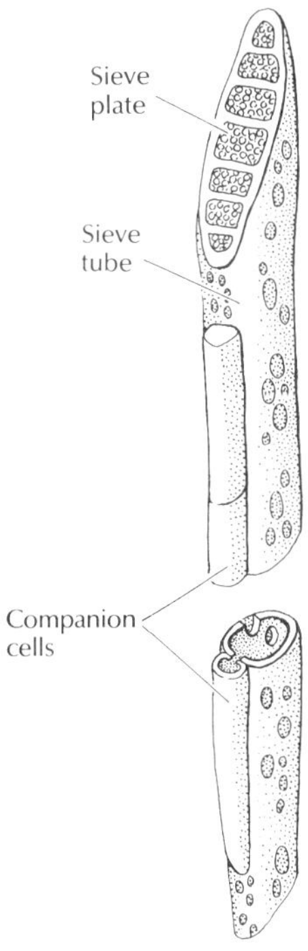

Sieve
plate
Sieve
tube
Companion
cells

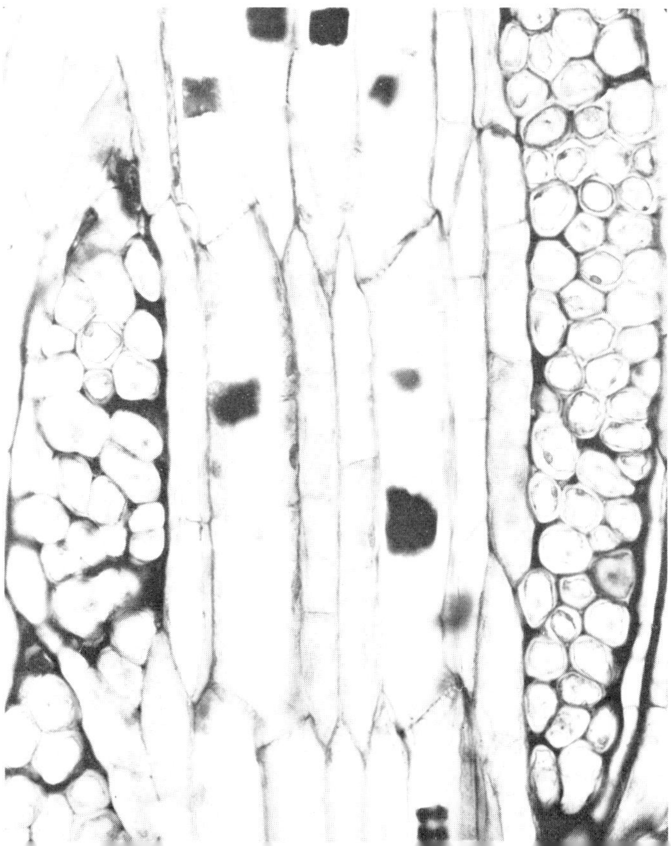

The Sap With help from aphids, it has been learned that the sap carries sugars in concentrations of 10 to 30 percent. It may also have sugar alcohols, vitamins, amino acids, organic phosphorus compounds, and inorganic ions. The flow rate is remarkably high, occasionally more than 5 cubic millimeters (.008 cubic inch) of sap per hour from one stylet. The surprise is that each phloem cell has to be refilled anywhere from three to ten times per second, and that the sap can flow at about 100 centimeters (40 inches) per hour. Only an active system of transport, such as mass flow, could account for such fast movement.

Puncture a phloem cell and the sap flows out; it is traveling under positive pressure. The familiar tapping of sugar maples in the spring produces a sweet sap that is boiled down to make maple sugar, but this is not the sugar solution carried downward in the phloem. It is converted starch carried upward in phloem from the roots. Why is it made? Many plants flower before they produce new leaves and, in any event, development of leaves in the spring requires large amounts of energy before the leaves can start to photosynthesize. Sugar maples store up starch in their roots; in spring, the starch is turned into sugar, forming a sweet sap that rises in the tree and supplies energy for leaf development. This is the sap that is tapped for sugar making. Sap of other trees is also sweet, but for an unknown reason, the sugar maple produces copious amounts.

Loss of Sap Almost any plant will exude sap when cut or punctured, but almost as soon as the injury happens, events in the phloem start to prevent the loss. The pores in the end walls of the damaged phloem cells are almost immediately plugged with **callose,** a complex sugar and an effective seal. Callose will eventually plug the end walls of the phloem cells where a branch has been cut from a tree. (In the same fashion, sap from maple trees will not continue to flow from a hole; a new tap is needed.)

All this sugar-rich liquid, excess food produced by photosynthesis in leaves, flows through only a few layers of cells that ring the trunk of a tree. It can lose part of its phloem cells in various ways, and enough sap will still flow through the phloem that is left to keep the tree alive until new phloem cells and bark grow. But if phloem cells are lost all the way around its trunk by **girdling,** the tree will die as soon as the reserves of food in its roots are used up. A traditional method for killing trees, girdling works because new cambial cells cannot grow back and produce phloem cells in time to transport needed food supplies to the roots before the tree starves. The reverse is seen with hollow trees. With its center almost completely hollowed out, a tree can still live because enough xylem and phloem remain around the exterior to transport water and minerals to the top and food to the roots.

10.16 Trees may be killed by cutting away a ring of bark. This removes phloem and prevents the downward conduction of photosynthetic products. When the supply of starch to the root cells is depleted, they will die. The lack of water and minerals results in death of the leaves.

OTHER TRANSPORT SYSTEMS

Lateral Transport For simplicity's sake we have treated transport as if it went only up and down. That this cannot be the whole story becomes clear when xylem and phloem are partially destroyed. Cut a deep notch in a tree and water and nutrients continue to flow through the intact xylem and phloem. If the trunk is tapped right above and below the notch, water from the roots will come from above the notch and sugar from the leaves from below the notch. Since they have traversed the xylem and phloem bridge, those materials must be transported laterally.

In the thin regions on the lateral walls of the xylem elements are the pits, where the thin and porous primary wall of cellulose is exposed. Water can diffuse laterally through those pores as it ascends the tree in the stream of transpiration. This ability to detour is an adaptive advantage. Think of xylem sectors clogged by disease, fire, or mechanical injury. This partial disruption of water flow is not unusual, because water flows only in the relatively young (though dead) xylem, the so-called **sapwood** (see Figure 10.17). The living cells in the xylem (the rays) transport waste products of other living cells; these are deposited in and gradually fill up the older xylem elements forming the **heartwood.** The center of the tree becomes its garbage dump.

Lateral flow is vital when water in the stem freezes during a hard

10.17 The trunk of a large tree is mostly dead wood, through which little water is transported. Most water flows through the younger wood, which is toward the outside of the tree. The living cells of cambium and phloem are a barely noticeable layer beneath the cork.

winter. The freezing point of water is determined by its concentration of solutes (the principle behind antifreeze in car radiators and rock salt on icy sidewalks). Because water in the xylem is a salt solution, it takes below-freezing temperatures to make it freeze. When it finally freezes, the water contracts, sometimes snapping the thin threads of water in the xylem and interrupting continuity. Breaks in the water columns leave air pockets in the xylem elements, which keep the columns from rejoining when the ice melts. It appears, however, that vertical contact is not entirely cut off and that lateral movement reestablishes the columns.

What about the phloem? Like all the plant's living cells, those in the phloem are connected by cytoplasmic channels through the cell walls, the **plasmodesmata** (see Figure 10.18). Plasma membranes and cytoplasm of all the cells in the plant appear to be joined in this way, which explains how viruses can spread throughout a plant once they infect a cell. The plant, then, is not simply a pile of separate bricklike cells. Cells stay in one place and organelles stay in one cell, but all are joined by very thin strands of cytoplasm

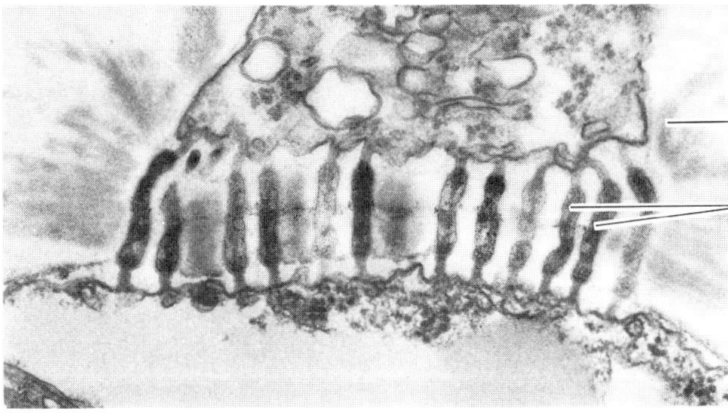

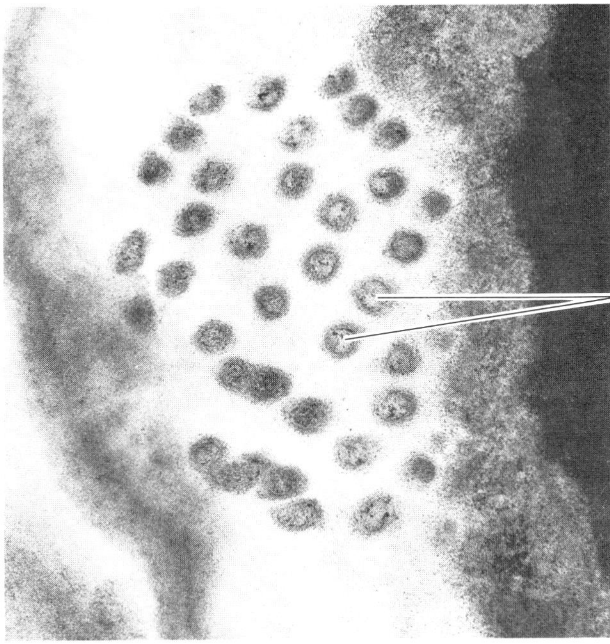

10.18 Living cells of plants are connected by cytoplasmic strands (called *plasmodesmata*) extending through pores in thin regions of cell walls.

bounded by plasmalemma. In the phloem, these connections permit lateral transport.

Sometimes cells become **plasmolyzed:** when they lose most of the water from their vacuoles because solute in the surrounding medium is highly concentrated, the plasmalemma loses contact with the wall as the vacuole shrinks and surrounds the glob of cytoplasm. Separation severs the plasmodesmata, but they reform when a favorable osmotic gradient returns and the cell becomes turgid. Like all membranes, the plasma membrane is dynamic and often acts almost like a liquid. In spite of their obvious importance, little is known about plasmodesmata.

Transport of Latex and Pitch Some compounds carried down the tree's trunk do not travel in the phloem cells; they are produced in living cells in the leaves or the trunk. The latex of

10.19 Latex in the rubber tree *(Hevea brasiliensis)* is produced and contained in a system of highly branched cells in the phloem region of the trunk.

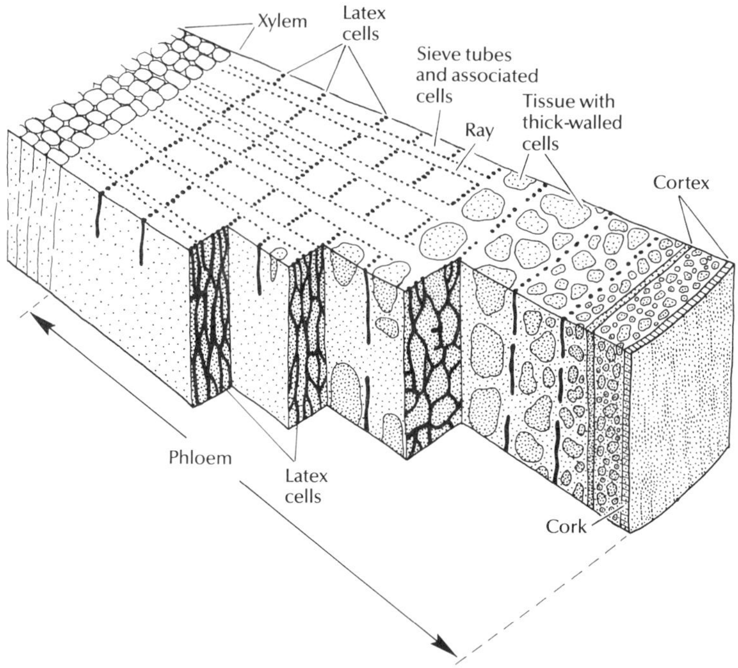

rubber trees is one product, an immeasurably useful compound that has survived the age of plastics. Other compounds are latex from the related poinsettia plant, the pine pitch that oozes from bark, and resin from tropical trees that make the beautifully clear substance, amber.

Here, too, the living cells that produce the substances form ducts through which their liquids flow (see Figure 10.19). But the function is different—not storage of food but probably protection. Cut a sliver of bark from a pine tree and you will see that the pitch that oozes out is not sweet with sugar, nor does it flow like sap. Instead it hardens over the wound, and in time the cork cambium will close the wound from the inside. Latex is often toxic and acts as a pesticide for herbivorous insects.

The production of latex by some plants is so copious it has been suggested that they be grown and harvested for the product, which is composed of hydrocarbons that could be used for fuels.

ENERGY AND PLANT INTERACTION—AGAIN

Transport always costs energy; so too does mining and concentrating minerals. Some of the energy produced by the mitochondria in the form of ATP is used for these jobs. It comes ultimately, of course, from photosynthesis, which can only occur when the plant has adequate light, carbon dioxide, and water. To photosynthesize, the plant must absorb CO_2, and it must take up from the soil both minerals and water.

For all these resources the plant may be competing with other plants or animals, including humans. Other plants may provide shade, depriving a plant of sunlight, and weeds decrease the productivity of crops by competing with the roots of the crop for both water and minerals.

Thus we see again the importance of energy and the interdependence of elements in an ecosystem. In the next section we will look more closely at how these interrelationships affect plants and their evolution.

Summary

Minerals, water, and energy-rich sugars are carried by two principal transport systems in plants: xylem primarily carrying minerals dissolved in water from roots to leaves, and phloem carrying sugars and other organic compounds from leaves to roots.

Water not only holds the dissolved minerals mined by the roots but also is the source of hydrogen for photosynthesis and is the solvent for the cell's chemistry. Transpiration is the process by which water is lost through the leaves. It is powered first of all by the sun: heat from solar radiation leads to loss of water by evaporation. The water is lost through pores—stomata—in the leaf's epidermis.

Photosynthesis cannot occur when the stomata are closed, since they provide the contact with the atmosphere that allows CO_2 to enter the leaf and O_2 to be released. Most leaf cells have some surface exposed to the intercellular space, which is continuous with the air when the stomata are open, and provides a pathway for diffusion of gases. Guard cells control the opening and closing of the stomata. When leaf cells have plenty of water, the guard cells are turgid and their adjacent cell walls separate, creating these pores.

Water vapor fills the intercellular spaces, so each cell is coated by a film of water. The epidermis and overlaying waxy cuticle protect the leaf from desiccation by hindering evaporation of this water vapor, but the stomata are holes in this covering, and water is lost. Some leaves have adaptations that hinder evaporation further: sunken stomata with hairs, waxy plates, or a reduced surface-to-volume ratio.

As water vapor is lost through the stomata, an osmotic gradient is created between the leaves' cells and the stream of water in xylem. Osmosis helps keep this stream of water moving. The upward-moving threads of water are maintained because of the forces of cohesion between water molecules and adhesion between the water and the walls of the xylem cells.

These thin cables of water are carried through dead xylem cells. The conducting cells in conifers are tracheids and angiosperms have tracheids and vessel elements that are more efficient for transporting water. These are dead xylem cells, one on top of the other, with holes at their ends; thus they form a tube. Both xylem and phloem extend into the leaves, forming the veins, which ensure that each cell is close to a source of energy, water, and minerals. Water flows only through recently dead xylem—the sapwood.

Water begins its trip through the plant at the root hairs, which take up water from the soil because of an osmotic gradient between their cells and the soil water. In addition, an osmotic gradient is maintained between root and xylem by the cells of the root's endodermis.

One hypothesis to explain how minerals are taken up and concentrated by the roots states that chemical compounds called carriers in the root cell membranes combine with mineral ions and transport the ions through the cell membranes to the cell vacuole. This process takes energy. Heat from the sun, causing evaporation, is the principal force involved in the transport of dissolved minerals. Mining the minerals, however, does cost the plant a great deal of energy; it comes to the roots as sucrose carried from the leaves in living phloem cells. These cells form sieve tubes—pipes with perforated end walls.

The sugar-rich solution, known as sap is thought to be transported as a result of mass flow—the roots at one end of a gradient keep using up the sugar produced at the other end. Both radioactive isotopes and aphids have been important tools for studying sap. It can flow at about 100 cm per hour and has sugars in concentrations of 10 to 30 percent.

Phloem also carries sugar produced from starch upward from the roots. The starch is stored in the roots, made into sugar in spring, then carried to the leaves where it supplies energy for leaves to develop.

Girdling can kill a tree by destroying cells through its whole circumference, because the cambium cannot produce phloem in time to transport food to the roots before the tree starves.

Food and water move laterally as well as vertically. In xylem, lateral transport occurs through the porous walls of the pits; in phloem, through plasmodesmata—channels of cytoplasm extending through the walls of different cells. If the columns of water in the xylem are disrupted because of freezing or injury to parts of the xylem, lateral transport can reestablish the flow of water.

Cells in the leaves and trunks also produce protective substances that are not transported through either xylem or phloem. Latex and pitch are examples.

Review Questions

1. Describe how the anatomy of many leaves enhances gas exchange and water loss.
2. How is transpiration related to the movement of minerals?
3. What is the relationship between water and mineral uptake in the roots?
4. Compare water- and food-conducting cells.
5. How does solar energy power both water and food transport?

6. Describe the function of these specialized cells: guard, endodermal, root hair, and sieve element.
7. How do water and food move laterally around the tree?
8. Herbaceous weeds in an orchard don't compete with trees for space, so why are they usually removed?

Evolution

of Plants

The structures and processes outlined in Part III apply to plants as we know them today, adpated to particular environments. How did they come to be this way, and how are they and their environments related? In many ways Part IV, on the evolution of plants, is the heart of this text. The most basic concept in biology is that all organisms are related through time; they evolve. Evolution takes place as the result of selective forces operating on variants that always occur in populations of organisms.

The cellular mechanisms of these variations are found in genetics, the subject of Chapter 11. By sexual reproduction, genetic variation is ensured and promoted. Both pollination and the dispersal of seeds are aspects of this process. Many plants also reproduce asexually, a method examined in Chapter 12.

Genetic variations are random, but the variations we see around us are not. Before these could be understood, new ideas about time and the meaning of fossils and, most important, a convincing theory of the mechanism of evolution were needed. These came in the nineteenth century, when scientists learned more about the history of earth, its age and fossils, and Charles Darwin developed an explanation of evolution—natural selection. Looking at the history of these ideas, we see that science is not simply a collection of numbers and facts but a body of ideas that are created, held, and changed by individuals, who are affected as we all are by personal history and culture. Chapter 13 traces the establishment of the concept of geological time and examines efforts to study plant life of the past. In Chapter 14 the principles of natural selection and many examples illustrating types of evolution, including coevolution of plants and animals, are provided.

With some familiarity about how plants work and the mechanisms involved in adaptation, we can look more closely in Chapter 15 at ecosystems, at the structure of plant communities and the environmental factors shaping them, including the role of humans in managing ecosystems. We end in a sense where we began, but it is hoped that, as a result of completing this cycle, your understanding of the structures, functions, and interrelationships of plants has been enriched.

from the

Origin of Species

Charles Darwin

SUMMARY OF CHAPTER IV

If under changing conditions of life organic beings present individual differences in almost every part of their structure, and this cannot be disputed; if there be, owing to their geometrical rate of increase, a severe struggle for life at some age, season, or year, and this certainly cannot be disputed; then, considering the infinite complexity of the relations of all organic beings to each other and to their conditions of life, causing an infinite diversity in structure, constitution, and habits, to be advantageous to them, it would be a most extraordinary fact if no variations had ever occurred useful to each being's own welfare, in the same manner as so many variations have occurred useful to man. But if variations useful to any organic being ever do occur, assuredly individuals thus characterised will have the best chance of being preserved in the struggle for life; and from the strong principle of inheritance, these will tend to produce offspring similarly characterised. This principle of preservation, or the survival of the fittest, I have called Natural Selection. It leads to the improvement of each creature in relation to its organic and inorganic conditions of life; and consequently, in most cases, to what must be regarded as an advance in organisation. Nevertheless, low and simple forms will long endure if well fitted for their simple conditions of life.

Natural selection, on the principle of qualities being inherited at corresponding ages, can modify the egg, seed, or young, as easily as the adult. Amongst many animals, sexual selection will have given its aid to ordinary selection, by assuring to the most vigorous and best adapted males the greatest number of offspring. Sexual selection will also give characters useful to the males alone, in their struggles or rivalry with other males; and these characters will be transmitted to one sex or to both sexes, according to the form of inheritance which prevails.

Whether natural selection has really thus acted in adapting the various forms of life to their several conditions and stations, must be judged by the general tenor and balance of evidence given in the following chapters. But we have already seen how it entails extinction; and how largely extinction has acted in the world's history, geology plainly declares. Natural selection, also leads to divergence of character; for the more organic beings diverge in structure, habits, and constitution, by so much the more can a large number be supported on the area—of which we see proof by looking to the inhabitants of any small spot, and to the productions naturalised in foreign lands. Therefore, during the modification of the descendants of any one species, and during the incessant struggle of all species to increase in numbers, the more diversified the descendants become, the better will be their chance of success in the battle for life. Thus the small differences distinguishing varieties of the same species steadily tend to increase, till they equal the greater differences between species of the same genus, or even of distinct genera.

We have seen that it is the common, the widely-diffused and widely-ranging species, belonging to the larger genera within each class, which vary most; and these tend to transmit to their modified offspring that superiority which now makes them dominant in their own countries. Natural selection, as has just been remarked, leads to divergence of character and to much extinction of the less improved and intermediate forms of life. On these principles, the nature of the affinities, and the generally well-defined distinctions between the innumerable organic beings in each class throughout the world may be explained. It is a truly wonderful fact—the wonder of which we are apt to overlook from familiarity—that all animals and all plants throughout all time and space should be related to each other in groups, subordinate to groups, in the manner which we everywhere behold—namely, varieties of the same species most closely related, species of the same genus less closely and unequally related, forming sections and sub-genera, species of distinct genera much less closely related, and genera related in different degrees, forming sub-families, families, orders, sub-classes and classes. The several subordinate groups in any class cannot be ranked in a single file, but seem clustered round points, and these round other points, and so on in almost endless cycles. If species had been independently created, no explanation would have been possible of this kind of classification; but it is explained through inheritance and the complex action of natural selection, entailing extinction and divergence of character. . . .

The affinities of all the beings of the same class have sometimes been represented by a great tree. I believe this simile largely speaks the truth. The green and budding twigs may represent existing species; and those produced during former years may represent the long succession of extinct species. At each period of growth all the growing twigs have tried to branch out on all sides, and to overtop and kill the surrounding twigs and branches, in the same manner as species and groups of species have at all times overmastered other species in the great battle for life. The limbs divided into great branches, and these into lesser and lesser branches, were themselves once, when the tree was young, budding twigs, and this connection of the former and present buds by ramifying branches may well represent the classification of all extinct and living species in groups subordinate to groups. Of the many twigs which flourished when the tree was a mere bush, only two or three, now grown into great branches, yet survive and bear the other branches; so with the species which lived during long-past geological periods, very few have left living and modified descendants. From the first growth of the tree, many a limb and branch has decayed and dropped off; and these fallen branches of various sizes may represent those whole orders, families, and genera which have now no living representatives, and which are known to us only in a fossil state. As we here and there see a thin straggling branch springing from a fork low down in a tree, and which by some chance has been favoured and is still alive on its summit, so we occasionally see an animal like the Ornithorhynchus or Lepidosiren, which in some small degree connects by its affinities two large branches of life, and which has apparently been saved from fatal competition by having inhabited a protected station. As buds give rise by growth to fresh buds, and these, if vigorous, branch out and overtop on all sides many a feebler branch, so by generation I believe it has been with the great Tree of Life, which fills with its dead and broken branches the crust of the earth, and covers the surface with its ever-branching and beautiful ramifications.

CHAPTER 11

Genes and Sexual Reproduction

Every one of earth's known organisms reproduces itself either sexually or asexually. In **sexual reproduction** gametes (often termed *male* and *female*) unite; the offspring of this union will carry genetic characteristics from both gametes. In **asexual reproduction** the individual reproduces by developing part of itself into a new organism. A bud (an "eye") of a potato, for example, will grow into a new potato plant. There is no union of gametes, no fertilization. The offspring of asexual reproduction will carry the genetic characteristics of its one parent.

Both methods can have advantages and disadvantages for a species' survival. The history of a popular flowering house plant, a species of begonia, illustrates both, as well as showing more generally some aspects of plant evolution. When discovered in the Andes in the middle of the last century, this begonia had small yellow flowers with four petals. Today the species has so many varieties, some with massive blossoms, that it is hard to believe that a little yellow bloom was their progenitor. Some look like roses, some like camelias, some like chrysanthemums, and they come in a rainbow of colors.

The begonia is an example of manmade evolution, possible because any sexually reproducing plant will produce offspring differing in one or more ways from its parents. In a series of sexually reproduced begonias, one may suddenly appear with a red flower instead of a yellow one. This one red plant can be "frozen" in an evolutionary sense by asexual reproduction: cuttings from it can be used to produce plants that will develop only red flowers. Now that red begonias are available, this new variety can be allowed to reproduce sexually, and here, among a flock of red begonias, other variations will appear, such as a plant with leaves of a different shape. This plant too can be made to reproduce itself from cuttings, and we will have a begonia with two characteristics different from the original small yellow begonia plant. All these varieties can be crossed with each other so that, in the end, there is a whole series of variants.

THE MECHANISM OF VARIATION

The begonia's appearance is its **phenotype,** only partly determined by the genetic characteristics passed on in reproduction. If organisms with the same inherited characteristics, the same **genotype,** are grown under different conditions, they can look and act quite differently, that is, have different phenotypes. In this chapter, however, we will be concerned only with the similarities and differences coming from inherited characteristics.

Genes are the carriers of inherited characteristics: they determine an individual's potential. All individuals produced by asexual reproduction are identical to the parent, because only one set of genes is involved. (In a way, identical twins are like the products of asexual reproduction. They are the result of the division of one fertilized egg that produces two identical individuals.) Asexual reproduction freezes evolution in a sense, since it prevents genetic variation that always results from sexual reproduction.

In sexual reproduction the offspring has characteristics different from those of the parent or parents, as well as from other offspring, because genes from both parents are intermixed. All the peas in a pod may look alike, but no two plants grown from seeds in those pods will be exactly similar. Children in a family are always different from each other, except for identical twins.

The reshuffling of genes by sexual reproduction produces genetic variations, enabling organisms to adapt to new environments. In any given environment, however, many of these varieties will be less well adapted than others. In contrast, if an organism is highly successful in its environment, asexual reproduction maintains a well-adapted population of organisms, because it results in many identical, well-adapted copies. Both sexual and asexual reproduction have advantages, and plants that can reproduce both ways have the best of both worlds. The plant world has a good many of these, including many common weeds such as the dandelion.

How do variations happen and how are genetic characteristics passed on? We will have to step back in time halfway through the nineteenth century to a most fascinating story in the history of science. An Austrian monk named Gregor Mendel (1822–1884), working alone, made a discovery almost as momentous as Darwin's

11.1 Genetic variation is illustrated by these begonias. They are all variants of a plant with small yellow blossoms found in the middle of the last century in South America. New variants are selected and maintained by asexual reproduction.

11.2 Johann Gregor Mendel (1822–1884) was the first to demonstrate how physical units—what we now call *genes*—control inheritance. Mendel's scientific career was brief, making his contribution all the more remarkable. He became abbot of his monastery and, like so many able scientists, became engulfed by the tide of administrative duties.

theory of evolution, only to have it ignored through his lifetime and rediscovered thirty-five years after he reported it.

Mendel's First Experiments: Dominant and Recessive Traits Johann Mendel was the son of a peasant fruit farmer living in what is now Czechoslovakia. Since the family was too poor to give him a complete university education, he entered an Augustinian monastery at age 21, where he took the name Gregor. He was able to spend some time on his own interests, and he put this time to use in a systematic series of experiments in the monastery garden.

Looking for variations among successive generations of a "true-breeding" plant, one that does not exhibit major variations in succeeding generations, Mendel chose the garden pea *(Pisum sativum)*, a

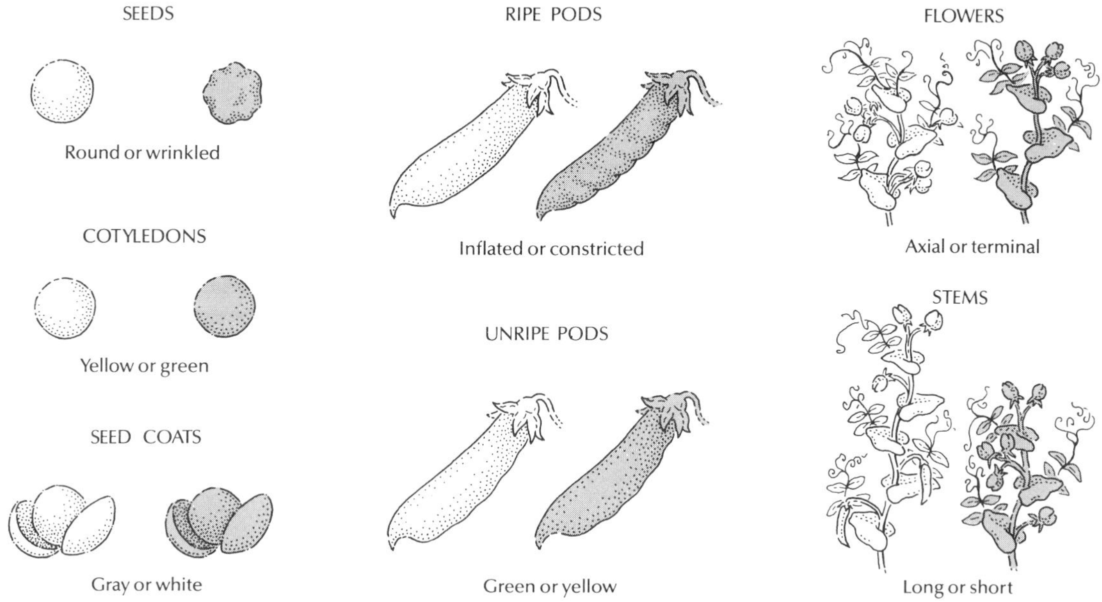

11.3 Mendel selected seven varying characteristics of pea plants for his experiments on inheritance. He was fortunate that the characters he selected were all on different chromosomes because his results would have been confusing under other circumstances.

self-pollinating plant. Each flower could easily be protected from pollen and resultant fertilization from other flowers, which would obscure the results of his experiments. He chose to follow only seven characteristics, or **traits,** an intelligent move without which he would shortly have begun to flounder, like so many of his predecessors, in a forest of unexplainably varying characteristics.

Once his pea vines had flowered, Mendel began carefully controlled breeding experiments among plants showing the characteristics he had selected. For example, he crossed plants with only wrinkled seeds with plants that had round smooth seeds, carrying out the **crosses** (exchanges of pollen) himself. Stamens were removed from each blossom to prevent self-pollination and outside pollination was prevented by covering the blossoms with little bags.

The results were clear-cut (see Figure 11.4). All the offspring—the **hybrid plants**—had round seeds; the wrinkled seed trait possessed by one parent seemed to have disappeared. He got similar results using plants with the other characteristics he had chosen to study. Tall plants bred with short plants produced only seed that grew into tall plants. He now labeled the traits **dominant** or **recessive**—*dominant* if the trait appeared in the hybrid, *recessive* if it did not. In the round seed-wrinkled seed experiment, he labeled the round seed trait *dominant,* the wrinkled seed *recessive.*

Next Mendel went on to find out what would happen if these hybrids were self-pollinated. The offspring showed both traits, that is, round seeds and wrinkled seeds, but in unequal numbers. Those with the dominant trait, such as round seeds, outnumbered the recessive almost three to one (Figure 11.4).

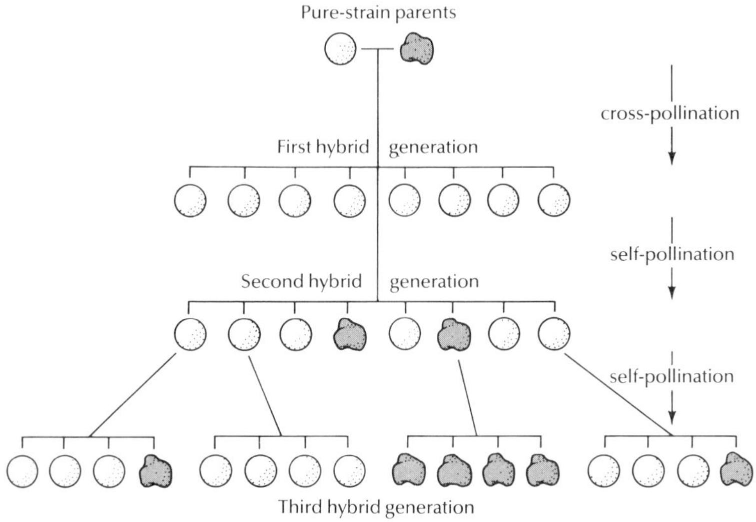

11.4 When Mendel crossed peas with round seeds with peas with wrinkled seeds, the first generation produced only round seeds. When plants from these seeds were allowed to self-pollinate, they produced both kinds of seeds. The ratio was 3:1. Wrinkled seeds from this generation produced only plants with wrinkled seeds, while round seeds produced either round seeded plants or a mixture of round and wrinkled.

The plants that grew from these seeds with the 3:1 ratio were self-fertilized, and the resulting seeds were examined. Wrinkled seeds produced only wrinkled seeds, continuing to do so for many generations. Mendel had produced a **true-breeding** variety. The round seeds from the second generation, however, produced plants with both round and wrinkled seeds and plants with only round seeds. The ratio was 2:1; that is, for every two of the round seeds that produced plants with both types of seed, one round seed grew into a plant with only round seeds, as Figure 11.4 shows. Mendel concluded that the recessive trait was still there, even though it could be concealed in plants that seemed to bear only round seeds. He could not tell which kind they would produce, but clearly both dominant and recessive traits were carried by these plants.

Mendel now decided that the hereditary characteristics of his pea plants were always determined by discrete (separate) factors that occurred as pairs in the parent plants. When the sex cells were formed in the flower, he reasoned, these pairs were separated, producing two kinds of gametes, each containing one factor of the pair. If the two factors were the same, the plant would breed true when bred to itself; that is, the offspring would be identical to the parent. But if the two factors were different, then the offspring would be different following self-crossing.

These alternative forms of a character, such as seed-coat texture, are now known as **alleles.** The determinant for the character is a **gene.** In the offspring of a round-seeded plant, round seeds and wrinkled seeds are alleles of the same gene. By the same reasoning, a plant with both alleles for one gene is said to be **heterozygous** (*hetero,* "different"; *zygous,* "zygote"). A plant that has the same alleles for a gene, and therefore breeds true, is **homozygous** (*homo,* "same"; *zygous,* "zygote").

Mendel analyzed the results of his experiments. For clarity he labeled the dominant trait *A*, and the recessive trait *a*. He proved that when plants with *A* gametes were crossed, only *AA* would result (see Figure 11.5). Likewise, when plants with *a* gametes were crossed, only *aa* resulted and the recessive trait would appear; these plants too were homozygous. Homozygous plants bred to each other always produced the same traits. But when plants with *A* gametes were crossed with plants with *a* gametes, then a heterozygous *Aa* plant was produced. In this case there are three genotypes—the two homozygous genotypes, *AA*, *aa*, and one heterozygous genotype, *Aa*—but there are only two phenotypes, round seeds and wrinkled seeds. Two of the genotypes, *AA* and *Aa*, yield the same phenotype of round seeds because *A* dominates in determining the appearance of the seed coat.

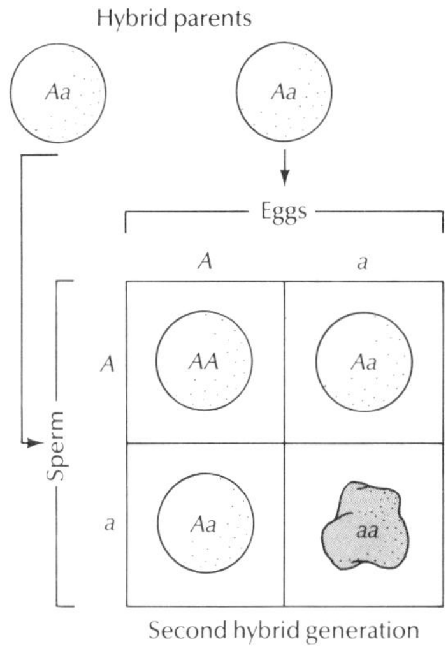

11.5 Hybrids of plants with round and wrinkled seeds produced only round seeds. Self-pollinated plants from these seeds yielded round and wrinkled seeds (3:1 ratio).

Experiments with Several Characteristics The next question was, what would happen when two dominant characteristics were crossed with two recessive ones? What would happen, for instance, if a pea plant with round and yellow seeds were crossed with one that had wrinkled and green seeds? Roundness and yellowness were the dominant traits. What would the offspring be like?

Guided by his first experiments, Mendel got what he expected. The first-generation offspring had round yellow seeds, the dominant characteristics. And when he allowed this generation to produce offspring, he found four kinds of seed: round yellow, wrinkled yellow, round green, and wrinkled green.

Out of the 556 seeds, he counted 315 round yellow seeds, 101 wrinkled yellow; 108 round green, and 32 wrinkled green. The ratios of the characters were almost exactly 9:3:3:1 (see Figure 11.6).

Mendel added another characteristic, crossbreeding pea plants that bore the traits for producing round yellow seeds having grayish seed coats with plants making wrinkled green seeds with white seed coats. Once again he predicted the results for the second generation, and again he was right: he got every possible combination of the inherited traits. In other words, one characteristic such as texture of seed coat could produce 3 combinations: *AA* (round), *Aa* (round), and *aa* (wrinkled). Each pair of alleles gives 3 possible combinations, so that if n = the number of pairs of alleles, any cross will give 3^n combinations. In the cross with 3 characteristics (seed coat texture, seed color, seed coat color), the number of combinations in a cross in which each plant has the 3 characteristics is 3^3 or $3 \times 3 \times 3 = 27$. Crosses with only 10 characteristics make 60,000 combinations. It is

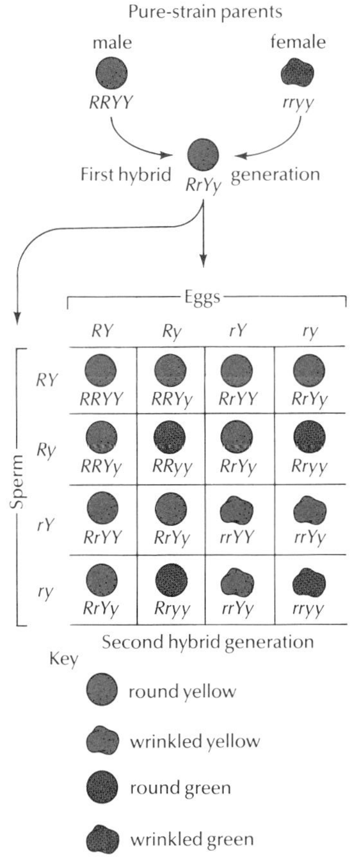

11.6 Mendel's crosses demonstrated that the hybrid of two plants, each of which was homozygous for two characteristics only, showed the dominant characteristics. When the hybrid reproduced, the offspring exhibited every possible combination of the four characteristics. The best explanation for this result was that the characteristics were independently distributed to the gametes of the hybrid.

intriguing to think of the combinations possible with the hundreds of charcteristics in the genes of any species of plant.

Mendel's Laws of Heredity These simple experiments provided strong evidence that plants have particulate units of heredity. Mendel found the key to heredity, and with it a vital clue in explaining evolution. The Mendelian laws are frequently stated as:

1. The heredity mechanism consists of the transmission of a large number of paired characters (these are the genes). Only one member of the pair is transmitted by each gamete. This is called the principle of segregation.
2. When the gametes are formed, the members of the paired genetic units form all possible combinations with each other. This is called the principle of independent assortment.

Mendel also worked on inheritance of flower color in beans, but his results from crossing beans with flowers of different colors were significantly unlike those from peas. He did not find simple dominant and recessive conditions. The first generation produced beans with intermediate colors, and the offspring of the hybrid showed a whole range of colors. He was confused, worried that his work with peas might be wrong. He stuck with the flower problem, however, and got systematic results with which he could present a hypothesis, again ahead of his time. He suggested that flower color may be controlled by multiple factors (genes) and not by two forms (alleles) of the same factor (gene). This complexity would account for the gradient of colors he observed in the second generation. It is now known that many phenotypic characteristics such as flower color are a result of multiple genes. Human skin color, for example, is determined by at least eight genes, and that is why it has so many gradations.

Reactions to Mendel Mendel's work was first brought to the attention of the scientific community in 1865 at a meeting of the Brünn Society for the Study of Natural Science. Apparently, none of the few people present at this historic occasion understood what he was talking about. The following year, his paper was published in the Society's *Proceedings,* but again it was ignored. Mendel subsequently became head of the monastery, devoted himself to administration, and died in scientific obscurity.

Not until thirty-five years after his epochal discoveries was the world finally prepared to acknowledge Mendel. Recognition, when it came, came all in a rush. Three botanists working in three countries rediscovered him independently as they searched for scientific publications related to their work, finding that Mendel already had the answer. Dutch botanist Hugo De Vries read Mendel's paper as a result of his own work on the evening primrose, and he reported his discovery of Mendel in 1900. Two months later the German, Carl Correns, reported; like Mendel, he had worked on inheritance in peas. The third report was by an Austrian, Erich von Tschermak, who also worked on peas.

By a quirk of history Correns was a student of Nägeli, a renowned botanist with whom Mendel corresponded about his crosses with peas but who failed to understand how significant Mendel's work was; he suggested that Mendel do more experiments. Tschermak was a grandson of the man who taught Mendel botany during his student days in Vienna. All these men went on to make important contributions to genetics, but De Vries is especially remembered for introducing the concept of mutation. (Mutations are changes in the genes such that different forms of a gene are produced. They will be discussed shortly.)

FINDING THE GENES

Mendel proved that the genetic determinants are particulate, but it took fifty years to establish precisely where the genetic material lies. Science often seems to move spasmodically; an original idea in one field goes unappreciated, leading nowhere, until other fields catch up with it. So it was with studies on inheritance. Mendel's laws were abstract, and they required correlation with concrete events in cells before their relevance was appreciated.

Three big ideas in biology were products of the nineteenth century. The biggest was natural selection, which will be discussed in Chapters 13 and 14. Second was Mendelian genetics, which eventually explained organismal variation, which is central to natural selection. The third was our concept of the cell, the only one of this trilogy to have a rich inheritance from the seventeenth and eighteenth centuries. The nineteenth century was biology's feast of giants; during our own time we have been gleaning scraps from the giants' table—a rich diet, but scraps nonetheless.

Discovering Cells Natural selection and the laws of inheritance are intellectual feats attributable to individual genius; not so our concept of the cell. It took many people working over several centuries to describe a complete picture of the cell.

Most cells can be seen only with a microscope, which was invented by Dutch spectacle makers near the end of the sixteenth century. But it was not until 1858 that German pathologist, Rudolf Virchow, summarized knowledge about the cell as the basic unit of life in a phrase: *omnis cellula e cellula* ("every cell from a cell"). He could say so only because of innumerable contributions by hundreds of investigators starting with Robert Hooke (1635–1703), who first applied the word *cell* to the compartments in a slice of bark from

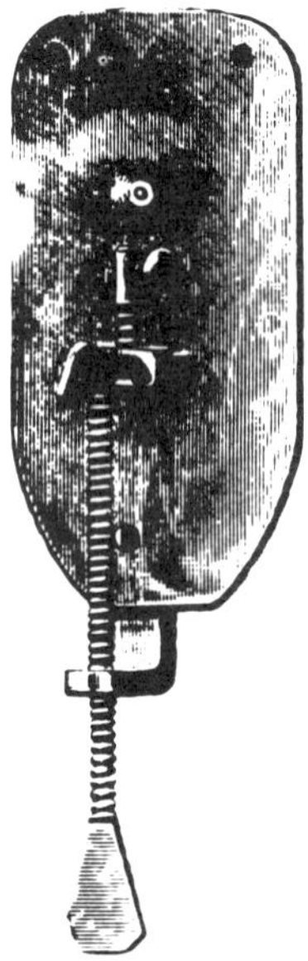

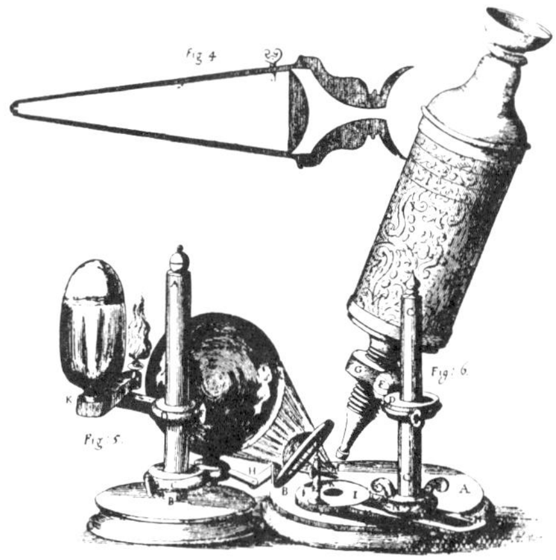

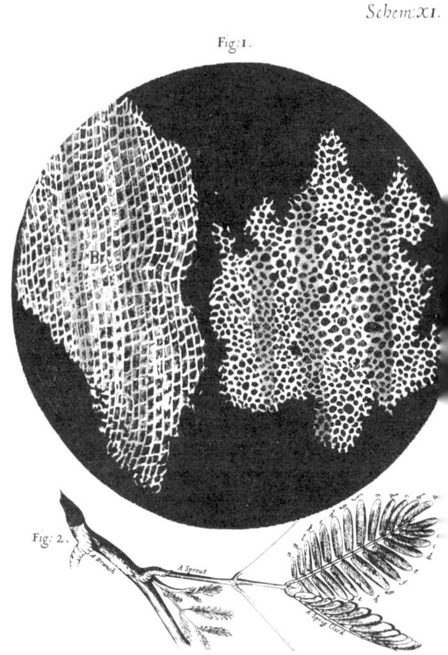

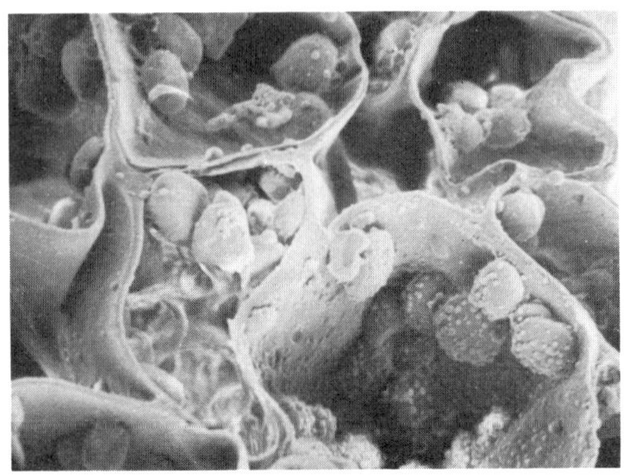

11.7 The first microscopes gave a view of the unseen world similar to that obtained by low power microscopes used today in introductory courses. Hooke's microscope (top, center) was mechanically very similar to them; Hooke's drawing of cells in a section of cork as seen through his microscope is shown (top, right). Leeuwenhoek's instrument (top, left) had only a single lens and had a moveable stage. A specimen was placed on the tip of the mount and moved in two directions by means of screws. A recent development in microscope technology is the scanning electron microscope (SEM) at right. The transmission electron microscope (TEM) provided a quantum jump in what could be seen in sectioned cells, and the SEM gives an even more remarkable depth of field. The three-dimensional view of leaf cells (center) is an example of the image that appears on the screen of the SEM. The big spheres around the insides of the walls are chloroplasts, and the numerous little spheres are mitochondria.

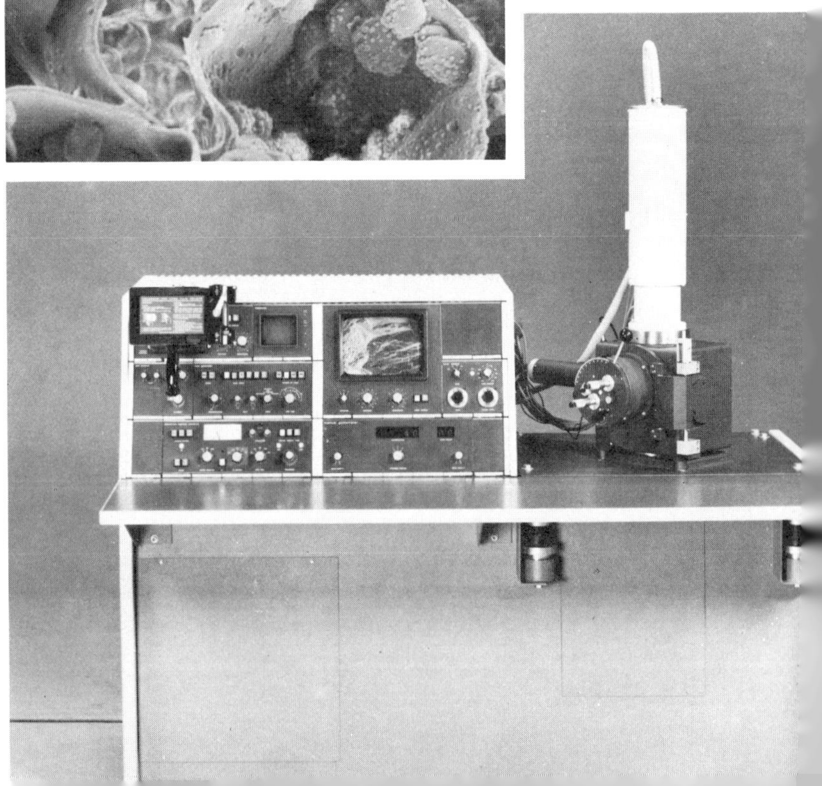

the cork oak (see Figure 11.7). His Dutch contemporary, Anton Leeuwenhoek, discovered bacteria and spermatozoa and a host of other things with the single lenses he made. The first systematic studies on cells were with plants, described by an Italian, Marcello Malpighi, and an Englishman, Nehemiah Grew, contemporaries of Hooke and Leeuwenhoek. In spite of this vigorous beginning, it took almost one hundred seventy-two years after Hooke's description for the cell theory to be formulated. Then Matthias Schleiden stated that all plants are composed of cells, and Theodor Schwann followed in 1839 with the general statement that the cell is the basic unit of both plants and animals.

The nucleus (shown in Figure 11.8) specifically interests geneticists, because it is where the chromosomes are located. English botanist Robert Brown discovered it in 1831. Smaller structures in cells are hard to see until stained, and the German synthetic dye industry had to develop before specific stains could be found for plant and animal tissues. In 1888 a number of dyes turned up that would stain threadlike bodies in the nucleus, the chromosomes ("colored bodies"). Then it was discovered that cells of each organism have a characteristic number of chromosomes: the cabbage, 18; frogs, 26; human beings, 46; dogs, 78; and goldfish, 94. And each

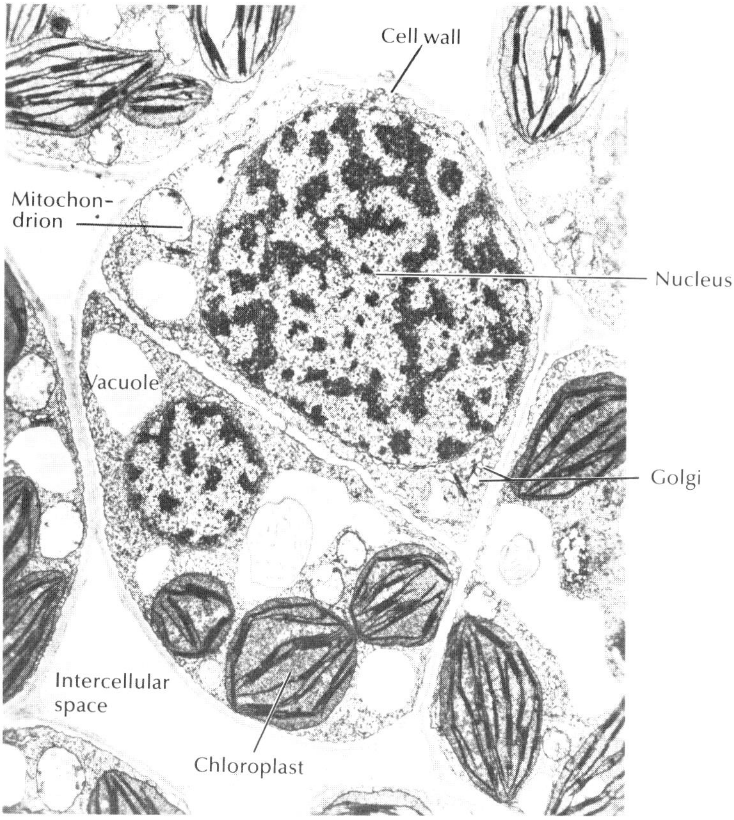

11.8 Products of a recently divided plant cell display sections of nuclei which are large relative to cell size, a characteristic of meristematic cells. The darker regions in the nucleus are the chromosome material.

chromosome seemed to split just before the cell divided, distributing the two products evenly to each of the new cells. This division was called **mitosis** (from the Greek word for "thread").

Mitosis Division of the plant cell has two distinct steps: division of the nucleus and formation of a transverse cell wall making two cells. The nucleus can divide without the cell's doing so. When conditions are right, cell divisions in bacteria can take only fifteen minutes, but in meristematic plant tissues, division usually takes an hour or more.

The first signal of division is the appearance of threads—the chromosomes—in the nucleus. These thicken and the nuclear membrane disappears, as you can see in Figure 11.9. This stage is called **prophase.** The now visible chromosomes thicken, shorten, and arrange themselves in a line along the middle of the cell. Now we can see that each chromosome is really double; each part of a pair is called a **chromatid** ("bearer of chromatin"). During this **metaphase** part of mitosis, very thin filaments appear in the cytoplasm, forming a spindle-shaped structure with the chromosomes attached to fibers at its thick part (see Figure 11.9). The chromatids separate and migrate to opposite poles of the spindle. This moving apart is the **anaphase.** Now the chromosomes lengthen, and a new nuclear membrane is formed. During this terminal part of mitosis, the **telophase,** a new cell membrane and wall are formed approximately at the point where the chromosomes lined up during metaphase. The new wall is made by deposition of cellulose all across the transverse plane.

The daughter cells formed by mitosis are smaller than the parent was just before cell division but otherwise they are exactly the same as the parent cell. Plant and animal tissues are constructed by mitosis, and it is the basis of asexual reproduction. When gametes are produced, a variant of mitosis occurs.

Meiosis "Decrease" is the literal meaning of **meiosis,** and it describes a distribution of chromosomes in which the daughter cells end with only half the chromosomes possessed by the parent cell. The daughter cells are called **haploid** (N), compared to the **diploid** (2N) cells of the parent cell.

Meiosis has two parts. First, **homologous** chromosomes (similar chromosomes from each parent) pair and assemble as if in metaphase during mitosis. (Remember that in mitosis the homologous chromosomes do not pair and at metaphase are already double.) All the chromosomes then double, and each double chromosome of a pair moves to opposite poles of the cell. Each daughter cell has an equal complement of double chromosomes, including one from each homologous pair. But the separation is random, so all the paternal chromosomes are not in one daughter cell and all the ma-

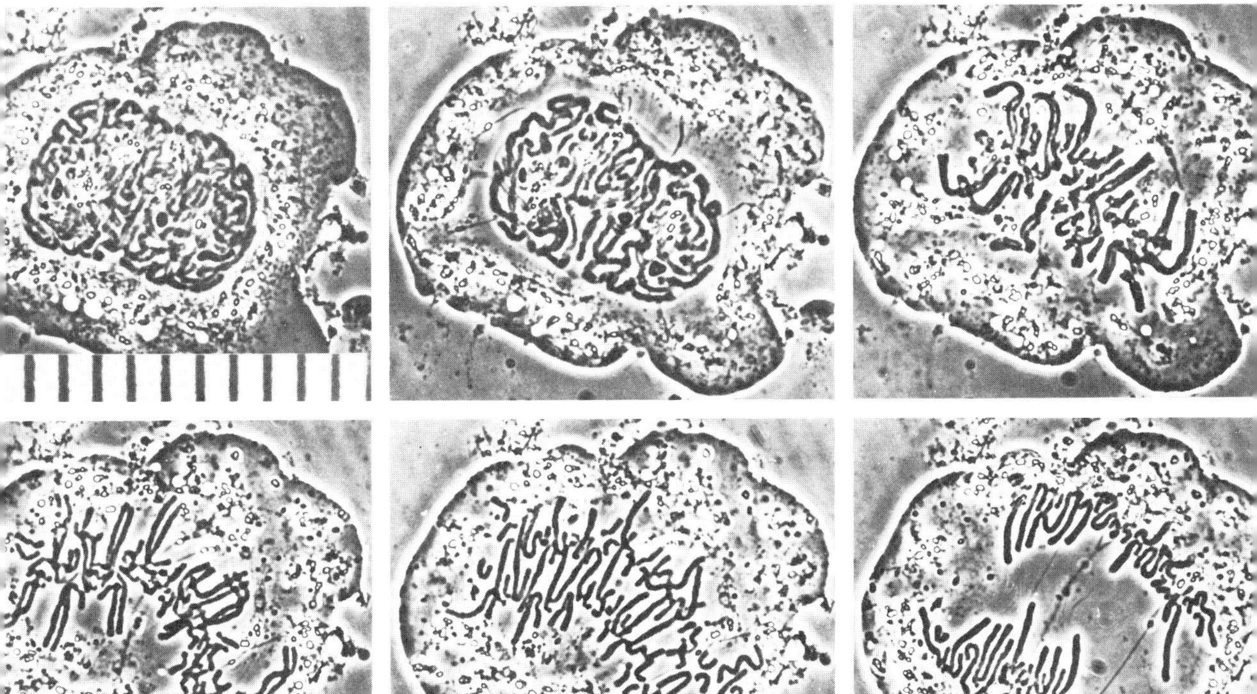

11.9 Mitosis is a continuous process, as illustrated in these photos. It is not possible to pinpoint exactly where one phase of the mitotic cycle ends and another begins. Prophase is the period when the chromosomes become visible and the nuclear envelope is lost. Metaphase is the period when the chromosomes form a row in the middle of the cell; anaphase is the period when the separated chromatids move apart; and telophase is the period when the nucleus reforms and a new cell wall is initiated. In the first photo above, a scale of measurement is shown: the distance from one black bar to the next is 10 microns.

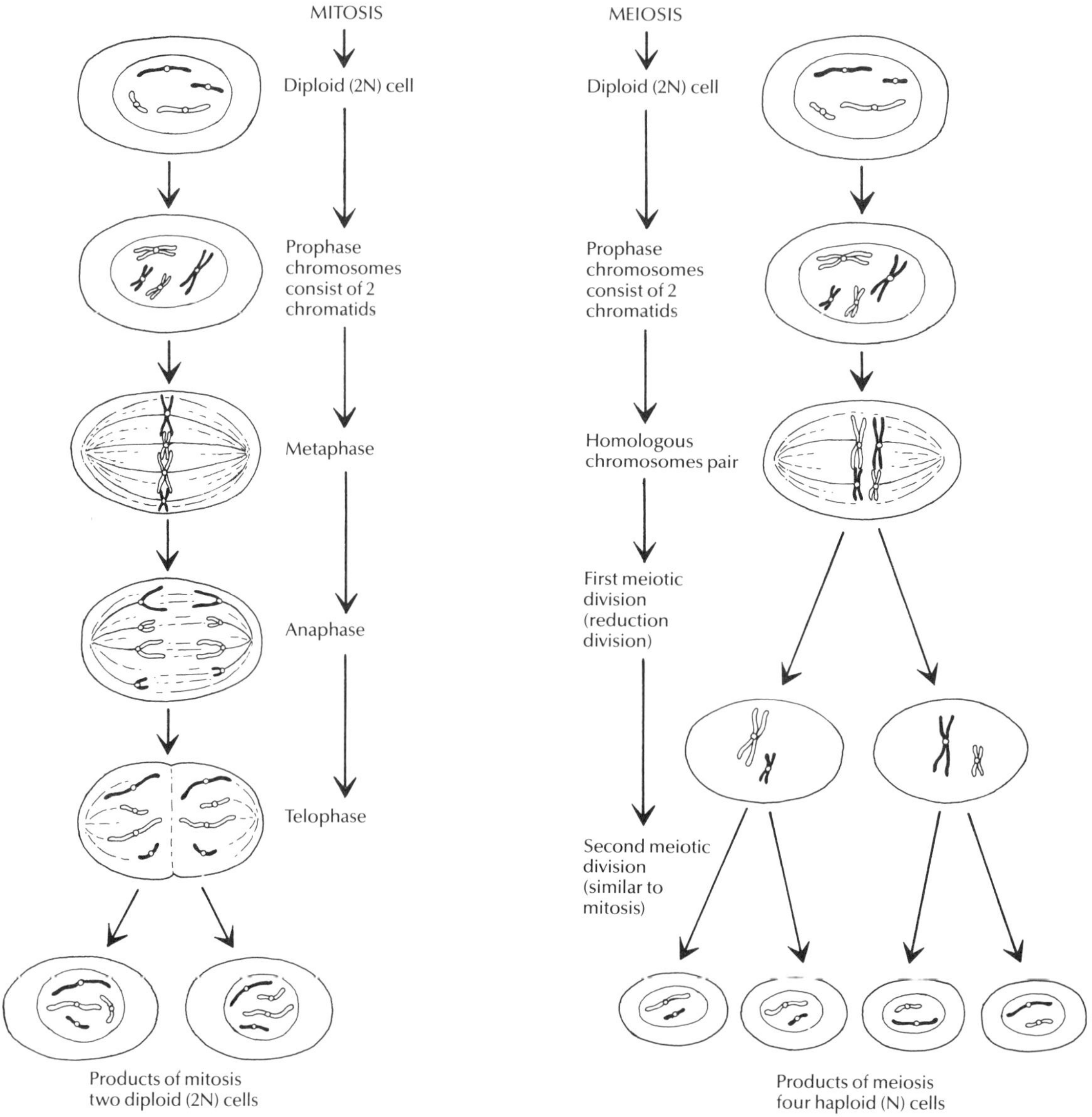

11.10 Mitosis provides each new nucleus with the same number and combination of chromosomes as the original nucleus. It is accompanied by one cell division. Meiosis consists of two divisions of nuclei which produce nuclei with half the number and a different combination of chromosomes as compared with the original nucleus.

ternal chromosomes in another. Each chromosome is still double (with two chromatids); the first division of meiosis leaves only half as many chromosomes in each daughter cell as were in the parent cell before division. Thus this first division of meiosis is a **reduction division** (see Figure 11.10).

The second meiotic division is much like mitosis in its phases and in distributing the same number of chromosomes to daughter cells. There is no reduction in number of chromosomes during this division, although the number of chromatids is halved. Each cell that

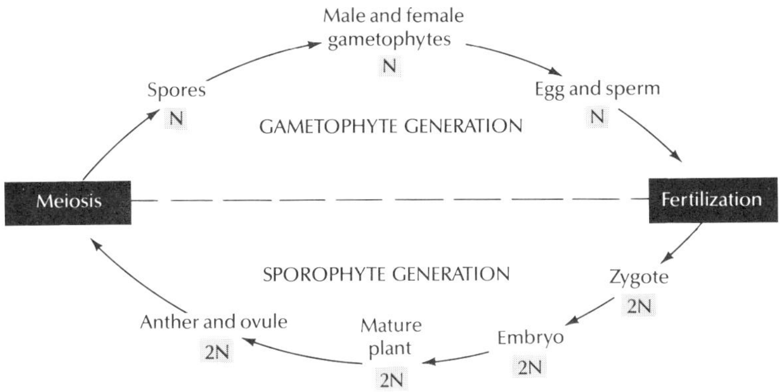

11.11 Green plants produce gametophyte and sporophyte generations. Each division has variations on this theme. The crossroads separating each generation are meiosis and fertilization. The haploid gametophyte generation begins with spores, the products of meiosis. Meiosis in flowering plants occurs in anthers and ovules and produces, respectively, pollen and a microscopic female gametophyte buried in the ovule tissue. These two structures produce egg and sperm and fertilization initiates the diploid sporophyte generation.

undergoes meiosis produces four daughter cells with half as many chromosomes as in the original cell, as can be seen in Figure 11.10. Because the chromosomes have been randomly assorted during meiosis, the genotype of each of the four meiotic daughter cells differs from the original cell.

In plants the products of meiosis are called **spores,** and in animals they are **gametes.** Fern and moss spores are products of meiosis, as are the cells in gymnosperms and angiosperms that produce pollen grains and those in the ovules of flowers that produce the female gametophyte. Thus in all green plants we find an alternation between stages of haploid cells (N) and diploid cells (2N)—the alternation of generations discussed in Chapter 3. Gametophyte cells, and therefore gametes, have half as many chromosomes as cells of the sporophyte generation, which are diploid (see Figure 11.11).

A great deal about the movement of chromosomes during mitosis and meiosis is not understood. A particularly tantalizing puzzle is how homologous chromosomes find each other at metaphase of the first meiotic division: some ferns have about 1800 chromosomes—so 900 chromosomes have to find their partners.

Meiosis and Variation Around 1885 several prominent biologists had concluded that the threads in the nucleus now called chromosomes were carriers of genes. By the time Mendel was rediscovered in 1900, the details of meiosis were pretty well known, and it took just two years for an American, W. S. Sutton, to propose that the Mendelian laws of the distribution of inherited factors could be explained by the distribution of chromosomes during meiosis. Fourteen years more went by, however, before a group of Americans working with fruit flies (*Drosophila melanogaster*) published the conclusive proof that chromosomes carry the genetic material. They had matched the occurrence of a genetic trait in offspring with the appearance of a chromosome of distinctive shape in parents.

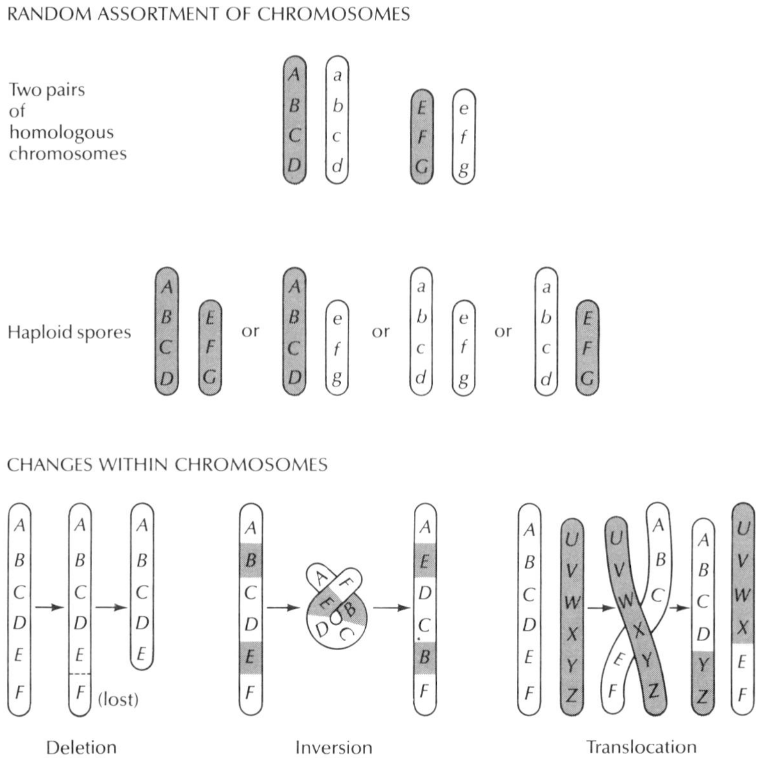

11.12 Meiosis provides for genetic variation by the process of chromosome shuffling during the reduction division. The genetic lottery continues during fertilization when new combinations of chromosomes are formed. Genetic variation also takes place as a result of changes within chromosomes. Most genetic characteristics of organisms are determined by groups of genes, and these changes produce new groups of linked genes which result in phenotypic variation.

Meiosis creates new genetic combinations by assembling a random assortment of chromosomes and hence of genes. These new combinations provide the constant variation observed in every species. This source of variation is replenished by **mutations,** heritable changes in genes (see Figure 11.12). Meiosis, then, provides the genetic fuel for evolution in multicellular organisms. In other words, sexual reproduction creates the genetic variants that enable a species to adapt to changing conditions. (Much more will be said about the selection of genetic variants in Chapter 14.)

Another aspect of meiosis should be understood. In addition to genetic variation from random assortment, variation results from recombination of genes. When the homologous chromosomes pair and then double during the first meiotic division, arms on the double chromosomes of one parental type entangle with those of the other. When they disentangle, an exchange of segments of chromosomes frequently occurs (see Figure 11.13). Imagine touching elbows with your neighbor and simultaneously exchanging forearms. The swap is called **crossing-over** and obviously mixes up the genes on a chromosome. Meiosis is like mixing tickets in a genetic lottery. Putting together two lottery tickets pulled from the basket at random is akin to fertilization.

Mendel was lucky in choosing his characters, and he made the most of it. Because the characters were carried by different chromosomes, the complexities of crossing-over did not jeopardize his results. If his characters had been on the same chromosome, they would not have been assorted randomly, and he would not have observed independent segregation in new generations. His results would not have fit into nice ratios, and frustrations and doubts would have assailed him as in his work on the color of bean flowers. He could not have explained **linked genes** (those found on the same chromosome) without knowing about chromosomes. His luck held, but with three others reporting almost simultaneous discovery of the basic laws of genetics in 1900, science might not have suffered if he had failed.

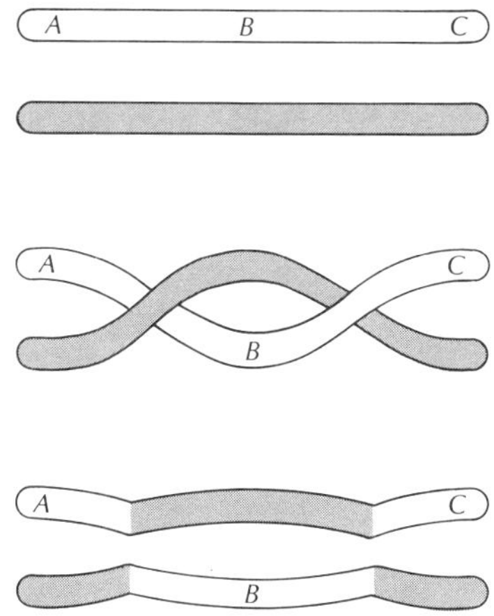

11.13 *A, B,* and *C* indicate genes on the chromosome. At the top are two homologous chromosomes (each in fact would consist of two chromatids). The center drawing illustrates the pairing of the two chromosomes; at this stage of meiosis, the arms of one entangle with the other. When they disentangle at a later stage, some segments of the chromosomes may be exchanged, as shown in the bottom drawing. Each of the two chromosomes now has a different complement of genes.

THE GENE AND ITS FUNCTIONS

With magnification we can see how chromosomes are duplicated and distributed and we know they are the bodies that carry genetic information. This much was known by the start of World War I. The next subject that challenged researchers was identifying the chemical structure of the genetic material, how it is duplicated, and how it controls the cell's characteristics and those of whole organisms. In the quarter of a century after World War II, the golden age of twentieth-century biology, answers were found.

Genes and Nucleic Acids One of the primary components in chromosomes had been discovered in cell nuclei as this century began—a class of compounds called **nucleic acids.** Later, proteins were found to be the other main part. Proteins are polymers made up of coiled and connected strings of amino acids. Much of the cell's substance is protein, and more important still, the enzymes making the cell function are proteins. Nucleic acids are made up of strings of compounds called **nucleotides.** These are large and fairly complex molecules composed of a phosphate group, a five-carbon sugar, and a compound rich in nitrogen called a **base.** We are specifically interested in the bases **adenine, thymine, cytosine, guanine,** and **uracil.** The first four are components of a nucleic acid found in the chromosomes: **deoxyribonucleic acid,** or **DNA.** The other nucleic acid is **ribose nucleic acid,** or **RNA,** very similar to DNA except that the sugar part (ribose) of the nucleotide has one more oxygen molecule, and it has the base uracil instead of thymine. RNA is found in both the nucleus and the cytoplasm.

Though chromosomes are made of both protein and DNA, proteins seemed more likely to be the genetic material because they are more complex. But Oswald T. Avery, Colin M. MacLeod, and Maclyn McCarty of the Rockefeller Institute in New York clearly showed that they were not, reporting in 1944 that DNA from one genetic strain of bacteria could change the genotype of another.

Like most important discoveries, this one was built on earlier work that demonstrated the basic phenomenon but did not quite provide a full explanation. English physician Frederick Griffith had reported in 1928 on experiments with *Pneumococcus,* the bacteria that cause pneumonia. He had two strains, one **pathogenic** ("disease-causing") and the other not. He injected small amounts of the harmless strain and large amounts of the pathogenic strain (which had been heat-killed) into mice. From the blood of the many mice that died, he isolated pathogenic *Pneumococcus.* But how could a dead pathogen still cause the disease? Years earlier, Louis Pasteur had disproved spontaneous generation of organisms (bacteria are not produced from spoiled soup), so that the best explanation was that something from the dead bacterial cells induced the living bacteria to become pathogenic.

The Rockefeller group conducted a similar but more refined experiment. They separated the heat-killed pathogenic strain into chemical fractions and incorporated the fractions into culture media on which they inoculated the nonpathogenic strain. Sure enough, when they included DNA from the pathogenic strain in the medium, the harmless bacteria were transformed to the pathogenic strain and transferred this character to all their offspring. Mendel's factors, the experiment showed, were DNA, but how could this fairly simple molecule control the destiny of cells?

Replication of DNA The DNA molecule is a helix, a spiral shape like the threads of a screw, a spiral staircase, a pretwisted telephone cord, or the stems of ivy. What is more, DNA is a double helix with two strings of nucleotides arranged in parallel (see Figure 11.14). The phosphates and sugars connect to each other and form a backbone while the bases jut out toward each other at right angles to this backbone. The "stairs" of this molecular staircase are formed by the linking of the bases by hydrogen bonds. Each kind of base will form a link with only one other kind of base: adenine bonds only with thymine, and guanine only with cytosine. The sequence of these linked pairs is variable, permitting an incredibly large number of sequences in the molecule's structure.

Chromosomes are replicated in the nucleus, so DNA must also replicate. In 1953 an American, James Watson, and an Englishman, Francis Crick, collaborators at Cambridge University, presented a model of the structure of DNA and how it replicates. They proposed that the double-stranded DNA unzips up the middle of the helix; as the two strands separate, free nucleotides in the nucleus pair up with their complementary nucleotides in the two halves of the unfastened "zipper" and are attached in sequence, as depicted in Figure 11.14. Here was a model explaining the unique self-replication of DNA and a starting place for a genetic code. It also brought Watson and Crick a Nobel Prize. (The sociology of this quest for the structure of Mendel's "factors" is nearly as gripping as the science, as told in Watson's popular *The Double Helix.*)

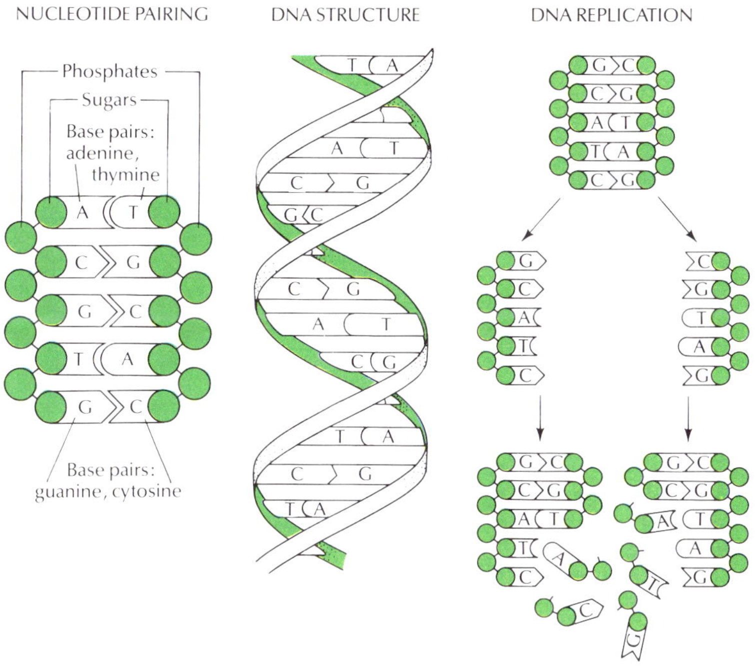

11.14 The nucleic acid DNA is the stuff of genes. The molecule consists of two strings of nucleotides arranged in a spiral (helix). The two spirals are connected by complementary bases of the respective nucleotides to form a double helix. The molecule replicates by unzipping at the middle where the bases are linked, and each single helix serves as a template for the attachment of complementary nucleotides. The double helix is reformed.

DNA and Proteins Another layer was removed from the scientific onion. But how does DNA provide for all the information required to build and run cells and organisms? And how are mutations produced? The key is to link DNA with proteins, because the synthesis of all the compounds of all the structures in cells is controlled by proteins—enzymes: whatever controls enzyme synthesis controls the nature of the cell.

In 1940 George Beadle and Edward Tatum, then of Stanford University, brought DNA and proteins together. They exposed the bread mold *Neurospora* to X rays and produced mutants that would not synthesize some vitamins and amino acids required for growth. Crossing the mutants with normal bread molds gave progeny (offspring) with predicted genotypes, just as Mendel obtained with his peas. Beadle and Tatum found from these experiments that the mutations were due to alterations of single genes, and they resulted in the blocking of single enzymes required to produce an amino acid needed for the growth of the bread mold. The conclusion: each gene in a chromosome is a segment of DNA responsible for synthesizing one enzyme. Remember dwarf corn (page 204): it is made that way because a mutation of one gene blocks the synthesis of an enzyme required to synthesize the growth substance gibberellic acid. Normal corn's phenotype can be produced by applying gibberellic acid to the dwarf mutants.

Beadle and Tatum had established the connection between DNA and protein, but how does DNA direct the production of proteins?

What is the "code" for producing one protein rather than another? An important step toward the answer came ten years after the original *Neurospora* work. Biochemists Linus Pauling of the United States (double Nobel laureate and vitamin C advocate) and V. M. Ingram of England found the structure and the sequence of amino acids in normal blood hemoglobin and in hemoglobin associated with sickle-cell anemia, a disease caused by decreased ability of the hemoglobin in human blood to carry oxygen to the tissues. The disease is the result of a single gene, which is found frequently in populations in Africa and in parts of Asia. The homozygous recessive genotype often kills early in the victim's life, but the heterozygous genotype gives resistance to the malaria common in these areas; the homozygous dominant genotype is susceptible to malaria, making the gene's presence in malaria regions an example of selection of variants adapted to a particular environment (selection will be discussed further in Chapter 14). With malaria missing as a selective force in the United States, its incidence in the gene pool will decrease.

The hemoglobin responsible for sickle-cell anemia, Pauling and Ingram discovered, differs from normal hemoglobin only in the position of one amino acid. It was concluded that a gene can control the placement of amino acids in the chains constituting proteins.

The Genetic Code Twenty amino acids are found in proteins. The kind of amino acids, their position, and the manner in which the polymer chains are folded and connected determine the properties of proteins. The combinations made possible by these variables are almost infinite. But how are these twenty amino acids selected and placed in proteins by only four bases? (Remember that the sugar and phosphate parts of the nucleotides are not involved in the pairing so important in DNA replication.)

Before this genetic code can be understood, attention must be paid to the other nucleic acid, RNA. The structure of RNA is very much like DNA. (In plant viruses RNA is the self-replicating genetic material; DNA is absent.) In addition to a difference in the ribose sugar, RNA has the base uracil instead of thymine. Uracil bonds with its complementary base adenine. Single strands of DNA act as a template or pattern for RNA synthesis, just as they do for DNA replication, dictating the sequence of bases in RNA. There is an important difference: the single strands of RNA separate from the DNA templates, leave the nucleus through pores in the membranes, and enter the cytoplasm (see Figure 11.15). The RNA acts as a messenger, carrying information from nucleus to cytoplasm.

Once in the cytoplasm, this **messenger RNA** is linked to the surfaces of small bodies in the cytoplasm, called **ribosomes** (see Figure 11.16). These are about 0.02 micron in diameter (1000 microns is 1 millimeter, or 0.04 inch) and are made of RNA and pro-

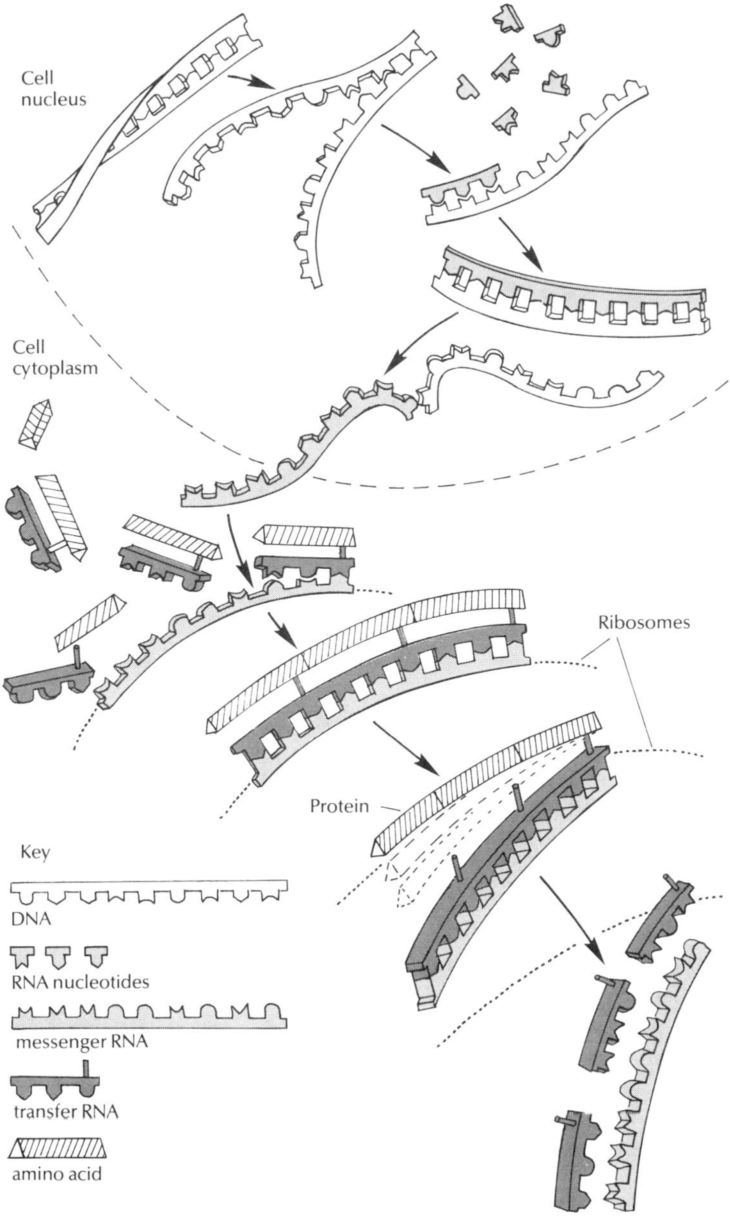

11.15 Protein synthesis takes place in the cell cytoplasm, but the events leading to it are initiated in the nucleus. Single strands of DNA are templates for the synthesis of RNA. This RNA slips through pores in the membranes surrounding the nucleus and it is the messenger carrying the genetic code to the cytoplasm. Messenger RNA attaches to the surface of ribosomes in the cytoplasm where it serves as a dock for the transfer RNA with its amino acid cargo. Each kind of transfer RNA carries only one kind of amino acid. Bonding between complementary bases of the messenger and transfer RNA determines the kind and location of amino acids in a protein chain. Cell and organism structure and function are regulated by enzymes, so genetic control is expressed through protein synthesis.

tein. Fabricated in the nucleus, they migrate to the cytoplasm. Proteins are formed on the surfaces of the ribosomes, where the messenger RNA sits with its reactive bases free in the cytoplasm somewhat like antennas. Yet another type of RNA, consisting of small segments about seventy-seven nucleotides long, is free in the cytoplasm. Each of these small chunks of RNA will form an attachment with only one of the twenty amino acids. It is called **transfer**

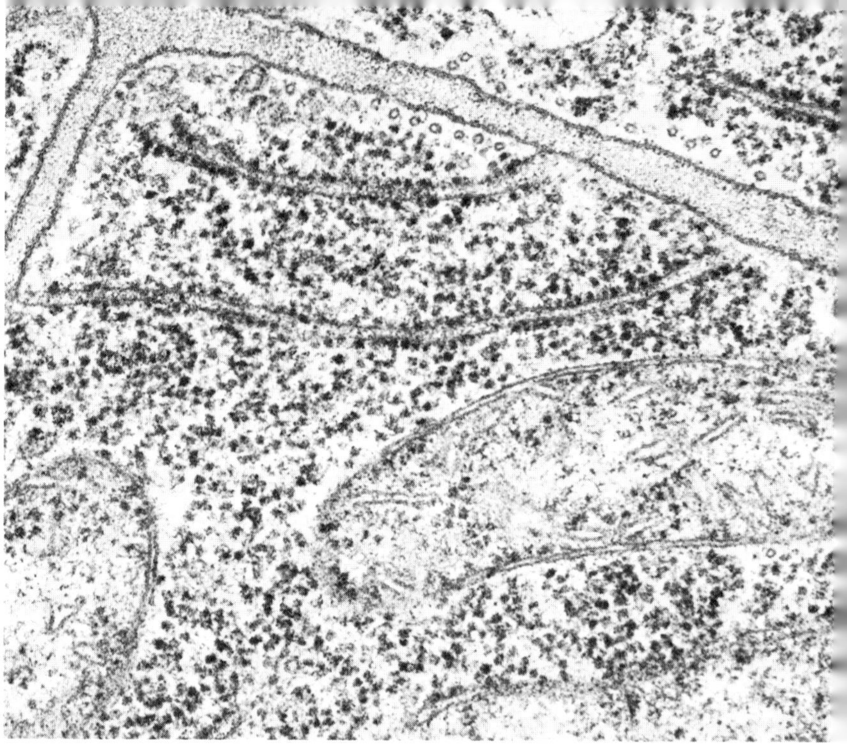

11.16 Cytoplasm of all cells is filled with ribosomes which can be seen only with the electron microscope. Protein synthesis occurs on the surface of ribosomes. This root tip cell is magnified × 53,500.

RNA, because it captures amino acids and transports them to the messenger RNA on ribosome surfaces. (The sequence of events is depicted in Figure 11.15.) Each transfer RNA has three exposed bases that bond with their complementary bases on the messenger RNA. The amino acid attached to the transfer RNA bonds with its neighbors on each side, and a protein is born. The protein then detaches from the transfer RNA, which in turn detaches from the messenger RNA and returns to the cytoplasm, where it bumps into another of its kind of amino acid and hauls it back to a ribosome.

Cells with active protein synthesis have so many ribosomes that this matching and attaching goes quickly. This molecular dance unites many partners with complementary costumes. But any transfer RNA can dance with only one kind of amino acid, and because the three bases of each transfer RNA attach only to the complementary bases on the messenger RNA, the secret of the genetic code is to be found in this *ménage à trois.*

The code was cracked in 1960 by Marshall Nirenberg and Heinrich Matthaei at the National Institute of Health in Bethesda, Maryland. They used synthetic RNA containing only nucleotides with the base uracil to make protein. The protein they produced contained only the amino acid phenylalanine. RNA with a triplet of uracils (UUU) coded for the amino acid phenylalanine, so the matching DNA code letters were AAA (for 3 Adenine which bonds with uracil).

In the next few years, intense activity in many laboratories deciphered the code. Because the four bases of RNA form code words in groups of three or triplets, there are 4^3 or 64 possible words. There are only twenty amino acids in proteins, and so the code has duplications. One amino acid can be specified by more than one triplet of messenger RNA: the messenger RNA words for phenylalanine are either UUU or UUC, and for glycine, GGU, GGC, GGA, or GGG (see Table 11.1). Some triplets are not words for amino acids but instructions to stop or start synthesizing a protein.

Table 11.1 Selected Amino Acids with Corresponding Messenger RNA and Transfer RNA Triplets

Amino Acid	Messenger RNA	Transfer RNA
Alanine	GCU	CGA
	GCC	CGG
	GCA	CGU
	GCG	CGC
Arginine	AGA	UCU
	AGG	UCC
	CGU	GCA
	CGC	GCG
	CGA	GCU
	CGG	GCC
Asparagine	AAU	UUA
	AAC	UUG
Aspartic Acid	GAU	CUA
	GAC	CUG
Lysine	AAA	UUU
	AAG	UUC
Phenylalanine	UUU	AAA
	UUC	AAG
Tryptophan	UGG	ACC

Back in 1865 Mendel had reported to the Brünn Society for the Study of Natural Science that inheritance can be explained by particulate units assorted randomly and segregated independently in the gametes of pea plants. A hundred years later the composition of genes and the molecular language responsible for expressing genetic information were discovered. Mendel's theory was verified. Species that starred in the story went from pea plants to *Escherichia coli*, a bacterial species of the human intestine. Discoveries came out of studies on fruit flies, corn, bread mold, and human beings. In every one, a gene is *a triplet of nucleotides,* the part of a DNA molecule that codes for placement of an amino acid in a protein chain.

Mutations Any rearrangement or deletion of a DNA triplet will alter the code and create a mutation. Let us say a DNA triplet is GGG; this triplet means the messenger RNA triplet will be CCC, a code word for the amino acid proline. If the DNA triplet is altered to GTG, the corresponding messenger RNA will have CAC, a code word for the amino acid histidine. This would lead to synthesis of a different protein and possibly a significant change in cell metabolism. And this alteration shows how X rays produce mutations. Striking just one part of a molecule, such as the base of a nucleotide in DNA, X rays can either destroy it or alter the molecular structure. The bases, after all, are very similar in structure. Changing one letter of a code word can cause a mutation. Also, the DNA synthesizing machinery occasionally makes errors, such as inserting an incorrect base. Only changes in the genes of the sex cells, however, can be passed on to the next generation, explaining why technicians protect the gonads of a patient with lead shields during exposure to X rays.

Much less intense than X rays, the cosmic rays that constantly bombard earth can also cause mutations. Elements in the earth's crust steadily give off radioactivity. It is quite possible that these constant streams of radiation contribute to the continuous mutation observed in all organisms. A substantial rise in radioactive emissions in the environment could lead to higher rates of mutation. Many mutations are potentially harmful, and this is the basis for concern about radioactive fallout from nuclear explosions and leakage from atomic reactors or their refuse.

Genetic Engineering Much that we know about the genetic material and the genetic code has been gathered in work with bacteria, in which DNA forms large molecules in the cell instead of being packaged with protein in chromosomes wrapped up in the nucleus. Today new genetic information can be introduced into bacterial cells, transforming their genotype. It is possiblè to synthesize a bacterial gene and have it work; this type of genetic engineering is a reality.

How significant is this news? Consider the possibility of producing rare and costly compounds very cheaply by bacterial mass production. Insulin is a protein necessary in treating diabetes; it is obtained from animals. Could not the appropriate genes be synthesized from DNA and introduced into bacteria? The resulting strain of bacteria could be grown in large tanks to manufacture limitless amounts of insulin at very low cost; as has been done on an experimental basis. This type of genetic engineering might have immeasurable economic and social benefits.

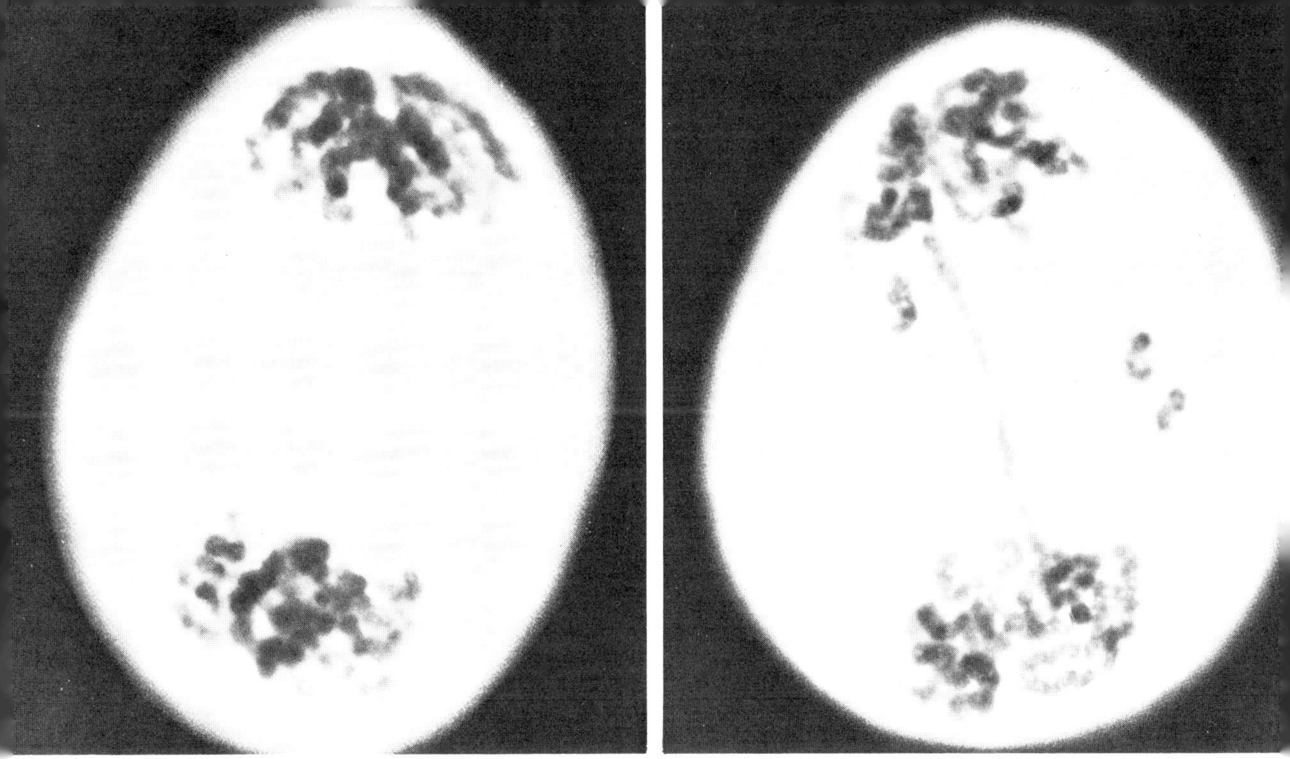

11.17 (Left) Normal chromosomes during cell division in a *Trillium* separate and move to the ends of the cell. (Right) A *Trillium* cell irradiated with X rays in the early stages of meiosis has an abnormal division of chromosomes. This is one way in which radiation can induce genetic mutations, which usually result in abnormal cells.

Like most human activities, though, the idea has a darker side. Many bacteria and viruses are themselves pathogenic, and introducing foreign DNA into bacteria might generate new pathogens or make established ones more virulent. Could it happen? Would the new strains compete with wild strains outside the laboratory? Certain answers to these questions are not known; and debate thunders on about continuing experimentation in this field of genetics. Will the almost certain benefits be balanced by dangerous new creations? Is it an acceptable risk or a Faustian bargain? This controversy is but one of many involving science, ethics, and politics.

Genetic engineering of the same type with animals and plants is some time off. Most characteristics of these organisms are controlled by the complex interactions of many genes, and we know little about them. We still do not know how DNA is really packaged in chromosomes, for example, and our ignorance about cell development is vast. Knowledge has expanded unbelievably in fifty years, however, and if understanding of genetic control increases as fast over the next fifty, you will face hard decisions about certain types of genetic engineering in multicellular organisms, perhaps even in humans.

DNA outside the Nucleus All DNA does not live in the nucleus, and the nucleus is not alone in replicating itself and directing other activities in the cell. The mitochondria and chloroplasts in plant cells also have DNA and, like the nucleus, are self-replicating.

The chloroplasts are passed on only through the egg cell in the gametophyte—an example of **maternal inheritance** in which genes are passed only through the female. Maternal inheritance usually is related to pigmentation and does not follow Mendelian patterns of heredity.

Like bacteria, chloroplasts and mitochondria divide by **fission,** or pinching into two parts, and their strands of DNA are free within the organelle. The quantity of DNA is similar to that in bacteria, and these organelles also have their own RNA and ribosomes. Still, most of their protein synthesis is coded for by nuclear DNA.

Because of these characteristics, it has been suggested that mitochondria and chloroplasts are descended from microbes that were engulfed by primitive cells and have lived since then as symbionts. Opinions differ on the validity of this theory.

POLLINATION

We have seen how information is transferred from generation to generation and how this information is generated by the genetic code. A fuller understanding of inheritance in plants requires that we now go from the cellular level to a larger scale: the plant living in its environment.

In sexual reproduction gametes produced by meiosis are united, but first the gametes or gamete-bearing structures must be brought close together. This is done in gymnosperms and angiosperms by **pollination,** the process discussed in Chapter 3. In angiosperms it consists of transporting pollen from the anthers of a flower to the stigma of a flower's pistil, allowing fertilization to occur. When the stigma is receptive, its surface cells excrete a sticky substance of sugars and other compounds, which holds the pollen grain, induces the pollen tube to emerge, and nourishes its initial growth. Stigmas of some plants even have two "flaps" (see Figure 11.18). When pollen is deposited, they close and encase it in a humid, nutrition-rich chamber.

Pollen can be transferred in a seemingly endless variety of ways, but we can divide these into a few basic groups. One group is self-pollinating, like Mendel's peas, and another requires a pollination agent, a **vector.** The latter can, in turn, be divided into plants employing either physical agents or organisms as vectors.

Self-Pollination Many plants have flowers that are **bisexual:** they have both stamens and a pistil or pistils. Often these flowers exhibit self-pollination: pollen from the anthers falls onto or is transported to the stigma of the same flower. Many other plants have **unisexual** flowers, flowers with only stamens or pistils but not both. Male and female flowers can occur together on one plant, or

the plant may have only one of a kind and not be self-pollinating. Some fruit trees, such as the mulberry, have flowers of only one sex on each plant, so unless you have a male reasonably close to a female tree, you will obtain no fruit. A fruitless mulberry has only male flowers.

In many plants complicated mechanisms ensure that some **outcrossing,** some transfer of pollen between plants, occurs rather than self-pollination (see Figure 11.19). Outcrossing increases the mixing of genotypes, which improves the possibility of new genetic combinations. The enriched **gene pool** (the total genetic information in a population) that results provides populations with a greater chance to exploit new environments and to adapt to changes in their present environment. It is not uncommon, for example, for self-pollination in bisexual flowers to be prevented because the pistil is taller than the stamens or because the pollen is released before or after the stigma of the pistil is receptive. Pollination between two different plants is then needed for a high rate of fertilization. In many cases there is a chemical incompatibility between pollen and stigma of the same flower and fertilization is prevented or inhibited. This is seen

11.18 Variation in stigma structure is endless. Some have two flaps, one of which is sensitive to touch. When pollen is brushed on the surface, the flaps snap shut. This snug and moist embrace provides fine conditions for pollen germination.

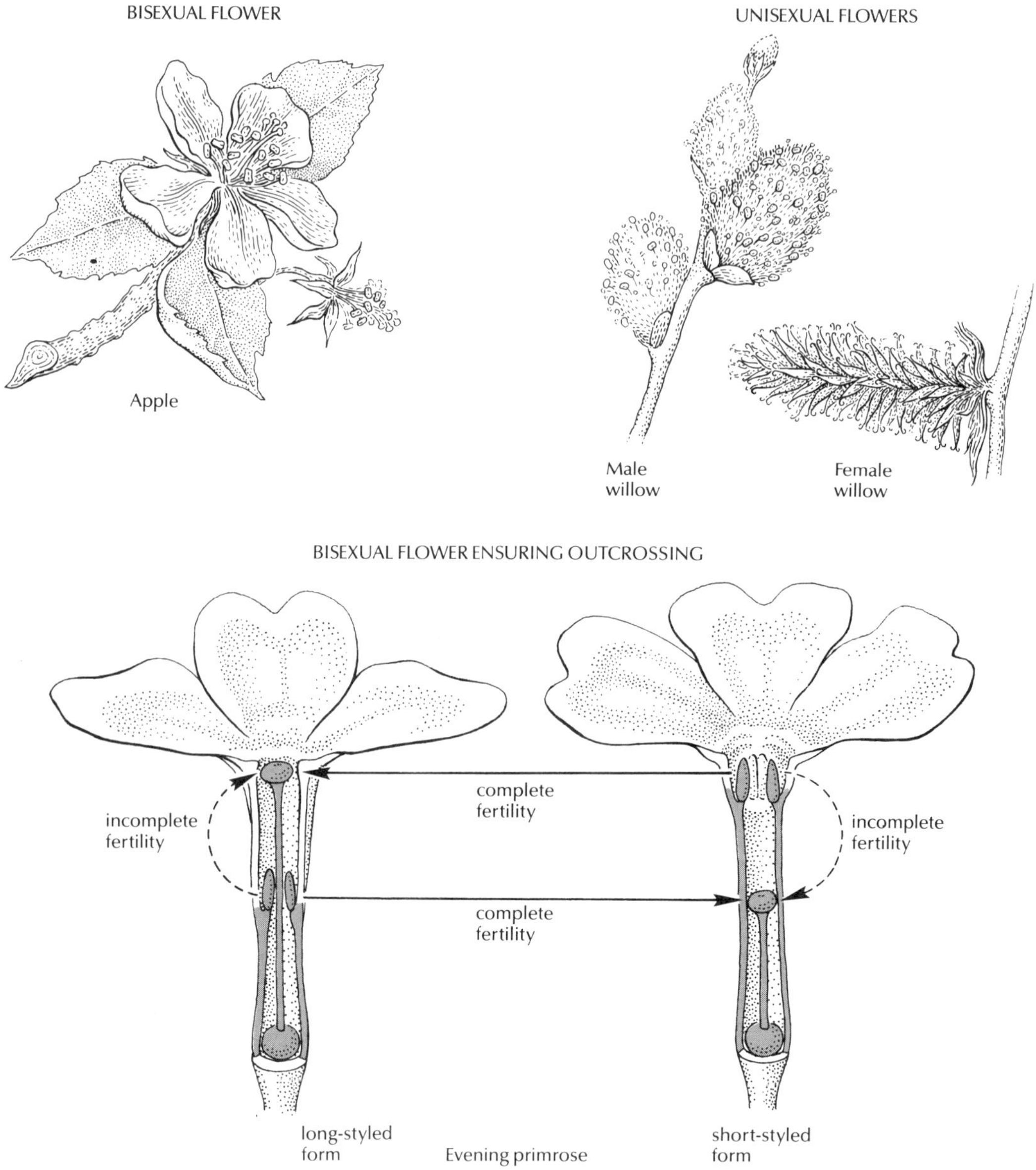

11.19 Self-pollination is the rule in bisexual flowers, while unisexual flowers are an effective mechanism for cross-pollination. Bisexual flowers, however, frequently exhibit adaptations which promote outcrossing. The evening primrose is an example. Relative lengths of pistil and stamens differ in different forms of the flower. These features and incompatibility between stigma and pollen of the same flower result in a breeding system favoring outcrossing.

in some fruit trees. Several cherry trees together, particularly of different varieties, will usually produce more fruit on each tree than a single tree standing alone will.

Physical Vectors Water and wind are the most common physical agents of pollination between plants. Pollen is carried among many aquatic angiosperms in both fresh and salt water by currents and waves. The gymnosperms (see Chapter 4) have no flowers, but their pollen, produced in cones, is transported to female cones by wind. All the grasses are wind-pollinated, as are many trees, such as beeches, birches, alders, and willows.

Not only are wind-pollinated flowers not showy, their floral parts are so reduced that many people do not know that grasses have flowers. Pollination by wind is a haphazard business, requiring copious production of pollen, and the plant's energy goes into producing pollen, not large showy flowers. The grasses have evolved recently and are still changing rapidly, so wind-pollination is neither primitive nor ineffectual.

Organisms as Vectors The evolution of the flower is a story of the coevolution of plants and animals (discussed further in Chapter 14). Many animals serve as pollination vectors; some, like the honeybee, are specifically adapted to life as a vector. Other insects, such as beetles, flies, and midges, are often incidental vectors. Beetles, for example, are always scrounging around for food, and while eating flower parts they can be dusted with pollen, later scraping some off as they contact stigmas. Birds such as the hummingbird act as vectors while seeking nectar at the base of tubular flowers. (Hummingbirds also catch insects, so their diet is not exclusively nectar.) Many bat species are strict vegetarians (sorry, Dracula) and pollinate night-blooming flowers. They eat nectar and fleshy flower parts and transfer pollen from flower to flower. Some flowers are pollinated by night-flying moths; their flowers are usually white with a heavy musty aroma. Aroma also attracts daytime vectors, particularly flies, which go for flowers that smell like rotten meat or even fecal matter. (See also Figure 11.20, page 338.)

HYBRIDIZATION AND PLANT BREEDING

Hybrid corn is an excellent example of the practical use of Mendelian genetics. Corn is easy to cross because it has separate male and female flowers. Because the stamens are carried in the tassels, corn can be hybridized by simply removing the tassels so that pollination occurs from neighboring plants. Corn can be crossed, double-crossed, and even triple-crossed to produce a truly rich gene pool from which to choose desirable qualities (see Figure 11.21). These range from the hardy, flinty types used for animal fodder in some parts of the world to the sweet, tender varieties we cherish for the table, to popcorn.

Hybrid Vigor Corn also illustrates that hybrids are often more vigorous than their parents, a condition called hybrid vigor, or **heterosis.** The average yield of corn has been increased

11.20 (Top, left) Grass flowers are wind pollinated. Large amounts of pollen are characteristic of wind pollinated flowers and these timothy *(Phleum pratense)* flowers have prominent stems but lack showy flower parts. Bats are numerous in the tropical and subtropical regions, and they are pollinators for many plants. This bat (top, right) is feeding on the flower of a century plant *(Agave palmeri)* and in the process will become well dusted with pollen which will be transferred to receptive stigmas as the bat moves among the flowers. Butterflies and moths feed on nectar and in the process serve as pollinators. A butterfly (inset, right) will transfer pollen as it flits from flower to flower and sips nectar. A number of flowering plants live in the ocean. Eelgrass (right) is a monocot, but not a grass, and is common in shallow waters with sandy bottom. It produces small flowers and pollen is transferred by water movement.

from 25 bushels per acre in the 1930s to 200 bushels regularly produced today as a result of hybridization, and further increases are likely. Forty-eight or so kernels per ear were found in the earliest known crops of domesticated corn; hybrid varieties now usually produce ten to twenty times that many, or five hundred to a thousand kernels per ear.

Heterosis has commercial value too in many types of vegetables, with tomatoes being a well-known example. The first generation of hybridized tomatoes will have a great variety of inferior plants, but crossbreeding plants selected for a certain size or sweetness of fruit can yield a variety with vigorous growth and desired fruit characteristics. This is a clear example of manmade evolution.

Hybridization can, of course, occur among animals. A most familiar product is the mule. Offspring of the union between a male donkey (Jack) and a female horse (mare), the mule demonstrates as well as corn the increased vigor that hybrid strains so often show compared with their parents. Mules are stronger than donkeys and many horses, and they can go longer without food and water than horses can. It was no accident that they pulled many a covered wagon on the frontier and were favorite riding and draft animals.

Like many hybrid plants, mules are usually infertile because normal meiosis is blocked and viable gametes are rarely produced. But **interspecific crosses** in which, say, one species of wheat crosses with another are common among plants in nature and certainly have been important in plant evolution. The resultant hybrids then crossbreed between themselves and even with their parents, creating a genetic soup, or **hybrid swarm,** that can provide a pool for evolution of new species.

Applications of Plant Genetics In 1933 the state of Iowa, America's greatest producer of corn, grew only 1 percent of its crop in hybrid corn. Ten years later, 99 percent of the crop was hybrid. These statistics make headlines, but behind the headlines lies the unheralded work of plant geneticists who for half a century or more have been working with genetic facts first learned in a little monastery garden in Austria. Those "units of heredity" that Mendel proposed as he watched the appearing and disappearing traits of generations of garden peas are today much manipulated by geneticists.

Synthetic hybrids of crop plants have a tremendous effect on agriculture, but natural hybridization of pathogens also has profound implications for the world's food supply. The fungus *Puccinia graminis,* for example, causes wheat rust. New wheat hybrids are selected to be resistant to wheat rust, but new genetic strains of rust overcome the resistance of the wheat varieties. New wheat varieties resistant to the rust must then be produced. As a result, growers of food are in an unending race against pathogens, seeking to develop

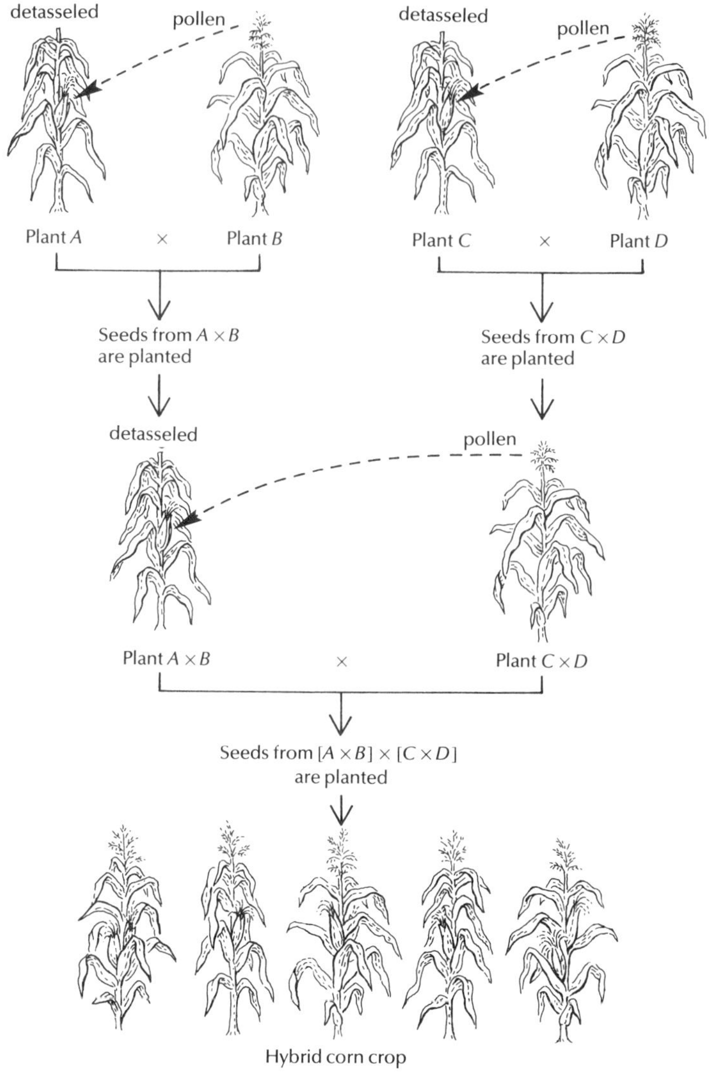

11.21 Hybrid corn is a major reason why so many persons can be fed by so few farmers in the United States. Hybrid corn seed is produced by crossing two hybrid strains. Since corn has both male and female flowers on the same plant, self-pollination is ruled out by either removing corn tassels which bear the male flowers or by using varieties bred for male sterility.

new genetic varieties of crop strains that will be resistant to new genetic varieties of pathogens.

Culturing of single cells, which we will look at more closely in Chapter 12, is the latest and perhaps most exciting development in plant genetics. It has already been possible to induce fusion between two cells from totally foreign plants, like soybean and barley. No one has succeeded in raising such a bastard plant to maturity, but the possibility is certainly there, and commercial applications of this type of intergeneric union are marvelous to contemplate (see page 358).

Triticale The potential of intergeneric breeding is shown by a new cultured genus of plant called *Triticale*, created by a cross between wheat (genus *Triticum*) and rye (genus *Secale*). The offspring has characteristics of both parents, combining the high protein content of wheat with the high lysine content of rye. Lysine is an amino acid essential to human beings, too little of which is supplied by protein from cereals. *Triticale* adapts well to cold weather and light, sandy, or arid soils, and some of its advanced varieties seem to have resistance to wheat rust. It is the first synthetic cereal grain that can compete successfully with the cereals grown by farmers for a thousand years. Its grains are also larger than those of wheat and plumper than those of rye. Now grown in fifty-two countries in environments from the Canadian prairies to the Himalayan foothills, *Triticale* everywhere has given yields that regularly surpass those of normal wheat crops.

Triticale was discovered in 1889 in Germany, where a botanist named Rimpau observed it growing spontaneously. The plants were infertile, and nothing could be done about breeding them; for almost forty years *Triticale* remained a tantalizing curiosity. Then in 1937 a Frenchman, Pierre Givaudon, found that by treating *Triticale* seedlings with colchicine, an alkaloid compound (commonly used to treat gout) derived from the autumn crocus, their chromosomes could be made to double. The first-generation *Triticale* hybrids had been infertile because they had only one set of chromosomes from each parent, and these could not combine during meiosis; normal gametes were not formed. The colchicine treatments doubled the chromosome numbers, so that each pair of wheat chromosomes could pair during metaphase of meiosis with wheat chromosomes, and rye chromosomes with rye. Gametes were formed and a fertile plant was produced.

In 1954 an extensive *Triticale* research program was set up in Canada by the University of Manitoba and the International Maize and Wheat Center of Mexico. Over the years this program has grown many varieties of *Triticale*. Mexican researchers, working with a large and diverse pool of wheat genes, combined their efforts with Canadian scientists, who were concentrating on new techniques for increasing the number of viable crosses. This was necessary because *Triticale* did not have many features required for economic success. The story of these efforts is engrossing, as told by Norman E. Borlaug, Nobel Prize-winning scientist in charge of the Mexican end of the project:

> I must tell you that the largest and most important step toward making the breakthrough in *Triticale* improvement was executed by Mother Nature herself, one early dawn March morning in 1967 in Ciudad Obregon, Sonora, while scientific man was still in bed. One

> promiscuous, venturesome stray wheat pollen grain with a potent and valuable "genetic load" from the nearby wheat breeding plots floated across the road under cover of darkness and fertilized a sad but permissive, tall, sterile, degenerate *Triticale* plant.

A year and two generations later, scientists identified several unusually promising plants among the hybrids. Their genetic makeup clearly told the value of the illicit stray wheat pollen grain. The *Triticale* hybrids indicated that the wheat pollen had carried genes for a dwarfed growth habit and partial photoperiodic insensitivity. It had completely overcome the barrier of sterility that for decades had inhibited progress in improving *Triticale.* "This seems to me," Borlaug concluded, "to be nature's way of telling scientists not to become too arrogant."

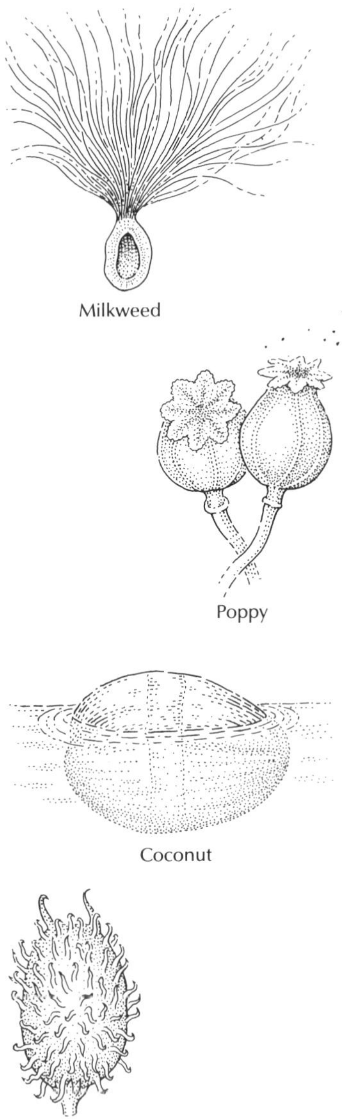

11.22 Fruits and seeds are dispersed in many ways. These range from the milkweed seeds, dispersed by wind because of their long silky hairs, to the floating coconut fruit. Cocklebur fruits catch rides on fur and clothes, and poppy seeds are shaken out of the fruit capsule.

SEED DISPERSAL

It is simple enough to think of a seed as a small package formed to protect and nourish the young plant, as we saw in Chapter 6, but that is only half the story. The young plant usually has to get somewhere so that it can grow apart from its parent. Where the seeds are scattered will affect not only the individual plant's survival but also the distribution of a population. Plants exhibit many mechanisms promoting wide dispersal of their seeds (see Figure 11.22).

For some plants, fruits surrounding the seeds are the vehicles for their dispersal. Birds and other animals ingest the seeds along with the tasty pulp of the fruit and later excrete the undigested seeds, distributing them widely and adding to the seedlings some handy fertilizer. Wind disperses seeds as well as pollen, and some plants have specialized structures adapted to dispersal by air. After the dandelion has flowered, it forms wonderfully delicate, round parasols of silvery threads, each with a seed attached. A slight puff sends these flying, to drift along wind-borne, until they fall to earth. If the landing place is suitable, they may start germinating. Maple seeds have leaflike appendages that wing them away on a breeze like tiny helicopters, spinning as they fall and traveling quite a distance, even in the forest.

Nut seeds are in particularly tough packages that can tolerate very distant dispersal. Squirrels carry them off to a cache, hoarding them against winter shortages of food, and people harvest them and carry them to distant markets. Nuts can lie on the littered forest floor for years before germinating to make another tree.

Berries, like flowers, can be brightly colored to attract birds, bears, squirrels, and other animals. The species react to different colors, and birds are particularly attracted to red. Distinctive odors may be

11.23 Coconut palms are seaside residents throughout the tropics. Their worldwide distribution is testimony to the effectiveness of waterborne dispersal. This coconut seedling demonstrates how castaways can colonize new areas.

the attraction held out by others. In Southeast Asia the durian (*Durio zibethinus*) is extraordinary in its odoriferous attractiveness. The fruit smells much like rotten meat, bringing flies in swarms. Yet even people relish its delicious white creamy pulp.

Bears eating blackberries, mice eating grain, squirrels eating and burying nuts, all these animals help spread seeds. And plants have still other ways of getting their seeds around. Burrs are fruits with hooks that cling to animal fur, perhaps to be carried miles away. Mud traps seeds, which stick to the feet of birds and are carried away. Coconuts, which often grow next to a beach, fall into the sea and drift miles before washing up on another beach where they can take hold and grow; they have spread around the tropical parts of the globe. When the milkweed seedpod has dried up, it bursts open, sending the seeds, each with a tuft of silky hairs, to float away on even the most vagrant of breezes.

Seed dispersal illustrates the complex interaction between plant reproduction and genetic variation and between plants and their environment. For example, wide dispersal favors the enlargement of a population of perennials because the chances that progeny will survive competition with parent plants is small. A genotype providing tolerance for a wide range of growing conditions might also produce abundant progeny but with little food reserves in the seed. Genotypes resulting in seeds with little tolerance for environmental extremes might have to provide for prolific seed production and wide dispersal if extinction is not to ensue. A plant with redder berries than another plant would be more attractive to birds and ensure greater chances for seed dispersal.

The test of successful genotypes is the production of progeny. Since the environment is constantly changing, it is only by the production of new genotypes as a result of sexual reproduction that the lottery of survival continues to be played.

Summary

Plants may reproduce either sexually or asexually. In sexual reproduction gametes from two parents fuse, producing a new individual unlike either parent; sexual reproduction is thus the source of the genetic variation that fuels evolution. In asexual reproduction, a new plant grows from one parent and is identical to it. If the parent is well-adapted to its environment, asexual reproduction has the advantage of perpetuating the well-adapted genotype.

In the middle of the nineteenth century, Gregor Mendel studied how traits are passed from parents to offspring. Alternative forms of a trait are called alleles; determinants of the traits are genes. From experiments with peas Mendel laid the foundation for two laws basic to our understanding of genetics: (1) Segregation—large numbers of paired genes are passed on to offspring; the gamete of each parent transmits one member of each of these pairs. (2) Independent assortment—when the gametes are formed, genes from mother and father are independently assorted; that is, members of the paired genetic units form all possible combinations with each other.

If an organism receives the same allele of a gene from each parent's gamete, its genotype (genetic constitution) will be homozygous for that trait; if it receives two alleles, it is heterozygous. An allele is said to be dominant if it will be expressed when the genotype is heterozygous for that trait; it is called recessive if the trait is not expressed when the genotype is heterozygous. Some traits, such as flower color, are determined by multiple genes, so one finds not simple dominant and recessive traits but many gradations.

Understanding the mechanisms by which Mendel's laws operate required a concept of the cell, which developed over many centuries, and detailed knowledge of how the cell operates, which has been gained only during recent decades.

Chromosomes within the nuclei of cells carry the genes, and each type of organism has a characteristic number of chromosomes. In mitosis (the process which produces all plant cells except spores), the chromosomes duplicate, separate, and migrate to opposite ends of the cell, and new membranes and cell walls develop. Thus two cells, identical to the parent cell except in size, are formed.

Division of sex cells, however, differs. It is called meiosis, reduction division. In its first stages homologous chromosomes (those derived from each parent for the same trait) pair up and double. Each doubled chromosome of a pair moves to opposite ends of the cell, but the separation is random—some chromosomes going to one end are from the mother and some from the father—and the doubled chromosomes (chromatids) do not themselves separate. Two cells are formed, each with half as many chromosomes as were in the parent cell. These cells then undergo another division much like mitosis but without further duplication of chromosomes. Thus meiosis consists of two divisions with just one doubling of chromosomes. The result is four daughter cells, each with half as many chromosomes as were in the original cell. These are called haploid cells; all other cells in an organism are diploid. When haploid gametes fuse, a diploid zygote is formed. Thus in a plant's alternation of generations one generation, the gametophyte, is haploid and the other generation, the sporophyte, is diploid.

The random assortment of chromosomes early in meiosis gives the independent assortment that Mendel's laws predicted. Genetic variability is also produced by rearrangements of chromosomes such as occur in crossing-over: arms on the doubled chromosomes entangle and, when they disentangle, segments of chromosomes may be exchanged. Mutations are heritable changes within the genes.

The chromosomes are made up of proteins and deoxyribonucleic acid (DNA). The DNA takes the form of a spiral staircase or helix with parallel strings of nucleotides, which are phosphates and sugars connected to each other to form the sides of the spiral and bases (adenine, thymine, guanine, and cytosine) forming the "stairs." When chromosomes duplicate, DNA does too: the double spiral unzips and free nucleotides in the nucleus are picked up and matched to each half of the unzipped spiral. Each of the four bases will bond with only one of the others, so the two spirals of DNA are identical to the parent DNA.

Each gene is a segment of DNA. It is the key to heredity because DNA directs the synthesis of enzymes and all other proteins. It does this in concert with another nucleic acid, ribonucleic acid (RNA), which differs from DNA in the sugar it contains and in one of its bases.

DNA acts as a template for synthesizing RNA much as it does for replicating itself. After single strands of RNA are made by DNA they leave the nucleus and link up to ribosomes in the cytoplasm. This kind of RNA, known as messenger RNA (mRNA), is not the only kind; within the cytoplasm are smaller segments known as transfer RNA (tRNA). Each transfer RNA will bond with only one amino acid. It picks up an amino acid in the cytoplasm and carries it to the ribosomes; here tRNA bonds with mRNA while its passenger amino acid bonds with neighboring amino acids that were carried by other segments of tRNA. By this linking of amino acids a protein is made.

The basic character of a protein is determined by the sequence of bases in the nucleic acids; the four bases of RNA form code words. Some code words "say" a particular amino acid and some are punctuation, giving instructions to begin or to stop synthesizing a protein. Any arrangement or deletion of a DNA triplet will alter the RNA's code, leading to synthesis of a different protein. X rays can create mutations by destroying a nucleotide's base or altering its structure; DNA itself creates mutations by making errors in replicating itself or in synthesizing RNA.

This knowledge of the genetic code has widened the possibilities of genetic engineering. Even now DNA can be introduced into bacterial cells, transforming their genotypes. This type of genetic engineering with plants and animals, however, will probably not be possible for some time, if ever.

Mitochondria and chloroplasts of plants also have DNA and, like the nucleus, are self-replicating. Chloroplasts provide an example of maternal inheritance: they are passed on only through the egg cell and do not follow Mendelian patterns of heredity.

The gametes are united in gymnosperms and angiosperms as a result of pollination and subsequent fertilization. Some flowers are bisexual, and some of these are self-pollinating. Many flowers, however, have mechanisms ensuring that outcrossing and not self-pollination occurs. Outcrossing increases the mixing of genotypes and the chances that new genetic combinations will be formed.

Many flowers require pollinating vectors—wind, water, insects, or animals. Wind-pollinated flowers are usually unobtrusive but produce a great deal of pollen. The structures of both flowers and their pollinators often show that they have adapted to a life of interdependence.

Comparing the yield of hybrid corn with the yield of its ancestors dramatically illustrates heterosis; that is, hybrids are often more vigorous than their parents. Crosses among species are common among plants, and the resulting hybrids may breed with other hybrids or with the parent population, creating what is known as a hybrid swarm. Two genera—wheat (*Triticum*) and rye (*Secale*)—have even been crossed, creating the plant known as *Triticale,* which has the high protein characteristic of wheat and high lysine content like rye.

Plants also have mechanisms to disperse their seeds widely; fruits are sometimes the agents for this dispersal. Some fruits and seeds are attractive to animals, which help disperse them; some have wings or burrs; some, such as coconuts, can drift in water. Wide dispersal encourages the growth of populations of perennials because it takes the offspring out of competition with the parent population.

Review Questions

1. Why is sexual reproduction such an important biological concept?
2. Why was Mendel fortunate in his choice of the pea as an experimental organism?

3. Why are Mendel's laws important?
4. If mitosis and meiosis represented platforms of political parties, which would describe a party in favor of the status quo and which would describe a radical party? Why?
5. Describe the relationship between meiosis and organismal variation.
6. What is a gene and how do genes actually determine phenotypes?
7. How do mutations occur?
8. Explain the advantages and disadvantages of the most frequently used pollination mechanisms of angiosperms.
9. Debate the proposition that ancient civilizations arose because of plant hybridization and that modern civilizations depend on the knowledge and application of plant hybridization.
10. Why can seed dispersal be thought of as an extension of the outcome of meiosis?

CHAPTER 12

Plant Propagation

Many plants bypass the sexual cycle. Instead, by asexual reproduction a new plant grows directly from part of the parent. In sexual reproduction meoisis and fertilization increase genetic variability among the offspring, but asexual reproduction produces plants identical to the parent. All the plants that are derived from one individual in this way are called **clones.** Long ago, farmers began using asexual reproduction for artificial breeding. With this method they and horticulturalists and gardeners can produce plants identical to the desirable one they started with.

NATURAL FORMS OF ASEXUAL REPRODUCTION

Most plants can reproduce sexually; many can reproduce asexually as well. Most blue green algae, many bacteria, and some fungi, however, reproduce only by asexual means, including fission, in which a one-celled organism divides into two. Filamentous algae and even moss and liverworts can undergo **fragmentation,** in which the plant breaks into parts, each of which continues to grow. Many plants form spores, specialized structures that can grow into new plants. Here we will examine other methods of asexual reproduction, also known as **vegetative reproduction,** particularly those found in flowering plants. Methods vary from species to species; we will see that the new plant might be formed from the stem, root, leaf, bud, or even flower of the parent.

From Stems Some plants like strawberries *(Fragaria)* send out long **runners** (stems growing horizontally along the ground) that stretch far from the parent plant, taking root here and there to start independent plants. Ivy plants climb walls or grow along the ground in much this way. Other plants, including many grasses, have **rhizomes,** underground stems from which new plants can rise.

Tubers are food storage organs that are enlarged apical parts of underground stems (see page 350). Both in nature and in agriculture, tubers are often used to propagate plants asexually. The white or Irish potato *(Solanum tuberosum)* is a tuber, and its "eyes" are nodes with much reduced leaves and axillary buds. You can observe that they form a spiral. Planting pieces of potatoes with eyes is essentially the same as reproducing a plant via stem cuttings. Tuberous begonias can be propagated in a similar way.

Another enlarged underground stem stores food: the **corm.** Often called a bulb, a corm is really a short, swollen, vertical shoot with buds on the top and roots growing from the bottom. Gladiolus and crocus have corms, and each year buds at the top of the corm grow into new shoots. In the humid tropics large starch-filled corms of the taro are an important source of food. (The poi of the Hawaiian Islands is a good example.)

From Roots Still another form of vegetative reproduction is observed in banyan *(Ficus bengalensis)* and mangrove *(Rhizophora mangle)* trees. Both trees have roots that come down through the air from branches. Once they reach the ground, the roots grow downward into the soil. The trunk of the parent tree may die, but the lateral branches with their strong roots will produce a whole new tree. Banyans and mangroves may spread far indeed, an individual ancestor producing a large grove; swamp mangroves reclaim many acres of tropical coastal wetlands by dropping roots through water and gradually trapping soil. The morning glory *(Ipomoea)* also starts new plants from its roots, spreading very widely as it forms buds on its roots from which new morning glories grow. In the wild a dense thicket forms, covering large areas of ground. Sweet potatoes *(Ipomoea batatas)* too are roots that will produce buds and form new plants.

From Leaves and Buds Specialized underground shoots with leaves that have fleshy bases are **bulbs.** The onions *(Allium cepa)* we use as food are bulbs, and so too are tulips *(Tulipa).* New bulbs form lateral buds at the base of a bulb leaf, and each bulb is capable of producing a whole new plant. Food for its initial growth is stored in the thickened leaves that form the bulb.

Some plants like *Bryophyllum* form buds on the edges of their leaves (see Figure 12.2). These buds will grow into new plants, dropping off the leaf when they are mature enough to survive on their own.

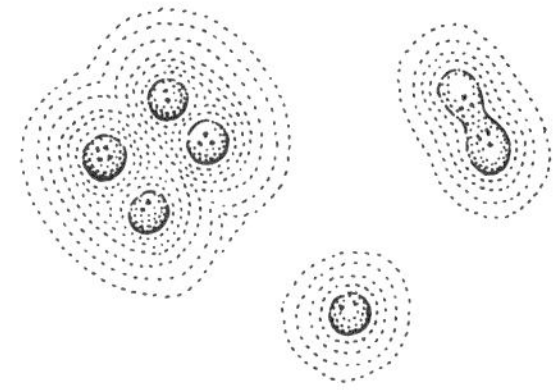

Fission

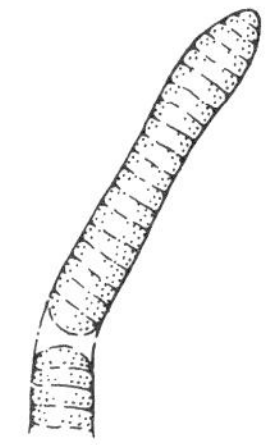

Fragmentation

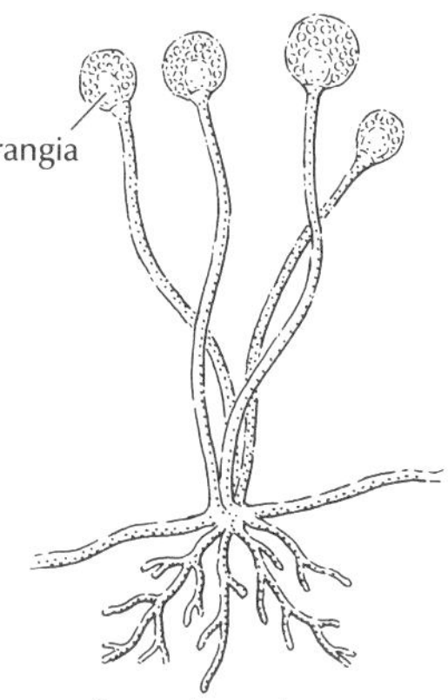

Spore formation

12.1 Forms of asexual reproduction are illustrated by fission and fragmentation in blue green algae. *Fission* is cell division and *fragmentation* is the breaking apart of a multicellular form. Fungi, such as bread mold, commonly reproduce by forming specialized cells called *spores.* These spores are products of mitosis.

12.2 ASEXUAL PROPAGATION IN NATURE

Tubers, corms, and rhizomes (below) are all modified stems, and they are all used to propagate plants asexually. Tubers are enlarged portions of underground stems, familiar to all in the form of the white potato. Begonias also have tubers and they are the means by which these plants are commonly propagated (bottom). A corm is a short, thick stem and is possessed by the crocus and gladiolus. New corms form from lateral buds on the old corm. They are modified lateral branches and each one will form a new plant. Rhizomes are underground stems, and sections with buds and roots can be used to start new plants. The Canna illustrated here is an example of a plant propagated by means of rhizomes (below, right). The roots growing from this mangrove tree along the coast of Puerto Rico (facing page, bottom) will implant themselves in the bottom mud or sand, and in time a whole new forest of mangroves will be produced from the original tree. Bulbs (facing page, top left) are shoots with thickened leaves and each bulb will produce lateral bulbs which can be separated and used to propagate plants like daffodils, tulips, onions, garlic, and leeks. *Bryophyllum* (facing page, top right) is perhaps the most dramatic example of asexual reproduction by leaves. Buds form in notches of the fleshy leaves, and miniature plants with roots are produced. They drop from the leaves and form new plants. A single plant can produce great quantities of these plantlets.

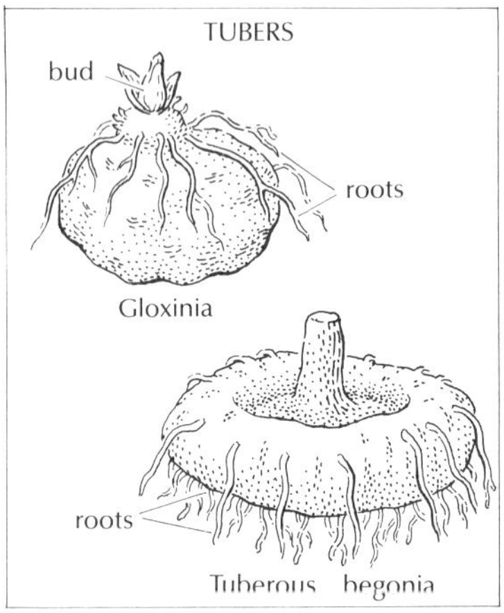

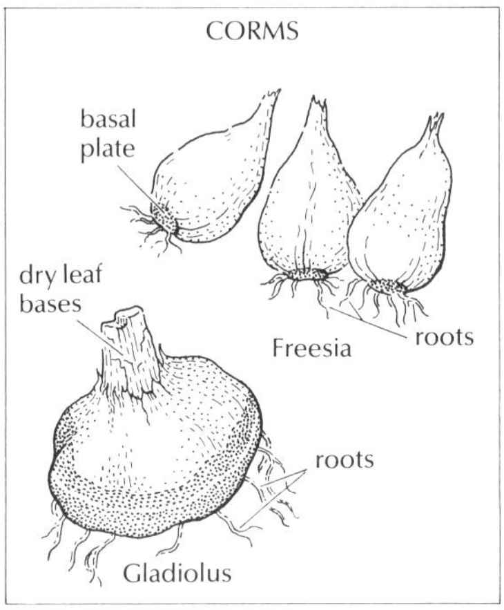

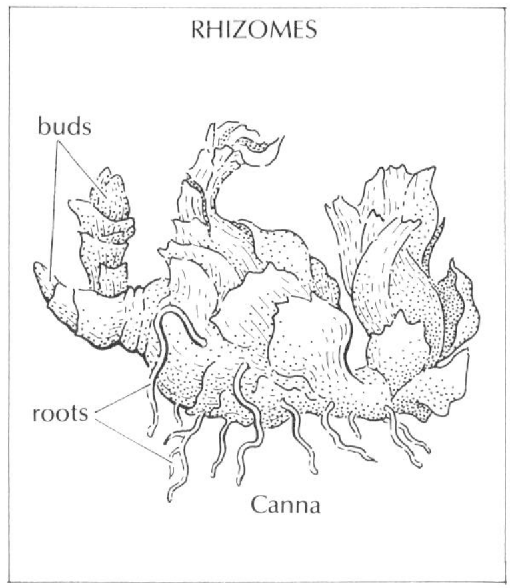

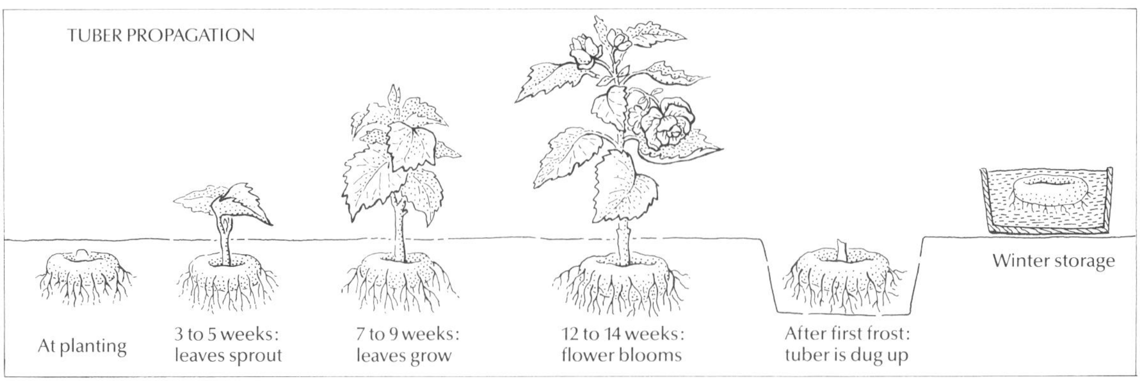

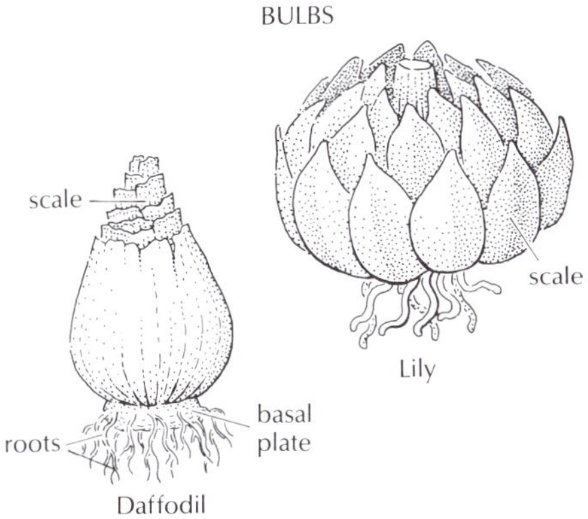
BULBS
scale
scale
Lily
roots
basal plate
Daffodil

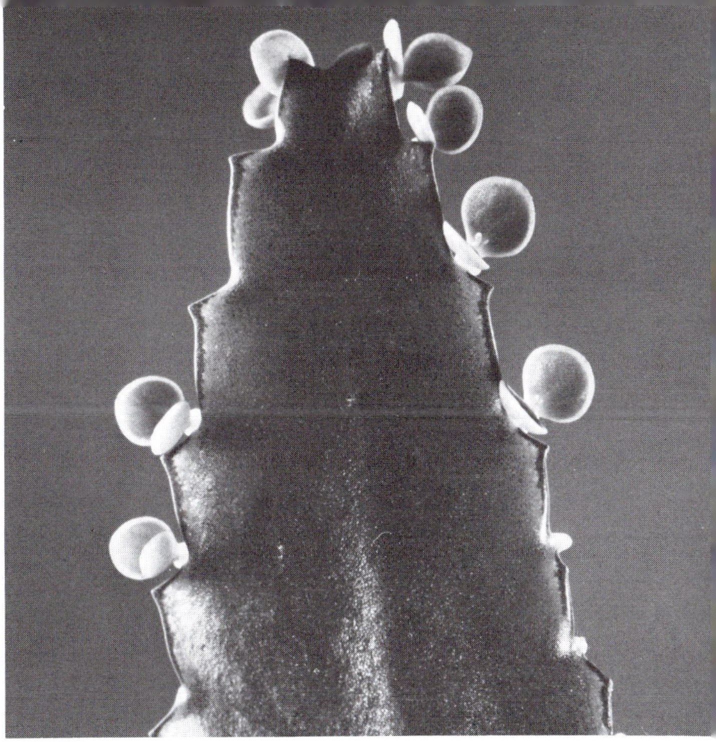

Aquatic plants like duckweed *(Lemna minor)* and water hyacinth *(Eichhornia crassipes)* can reproduce from lateral buds at the base of leaves. After a while these plants break off and drift away to grow into mature plants on their own. The century plant *(Agave americana)* dies after it flowers, but new plants developed from axillary buds of the older leaves produce a new generation.

Asexual Embryos Embryos of some plants are not products of a fertilized egg; instead they result from division of an unfertilized egg cell. **Apomixis** is the term for asexually produced embryos. In most of these the female gametophyte is derived from a diploid (2N) cell that has not undergone meiosis. Since the egg is diploid, the asexual embryo is diploid. These are called **parthenogenic** (*partheno,* "virgin"; *genic,* "birth") embryos. Sometimes other cells of the female gametophyte can develop into embryos.

Adventitious roots and buds (those arising from an unusual place) have been mentioned in earlier chapters, and there are also adventive embryos, which develop from cells of the ovule, not cells of the female gametophyte. In dandelions and citrus trees like oranges and lemons, such embryos are very common. In the latter there are often a number of adventive embryos in one seed.

Embryos are like buds (Chapter 6), and the steps leading to the production of adventive embryos are much like those leading to the production of adventitious buds such as those on a begonia leaf, in which one cell of the leaf epidermis divides by mitosis and forms a bud (see page 355). To see this vegetative reproduction from a single epidermal cell, try a home experiment with flax seed. Many drug-

12.3 Roots produced from stems and leaves are called *adventitious.* The ability to produce adventitious roots is the basis for much natural and human-induced plant propagation. These adventitious roots are characteristic of the tropical screw pine *(Pandanus).*

stores and supermarkets carry the seed, a traditional remedy for colic in babies. Plant the seeds in sand or soil or lay them on a damp towel. When they have germinated, let the seedlings grow about three quarters of an inch tall, then cut them off just below the cotyledons (Chapter 6). This part of the stem is the **hypocotyl** ("below the cotyledon"). In a week or so one or more buds will grow from the hypocotyl. The buds will form complete new plants, begun from just one cell in the epidermis of the hypocotyl.

It is fairly common for fruit to develop without sexual reproduction and the formation of seed; bananas are good examples. And many kinds of flowering plants have apomictic seeds. The evolutionary significance of apomixis is that it reproduces genotypes that have advantages for living in their particular habitat. The method can have disadvantages: consider what might happen to plants that reproduce only by apomictic seeds if their environment changes.

HORTICULTURAL PROPAGATION

Genetic crosses can produce new varieties of plants, but when a desirable type has been made, how is it to be preserved? Crossbreeding such hybrids will not work; it will simply produce additional variations. To keep the desired traits, horticulturalists propagate the plants asexually; therein lies the practical value of asexual reproduction.

Any plant that sends out runners, such as the strawberry, can be made to propagate simply by letting the runner root in another pot alongside. When the runner has produced new leaves, the new plant can be separated from the original by cutting the runner.

Adventitious roots (Figure 12.3) are very common in a number of plants. In the vegetable garden roots grow naturally out of stems of corn and tomato plants. Cut a twig from a willow and put the end of it in water. In a week or so it will develop roots, buds will break, and the stick will form a new willow tree. Bean plants will do the same if you cut the roots off young plants. If the ends of the stems are put in water, they will not die; they will simply grow new roots in a few days. The production of roots from stems is the basis for much plant propagation.

Selected varieties of plants such as roses, carnations, African violets, and many others are propagated for commercial use by rooting stem sections. These may be treated with growth regulators, such as the synthetic auxins discussed in Chapter 7, to stimulate rooting. Buy a rose at the florist's shop and you are buying a flower shaped into the beautiful thing it is by horticulturalists during years of patient work. Having asexual reproduction as a tool, growers know they will get an identical copy of that rose time after time.

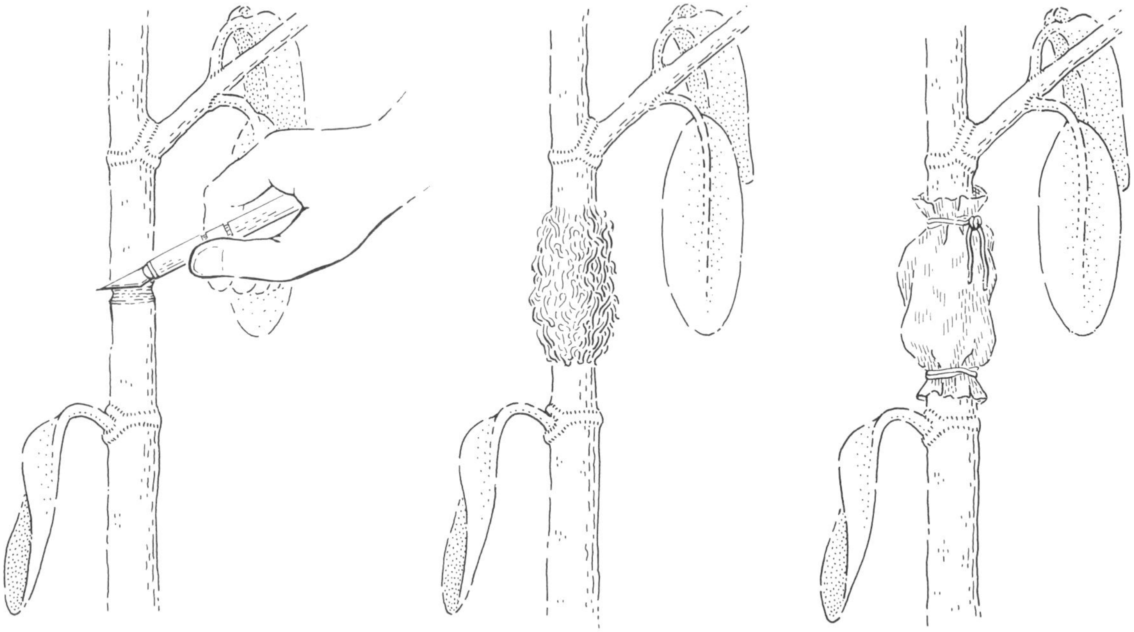

12.4 Air layering is a time-honored method for obtaining adventitious roots, and hence a new plant. Bark is cut or removed around the stem, and a hormone-containing rooting compound is applied. A moisture-holding material, such as sphagnum moss, is applied; and it is all wrapped up with a layer of polyethylene, which permits the diffusion of gases but not water. In time, roots appear, the stem is cut beneath the rooted area, and the rooted branch is planted.

Another technique for asexual reproduction is **air layering,** used chiefly with woody plants like hibiscus or camellias, and with house plants like the philodendron. Small slits are cut into the stem, just deep enough to go through the bark; these are then dusted with a rooting powder (a synthetic auxin). The segment of stem is wrapped in damp *Sphagnum* moss, which is held in place by a clear plastic bag secured with string. Within a few weeks, tips of small roots will have grown out through the moss. The plastic bag can now be removed, the plant cut below the ball of moss and roots and potted, moss and all, to grow into a replica of the plant from which it was derived.

Horticulturists propagate begonias, gloxinias, and African violets by asexual reproduction from leaves. The rex begonia can be propagated by cutting across the larger veins in the leaf and laying the leaf on potting soil. In a few weeks newly formed buds at the margins of the slits will turn into plantlets. Later they can be transferred to pots of their own where they will grow into replicas of the plants from which they were grown. Rex begonia leaves can also be cut into sections, each including a piece of the stem at the base of the leaf. These will grow roots in potting soil and from them new plants can be grown.

Grafting is an ancient method of growing new plants without going through the sexual reproductive cycle. Essentially, grafting joins a bud or stem section of one plant to the rooted stem of

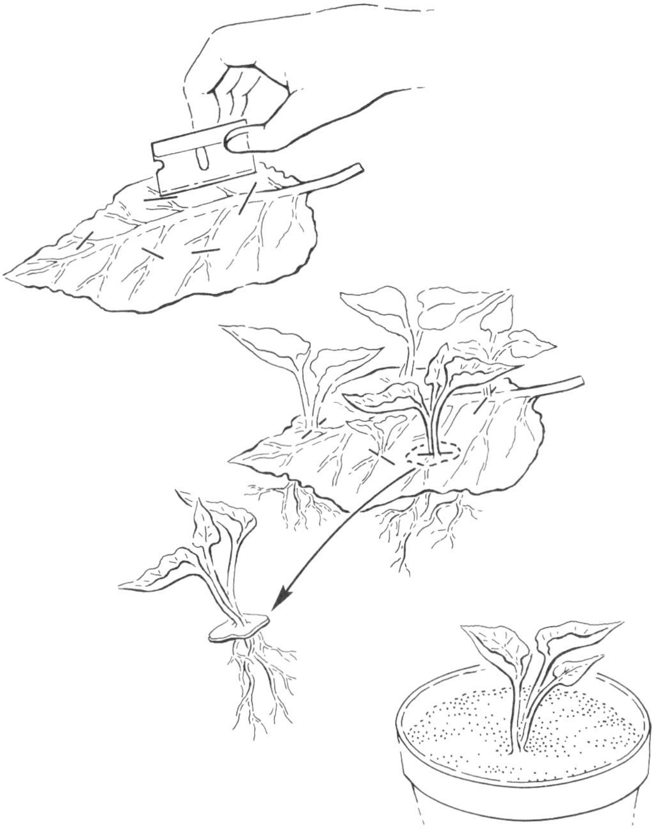

12.5 Leaves of many plants can be induced to form buds and roots. A well-known example is the *Rex begonia.* When the veins of a leaf are cut and it is placed on a moist surface, buds and roots form near the cuts. Each isolated plantlet will develop into a new plant.

another; the newly attached grafted section becomes part of the rooted plant. Although grafting will succeed only if the two plants are closely related, the host and donor tissues do not influence each other. There is no interchange of genetic material. Thus one apple tree rootstock can carry many varieties of apples on branches that grew from grafted buds.

A fruit or nut tree that you buy at a nursery will almost certainly be grafted. Its trunk and branches were produced by a bud from a high-producing commercial donor grafted onto the stem of a young host plant of a low-producing but hardy variety. When the grafted bud grows into a shoot, the original shoot is cut back to the grafted shoot, leaving a tree with the aboveground parts of one genotype united to roots of a different genotype. When the tree grows, only a swelling near the base of the trunk marks where the graft was made. These grafts are advantageous because the roots of wild varieties are more resistant to disease, pests, and drought than those of the plant breeder's high-producing creations.

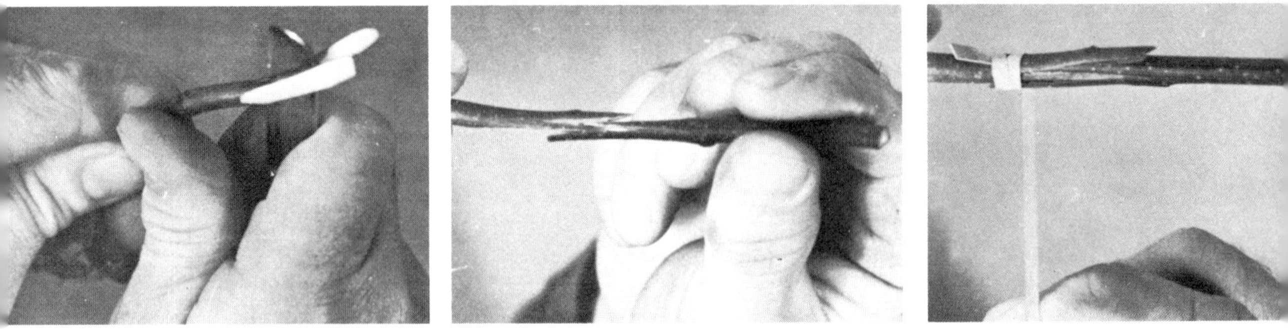

12.6 Buds and pieces of stem can be grafted to the intact stem of another plant of the same species. When a stem cutting is used, it is called a *scion* (a descendent) and the stem to which it is grafted is called a stock. A splice graft is illustrated here. One end of the scion is split and the outer portion of one side is trimmed so as to form a wedge. The end of the stock is cut in the same manner, and the part of the scion with the two cut surfaces is inserted into the stock so that six cut surfaces are in contact. A union of the tissues occurs when the cambium layers are in contact. The scion and stock are held in place with rubber strips until cambial growth results in a union.

NEW DIRECTIONS

Agriculture would be far poorer without asexual reproduction. Vegetative methods propagate commercial plants as diverse as sugarcane, potatoes, grapes, pineapple, bananas, tea, cocoa, asparagus, onions, and all fruit and nut crops. Although the technique has been used in agriculture for thousands of years, a new technology has begun a profound change: plant **tissue culture,** which essentially permits gardening in a test tube. Plant cells, tissues, or organs can be grown in a sterile environment on an artificial medium containing minerals, sugar, vitamins, amino acids, and growth substances, such as auxins and cytokinins. Buds and roots can often be induced at will by specific chemical treatment, allowing unlimited cloning from a clump of cells or even a single cell (see Figure 12.7). Cytokinins are particularly effective in bud induction. The first successful tissue cultures were produced in France in the early 1930s, and by 1960 laboratories in the United States had successfully grown entire plants from a single cultured cell. This technique was a product of basic research, and as is so often the case, no one knew what use would be made of it.

Uses turned up, however. Tissue culture revolutionized the orchid industry in the mid 1960s. It takes years to propagate a new orchid hybrid by stem cuttings and then to build up a commercial stock. Now a lateral bud from a new orchid can be placed in a culture medium, and it will make thousands of small plants in a few months. With the old method hybrid orchids were a luxury, but use of tissue culture dropped the price substantially.

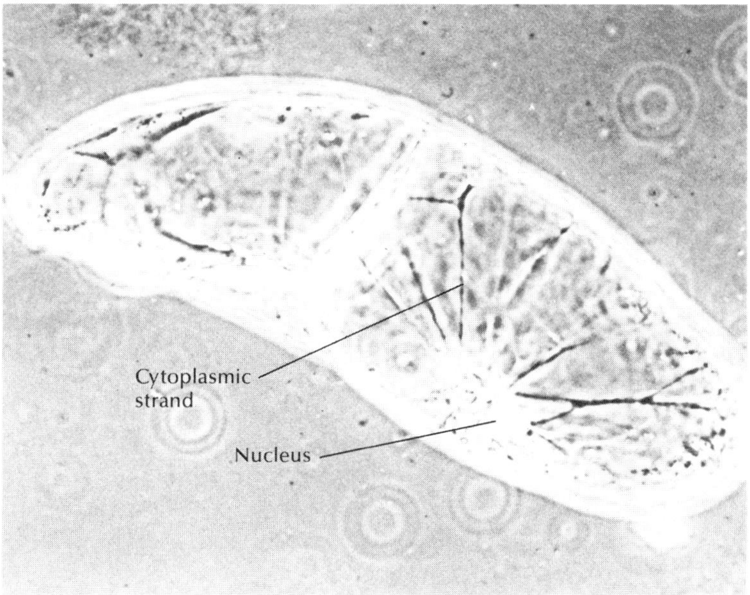

12.7 The ancient practice of cloning plants entered a new age with the advent of test-tube horticulture. Plant tissue culture techniques allow growth of cells and tissues in sterile and defined chemical media. Like many other plants, tobacco cells can be grown in liquid media. When grown on solid media, cells form clumps of tissue called *callus* (below, left). Adjusting the levels of auxin and cytokinin results in the formation of small plants (below, right). One flask can produce a small field of tobacco plants. Similar techniques have revolutionized the orchid and house plant industries and are increasingly important in the breeding and propagation of food plants.

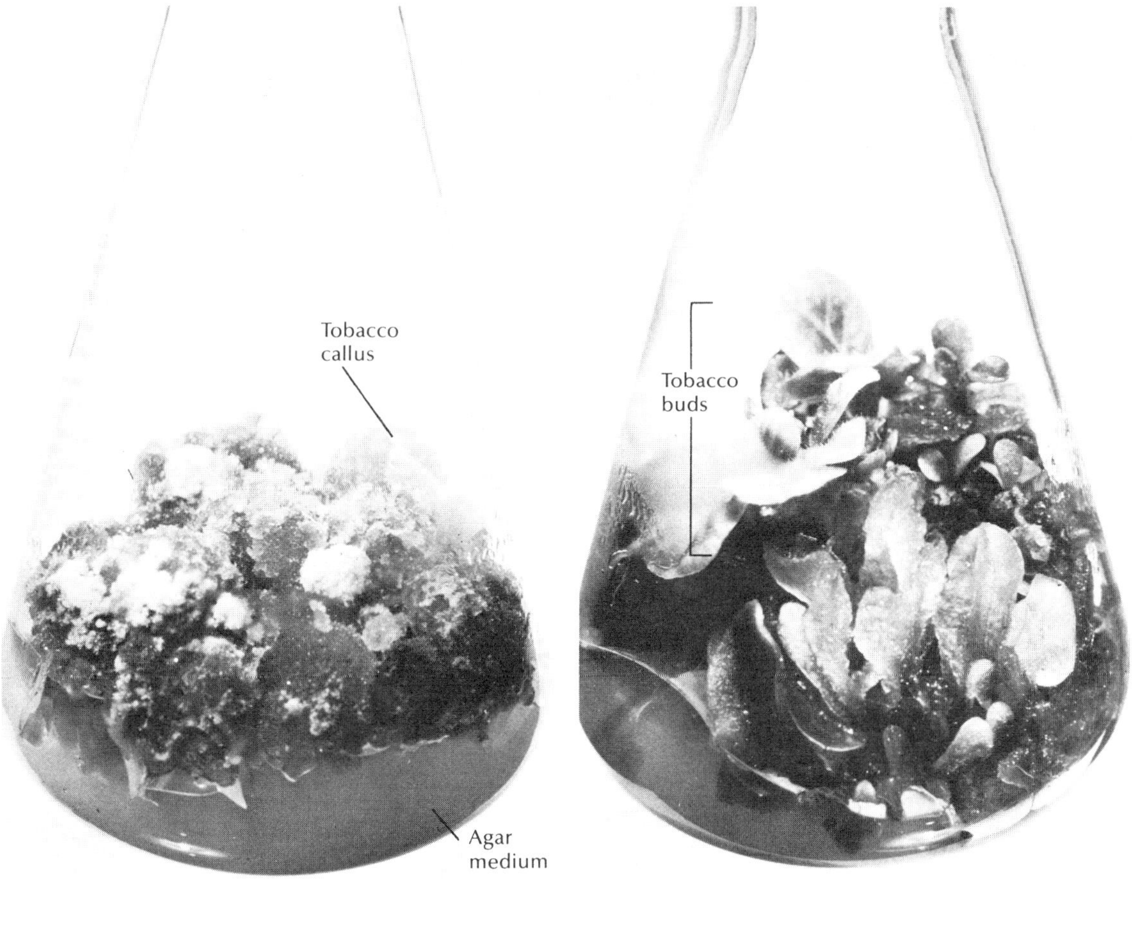

Similar techniques have been applied to the pineapple industry and to asparagus growing as well as to the flower industry. Apical meristems of commercially grown flowers such as carnations are cultured and regenerate new, virus-free plants, because the virus is not found in the rapidly dividing cells of the meristem. New varieties of chrysanthemum have been produced as a result of culturing cells from a single flower petal that had a new color caused by a mutation. Chemical treatment induces this cultured tissue to form buds and roots, and a new variety of chrysanthemum is ready for the market. A number of houseplants are now being propagated on a large scale and in a fairly small space using tissue culture techniques.

Tissue cultures have another application in agriculture: producing new varieties by treating cultured cells with **mutagenic** ("mutation-causing") **agents.** Certain chemicals and irradiation by radioactive elements can cause extensive mutation in cells by altering the DNA (see page 333); the cells can then be induced to form plants with new genotypes. This method is being used in Hawaii to produce new varieties of sugarcane, which are then field-tested and selected in the conventional manner. Such treatment of cultured tissue bypasses much of the expense and time it takes to produce new varieties by actually crossing varieties. It is a direct method of creating mutants and is being tried with other plants as well.

The possibility of using tissue culture techniques to breed haploid plants is particularly exciting. Haploid (N) plants are sterile because they contain only one set of chromosomes and cannot go through normal meiosis. Pollen grains from tomato and tobacco plants have been cultured and buds induced from the cultured cells have formed haploid plants. Although the plants do not form normal seed, they can be propagated by stem cuttings or from buds induced from tissue cultured from these plants. These haploid plants interest plant breeders because expression of the genotype is not masked by the heterozygosity that results when dominant and recessive alleles of genes are present. The haploid plants can then be made diploid by treating the flower parts with colchicine, a drug that induces chromosomes to double. The plant's genotype is then homozygous, and it can be used with great effect in breeding programs. New improved commercial varieties of rice and barley have been produced in this manner.

Using haploid plants for plant breeding is one result of applying tissue culture techniques, but it is only part of the story; **somatic cell hybridization** may give still larger benefits. This technique involves fusion of somatic cells—the cells of an organism's body, not its gametes—and is much like fertilization. Many researchers are working on this technique with animal cells. They have induced cultured cells of human and mouse to fuse into a cell that continues to divide, forming a population of cells with a mixed human-mouse

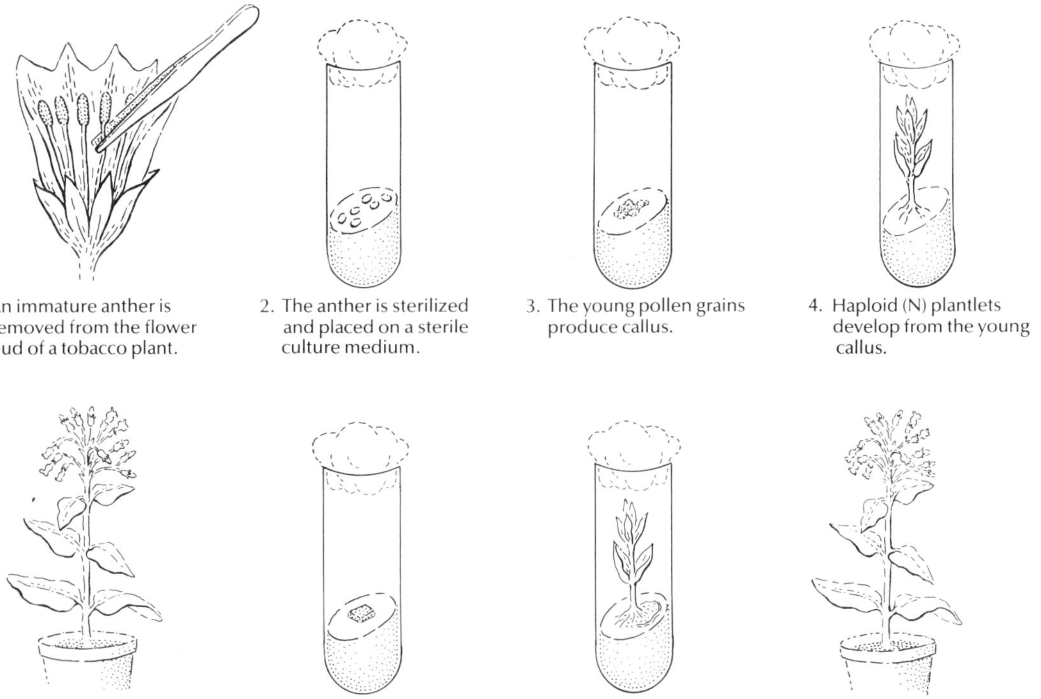

1. An immature anther is removed from the flower bud of a tobacco plant.

2. The anther is sterilized and placed on a sterile culture medium.

3. The young pollen grains produce callus.

4. Haploid (N) plantlets develop from the young callus.

5. The haploid plant is grown in soil, but it produces sterile flowers.

6. Tissue from the pith is excised and placed in a culture, as in the second stage. The pith is treated with colchicine, producing 2N cells.

7. Many haploid plants are produced.

8. The haploid plants are treated with colchicine and produce seeds which produce diploid plants.

12.8 The ability to induce buds from cultured tissue has made possible the production of haploid flowering plants. Tissue derived from haploid pollen is induced to form buds. The haploid plants are sterile but cells from these can be transformed to the diploid state by treatment with the drug colchicine. Plants induced from this tissue are diploid and form normal seed. They can be used to great advantage in breeding programs because they are homozygous for genetic characters.

genotype. Animal tissue cultures do not form organs or organisms, but plant tissue cultures do. In cultures of plant tissue, somatic cells do not normally hybridize because the rigid walls of plant cells prevent contact and fusion of plasma membranes. Now, however, naked protoplasts can be made by dissolving their cell walls with enzymes derived from fungi and snails. If conditions are right, the naked protoplasts will fuse with each other, and their nuclei will fuse as well, just as in normal fertilization of sex cells. These cells quickly produce new cell walls and divide to form a mass of cells (see Figure 12.9). Buds can be induced from the hybrid cultured cells, and they grow into hybrid plants.

Interspecific (between species) crosses have been attempted with some success. Hybrid tobacco plants made by fusing protoplasts from different species have been grown. **Intergeneric** (between genera) **crosses** have also been attempted. In this combination cells have been reported to fuse, but the nuclei do not.

A question behind all these experiments in somatic cell hybridization is, how can their results be applied to agriculture? There are

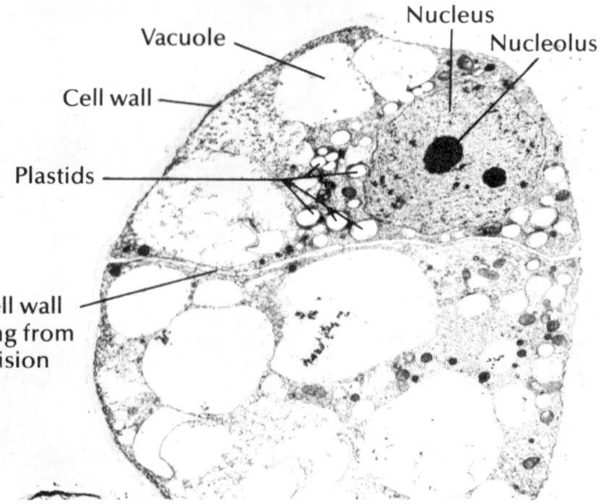

12.9 Plant cell protoplasts can be produced by removal of cell walls with enzymes. Cells of the same or closely related species will fuse and form a cell wall, divide, and eventually form a plant which is the same as a hybrid produced by normal sexual reproduction. Hybridization of protoplasts from haploid plants offers great promise for crop improvement. The two sugar cane cells shown here are the product of a protoplast which formed a new cell wall in a matter of hours and then divided. These cells can form a callus.

many barriers to fertilization between species of the same genus, and particularly between species of different genera, like those hindering formation of the wheat-rye hybrid *Triticale.* Some, maybe most, of these barriers can be bypassed by hybridizing protoplasts, making way for heretofore impossible genetic combinations. We can at least think of what would happen if a hybrid of a legume and a grain such as wheat or corn is produced, though it might well be impossible. A plant that could give high yields of grain and also carry nitrogen-fixing root nodules like those of many legumes would change world economics and politics. (Recall both the necessity and high cost of fixing nitrogen for intensive agriculture discussed in Chapter 9.) Before we get carried away, however, we should note that a grain-legume hybrid, if produced, might have the root characters of wheat and the shoot characters of a legume, and be little more than a curiosity.

Hybrids made by fusion of haploid cells are likely to be of more immediate use. Haploid cells would be like gametes in creating a diploid hybrid, but unlike gametes in that the genotype of the somatic cells would not be shuffled by meiosis; the plants would be duplicates. With fuller knowledge of the genotypes of these cells, breeders could manipulate crosses more surely and predict genotypes more accurately than they can with conventional plant breeding.

In one way or another, human experimenters have conducted genetic engineering among plants for some ten thousand years, but we have begun to understand the mechanics behind this genetic manipulation—the workings of the genes—for only a half century. The next fifty years will be more exciting still, as researchers manipulate plant cells in ever more sophisticated ways. Research on tissue culture took thirty years to produce a commercial application, but the doors are now opening, and future basic and applied research will be loaded with surprises.

Summary

Asexual or vegetative reproduction produces plants identical to the parent. Meiosis is not involved, nor fusion of gametes. A new plant—a clone—grows directly from the parent. Most blue green algae, many bacteria, and some fungi reproduce only asexually, by fission, fragmentation, and formation of spores that grow into new plants.

Numerous methods of asexual propagation occur naturally in flowering plants. New plants may be formed from several specialized stem structures: (1) from runners, stems growing horizontally that extend far from the parent plant and take root; (2) from rhizomes, underground stems; (3) from tubers, enlarged parts of underground stems; they are also food-storage organs; (4) from corms, food-storage organs that are swollen underground shoots with buds on the top and roots on the bottom; new shoots grow from the buds.

New plants may also grow from the roots of some species. In banyan and mangrove trees, for example, roots grow down from lateral branches, reaching into the soil and producing a new tree.

Some specialized underground shoots form bulbs—leaves with fleshy bases, which form lateral buds producing a whole new plant. Other species form buds on the edges of their leaves. These grow into new plants, dropping off when mature. And some aquatic plants grow buds at the bases of their leaves; these too grow into new plants and then break off, drifting away.

Embryos may also form asexually, by apomixis—division of an unfertilized egg cell. They may be adventitious, developing from cells of an ovule. Many kinds of flowering plants have apomictic seeds.

Asexual reproduction allows horticulturalists to maintain desirable traits in the plants they breed. To do so, they duplicate nature's own methods, using runners and rooting segments of adventitious roots; stem sections with nodes will also form a new plant if induced to form roots.

Air layering is another technique: small slits are cut in the stem, dusted with a rooting powder, and wrapped in *Sphagnum* moss. Small roots will grow into the moss, which can then be planted.

Some plants may be propagated by manipulating the leaves: large veins are cut, the leaf is put in soil, and buds from which plants can be grown form at the edges of the leaf. If a *Rex begonia* leaf is cut into sections, each including a piece of the leaf's stem, roots and then new plants will grow from the sections.

An ancient method for growing new plants is grafting: the bud or stem of one plant is joined to the roots of another.

New techniques of asexual reproduction promise not only horticultural and agricultural improvements but scientific advances as well. These include tissue culture techniques, the breeding of haploid plants, and somatic cell hybridization.

In reproduction by tissue culture, cells, tissues, or organs are grown on a medium containing sugar, vitamins, amino acids, and growth substances such as auxins and cytokinins. The technique allows unlimited cloning from one or many cells. If the cells are treated with mutagenic agents, new varieties of plants may be produced.

Pollen grains have been cultured, producing haploid (N) plants. They may then be propagated by stem cuttings or tissue cultures. Or the plant may be treated with colchicine, which induces the chromosomes to double, so the plant is diploid. It is also homozygous, so the genotype is clearly expressed.

Normally a plant's somatic cells (all cells except gametes) do not hybridize because of their rigid cell walls. But enzymes can be used to dissolve the walls; under certain conditions, the protoplasts and the nuclei of the cells will then fuse. These cells form new walls, divide, and produce a mass of cells. Buds and new plants can be induced to grow from these.

Scientists have had some success in fusing cells from different species, but so far nuclei have not fused when intergeneric crosses have been attempted. Breeding haploid plants is likely to be of more immediate use than somatic cell hybridization; knowing more about the genotype of haploid plants, scientists or horticulturalists can manipulate breeding programs more effectively.

Review Questions

1. Some plants possess only asexual means of reproduction and individuals within populations of these plants vary. How does this happen?
2. One new hybrid rose plant is produced. Soon, exact copies can be seen in nurseries and supermarkets throughout the country. How is this accomplished?
3. The term "cloning" is much in the news, but many people do not realize that every time they eat a meal, walk through a garden, or tend their houseplants they are experiencing the results of cloning. List your daily contacts with this technique.
4. Why are plant breeders excited about haploid plants?
5. Can you think of applications of plant tissue culture techniques not covered in this chapter?

CHAPTER 13

Time and Diversity

Biological diversity is a fact. From the work of Mendel and those who followed him, we know that characteristics are passed from generation to generation by the genes, and we know how offspring come to vary from their parents. But understanding the genetic mechanisms of variation does not explain how organisms came to have their present diverse forms. Nineteenth-century scientists had to develop two further great ideas before the immense changes earth's organisms have undergone could be understood: a new concept of time and a new theory of evolution.

THE PROBLEM OF TIME

The idea that the earth is very old is recent, and the concept of almost limitless time represents a major revolution in human thought. A hundred and fifty years ago, geologic time stretching billions of years was unimaginable to most. Certainly curiosity about earth's age was plentiful, but no one had facts on which to build estimates: little was known about the origin of rocks, about stratification, about fossils and the record they can give us. So it is understandable that those in the Western World who first tried to date the earth's beginning should have gone to written history first and the Bible foremost.

13.1 A biblical view of the creation of living things is illustrated in this painting by Jan Breughel the Elder. Of particular interest is the simultaneous appearance of wild and domesticated animals, fanciful creatures like unicorns, and unlikely sea monsters.

Catastrophe and Creation People once thought of the earth as the center of the universe, a vast, immobile mass of matter made of the four primary elements: fire, air, water, and the earth. Beyond was space, in which were suspended the sun—giver of heat and light—the moon, and all the stars. The Bible provided an explanation for how these things came to be, and its theory of creation was accepted and persisted for hundreds of years.

Halfway through the seventeenth century, a famous attempt to calculate earth's age was made by Irish Archbishop James Ussher. He based his calculations on the theory of creation and the Bible. Counting the number of generations after Adam, Ussher estimated that the world had been created at precisely 9 A.M. on Sunday, October 23, 4004 B.C.; earth therefore was about 6000 years old.

Assuming this date to be correct, there had to be some explanation for how during this short period everything on earth had come to be—its oceans, mountains, valleys, deserts, and all the living things. Scientific thinkers at that time could come no closer to accommodating all these visible, tangible things than to postulate a series of tumultuous, catastrophic events that had left in their wake earth's physical features.

But this idea that catastrophes had overridden all natural laws became more and more questionable as knowledge grew, and by the eighteenth century more rational explanations were offered. A Scotsman named James Hutton (1726–1797) was prominent in proposing a theory contradicting the earlier catastrophic suppositions.

Uniformitarianism and Neptunism Hutton was a remarkably perceptive man, and Scotland is well stocked with material for a geologist with a searching and logical mind. In the weathered rocks, deep valleys, and rugged coastline of his land, he saw evidence of basic processes for which the catastrophic theory made no allowance. Erosion, for example, was to be seen in the exposed rocks of every ancient Scottish hill, and there was visible evidence of volcanic activity. Evidence lay too in the strata of rock formations, which, apparently folded one upon the other, told of enormous movements in the earth's crust long ago.

Hutton's observations led him to formulate a theory about the earth quite different from the prevailing one. As he saw it, this was no immobile sphere formed a long time before and never changed but a dynamic living mass within which great forces had uplifted the beds of oceans and created new mountains and lands, even as erosion was wearing older lands and mountains away. "From the top of the mountain to the shore of the sea," he wrote in 1775, "everything is in a state of change. The earth has a state of growth and augmentation; it has another state, which is that of diminution and decay. The world is thus destroyed in one part, but it is renewed in another."

Hutton's interpretation was influential, but it lacked an estimate of how long such vast movements might have taken to give earth the physical features he observed and analyzed. Hutton himself saw this weakness and even realized that, somehow, geologic time might be gauged by observing earth's visible structures. As he put it, all the principles are before our eyes, "from whence we may reason back into the boundless mass of time already elapsed." But neither he nor anyone else understood how such reasoning back might be effected.

Hutton's point of view came to be known as **uniformitarianism,** and it dominated scientific thought for years. This theory stated that past geological changes were due to the same forces producing changes in the modern world. But old concepts die hard, and uniformitarianism did not sweep the catastrophic theory aside in a moment. Instead, some tried to adapt catastrophism in order to account for the evidence of earth's evolution on which Hutton built his theory. One of the most convincing efforts came from an influential German teacher, Abraham Werner (1750–1817).

Werner's theory took account of the stratification of rocks (which is obvious evidence of some sort of evolutionary process), but he chose to link this scientific observation with the Great Flood of

13.2 Charles Lyell (1797–1875) has been called the father of modern geology. His *Principles of Geology* was a treasured possession of young Charles Darwin aboard the *Beagle*. Lyell became a close friend and advisor to Darwin following the latter's return to England, and Darwin's theory owes much to Lyell's genius.

Noah, one of the Bible's oldest and most widely accepted stories. As Werner saw it, at one time the entire world was covered with water, and in this huge primeval ocean a definite sequence of rocks was precipitated. When the water went away, leaving only the oceans and lakes as he knew them, the rocks, valleys, and all other features of the landscape were left behind. This theory, long supported against Hutton's uniformitarianism, was known as **Neptunism** (from Neptune, Roman god of the sea).

Time without Measure Neptunism, like uniformitarianism, lacked a convincing explanation for geologic time. In the early 1830s Charles Lyell (1797–1875), born in England in the year Hutton died, provided strong evidence against the catastrophic theory of creation and established uniformitarianism as the accepted theory. Lyell, later one of Charles Darwin's principal sponsors, was to become the father of modern geology by proving that earth's history can be interpreted by natural events still going on today, and it was he who forcefully presented the theory of unlimited time. Lyell's theories put in rational order most of the grand discoveries of the century in which he lived, including Darwin's theory of the origin of species, because his scheme did allow time for evolutionary processes to take place.

But telling geologic time was still the great barrier, and there seemed to be no dissolving it. Hutton and Lyell based their theories on visible evidence in the oldest thing on earth, rock, but they could not see a way of deriving a geologic calendar from rocks. Hutton confessed that in them he saw "no vestige of a beginning—no prospect of an end," and Lyell had to admit that "the same conclusion seems to me to hold true." Without some way of differentiating ages of rocks, ascertaining what they had to tell, the idea seemed inescapable that earth is repeating, with machinelike regularity, the same things that had gone on through endless time. It was dynamic, yes, but did not possess a time scale, so far as anyone could tell. Time still was impenetrable.

A Modern Calendar Development of a geologic clock began when scientists recognized that there are differences among rocks, that some are older than others. Much evidence came from observations associated with the building of the great English network of canals early in the nineteenth century, evidence that greatly influenced Lyell. It climaxed (for the time being at least) when **radioactivity** (the spontaneous breaking apart of atomic nuclei) was discovered, which made it possible to calculate accurately the age of certain kinds of rocks (see page 374). Discoveries showed that Lyell's uniformitarianism was right in picturing earth as dynamic, not static. He correctly thought that processes observable today were at work in the past, but he was wrong in thinking that these are always the same. Today it is known that they are not.

A key to the concept of geologic activity and organic evolution is time. How can we visualize it? Historian Hendrik van Loon described its enormity in a very dramatic, if not quite satisfactory, way when he wrote, in *The Story of Mankind:*

> High up in the North in the land called Svithjod, there stands a rock. It is a hundred miles high and a hundred miles wide. Once every thousand years a little bird comes to this rock to sharpen its beak.
>
> When the rock has thus been worn away, then a single second of eternity will have gone by.

For some this image is helpful, but another may give a better notion of temporal relationships. Imagine that all of time, from the moment when the earth's crust first formed right down to this moment, is compressed into one calendar year. Here is what geologic time and its events look like on this calendar.

The earth's crust begins to form on January 1; the oldest rocks known date from about the middle of March. In May the first living things appear in the sea. By late November land plants and animals have appeared, and in early December the vast swamps that formed the Pennsylvanian coal deposits have appeared, flourished, and disappeared after four days. By mid-December dinosaurs have begun to roam the earth; by December 26, just as the Rocky Mountains begin to rise up in the American West, they become extinct. In the evening of December 31, the first humanlike creatures have begun to appear. One minute and fifteen seconds before midnight on December 31, the glaciers of the last great Ice Age begin to retreat in America and Northern Europe. The Roman Empire rises, flourishes for five seconds, and disappears in the waning seconds of our year. Three seconds before midnight, Columbus discovers America. In the final fractions of the last second of this year, which represents earth's entire history, James Hutton begins to write his theory of uniformitarianism.

In time as we know it, this one imaginary year stands for 4.5 to 5 billion years, or perhaps more; with all the sophisticated knowledge available to us, earth's age is now calculated as 4.5 billion years. No rocks exposed on earth's surface are known to have survived the first billion years of earth's existence.

CLASSIFYING TIME

The fossil record is the best and most persuasive evidence we have that organisms have changed over time. Fossils are the true benchmarks in the calendar of life on earth, even though the fossil record fades out before we come really close to the origin of living things. But because Western peoples have traditionally equated time with human history, it was not until the nineteenth century that scientists began to understand the fossil record. In the

13.3 FOSSILS

Fossils are the pages of a guidebook to the history of life on earth. They are found in stratified sediments formed in ancient bodies of water. Remains of organisms at the bottom of the lake in the drawing (below, left) were covered by river-borne sand and gravel which eventually filled the lake. Other organic remains were then covered by glacier-borne soil, sand, and rocks. Much of our knowledge of plants growing in a region in prehistoric times, and hence knowledge of the region's climate, is obtained by fossil pollen in lake bottom sediments and bogs. Pollen is collected with a hollow collecting core which is driven into the earth by clockwise rotations (below, right). A sample is collected by reversing the direction of rotation. This opens an inner container and causes a flange on the side of the core to push a sample into the container. Stromatolites are hardened sediment cemented together by blue green algae. These stromatolites from Minnesota (facing page, top) are 2 billion years old and were formed in the Precambrian era. Coal is fossilized plant remains, and the form of plants of the Paleozoic era are often found in coal. This piece of coal from Illinois has a compressed leaf of *Neuropteris* (facing page, middle) a member of the seed ferns, the extinct gymnosperms (see Chapter 13, Supplement, Early Land Plants). This fossil dates from the beginning of the Permian period, 280 million years ago. A fossil within a fossil is represented by an ant within amber (facing page, bottom left). One hundred million years ago in the Cretaceous period of the Mesozoic era, a wayward ant was entrapped in the sticky resin from a tree. The resin sealed the ant in a plastic tomb, and over the ages the resin changed to what we call *amber.* A birch tree leaf (facing page, bottom right), compressed in sediments at the bottom of an Oregon lake until only a thin layer of carbon remains, still bears as detailed likeness of its original form. The tree which dropped this leaf was growing at the edge of the lake 30 million years ago in the Oligocene epoch of the Tertiary period of the Cenozoic era.

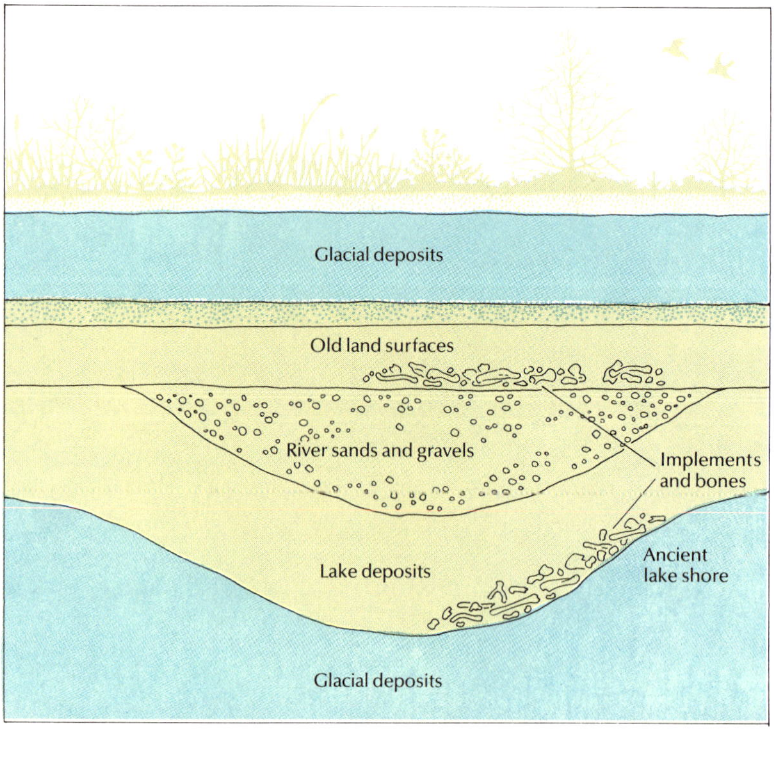

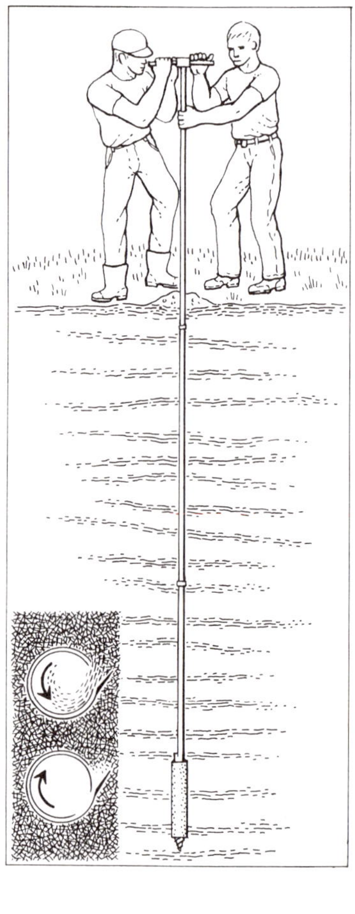

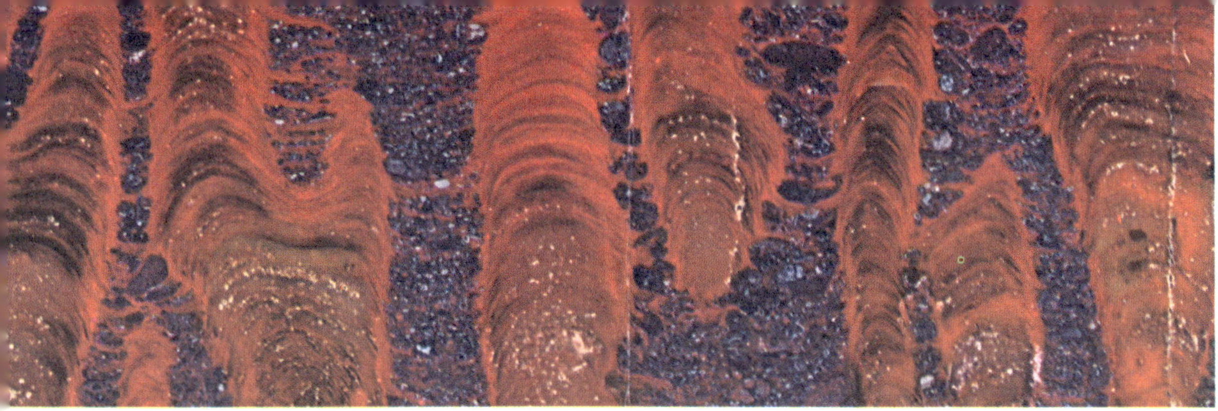

latter part of the seventeenth century, an elephant skeleton was discovered in Italy, and so immersed were scholars in their traditional ways of measuring time, they concluded that it must be the remains of one of the war elephants with which Hannibal had crossed the Alps. There were records in Italy showing that Hannibal had used elephants, but nothing hinted that they had ever lived there, and the skeleton was not recognized for the fossil it was.

Formation of Fossils There are many kinds of remains or fossils of organisms, and the definition of what constitutes a fossil is often arbitrary. Our word *fossil* traces back through French to the Latin verb *fodere,* "to dig." It was originally used for anything dug from earth or rock, including gems and other inorganic articles. A major problem in biology was that of relating fossils with organisms, and the first tentative relationship was made between what had traditionally been called "tongue stones" and shark's teeth. This connection was made in 1588, and the fact that these two objects were the same was only proven a hundred years later. The best naturalists were concerned with this problem of relating fossils to living organisms until principles of paleontology were developed in the nineteenth century.

Fossils were created in many ways; scientists have even been able to find traces of soft-bodied life forms such as ancient algae. They were preserved by the precipitation of silica, a process so gentle that little detail was lost. Similarly, ancient insects have been preserved in amber, a material that originally was resin oozing out of a tree. If an insect got caught in the resin, it was trapped for all time: the resin gradually flowed over it until it was completely covered. The resin was buried and solidified into a clear solid mass called amber, which was found thousands of years later.

Other organisms were trapped in marshes or shallow waters that were later covered by sediment, which became rock. Under immense pressure and heat, this former vegetation turned to coal, and in the coal can be found fossil imprints of leaves, with veins still clearly defined. In tar pits paleontologists have found the bones of ancient animals that were trapped and died there. The tar itself came from plants, so you might say plant fossils caught animals. Footprints have been found and even dung, hardened into formations called coprolites, which yield valuable information on the feeding habits of the animals that, long ago, dropped them.

The walls of plant pollen are almost indestructible and are commonly found fossilized in former marshes and lake bottoms. When they can be identified and dated, they are priceless carriers of facts about the kinds of plants that grew in ancient times. From this knowledge the region's climate, even its appearance, can be reconstructed.

All we know about the history of life comes from such samples of once-living populations. Of course each sample is limited to the plants and animals most likely to be fossilized: those which lived in wetlands such as the shores of ancient lakes, rivers, and seas, and the bottoms of shallow bodies of water. In arid regions like deserts, the chances of being fossilized were small.

We must handle fossils as we do living plants and animals in order to make sense of their relationships to each other: that is, they need classifying. They can be as hard to tell apart as a heap of kernels from several varieties of corn, to say nothing of the difficulty of grouping them meaningfully. Even determining if parts of plants found separately in rock belong to the same plant is not always easy. Leaves and reproductive structures of pieces from a specimen that turned out to be all from the same plant have accidentally been classified as parts of different organisms. When fossils are accurately identified and classified, how can we reconstruct their history?

Geologic Time It was the classification of geologic time that laid the basis for assigning dates to fossils. The strata (layers) in sedimentary rock on the earth's surface were analyzed and from their thickness and sequence huge periods on the geologic calendar were defined. The relative ages of these rocks are compared on a vertical scale: those found farthest down are the oldest (unless there is evidence of folding or faulting, which might have pushed an older layer of rock up to overlap a younger one). The fossil organisms in each stratum can therefore be compared for age. Those in the lowest stratum of rock are oldest; those in the highest are most recent (see Figure 13.3, pages 368–369).

On this vertical scale of time the two largest divisions are called **aeons:** the Cryptozoic, which covers the nearly 3 billion years from the time earth began to form until our knowledge of it begins, and the Phanerozoic, which covers everything since. The Phanerozoic is divided into three great subdivisions called **eras:** the Paleozoic (ancient animals), covering about 370 million years; the Mesozoic (middle animals), with 167 million years; and the Cenozoic (recent animals), covering some 63 million years. These eras are in turn divided into **periods** (see Figure 13.4).

Strata containing similar fossil populations were arranged into a relative time scale. The classification scheme is based in the first instance on animal life, as can be seen in the names of the eras; the relative time scale would often be quite different if it were based on the fossil record of plants. The method of classifying strata is similar to that used by Linnaeus in classifying organisms: a stratum is described, labeled as a type specimen for that particular type of rock, and then similar rock layers are classified by comparing them to the type specimen.

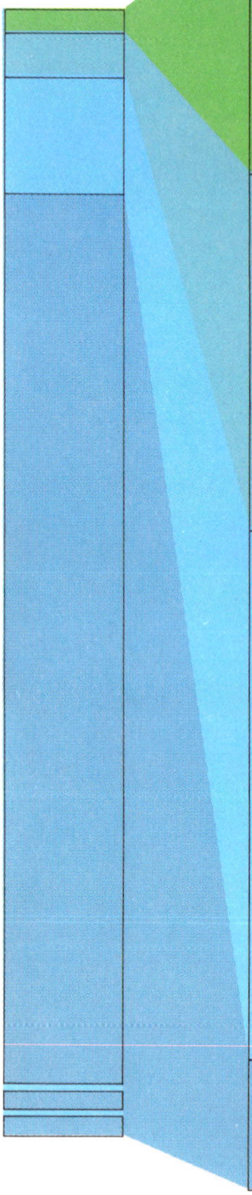

Era	Period	Years before present	Biological features
Cenozoic Present to 63 million years ago	Quarternary		
	Holocene epoch	present to 10,000	
	Pleistocene epoch	10,000–2 million	First true humans: *Homo erectus*
	Tertiary		First hominids; herbaceous plants more abundant
	Pliocene epoch	2–10 million	
	Miocene epoch	10–25 million	Grasslands spread as forests contract
	Oligocene epoch	25–40 million	
	Eocene epoch	40–60 million	
	Paleocene epoch	60–63 million	First primates: prosimians
Mesozoic 63–230 million years ago	Cretaceous	63–130 million	Angiosperms appear and become abundant; modern conifers arise Last dinosaurs
	Jurassic	130–180 million	Gymnosperms, especially cycads, and ferns dominate First birds Giant reptiles reign First mammals
	Triassic	180–230 million	Forests of gymnosperms and ferns spread Early dinosaurs appear
Paleozoic 230–600 million years ago	Permian	230–270 million	Conifers, cycads, and ginkgos appear; earlier types of forests wane
	Carboniferous (Pennsylvanian and Mississippian Periods)	270–345 million	First gymnosperms Reptiles and insects appear
	Devonian	345–400 million	First large vascular plants Modern groups of algae and fungi appear First amphibians and bony fish
	Silurian	400–440 million	First vascular land plants First fish with jaws
	Ordovician	440–500 million	Plants begin to invade land First vertebrates: armored fish without jaws
	Cambrian	500–600 million	First marine plants First shell-bearing animals
Precambrian 600 million–4.5 billion years ago		600 million–4.5 billion	First living things: blue green algae and bacteria; eukaryotic cells and multicellular organisms appear

13.4 The scale of geologic time must always form the background for considerations of organic evolution. The most dramatic feature of this illustration is the short time, relative to the age of the earth, that life has been on land.

Thus Adam Sedgwick, a nineteenth-century geologist, gave the name *Cambrian* to fossil-bearing dark sandstone rocks in Wales (the Roman name for Wales was Cambria). (Geology professor at Cambridge University, Sedgwick had associated with the student Darwin.) Elsewhere in the world, strata were assigned to the Cambrian by comparing them with the type of rock and fossils found in Wales.

The Ordovician and Silurian strata, named for ancient Welsh tribes, were assigned later dates because they occurred above the Cambrian level. The Devonian strata were named for Devonshire in England, where they were first classified; the Carboniferous period was named for the coal-bearing and fossil-rich strata first explored in England; the Permian was named for the province of Perm in Russia; the Triassic for strata with three subdivisions; and the Jurassic for strata in the Jura Mountains of Switzerland. The Cretaceous was named for chalk strata in central France.

Subdivisions in the Cenozoic, the most recent, era are known as **epochs,** defined according to the percentage of living or extinct mollusk shells (snails and others) in their strata. This classification was initiated by Sir Charles Lyell, who in 1833 defined the Eocene,

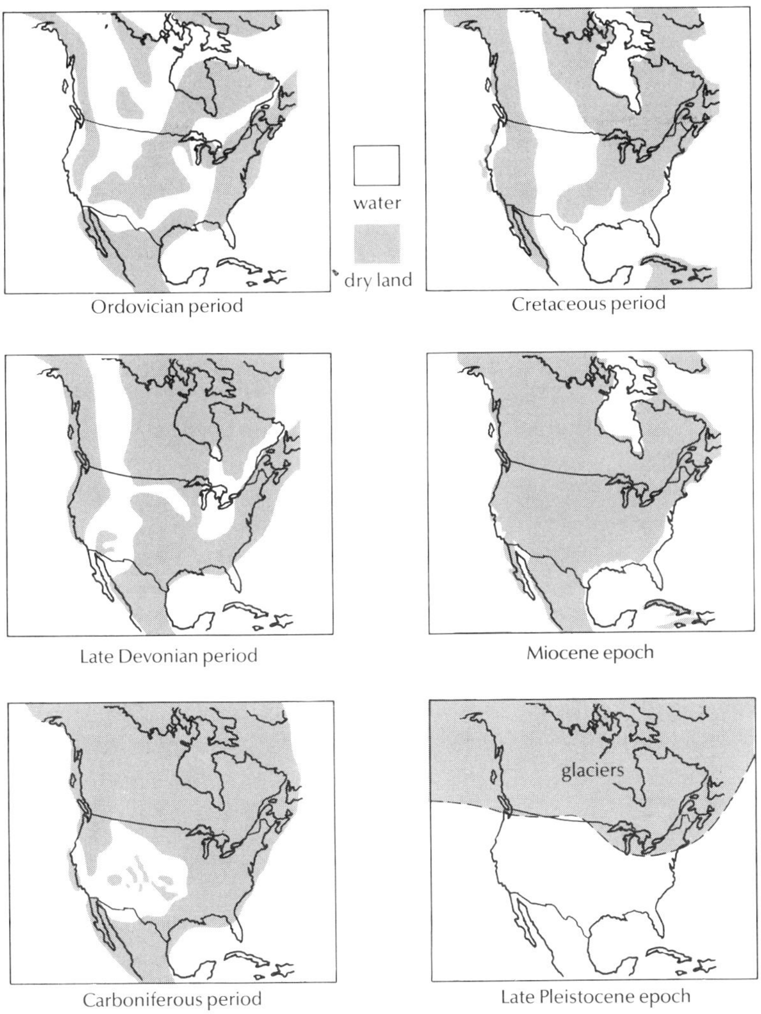

13.5 The distribution of plants and their fossil remains has been greatly influenced by the ebb and flow of oceans (clear areas) over the continents. The North American continent, to the extent that we now recognize it, emerged from the oceans only about 25 million years ago. Much of it was then swallowed up by glaciers until near historic times. The message from geologic time is that change is constant and frequently drastic.

Miocene, and Pliocene series. The Oligocene was added in 1854 by the German August von Beyrich, and the Paleocene by Wilhelm Schimpfer in 1874.

Absolute Time: Radioactive Dating This classification of geologic time and organisms, even with its continuing refinements, is arbitrary and unwieldy. The table of geologic time is no more than a relative scale based on relative types and abundance of fossils; its times are anything but absolute. To classify the Triassic as lasting from "about" 230 million to "about" 181 million years ago is to overlook a mere moment 49 million years long. But can geologic time be told on an absolute scale?

While the geologic table was being assembled, little was known of absolute time. Darwin made it clear that his theory depended on vast periods of time. He could not prove it, but he proposed that the world is much older than anyone then thought. He was bothered by the calculations with which Lord Kelvin, renowned English physicist, in 1865 "proved" the world to be only 20 to 40 million years old. Kelvin claimed that his quantitative methods refuted Darwin's theory of natural selection, but his calculations were figured from wrong assumptions.

Since Lord Kelvin's time, the age of the earth has been extended by over a hundred times. The new age is arrived at by measuring the rate at which radioactive **isotopes** (atoms with the same number of protons but different numbers of neutrons) of elements in rocks decay into other forms. This incredibly valuable tool became available only in this century as scientists learned about radioactive materials and understood how they decay. All radioactive isotopes have a constant rate of decay and a known **half-life** (the time it takes for half the radioactivity in the sample to decay). Uranium 235 decays to form lead 206 and has a half-life of 4.51×10^9 years (45,100,000 years). Carbon 14, the favorite isotope for dating human artifacts, has a half-life of 5,730 years, but that is too short for dating most fossils because little carbon 14 is left in fossils older than 50,000 years or so.

How do these time clocks work? They are based on the speed at which isotopes produced by decay accumulate. Potassium 40 decays into its daughter element, argon 40, and has a half-life of 1.3×10^9 years (1.3 billion years). In 1.3 billion years the amount of potassium 40 in a fossil will decrease by half; in the next 1.3 billion years it will decrease by half again to one-quarter the original amount, and so on. Determine the proportion of argon 40 to potassium 40 in the fossil, and, because the rate of decay is constant, the age of the rock can be established.

A number of elements have proved to be effective in dating fossils. Rubidium 87 was the tool that established the age of the earliest microfossils (see page 375). It decays into the daughter element strontium 87 and has a half-life of 4.7×10^{10} years. Carbon 14 dating is done by measuring how much of that element has been lost (see Figure 13.6). It decays to carbon 12. There is a constant amount of carbon 14 in the atmosphere since it is formed from nitrogen 14 by cosmic rays. It is incorporated into organic compounds by photosynthesis, and the amount of carbon 14 in a remnant of once-living tissue can be compared to that of the atmosphere. There is only half the amount of carbon 14 relative to carbon 12 in a 6000-year-old bone as there is in the atmosphere.

With this type of geological clock, scientists have been able to date rocks 3.6 billion years old. These rocks, granite gneiss (a metamorphic rock with its minerals in layers) formations near Dodoma in Tanzania, so far predate any life forms on earth. But with geological discoveries coming so swiftly, even this may prove to be a threshold beyond which evidence may be discovered carrying us still further back into the deep mysteries of the Cryptozoic epoch.

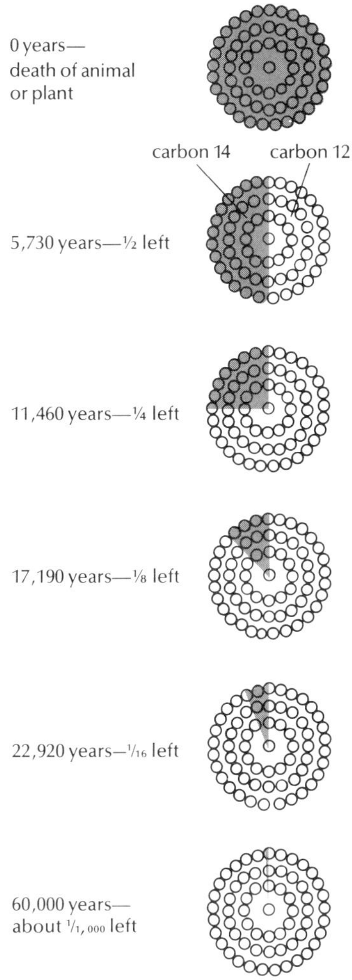

13.6 The carbon in all organisms, living and dead, is derived from the CO_2 obtained from the atmosphere during photosynthesis. A small percentage of this carbon is the isotope carbon 14. Carbon 14 decays to carbon 12 at a known rate. Therefore, by comparing the amount of carbon 14 in organic material with the amount in the atmosphere, it is possible to date the age of bones, wood, leather, natural fibers, and anything else derived from living organisms. Since only half the carbon is left after each 5730 years, carbon dating cannot be used for material over 60,000 years. This means that the technique is more useful to archaeologists than to many paleontologists.

LIFE'S BEGINNINGS

Although the fossil record of life on earth is incomplete—probably only between 1 and 10 percent of all species that ever lived are known to us—we can make intelligent deductions about life during the great span of geologic time. Think back to van Loon's imaginary second of eternity: the time it takes a little bird to wear away a mountain a hundred miles high and wide by sharpening its beak there once every thousand years. We are speaking here about almost unimaginable spans of time. The Precambrian began about 5 billion years ago and lasted for some 3.9 billion years, and yet scientists have been able to look back across this immensity of time to find what they believe to have been among the first living organisms in earth's history.

The First Microfossils These first life forms were simple one-celled organisms. Their remains are called **microfossils,** because with a diameter of only 5 to 15 micrometers, they can be seen only with a microscope. Some looked like filaments 2 to 13 micrometers in diameter and up to 90 micrometers in length; others were spheroids. Though there is still some learned dispute about them, their age is established at 3.4 billion years.

The first microfossils were found by a University of Wisconsin geology professor, the late Stanley Tyler, who was working as consultant for a mining company, classifying rocks. On the Canadian shore of Lake Superior, Tyler discovered strange objects in a hard flintlike rock known as chert, a formation he dated at about 1.9 billion years. These were dome-shaped structures within the rock,

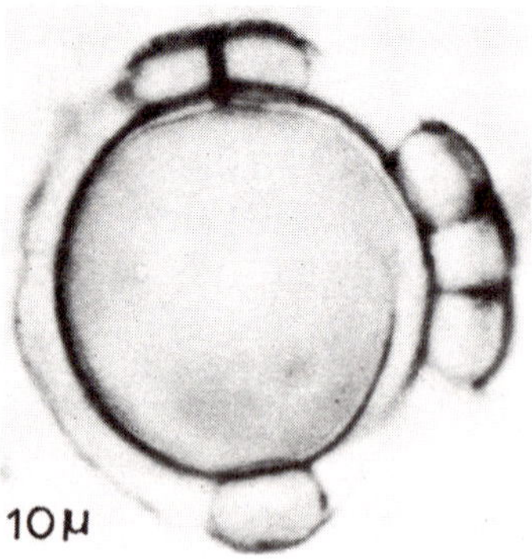

13.7 This sporelike body is a Precambrian fossil 2 billion years old, found with remains of many other microorganisms in the Gunflint Chert rock formations in southwestern Ontario. They rank among the oldest known fossils.

some of them (about 20 feet) in diameter. Because it was difficult to see how these structures could have originated in the normal way of inorganic rocks, Tyler decided to test the notion that they might be organic and went to Professor Elso Barghoorn at Harvard University. Barghoorn studied slides of slices cut out of the dome-shaped rock formations and discovered under the microscope the fossil remains of a rich variety of lifelike forms that looked very much like modern bacteria and blue green algae.

This exciting news was published in 1954, but more conclusive proof was required that these strange things had been living creatures. One way to check was to test for carbon content. All living things, past and present, contain carbon, and so indeed did these microfossils, as shown when they were extracted from the rock with acids and heated to see if they would burn (they did).

But was this carbon from something that had been alive? A further, very sophisticated test was necessary. Modern photosynthetic organisms, including blue green algae, incorporate more carbon of molecular weight 12 (^{12}C) than they do of another isotope of carbon with a molecular weight of 13 (^{13}C). A true organic fossil would, therefore, show a higher proportion of ^{12}C, and so indeed did the gunflint chert fossils. Although some of the structures had little

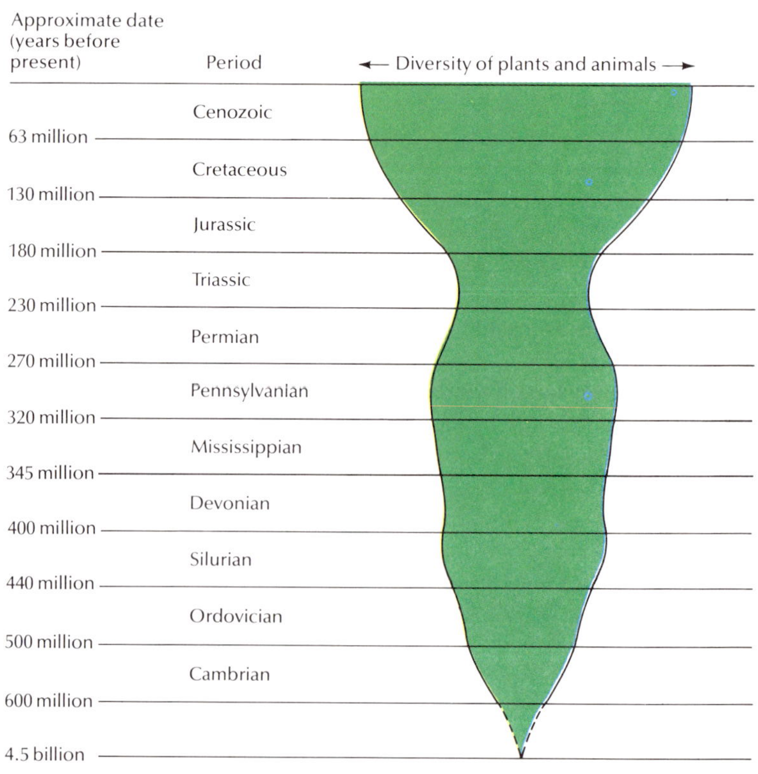

13.8 Diversity of organisms through time, as determined by the fossil record, is reflected by the width of the column. The constriction during the Permian and Triassic Periods represents a time of widespread extinctions presumably influenced by major environmental changes.

13.9 Increased specialization in the evolution of the flowering plants is illustrated by these representatives of four families. The magnolias (top, left) are an ancient angiosperm family, and the flowers are unspecialized in structure and pollination vectors. The composites, orchids (bottom, left) and grasses (bottom, right), are the largest families of flowering plants and have evolved relatively recently. Their flowers show a high degree of specialization. The gazania (top, right) is a daisylike plant from South America, possessing the ray and disc-flowers characteristic of many composites.

resemblance to any living organisms known today, the evidence that they were indeed early living forms was strong enough that the claim is now generally accepted by paleontologists.

Since then, life's beginnings have been rolled back still further. A rock formation in the eastern part of the Republic of South Africa had microfossils that lived about 3.4 billion years ago. These rods and spheroids are accepted widely, though not fully, as the oldest living structures yet found. And if the evidence for them is not quite strong enough, there are others that are accepted as having originated 2.6 to 2.7 billion years ago.

The First Land Plants The most ancient forms of life on earth probably acquired both energy and building materials for their cellular structures by consuming organic compounds in their environment, so they were heterotrophs. There is good evidence that suitable organic compounds, even proteins, could have been synthesized during lightning storms from gases such as methane and ammonia in the primitive atmosphere.

A major evolution took place when the first autotrophs appeared. These organisms took carbon from carbon dioxide in the air and produced oxygen from water and the development of an oxygen-rich atmosphere was correlated with the rapid appearance of diverse unicellular and multicellular organisms. All these evolutionary advances probably took something like 3 billion years, or about three-quarters of the age currently allotted to our planet.

About 600 million years ago, life began to display its current diversity. The oxygen-rich atmosphere resulted in formation of an ozone (0_3) layer, and because ozone as well as water absorbs life-killing ultraviolet radiation from the sun, land areas became favorable for colonization by organisms. Plants colonized the land and began to change their new terrestrial environment; their evolution in turn was influenced greatly by dramatic climatic and geologic events (see Chapter 15).

Measured by geologic time, evolution of life proceeded almost explosively. The diversity of organisms has been fairly constant for much of the last 400 million years. Exceptions are a period of mass extinction during the Triassic period, around 200 million years ago, perhaps as a result of major climatic changes, and an apparent expansion in diversity since then (see Figure 13.8, page 376). But diversification has affected living groups irregularly. Today there are far fewer varieties of reptiles than there were 400 million years ago, but there are many more varieties of mammals. The world had no flowering plants at all 200 million years ago, but many other vascular plants were present (see the Supplement at the end of this chapter); the record left by one nonflowering lowland and wetland fern goes back 300 million years or more. Now these ancient vascular plants are less diverse, but flowering plants have far more species and are now the dominant plant form.

Flowering Plants Much remains to be learned about the origins and forms of ancient flowering species. They seem to have originated principally in upland areas, such as the foothills of mountains, where the likelihood of being fossilized was low. However, we do know something about the changes angiosperms have undergone since they first appeared. Prominent among these is that as they evolved angiosperms developed increasingly specialized equipment for pollination.

The color of flowers, their shape, their arrangement on stems or in clusters, and the time they mature are frequently correlated with the structure, habits, and physiology of animal vectors, promoters of outcrossing. The evolution of these vector-mediated mechanisms of pollination attests to the significance of outcrossing for adaptations of plant populations.

Older families like the magnolias have unspecialized flowers; many of their parts are not adapted to specialized vectors. Such flowers are usually pollinated by beetles. But the largest and therefore most rapidly evolving angiosperm families like grasses, orchids, and **composites** (sunflowers, daisies) have highly specialized ways of getting pollinated (see Figure 13.9, page 377). The wind-pollinated grasses have clearly specialized flowers. A composite's "flower" is in fact many flowers arranged in a variety of complex ways. Their colors, usually found in the strap-shaped corollas of the flowers in the outer ring, are most often blues and yellows that bees and butterflies can see. Once an insect is on the flower head (think of a sunflower), at least some of the flowers are likely to receive pollen and set seed. Pollination mechanisms of orchids are extremely specialized (see Chapter 14). Unlike the specialized but still scattergun approach of the grasses, orchids demonstrate complex mechanisms involving a few or even just one species of vector.

From these examples we see both the complexity of sexual reproduction in flowering plants and how it is involved in producing genetic combinations in whole populations. Knowing these complexities, we can learn more about evolution and also apply them to practical work.

Extinction A lesson we can learn from the history of life is that most species that have lived are now extinct and that present-day species are a small percentage of those that lived in the past. But skepticism about extinction is still alive in many parts of the world. One tantalizing notion is that some ancient organisms, known to us only in fossil form, may still live somewhere in the world. What else could the Loch Ness monster be but an ancient form of dinosaur? What of the abominable snowman of the Himalayas? What of Bigfoot, the wild giant apelike creature constantly reported in the deep forests of Oregon and northern California?

We have no proof that any of these things exist, but not for lack of trying. Serious people have organized elaborate expeditions to hunt out all three of these supposed relics of primitive times, and they find ready audiences to support their belief that, in these cases at least, extinction did not take place, that survivors live on.

We have difficulty in thinking about and accepting both long periods of time and extinction. Take the ever-popular idea that because dinosaurs became extinct they were unsuccessful. But dinosaurs were reptiles made up of several orders and hundreds of species; many of the species became extinct, but new species were also evolving, and the group as a whole lived for about 250 million years. In contrast, humanlike primates first appeared only about 14 million years ago, and *Homo sapiens* is much younger. By any measure, then, the dinosaurs were extremely successful organisms.

Summary

For centuries the Western concepts of time and earth's origins were based on the Bible. In the seventeenth century Archbishop James Ussher even dated earth's creation: 9 A.M., October 23, 4004 B.C. The idea that the earth is very old, an idea that was necessary for the evolution of life to be understood, is recent.

For a time catastrophism prevailed as an explanation for earth's physical features; it held that a series of tumultuous catastrophes left the earth as we find it today. In the eighteenth century James Hutton challenged this view, asserting that the earth is dynamic not static and that past changes were due to the same forces that are producing geologic change today. His view, known as uniformitarianism, was supported in the nineteenth century by Charles Lyell, "father of modern geology." Lyell produced strong evidence against catastrophism, established uniformitarianism, and presented a theory of unlimited time.

Still it seemed that the earth, though dynamic, was not evolving; the same processes, it seemed, were simply being repeated, endlessly. And still there was no way to "tell" the earth's time.

Fossils provide a calendar for the history of life on earth. They may be formed in many ways: when silica precipitates, when insects are caught in resin, when organisms are trapped in shallow waters and covered by sediment. Fossilized dung and pollen grains provide much information about early plant life. Organisms are much more likely to be fossilized in wet areas than in arid regions.

The dating of fossils, however, depended on using a geologic clock. The first step in establishing such a clock came with the recognition that some rocks are older than others. The layers—strata—in sedimentary rocks were analyzed and geologic periods defined on the basis of their thickness and sequence. Generally, those rocks furthest down in a sequence are oldest, and the age of fossils found in different strata can therefore be compared. A geologic classification of aeons, eras, and periods was established by describing a stratum, establishing a type specimen, and then classifying similar strata by comparing them to the type specimen.

The resulting classification scheme is relative and fairly arbitrary. The relative types and abundance of fossils can be compared, but such a calendar does not tell us about absolute time, a feat that became possible when radioactive dating was discovered. Radioactive isotopes decay into other forms at a constant rate and have a known half-life (the time it takes for half the radioactivity in the sample to decay). Because the proportions of isotopes present in samples can be measured, they can also be dated. The relative proportions of potassium 40 and of argon 40, into which the potassium decays, and of carbon 14 and the element into which it decays, carbon 12, are frequently measured to date fossils.

Today earth is calculated to be 4.5 billion years old. The earliest life forms were one-celled organisms; remains of these, known as microfossils, have been found that are as much as 3.4 billion years old. These earliest organisms were probably heterotrophs. When autotrophs appeared, they helped to create an oxygen-rich atmosphere, which led to the development of the protective ozone layer. Only then was land safe for life as we know it.

About 600 million years ago life began to show its current diversity. Remains of one fern date back 300 million years or more. Little is known about the origin of flowering plants but they probably did not exist 200 million years ago. During their evolution the families of angiosperms developed increasingly specialized structures for pollination; the magnolias, for example, are an ancient family, and their flowers are unspecialized compared with those of the more recent grasses, orchids, and composites. Today angiosperms are the dominant plant form.

Probably only one to ten percent of the species that have lived on earth are known to us through the fossil record. Most species that have lived are now extinct.

Review Questions

1. Describe the various developments in the concept of almost limitless time.
2. How were the divisions of the geologic table determined?
3. If a piece of bark from Noah's Ark were dug out of the earth, would it be a fossil? What is a fossil?
4. Carbon 14 dating could be used to determine the age of the ark's bark, but it could not be used to date a dinosaur bone. Why?
5. Our understanding of organismal diversity, as determined by the fossil record, is probably biased in favor of certain groups. Why?
6. What are some probable reasons for so little knowledge about the origin of angiosperms?

SUPPLEMENT

Early Land Plants

Much of the evidence for the process of evolution comes from fossils found in the earth's crust, and such evidence indicates that the first organisms to live on land were early ancestors of vascular plants, which made the transition from water to land about 400 million years ago. It was 50 million years later that animals began to inhabit the land, an evolutionary transition most likely made possible because land living plants were already plentiful.

Although no fossil data exist that identify the first living plants, it is theorized that the first land plants evolved from green algae that lived in tidal pools or brackish waters. These plants evolved structures that permitted the acquisition of nutrients and water from the land, reproductive mechanisms suitable for procreation on land, and an external covering that prevented drying out.

Fossils do exist of plants that lived on land 350 to 400 million years ago that are probably like the ancestors of some or all modern land living

1 These representatives of early land plants date back to the Devonian period (about 405 million years ago), the Carboniferous period (345 million years), or the Cretaceous period (130 million years).

plants. These are called the *Psilophytales* and are represented by the genus *Rhynia,* of which there is fossil evidence of two species, one 2.3 centimeters (9 inches) high and another about twice that size.

Rhynia was a leafless plant with dichotomously branching stems. Some of the branches had sporangia containing spores. The stem was covered with a thick epidermis and contained a central core of tracheids surrounded by phloem. *Rhynia* possessed no roots but instead absorbed nutrients from the ground through filamentous structures called rhizoids.

There is disagreement among experts about whether *Rhynia* represents the single ancestor of all land plants or only one of the great subdivisions of vascular plants that arose from separate groups of green algae. Although more evidence will be required to resolve this question, it is apparent that before the end of the Devonian, about 350 million years ago, the great subdivisions of vascular plants existed on land.

Relatives of the club mosses, like *Lycopodium,* have existed since the beginning of the Devonian period of the Paleozoic era, or about 405 million years. These lycopods were large trees during the Carboniferous period, which began 345 million years ago. After being common plants for at least 65 million years, they disappeared. Today the division (Lycopodophyta) is represented by relatively few species.

These ancient lycopods contributed greatly to the coal deposits formed during the Carboniferous. A well-known fossil in this group is *Lepidodendron,* a large tree 35 meters (105 feet) high and one meter (3 feet) in diameter. It had a branched top and long leaves giving it a mop-like appearance. This sporophyte had sporangia arranged in cones. Presumably because of a change to a cooler and drier climate, this plant and its relatives were extinct by the end of the Permian period about 200 million years ago.

The division Arthrophyta is today represented by one genus, *Equisetum,*

with twenty-five species. These are the only survivors of a large group of plants that first appear in the fossil record at the beginning of the Devonian period, which ended the Paleozoic era. Many specimens of the fossil genus *Calamites* have been found and these 10 meter (30 feet) high trees coexisted in the coal forests of the Carboniferous with *Lepidodendron.* These giant "horsetails" were borne on rhizomes like modern *Equisetum.* While similar in appearance to modern "horsetails", it is difficult to know if the latter are descendants of *Calamites* or if their evolution was parallel.

A group of fossil plants called preferns date from the middle of the Devonian period. Their fronds look like modern ferns, but it is hard to distinguish between branches and leaves. There is a controversy about whether the preferns are more closely related to the division Psilophyta or to the ferns (Pterophyta). This group is an example of the problems of determining relationships of extinct plants.

The progymnosperms are a group of fossils that appear in late Devonian deposits, and they are considered ancestors of both the early gymnosperms, from which our modern conifers evolved, and the extinct seed ferns (Pteridospermophyta). A well-known progymnosperm is the genus *Archaeopteris.* A tree 20 meters (60 feet) in height, it had a crown of frond-like leaves. Sporangia were borne on some of the leaves, as in modern ferns.

The seed ferns or Pteridospermophyta are a completely extinct division. Known only from fossils, this group existed from the early Carboniferous to the Cretaceous period, which began 130 million years ago. They ranged from small plants to trees and had fern-like fronds. They are important to paleobotanists because they had seeds that resemble the seeds of cycads. This makes them the most ancient group of seed plants. It is thought that they descended from progymnosperms similar to those that gave rise to the conifers, and that the seed ferns were in turn the ancestors of angiosperms.

An extinct order of the conifers is called the Cordaitales, and they were ancestors of modern conifers. Like seed ferns, they probably evolved from progymnosperms. A member of this order, *Cordaites,* was a large tree with long, strap-shaped leaves. It possessed cone-like structures bearing seeds. This reproductive structure is obviously related to the cones of modern conifers. Unlike the situation with angiosperms, there is good fossil evidence for the origin and evolution of the conifers.

The fossil record of the origin and early evolution of flowering plants is scanty, but indications are that flowering plants appeared on the earth about 130 million years ago, most probably evolving from gymnosperm ancestors. Several groups of gymnosperms have been considered as possible ancestors of flowering plants. One common theory proposes that flowering plants arose from *cycads.* The identity of such early angiosperms is unknown, but many think they may be related to the magnolia family because of the structure of the magnolia flower. From such magnolia-like ancestors have evolved the myriad forms of flowering plants carpeting our present earth.

CHAPTER 14

Processes and Patterns of Plant Evolution

The biblical account of creation gave not only a theory of time but also the vision of a world created in accordance with a design that set aside for every living thing a place in the scheme of all things. A new concept of time cleared the way for a new vision of life itself.

People had long believed that everything that could have existed still did, continuing in that belief even when they were confronted with bones of extinct animals like the saber-toothed tigers and long-tusked elephants. Charles Darwin, among others, challenged that belief. In South America he found fossilized remains of giant browsing animals, unlike any living animals he had seen on that continent. He pondered long on how it was that they had left behind no remotely similar descendants. Even more puzzling was his discovery of an ancient type of horse that had died out long before the Spaniards repopulated the Americas with the modern species of horse. His later conclusion was that "no fact in the long history of the world is so startling as the wide and repeated extermination of its inhabitants."

It was difficult for the fact of extinction to be accepted by the scientific world. Why did extinction happen? Obviously it did happen, but from where did the organisms come to take the place of extinct species? Darwin formulated a theory to explain these changes. He showed us how to explain the past by looking at what happens to living organisms. This biological uniformitarianism extended the geological uniformitarianism of his teacher and friend, Charles Lyell.

THEORIES OF EVOLUTION

Charles Darwin did not "discover" evolution; the idea was ancient when he proposed it. Ancient Greek philosophers had discussed it, and the mutability of organisms, their capacity to change or be changed, was a favorite topic of scientific and intellectual circles in eighteenth-century Europe. What Darwin did was to formulate a theory to explain evolution. He also buttressed the concept of evolution with facts collected on his voyage on the *Beagle* and during the rest of his life (see the Supplement on page 417). As a result, he not only helped establish the fact of evolution; he also devised a theory to explain how it works.

14.1 Physician, poet, inventor, meteorologist, and educator—Erasmus Darwin was a man of genius and vigorous appetites. He was, like his grandson Charles, embroiled in controversy over his writings on evolution. Not the least of his accomplishments, his son married the daughter of the wealthy industrialist Josiah Wedgewood. The resulting addition to the family fortune permitted Charles to spend his life in scholarly seclusion.

Erasmus Darwin We can better comprehend evolution and its part in biology and our lives by tracing some of the ideas that led to it. The Darwin family gave much to biology, and Charles deserves most credit. But an extraordinary forebear, his grandfather Erasmus Darwin, prepared the way. A true genius, he proposed biological evolution as fact long before his illustrious grandson was born. Recognized as perhaps the finest English physician in his time, Erasmus Darwin was also an inventor, a scientist who came up with sound theories on plant nutrition (including an explanation of photosynthesis) and the first correct explanation of cloud formations, a student of psychology who linked mental disorders with physical illness, a proponent of theories on female education that should earn him honor in the history of the women's movement, and a poet of distinction and enormous output. In much of his poetry he tried to present science to the public. A Leonardo da Vinci of his time yet, perhaps because of his grandson's world-shaking work, he is nearly forgotten in modern education.

Erasmus Darwin lived from 1731 until 1802, packing into his life so much work and so many ideas that even in outline they are voluminous. Biological evolution was a part of his thinking for twenty years. He appreciated the competition for resources that characterizes all life, but he went further, constructing a philosophy of life based on his ideas of evolution. He saw delight in life, and he believed that this realization was shared by all living things, even plants. Believing this, he proposed that evolution is a struggle not just for survival of the fittest but for survival of the happiest. His philosophy of organic happiness embraced the highest as well as the lowest forms of life. The better adapted a creature is to its environment, he said, the happier it will be. Erasmus Darwin thought that the most happy triumphed in life. This relationship between evolution and happiness was echoed by Charles Darwin in his later years.

Erasmus Darwin's evolutionary theories were published in a treatise on animal life, *Zoonomia,* which roused almost as much outcry as did his grandson's work half a century later. Biological evolution is treated in only one chapter, but its premise, that all life evolved from primitive specks in the oceans into the world's many complex forms, was denounced as heresy, a denial of God's omnipotence, indeed close to treason. But Darwin stuck by his beliefs, restated in a long poem published after his death, *The Temple of Nature:*

> Organic Life beneath the shoreless waves
> Was born, and nurs'd in Ocean's pearly caves;
> First forms minute, unseen by spheric glass.
> Move on the mud, or pierce the watery mass;
> These, as successive generations bloom,
> New powers acquire, and larger limbs assume;
> Whence countless groups of vegetation spring,
> And breathing realms of feet, and fin, and wing.

Erasmus Darwin gathered no evidence to support his evolutionary ideas as his grandson did; and for all its unusual perceptiveness, much of his reasoning was flawed. Like many thinkers before and after who contributed ideas on the subject, he repeatedly stressed three principles now recognized as erroneous:

1. **Teleology,** a form of reasoning that finds a purpose and a design in nature.
2. Inherent activity of an organism that eventually changes it, as *use or disuse of its parts* changes those parts.
3. **Anthropomorphism,** ascribing human characteristics or motivations to an organism to explain its development or behavior.

None of these methods of thinking has died out despite their recognized error. Teleological thinking turns up in myths and in a great deal of popular science writing. The idea that activity will in time lead to organic change is still with us in popular and sometimes even learned explanations of biological phenomena. As for anthropomorphism, it is copiously dispensed in books for all ages, comic strips, television programs, and movies.

14.2 Jean Baptiste Lamarck (1744–1829) was an outstanding biologist whose life ended in poverty and tragedy. His ideas on evolution were part of a philosophical system to explain the unity of nature.

Lamarck The first comprehensive theory on the way evolution works was stated in 1809, the year of Charles Darwin's birth, and it contained all three of these misleading ideas. Its author was the Chevalier Jean-Baptiste de Monet de Lamarck, who lived from 1744 to 1829, acknowledged as one of the greatest scholars of the French flora and fauna of his time. It was Lamarck who coined the word *biology,* and his idea of evolution was only one aspect of a cosmic theory with which he explained events in the world and in the universe.

Lamarck believed that life had progressed from the lowest to the highest forms by an inherent drive in all living matter to achieve perfection. Thus he proposed an evolutionary ladder, an idea stretching back to Aristotle and Plato and unfortunately persisting to this day. At the top of the ladder are humans, most complex and most highly developed of all creatures, the form to which all other life mutely aspires. At the end of their days, humans return to the inorganic world; their dust starts the cycle over as it creates, by spontaneous generation, primitive organisms that begin anew the long climb up the scale.

Lamarck did not believe in extinction, and his theory had no place for competition for resources in which the fittest variants are most successful. Instead he found a wholesale transformation of organisms from one rung of the ladder to another. As environmental conditions change, new habits will form to deal with these changes, and these in turn, Lamarck thought, will lead to new actions and to change. The best-known example was the giraffe: changed conditions in the environment led first to shrubs, then to trees; browsing animals then had to change their habits and their actions; and the giraffe, as if pursuing the leaves that grew higher and higher with the trees, eventually had the long neck with which it could get food.

Lamarck is often unfairly portrayed as the progenitor of the belief in the "use and disuse" of organs as a factor in inheritance, and he is compared unfavorably with Charles Darwin. The idea was current at the time and was not the major part of his evolutionary thesis. Furthermore, Darwin too believed that use and disuse was a factor in heredity. Unlike Darwin, however, Lamarck did not believe in extinction and never accepted variability in species. Selection of variants that leave more offspring, the key to Darwin's theory on the mechanism of evolution, was never part of Lamarck's thinking.

Regardless of wrong ideas on biological evolution, Lamarck is outstanding in the history of biology. He published an extensive flora of France and was a well-known invertebrate zoologist. His ideas also keep cropping up today. Perhaps the most sensational recent neo-Lamarckian event was the rise of Soviet agronomist Trofim Denisovich Lysenko (1898–1976). A Rasputin of agriculture, Lysenko based his theories of genetics on Lamarckian environmental influence. Lysenko was able to convince Russian leaders, among them Stalin, that Mendel's genetic theories were false, and he quickly rose to dominate Russian agriculture like a czar, persuading Stalin to purge biologists who opposed him. Those trained during the Mendel-less period occupied many positions in the agricultural bureaucracy (and still do).

Lysenko's principal tool of persuasion was the claim that he could produce new crop plants that would turn frigid areas of the Soviet Union like Siberia into gardens of abundance. This event has never occurred, and partly because of Lysenko's disastrous errors, to feed its people the Soviet Union often has had to import huge quantities of grain from the United States and elsewhere, giving us a profound example of how biology used for political ends can affect world events.

Darwin's Revolution Fifty years after Lamarck's book was published, Charles Darwin published a contrasting view in his epochal *Origin of Species.* The book's full title tells his real message: *On the Origin of Species by Means of Natural Selection, or the Preservation of Favored Races in the Struggle for Life.* What Darwin presented was not a theory of evolution but an explanation of the mechanism of evolution. The difference is significant; his grandfather and many others had already stated the theory of the evolution of biological life. Charles Darwin published an explanation of how evolution takes place, an explanation solidly supported by observed facts.

14.3 Alfred Russel Wallace (1823–1913) was one of the great adventurers of all time. His theory of natural selection, developed independently, was basically identical to Darwin's, forcing Darwin into hurried publication of the *Origin of Species.* Younger than Darwin, self-educated, and poor, he was always in Darwin's shadow and probably did not receive the public attention he deserved.

Darwin's true contribution to modern scientific knowledge is sometimes confused because few people today, even biologists, have completely read the *Origin of Species.* It was published in 1859 as an "abstract" of a much longer work (see the Supplement on page 417). Darwin prepared this abstract after receiving an essay from the English naturalist Alfred Russel Wallace (1823–1913), who was working in the East Indies. (He sent it to Darwin because Darwin was a well-known naturalist.) In this essay Wallace presented the theory that Darwin had been developing for twenty years. What was Darwin to do? Wallace had not asked him to publish his essay but only to criticize it and forward it to others. But was not he, Darwin, honor-bound to have Wallace's work published, even if it did detract from his own? Agonizing over the decision, he wrote: "I would far rather burn my whole book than that he or any other man should think I had behaved in a paltry spirit."

He turned to his friends, the geologist Charles Lyell and botanist Sir Joseph Dalton Hooker, for advice. They arranged a compromise: both Wallace's paper and extracts of Darwin's work would be read before the prestigious Linnaean Society on July 1, 1858, by Lyell and Hooker. Wallace was so far away that his work was presented without his knowledge. Hooker had read Darwin's sketch on "Species" when it was written in 1844, so this evidence for priority was presented along with an extract of a letter from Darwin to American botanist Asa Gray, written in 1857, showing that Darwin had not changed his ideas on how species are formed. This presentation generated little interest from the members of the Society.

Darwin wrote quickly and *On the Origin of Species* appeared on November 24, 1859. The 1250 copies of the first edition sold out on the first day. The publication of *Origin* created a stir in scientific, religious, and lay circles. The famous debate between Darwin's "bulldog," Thomas Henry Huxley, and the Bishop of Oxford, Samuel Wilberforce, illustrates the vigor of the controversy.

14.4 Cartoonists enjoyed a field day with Darwin and his theory. This cartoon, appearing in the "London Sketch Book" (1874), depicts Darwin holding a mirror and explaining his theory to an apelike companion. The original caption quotes Shakespeare: "This is the ape of form" and "Four or five descents since."

This debate took place at the June 1860 meeting of the British Association for the Advancement of Science and was climaxed by the Bishop asking Huxley if it was through his grandfather or grandmother that he claimed to be descended from an ape. In response Huxley spoke on Darwin's ideas and facts supporting them; he concluded by stating that he would not be ashamed to have an ape for an ancestor, but he would be ashamed to have as an ancestor a brilliant man who used his gifts to attack scientific questions of which he was ignorant. The Oxford undergraduates clapped and stamped their feet, ladies fainted, and Captain Fitzroy, captain of the *Beagle,* was seen at the edge of the crowd holding a Bible over his head and shouting that it was the true word of God. Today the controversy has lost much of its heat, but the arguments of Bishop Wilberforce continue to surface in the unfortunate discussions about whether evolution should be taught in schools.

NATURAL SELECTION

What did Darwin say? The following are the basic elements of his theory of natural selection:

1. Organisms vary.
2. More organisms are produced than survive.
3. Therefore, there is a struggle for existence.
4. Therefore, more of those variants adapted to their environments will survive to reproduce.

These events result in a series of

variation → selection → variation → selection → variation → . . .
↘ extinction ↘ extinction

This is a simple statement full of meaning and the basis for our interpretation of the living world. Biology is perhaps unique among the sciences in that its basic idea can be understood readily by the nonscientist.

The Origin of Species was and remains the most important book in biology, perhaps in all science. It revolutionized our conception of the world we live in, and of ourselves. This intellectual revolution does not keep the principles of *The Origin of Species* from being ignored in many high school and even college biology courses.

Variations in Organisms Why individuals vary, Darwin could not say. Mendel's discoveries of the laws of genetics and their appreciation were still years away, and though Darwin's evidence supporting variation in species was massively documented, he could not explain it, although at times he offered the explanation that the use or disuse of parts might produce heritable modifications that would ultimately shape offspring. We now know that Darwin performed crosses on pea plants, with results similar to Mendel's, but he did not count the characters in such a way as to see the ratios. It is tempting to speculate on how biology might have changed if Darwin had completely broken the bank by discovering the mechanics of variation.

Darwin noticed variation in organisms during his voyage on the *Beagle,* and as he traveled through climatic zones on his way down the coast of South America, he was particularly impressed with the way closely related organisms succeeded one another. He came to recognize that variation is often a continuum, so that it becomes difficult in natural populations to distinguish clearly between varieties and species. On returning to England he began to investigate the selection of domestic varieties of plants and animals, and he found that breeders, who constantly selected certain types to maintain or change an animal breed or plant variety, considered well-established varieties to be of different species. This idea was, of course, fallacious: a Percheron is still a member of the species *Equus caballus* although it looks very different from that other member of the species, the racing thoroughbred. Darwin learned much by studying domestic organisms while developing his theory: the first chapter of *The Origin of Species* is titled "Variation under Domestication," a summary of material later published in two thick volumes.

Darwin also learned a great deal about individual variation in eight years of observing a group of living and fossil barnacles (see page 419). This study taught him how every part of an organism, not just the organism as a whole, may be subject to variation.

The Struggle for Existence That unchecked increases in population could jeopardize the livelihood of the human race was no news when Darwin considered it as part of evolution. In 1838 the economist Thomas Malthus published his *Essay on Population,* emphasizing the ability of human populations to multiply faster than the food supply. Darwin read the essay, was strongly influenced, and went on to incorporate competition for resources in his explanation of evolution. He presented many examples of reproductive abilities of organisms and what would happen if their numbers were unchecked. The important point, he said, is that their numbers *are* checked and the number surviving never equals the number reproduced. This results in a struggle for existence.

Selection for Survival Darwin pointed out that there are undeniable variations among species: some rabbits can run faster than others and therefore have a better chance to escape the fox and live to produce other rabbits that are faster runners; some leaf-eating insects that depend on camouflage for protection have a color more closely matching the leaves and therefore are less likely to be detected by predatory birds. On every level of life this eternal drama is enacted and reenacted; always, the individuals best adapted to their environment and better able to exploit it are the most likely to survive to procreate their kind.

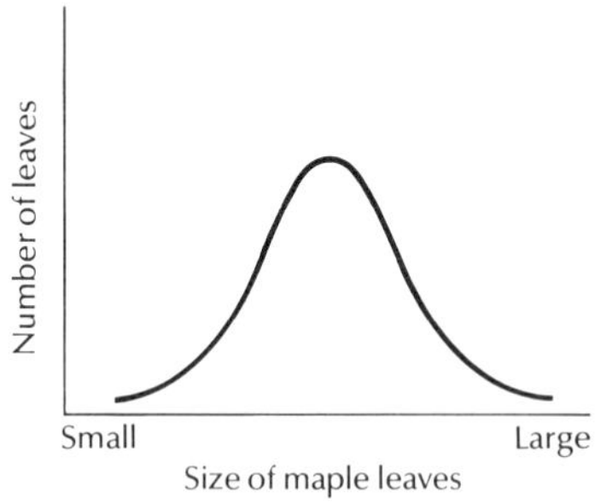

14.5 A graph plotting the size and number of maple leaves illustrates a normal curve. The height of humans or oak trees, or any measurable feature of a population, will result in a curve of similar shape.

Darwin's favorite illustrations to support his views were domestic plants and animals. Did not people themselves produce by selective breeding the qualities they desired or needed in, for example, cattle, sheep, pigeons and domestic plants? The same ancestral stock yielded many extreme variants. Could their similar ancestry and their present differences be denied?

As Darwin's theories grew more generally known and popularized, they were inevitably simplified, which led to error. The phrase "survival of the fittest" became a part of the English language but was usually misquoted. The heart of Darwin's theory was not simple survival but survival to reproduce and propagate a species. Again and again he said that no matter how many individuals are reproduced, relatively few survive to reach reproductive age. And yet, reaching that age is all-important.

Differential Survival Darwin, by his own admission, was a poor mathematician (as are many other biologists), but his postulates are really statistical statements. Much of statistics is about analyzing variation. In fact, variation itself is an idea arrived at statistically, and Darwin thought of a population as consisting of variants. Plot any measurable feature of a population, such as leaf size, and you will find yourself constructing a normal curve (see Figure 14.5), the same sort of curve used in establishing course grades. And there is a certain probability that any individual will fit somewhere on the curve.

Stating that more individuals are produced than survive is also a statistical expression, leading to **differential survival.** Because all organisms vary, even those that survive to the age at which they can reproduce do not have a uniform rate of reproduction. Their differential rate of reproduction can be expressed as a probability, which in turn means there will be a differential rate of survival.

Reproduction is not an all-or-none event, because even the less fit organisms in a population have a chance to reproduce. Reproduction of these less than wholly fit individuals, in turn, will influence evolution, which is really a matter of shifting percentages of genotypes in a population. There will be "fit" genotypes, "less fit" genotypes, and some losers. Therefore, it is possible to make predictions on how populations will evolve and to check those predictions.

DARWIN AND MENDEL

As a student Mendel too had trouble with his math exams, but he was able to analyze his experiments statistically. Like Darwin's, his postulates are statistical statements. Unlike Darwin's, however, his are unambiguously statistical, which may have been a reason why he was ignored for so long: scientists were not accustomed to explanations of the natural world in terms of probability; tradition held that interpretations must be based on certainty. Both Darwin and Mendel, therefore, were making a major break with old traditions. They dealt with the probability of events. This was a new way of thinking, always something difficult to sell, and it increased scientists' ability to make predictions. And it is predictive ability that determines the power of a theory.

Darwin combined in his theory of evolution both *inductive* and *deductive* elements. His proposition that organisms vary, and that more are produced than survive, was induced (based on direct observations). From these inductions he concluded, or deduced, that a competition for resources was going on and that the variants best adapted to survive this competition would be "selected" to continue the species by reproduction in a natural way.

This kind of deductive reasoning attracted a great deal of criticism, because it contradicted a well-established philosophical tradition that science should be based on inductions arrived at by direct observation. Deduction (reasoning from assumptions), most scientists felt, led all too easily to empty speculation.

In their original work both Darwin and that other great innovator who followed him, Gregor Mendel, used deductive reasoning to great effect. Both these giants of biology had been trained in theology. As a result, they were well acquainted with an intellectual tradition based on deduction. And since induction is difficult to apply in a field where so little can be directly observed, perhaps theology provided some of the essential intellectual tools both men needed to develop a viewpoint so different from prevailing theological thinking.

Darwin and Mendel are linked in another fundamental way. Darwin could not explain how successful traits are passed on to successive generations, exposing his theory of natural selection to growing criticism. When Mendel was rediscovered, geneticists were paying a lot of attention to mutations. They still felt that natural selection of variants had a minor part in evolution. The major factor, they believed, was sudden change introduced by mutation. Not until the 1930s did biologists realize, at last, that Darwin's theory of natural selection and Mendel's laws of genetics were fully compatible. Together the two form the basis of population genetics, a major science of today.

EVOLUTION IN ACTION

Genetics of Populations Put together genetic mechanisms, natural selection, and almost limitless time, and you have the raw materials to account for the great diversity of living things. Two questions pose themselves: How, more specifically, does diversity come about? And what are the limits of diversity?

Individuals do not evolve; populations do. It is variation and selection within the population that determine evolution. To better understand and interpret the history behind today's diversity, the population is, in effect, treated as an individual, and its gene pool is expressed in an individual way. The study of these is **population genetics,** in which the genotypes of the individuals in a population are combined to form aggregate genotypes for the population as a whole. If the frequency with which specific genes appear in a population changes markedly, we say that the population has evolved.

Because evolution results from shifts in the frequency of genes within a population, anything responsible for that shift can be considered an agent for natural selection. And because differential rates of reproduction are at the heart of the shifts in gene frequency, it can be said that natural selection is anything that promotes differential reproduction; that is, some members of a population leave more offspring than others. Natural selection can work in different ways, several of which we will look at next.

Varieties of Natural Selection Populations are almost by definition adapted to their environments; if they were not, they would not survive. Extreme variants will be selected against; they will not survive because of their extremeness. Let us say that in a population of plants those which produce seeds of a specified size will survive and those which produce larger or smaller seeds will not. Because extreme variants are selected against, natural selection can bring stability.

This **stabilizing selection** minimizes the extremes in the population, both the very large and very small seeds. Plants that produce seeds close to the norm for the population are the best adapted, and more of these survive (see Figure 14.6). Extreme seed sizes continuously reappear in the population because of normal variations, and they are continuously selected against.

Now, what if the environment changes so that large seeds are best adapted to the changed conditions? Gene frequency will then shift toward the large seeds. This directional selection will cause a shift in the norm or normal frequency. The population as a whole will have mostly plants with large seeds; it will have evolved.

Environmental changes may give a greater advantage to two or more types in the population, say to plants that produce large seeds

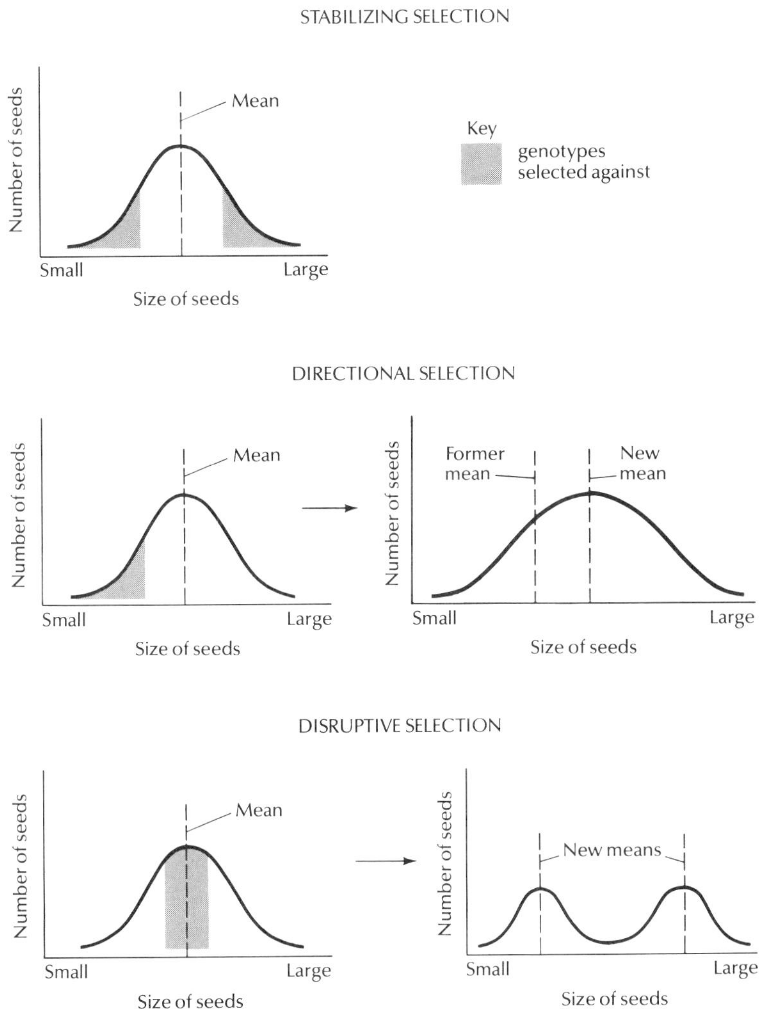

14.6 These curves reflect the effects of different forms of natural selection in a population. Stabilizing selection preserves mean values for characters. Directional selection results in a shift in the distribution of a character, while disruptive selection splits the population.

as well as to those producing small ones. This change will cause a shift in at least two directions. The norm will now be selected against, and the selective advantage will lie with the large-seed and the small-seed plants. This is called **disruptive selection**—it disrupts a stable pattern and leads to evolution in several new directions.

A graph illustrating these changes in selection is misleading because it must show that such shifts take place at one specific time, but they do not. Any shift in gene frequency is dynamic, and if the rate of change is slow, as it is likely to be, it would be difficult to notice a change in the population on a selected day. Redwood trees are a good example. They evolve at an extremely slow rate because they are an ancient group now restricted to a tight range of environmental conditions. They are only relics of much larger populations. The chance of their surviving something as drastic as a

14.7 A group of finchlike birds, frequently referred to as Darwin's finches, on the Galapagos Islands are classic examples of adaptive radiation. Closely related plants on the Galapagos illustrate the same phenomenon in as dramatic if less well-known fashion. (From *Island Life*, © Sherwin Carlquist.)

major shift in climate must be low. But the rate of change can be quite rapid. Some oak tree species are highly variable, which means that they are evolving rapidly.

There are many intermediates between groups that are designated as species, making it difficult to distinguish between species. Darwin himself maintained that the distinction between species often had to be arbitrary.

Adaptive Radiation Perhaps you remember the colonization of a newly formed volcanic island mentioned in Chapter 1. Now we can apply to that new environment an evolutionary concept, adaptive radiation.

On the newly formed island, a relatively small number of species have successfully survived the overseas journey to become colonizers. Many environmental conditions are different from those of their homeland. Some varieties not well adapted to their original environment are better adapted on the island. They do not face competition from related species, and predators and diseases may be absent. Directional or disruptive selection or both take place and new varieties evolve, which eventually may become higher taxonomic categories. New ecological niches open on the island as it is modified by weather and by other colonizers, particularly plants.

This is the sort of change that happened on the volcanic Galapagos Islands a million or so years before Darwin arrived there. Variants of the originally small number of colonizers adapted to a great many habitats. Considering the extensive variation in any population, the presence of many new and unoccupied habitats can result in rapid evolution. This process of organisms becoming adapted to new niches is called **adaptive radiation.** New variants exploit and increasingly adapt to new habitats by genetic variation.

Examples of Adaptive Radiation Darwin had seen adaptation with his own eyes during his stay on the Galapagos, in the amazingly varied finchlike birds (Figure 14.7) that could only have evolved from a common ancestor and had become adapted to the way of living the islands offered, and in the several varieties of lizards, both aquatic and terrestrial, which also appear to have common ancestry. Darwin's own description of these finches and what they told him about evolution is difficult to improve upon:

> The remaining land-birds form a most singular group of finches, related to each other in the structure of their beaks, short tails, form of body, and plumage: there are thirteen species, which Mr. Gould has divided into four sub-groups. All these species are peculiar to this archipelago; and so is the whole group, with the exception of one species of the sub-group Cactornis, lately brought from Bow Island, in the Low Archipelago. Of Cactornis, the two species may be often seen climbing about the flowers of the great cactus-trees; but all the other species of this group of finches, mingled together in flocks, feed on the dry and sterile ground of the lower districts. The males of all, or certainly of the greater number, are jet black; and the females (with perhaps one or two exceptions) are brown. The most curious fact is the perfect gradation in the size of the beaks in the different species of Geospiza, from one as large as that of a hawfinch to that of a chaffinch [Old World species], and (if Mr. Gould is right in including his sub-group, Certhidea, in the main group), even to that of a warbler. The largest beak in the genus Geospiza is shown in Fig. 1, and the smallest in Fig. 3; but instead of there being only one intermediate species, with a beak of the size shown in Fig. 2, there are no less than six species with insensibly graduated beaks. The beak of the sub-

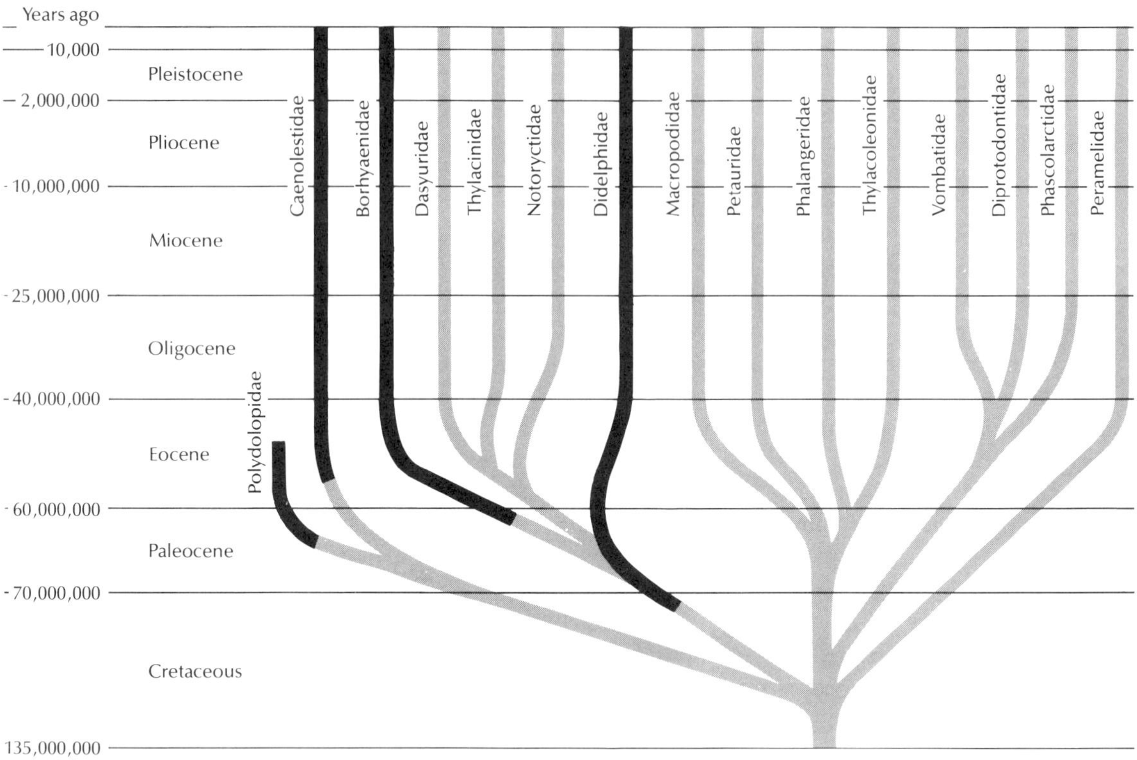

14.8 The marsupials of Australia are perhaps the most dramatic example of adaptive radiation. Small, mouse-like marsupials were probably the only mammals aboard when Australia began its continental drift. Through the ages, variants of this ancestral group became adapted to many habitats. Variation over millions of years has resulted in the diversity we organize into the many families listed on this chart.

group Certhidea is shown in Fig. 4. The beak of Cactornis is somewhat like that of a starling; and that of the fourth sub-group, Camarhynchus, is slightly parrot-shaped. Seeing this gradation and diversity of structure in one small, intimately related group of birds, one might really fancy that from an original paucity of birds in this archipelago, one species had been taken and modified for different ends.

Adaptive radiation describes what happened after colonies reached the Galapagos group from the nearest land, but how did the plant and animal populations now flourishing there reach the islands? Seeds, of course, might have floated there, and Darwin conducted experiments which demonstrated that some seeds could still germinate after several months in salt water. He also found that seeds eaten by birds survived their trip through the digestive tract and germinated. Adding these findings to the conjecture that seeds might have floated to the islands on driftwood rafts, or come in mud on birds' feet, Darwin felt he had satisfactory answers.

The Hawaiian birds called honeycreepers are also classic examples of adaptive radiation. Like the Galapagos, the Hawaiian Islands are volcanic, but they are even younger than the Galapagos. In less than a million years in some cases, honeycreepers have evolved into many new species and are adapted to many niches. Some look like woodpeckers, others like finches, and so on, though they have no relationship to the mainland versions of these birds.

Another example of adaptive radiation is the order of ancient mammals called marsupials (see Figure 14.8). A distinguishing characteristic of marsupials is that the offspring spend much of their early development in a pouch on the mother's belly, where they attach themselves tightly to nipples soon after birth. Marsupials were the only mammals in Australia when the continent separated from Southeast Asia (see the Supplement on page 460). A few marsupials live elsewhere, such as the opossum in North and South America, but they are relatively unspecialized relics of an ancient order that died out almost everywhere except in Australia.

Plants undergo adaptive radiation in the same way as animals. Darwin particularly noticed the extensive plant **endemism,** that is, restriction of a species to a particular area, in the Galapagos Islands. Especially impressive were plants of the composite family, such as daisies and sunflowers, which are generally herbaceous, but in the Galapagos they had evolved into large trees. Darwin described the unique features of the plants of the Galapagos and of the islands themselves in his popular *Journal of Researches,* describing his trip around the world on the *Beagle:*

> The botany of this group [the Galapagos] is fully as interesting as the zoology. Dr. J. Hooker will soon publish in the "Linnean Transactions" a full account of the Flora, and I am much indebted to him for the following details. Of flowering plants there are, as far as at present is known, 185 species, and 40 cryptogamic species, making together 225; of this number I was fortunate enough to bring home 193. Of the flowering plants, 100 are new species, and are probably confined to this archipelago. Dr. Hooker conceives that, of the plants not so confined, at least 10 species found near the cultivated group at Charles Island, have been imported. It is, I think, surprising that more American species have not been introduced naturally, considering that the distance is only between 500 and 600 miles from the continent; and that (according to Collnett, p. 58) drift-wood, bamboos, canes, and the nuts of a palm, are often washed on the south-eastern shores. The proportion of 100 flowering plants out of 185 (or 175 including the imported weeds) being new, is sufficient, I conceive, to make the Galapagos Archipelago a distinct botanical province; but this Flora is not nearly so peculiar as that of St. Helena, nor, as I am informed by Dr. Hooker, of Juan Fernandez. The peculiarity of the Galapageian Flora is best shown in certain families;—thus there are 21 species of Compositae, of which 20 are peculiar to this archipelago; these belong to twelve genera, and of these genera no less than ten are confined to the archipelago! Dr. Hooker informs me that the Flora has

14.9 Having Australia pretty much to themselves, the marsupials have filled many of the available ecological niches. Similar environments have similar selective pressures, so marsupials look and act like unrelated mammals occupying similar niches in other parts of the world. There are marsupials which resemble mice, rabbits, dogs, cats, and groundhogs, as well as many other mammals.

an undoubted Western American character; nor can he detect it in any affinity with that of the Pacific.

If this character were owing merely to immigrants from America, there would be little remarkable in it; but we see that a vast majority of all the land animals, and that more than half of the flowering plants, are aboriginal productions. It was most striking to be surrounded by new birds, new reptiles, new shells, new insects, and yet by innumerable trifling details of structure, and even by the tones of voice and plumage of the birds, to have the temperate plains of Patagonia, or the hot dry deserts of North Chile, vividly brought before my eyes. Why, on these small points of land, which within a late geological period must have been covered by the ocean, which are formed of basaltic lava, and therefore differ in geological character from the American continent, and which are placed under a peculiar climate,—why were their aboriginal inhabitants, associated, I may add, in different proportions both in kind and number from those on the continent, and therefore acting on each other in a different manner—why were they created on American types of organization? It is probable that the islands of the Cape de Verd group resemble, in all their physical conditions, far more closely the Galapagos Islands than these latter physically resemble the coast of America; yet the aboriginal inhabitants of the two groups are totally unlike; those of the Cape de Verd Islands bearing the impress of Africa, as the inhabitants of the Galapagos Archipelago are stamped with that of America.

I have not as yet noticed by far the most remarkable feature in the natural history of this archipelago; it is, that the different islands to a considerable extent are inhabited by a different set of beings. My attention was first called to this fact by the Vice-Governor, Mr. Lawson, declaring that the tortoises differed from the different islands, and that he could with certainty tell from which island any one was brought. I did not for some time pay sufficient attention to this statement, and I had already partially mingled together the collections from two of the islands. I never dreamed that islands, about fifty or sixty miles apart, and most of them in sight of each other, formed of precisely the same rocks, placed under a quite similar climate, rising to a nearly equal height, would have been differently tenanted; but we shall soon see that this is the case. It is the fate of most voyagers, no sooner to discover what is most interesting in any locality, than they are hurried from it, but I ought, perhaps, to be thankful that I obtained sufficient materials to establish this most remarkable fact in the distribution of organic beings.

If we now turn to the Flora, we shall find the aboriginal plants of the different islands wonderfully different. I give all the following results on the high authority of my friend Dr. J. Hooker. I may premise that I indiscriminately collected everything in flower on the different islands, and fortunately kept my collections separate. Too much confidence, however, must not be placed in the proportional results, as the small collections brought home by some other naturalists, though in some respects confirming the results, plainly show that much remains to be done in the botany of this group: the Leguminosae, moreover, have as yet been only approximately worked out:—

Name of Island	Total No. of Species	No. of Species found in other parts of the world	No of Species confined to the Galapagos Archipelago	No. confined to the one Island	No. of Species confined to the Galapagos Archipelago, but found on more than the one Island
James Island	71	33	38	30	8
Albemarle Island	46	18	26	22	4
Chatham Island	32	16	16	12	4
Charles Island	68	39 (or 29, if the probably imported plants be subtracted)	29	21	8

Hence we have the truly wonderful fact, that in James Island, of the thirty-eight Galapageian plants, or those found in no other part of the world, thirty are exclusively confined to this one island; and in Albemarle Island, of the twenty-six aboriginal Galapageian plants, twenty-two are confined to this one island, that is, only four are at present known to grow in the other islands of the archipelago; and so on, as shown in the above table, with the plants from Chatham and Charles Islands. This fact will, perhaps, be rendered even more striking, by giving a few illustrations:—thus, Scalesia, a remarkable arborescent genus of the Compositae, is confined to the archipelago: it has six species; one from Chatham, one from Albemarle, one from Charles Island, two from James Island, and the sixth from one of the three latter islands, but it is not known from which: not one of these six species grows on any two islands [see Figure 14.10]. Again, Euphorbia, a mundane or widely distributed genus, has here eight species, of which seven are confined to the archipelago, and not one found on any two islands: Acalypha and Borreria, both mundane genera, have respectively six and seven species, none of which have the same species on two islands, with the exception of one Borreria, which does occur on two islands. The species of the Compositae are particularly local; and Dr. Hooker has furnished me with several other most striking illustrations of the difference of the species on the different islands.

The same sort of adaptive radiation can be found in the Hawaiian Islands, where species and genera of plants look very much like species endemic to other regions, although they are in no way related. They have simply evolved into similar forms because of selective forces in similar habitats. They exploit environments that offer similar opportunities for living. And in Australia there are trees of the endemic genus *Eucalyptus*, which are adapted to rain forests, though others of the same genus thrive in the mountains right at the edge of the timberline (see Figure 14.11).

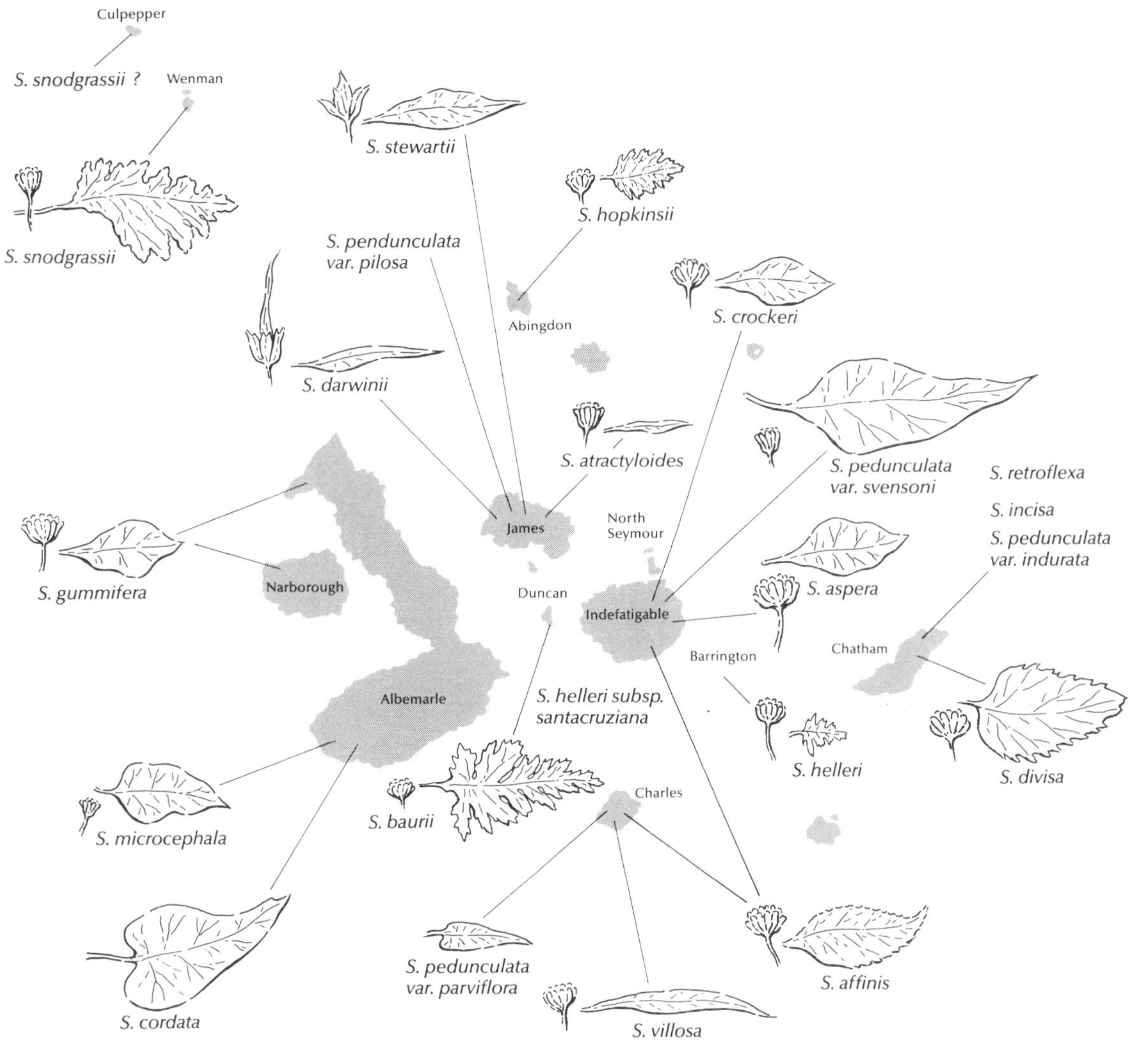

14.10 Adaptive radiation of the genus *Scalesia* was described in detail by Darwin. Members of the family Compositae, *Scalesia* looks like large woody sunflowers and is endemic to the Galapagos. Most of the species are restricted to specific islands.

Adaptive radiation is often apparent on islands, but there are other kinds of islands, biologically speaking, not isolated by water. California is a good example. It is isolated from the rest of the American continent by mountain ranges and deserts difficult for plants and animals to cross, and it has many habitats. Plants that colonized the region have therefore been able to evolve in isolation from other continental species. With little competition and many habitats, they evolved into many species, and today California has

Mountain ash River red gum

Snow gum

14.11 The *Eucalyptus* trees of Australia demonstrate adaptive radiation even more graphically than the marsupials. There are at least several hundred species, ranging from the 300-foot mountain ash of tropical regions to the snow gum, which survives snow and freezing temperatures.

more kinds of plants than any other region of the country, and a very large number of endemic plants. Mountaintops, deserts, valleys, and lakes can all be islands for isolated populations.

Convergent Evolution When different organisms occupy similar ecological niches, they tend to develop morphological similarities, a phenomenon known as **convergent evolution.** The cacti in our deserts are one example. Cacti are native to the New World and belong to the family Cactaceae, but Old World deserts have plants that look very similar to them. These, however, are members of a completely different family, the Euphorbiaceae (remember the poinsettia). Occupying similar habitats, they have faced similar selective pressures and so have developed morphologically in similar ways, as the photographs in Figure 14.12 demonstrate. This is called convergent evolution. There are members of the Euphorbiaceae in the New World, but they do not look like cacti. They occupy different ecological niches and, as a result, they are very different.

The ecological niche is a place in which an organism lives, but it also represents all conditions in the way an organism makes a living. We know that, to live, plants need light, carbon dioxide, water, selected minerals, and a definite range of temperature during at least part of the year. The ways in which these conditions can be satisfied are endless, from the desert (little water, lots of light, extremes of temperatures) to the arctic tundra (lots of water, less light, extremes of temperature). All offer ecological niches, and the kinds of plants that evolve and make a living in each are as endless in variety as the niches themselves.

Added to the enormous range of environmental opportunities are the constant changes in the environment itself, in ways and at a pace that our short lifetimes may keep us from discerning but that become significant over geologic ages. Evolution itself changes: no organism lives under unchanging selective pressures. Instead, different selective forces work on the constantly changing genetic recombinations represented by individual organisms. Look at the variety in the world and see that the variances in a population are, in truth, endless, that all around us there are constant genetic changes caused by mutations and compounded by genetic recombinations.

Hybridization Hybridization is a great agent in plant evolution, one that humans have taken advantage of to develop new species of plants suited to their own needs, such as the genus of grain *Triticale.*

You may remember that the promising hybrid of wheat and rye that led to *Triticale* had a problem (Chapter 11): the seeds of the hybrid were infertile. Treating the flowers with colchicine produced

14.12 (Top, left and right) Euphorbiaceae, (bottom, left) Asclepidaceae, (bottom, right) Cactaceae—these plants demonstrate convergent evolution. They all live in hot, arid regions and have similar selective pressures. Although they resemble each other, their flowers show they are not closely related. The flowers are short-lived and do not have the same selective pressures as the vegetative parts.

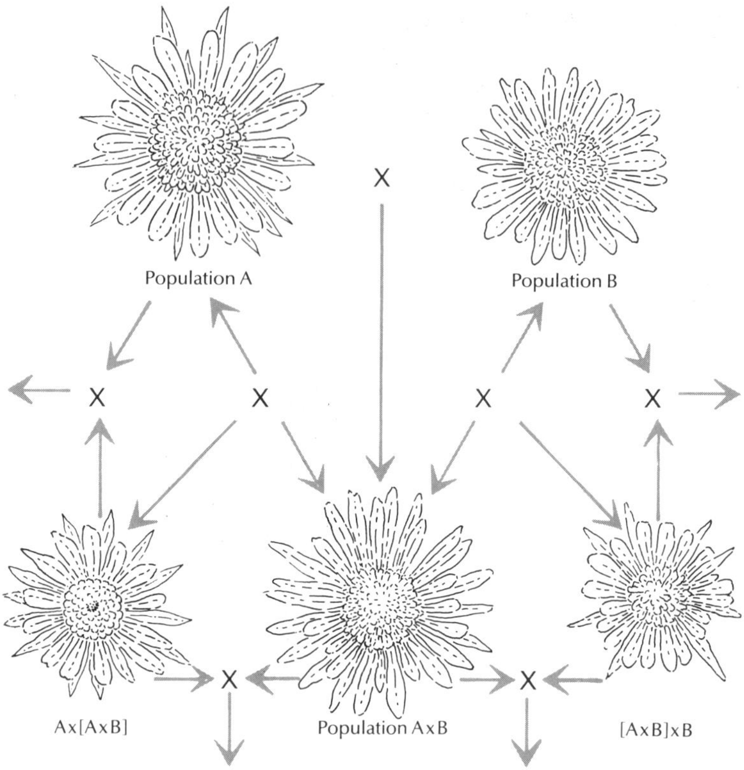

14.13 Hybridization between species of tarweeds *(Tragopogon)* illustrates how new species can be formed. Hybrids between members of populations A and B can backcross with populations A and B and form new hybrids which can produce new hybrids with other members of this "hybrid swarm." This process of new genetic combinations leading to new species is represented by the outward-pointing arrows.

fertile *Triticale* seed, and this form was now hybridized again with wheat, making a second generation that could be backcrossed with either the wheat or the rye parent, or even with entirely different varieties of wheat or rye. *Triticale-Triticale* crosses could also be made. The new genus, therefore, could be adapted to many habitats in which it had not been able to grow before. Each hybrid provided a source of other possible variants that could adapt to all manner of new environments.

When different plant populations hybridize, as they do all the time in nature, the hybrids themselves often hybridize again with the parent populations (see Figure 14.13). The probability of this happening is high, because parental populations are so much larger than the hybrid population. The latter might be only one or a few plants. This inbreeding can produce intermediates, which may have characteristics with which they can either exploit habitats not open to the original parent population or be **preadapted,** that is, have a genotype that by chance makes them more adapted to newly developing environments. This type of hybridization is called **introgression,** and it transfers genes between species adapted to very different habitats. Some of our wild sunflowers are well known examples of speciation resulting from introgressive hybridization.

COEVOLUTION

Of all the ways in which plants and animals have deeply influenced one another during their evolution, the most striking appears in the reproductive structure of the angiosperms, the flower. The flower holds the plant's reproductive organs, and, at the same time, its tissue, nectar, and pollen are food for animals.

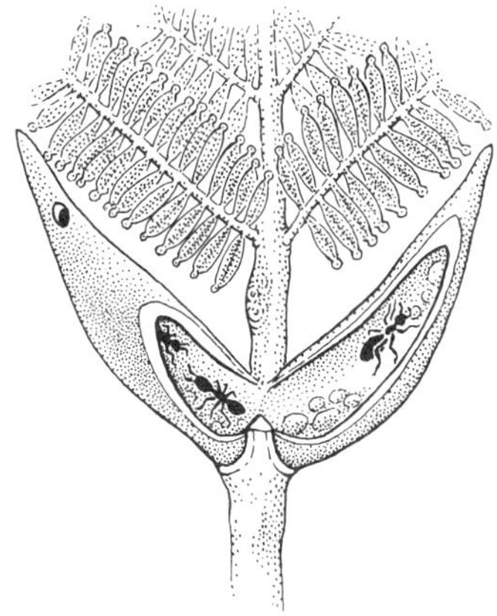

14.14 The ant acacias of Central and South America have thorns which house ant colonies. The ants attack herbivores and obtain food from specialized structures at the tips of leaflets. This symbiotic relationship benefits both parties and is called *mutualism.*

In Reproduction Flowers and their pollinators are often **coadapted**—they have developed a dependent relationship. Bees are the most visible of animals involved in this mutualism, but it is also demonstrated by other insects such as moths and even by some birds and mammals.

Bees and the flowering plants they pollinate began to evolve some 130 million years ago, when the first flowering plants appeared. The bees pollinate a wide range of flowers in many environments, and their whole complex social structure revolves around this activity, which gives them the raw material for making food. Some species of bees have gradually become restricted to visiting one type of flower; such as the orchid discussed below. Increasing specialization of flowers and pollinators has greatly diversified both plants and pollinators.

Such coadaptation sometimes has extraordinary results, as with the small yucca moth, which lives entirely on the blooms of the desert yucca tree. The yucca tree in turn depends completely on the yucca moth for pollination. Together the two have developed a classic case of *mutualism,* a relationship beneficial and vital to both. The yucca moth pollinates the yucca bloom; the bloom is the place for the moth to lay its eggs, at the base of the flower's ovary, and it even supplies food—its seeds—for the larvae after they hatch. The plant always has a surplus of seeds, so larval infestation does not keep it from reproducing itself. This extreme coadaptation has one problem: in a sense, all the genetic eggs are in one basket and hence are vulnerable.

In addition to attracting pollinators with food, some plants have developed colors, shapes, and odors with which they interject themselves into the life cycles of pollinators. The tiny *Ophrys* orchid of Europe, North Africa, and the Middle East is one of several species that have evolved a bizarre way of attracting their pollinator, a male wasp: the orchid's flower looks and apparently smells like a female wasp. The male wasp is attracted and assumes a copulating position on the flower, which is remarkably well adapted to this piece of deceit. The copulating wasp thereby loosens the orchid's pollen packets, which stick to its head or abdomen. Flying off, it carries the pollen to another orchid flower, where it repeats the act, this time inadvertently fertilizing the flower. Pseudocopulation of wasp with orchid is high specialization, and it shows the potential

14.15 COEVOLUTION

The relationship between primary consumers and plants is so basic that coevolution has been a constant theme. One of the most complete fossil histories of any organism is that of the horse (below, right, and bottom). The small forest-dwelling browser of the Eocene epoch (beginning 60 million years ago), *Eohippus,* differs considerably from the grass-eating plains dweller, *Equus,* arising in the Pleistocene epoch less than one million years ago. Changes in teeth, accompanied by changes in jaw and skull structure, were adaptations to grasses. Changes in position of eyes and toward long, speed-producing legs were adaptations to new predators. Underground stems of many grasses (below, left), adequate to sustain the plant while it grows again, are adaptations to both grazing animals and fires. The velvety soldier beetle (facing page, top right) has fine hairs covering its body that gather and transfer pollen. The sphinx moth (facing page, bottom) and the scarlet-tufted malachite sunbird (facing page, top middle) transfer pollen between flowers as they forage for nectar. The honey possum (facing page, right middle), a marsupial of Western Australia, gathers nectar and small insects. The honey possum, the sunbird, and the sphinx moth have coevolved with plants on which they feed and serve as pollinators. The most specialized example of coevolution is the relationship between a single plant and animal species. The yucca moth (facing page, top left) transfers pollen to the stigma of the yucca flower and lays eggs in the ovary. The larvae feed on the seeds which are the product of this pollination. This species of *Trichoceros* (facing page, photo, middle left), like a number of orchids, mimics the appearance of a fly to attract an insect, which pollinates the flower while it attempts to copulate with its presumed mate.

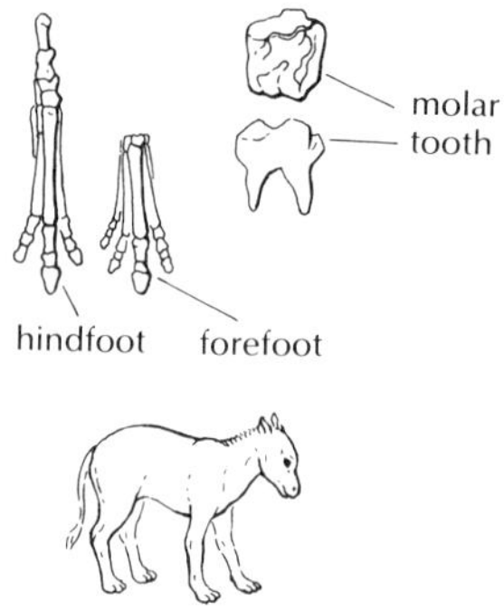

Eohippus: forest-dwelling browser
Eocene

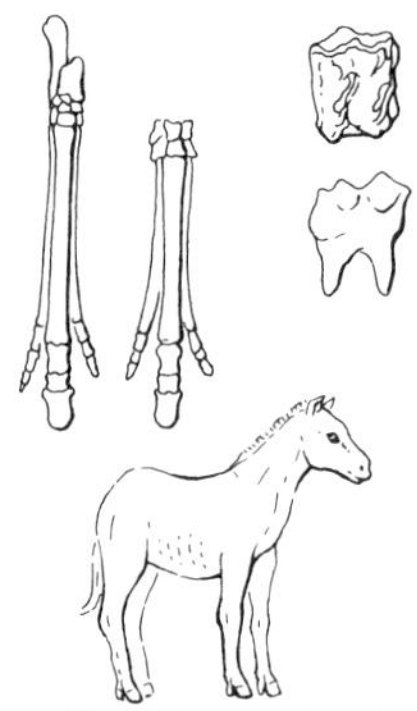

Mesohippus: forest-dwelling browser
Oligocene

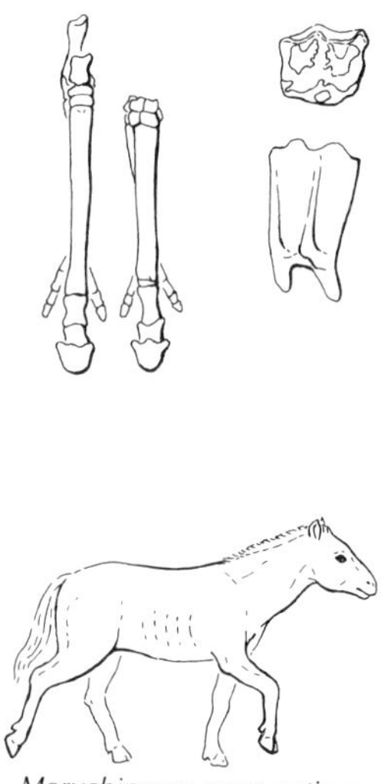

Merychippus: grass-eating plains-dweller
Miocene

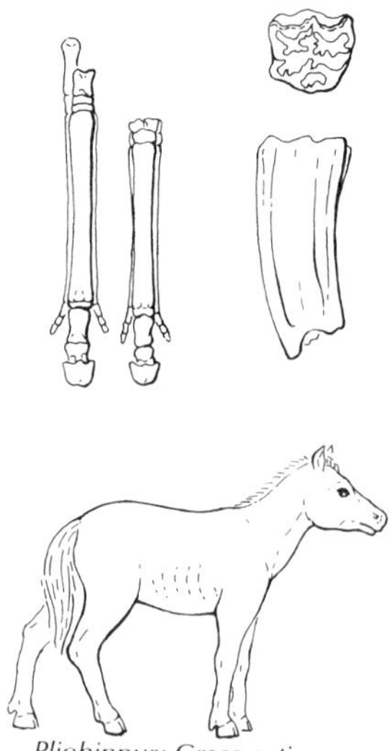

Pliohippus: Grass-eating plains-dweller
Pliocene

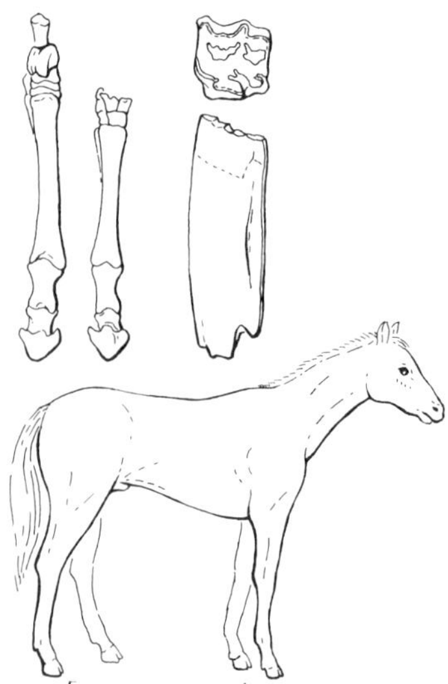

Equus: grass-eating plains-dweller
Pleistocene

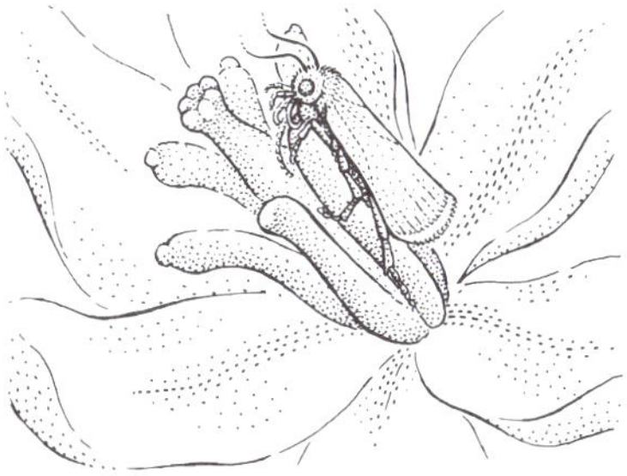

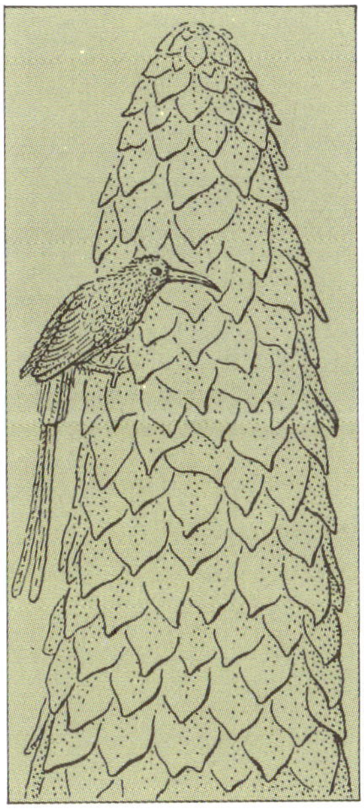

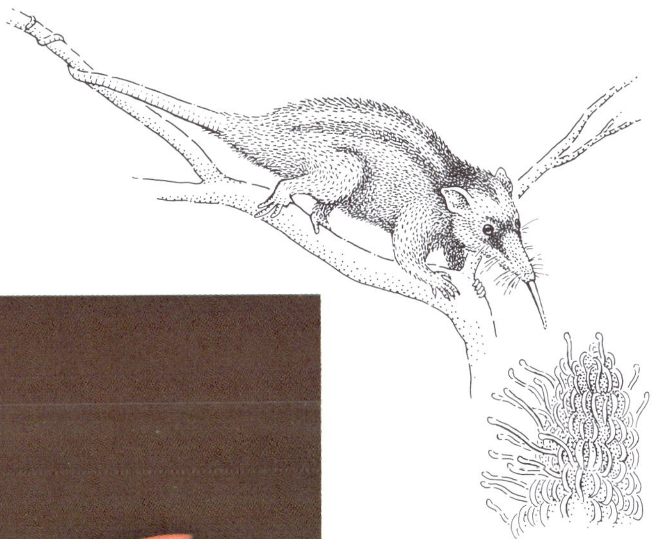

danger of a plant's depending on a specific pollinator: extinction of the pollinator could result in extinction of the orchid. The flower mimics the wasp and the plant's reproduction depends on it, but the advantage to the wasp is not clear.

Some flowers turn the tables, influencing reproduction of bees. They have odors to attract a solitary bee, which not only gets the flower pollinated but helps lone male and female bees find each other. (A human version might be the college or university, which fertilizes minds but also serves as an aggregation center for selecting mates.)

Defenses against Predators Butterflies and food plants provide another excellent example of coevolution. A great many plants wage chemical warfare for protection against predators. They produce a great array of toxic chemicals that discourage the many animals trying to eat them. Larvae of some species of butterflies have evolved immunity to the poisons of the plants on which they feed. If the butterfly should lay eggs on the wrong species, the mortality rate of larvae would be high, because the larvae would be poisoned as they fed. Some butterflies can identify hospitable plants by their leaf shape.

A constant natural struggle pits plants developing a poison against larvae developing tolerance for that poison. Like the tolerance of insects to pesticides, first one side is ahead and then the other. But another twist complicates this relationship. Larvae of a species may be immune to the poison of their food plant, yet they can store the poison in their bodies, making the adult butterfly distasteful to other animals that prey on it. In essence, the butterfly steals the plant's chemical defenses and uses them for its own ends.

Coevolution of this kind has all sorts of ramifications, some of them surprising. Females of some moth species will not lay eggs on a leaf that already has an egg on it, and some plants have lobes on the leaf stipules that look like eggs. The "egg mimic" decreases the incidence of egg laying and therefore the chance that larvae will eat the leaf. Conversely, some plants have **trichomes,** leaf hairs with sharp barbs, as a defense against larvae. Larvae of some species weave a web over the "bed of spikes" on the leaf surface, and nibble the leaf away at its edge.

One of the most dramatic examples of a mutualistic relationship between plant and insect is that of the ants and acacias of Central America. This acacia tree has hollow thorns and specialized leaflet tips that are modified into small bulblike structures rich in stored food. The ants live inside the thorns and feed on the leaflet tips. As if in exchange, they protect the tree from being eaten by herbivorous animals. Any animal that begins to feed on the leaves is immediately attacked by swarms of biting ants.

And another type of acacia tree in Africa that also harbors ants has thorns with swollen bases. Mantids, the insect group to which the praying mantis belongs, mimic these thorns, attracting the ants and feeding on them.

These acacia trees have not evolved chemical defenses against predators, probably because it takes energy to produce chemicals. Energy is also needed to produce food for the ants, but the ants provide a defense. Only so much of the budget can go for defense, and plants spend it in different ways.

Other Partnerships Familiarity may lead us to take such coevolutions as that of bees and flowers for granted, but many less obvious partnerships between plants and mammals are impressive. If an animal eats a type of plant, it must have the proper morphology to do so. Grazing animals must have teeth that will resist excessive wear from the grinding action of deposits of silica found in grass leaves. Human beings could not eat grass even if they could digest it, because these glasslike bits would soon wear our teeth down to the gums. In turn, the selective force imposed on the grass by grazers resulted in changes enabling the grass to survive predation by grazing animals. Grass leaves have basal meristems that keep young leaves growing when their tops are cropped. Many grass plants also have rhizomes that are not usually eaten by the grazers, so that the plant still has reserves of food and buds even though the leaves are clipped. Grasslands and many grazers evolved together, just as the bees and the flowers did.

This assumption can be tested because the evolution of grazing animals, particularly the horse, is very well documented. Horse fossils and fossil pollen also demonstrate concurrent evolution of plants with grazers in today's grassland regions.

The ancestor of the horse was about as big as a terrier, with four toes on each foot. Originally this creature lived in moist forests 50 million years ago. But then the climate changed, and the forests disappeared and were replaced by dry grasslands. The grasses began to evolve, and the horse along with them. The fossil record shows that when the grasslands were forming, the horse underwent very rapid evolution of the teeth and therefore the skull. Its hooves also changed: in the grasslands, speed was vital to escape predators, as was keen eyesight even while grazing.

As a result of its long association with the horse and other grazing animals, grass is preadapted for the lawnmower. Just as in the wild after a herd of grazing animals has passed through, grass grows back after we have mowed it. The lawnmower acts like an artificial horse or cow; it is a contemporary selective force that influences the breeding and selection of lawn grasses that are easy to cut. The grasses in turn have influenced the designs for lawnmowers.

Other evolutionary relationships of plants with animals are intriguing. Single-celled green algae live within the cells of animals as diverse as protozoans, hydra, sea anemones, marine snails, and giant clams. These host animals get the benefit of photosynthesis, and the algae obtain minerals and CO_2 from the host's respiration.

Another benefit goes to the sloth, a tree-dwelling mammal of South America that also plays host to green algae. The algae grow in grooves in the hairs of the sloth's fur, which helps to camouflage the slow animal among the trees while providing the algae a moist, protected habitat. Still another creature, a moth, joins in this mutualistic relationship. The moth lives in the sloth's fur, and its larvae feed on the algae. The feces of both moth and larvae provide nutrition for more algal growth, keeping the sloth's fur green and camouflaged.

A number of these relationships belong to the class of symbiosis called mutualism. Bees and many flowering plants are mutualistic in that each obtains something from the other. The ant-acacia relationship is also mutualistic. These benefits are costly, however, because the plant channels a lot of its energy into producing nectar, and the bees and hummingbirds have the problem of obtaining more energy from the high-energy but specialized food source than they expend in searching for it (hovering consumes a lot of energy). Likewise, the acacia trees devote energy to ant food, and the ants to fighting herbivores that do not directly threaten the ants (if the animal eats too many leaves it will cause a shortage of ant food).

Symbiotic relationships are often hard to define precisely because one form grades into another. When does mutualism drift into parasitism, and when does parasitism become a predator-prey relationship? A tapeworm or a louse is regarded as a clear-cut parasite, but what is an aphid or mealy bug sucking at the juices of the phloem? For that matter, is a herd of horses feeding on a grassland different from a flock of fleas sucking the juices of old Spot? What is important is that all organisms are interdependent, not the taxonomy we use to describe that interdependence.

Another fascinating relationship between plants and animals is **mimicry** of plants by animals (see Figure 14.16). Coevolution is not involved because the plant is a silent partner. Examples range from a slothlike mammal, the saki, which looks like a bunch of dried leaves, to fish mimicking algae, to lizards that resemble lichens. The most spectacular cases are provided by the insects (no wonder, because most animals are insects), with a treehopper that looks like thorns, moths like lichens, and katydids like dried leaves. Perhaps hardest to imagine are the flower mantids. These ferocious predators closely resemble flowers, with which they are associated, and are another example of adaptive radiation. There are even plants that appear to mimic stones: the pebblelike leaves of the African stone plants (the genus *Lithops*) are hard to distinguish from the surrounding rocks.

14.16 Mimicry has many forms. Some plants *(Lithops)* look like stones (top, left) and some animals mimic plant organs. Can you tell the tree hoppers (above) from the thorns? Flower mantids (left) are almost unbelievable; certainly at least to their victims.

The great range of variation in organisms shows the myriad ways in which organisms make their living. A trap always easy for people to fall into is ascribing a specific function to all the adaptations we observe. These interpretations are often teleological, because we infer purpose and direction to events when, in fact, there is none. The significance of many specific structures or colors is simply not known. Unless a causal relationship can be established between these structures and specific functions, it is best to refrain from speculation, which is more than likely to fall into anthropomorphic bias.

LIMITS OF DIVERSITY

Evolution, or a shift in gene frequency, works by selection of variants in a population. The theory of natural selection states that selective forces acting on natural variation in populations have created the diverse organisms seen throughout life's long stay on earth. The fossil record also reports that there are limits to diversity. What are these limits? Remember that diversity of organisms reflects diversity in ecological niches—in ways of capturing enough energy to make a living. The energy captured by the biosphere is finite, and it follows that the amount of diversity is also finite.

Many factors prevent full sharing of the energy pie. Limitations of temperature, nutrients, and moisture prevent plants from capturing all the energy that is available. Regions with the fewest limitations, like the tropics, have the most diversity of plant life and therefore animal life. Habitat variety also affects niche diversity, explaining why a region like California has so much more plant diversity than other regions with less varied geography and climate.

Summary

The idea that life evolves goes back at least to the ancient Greeks, but not until the nineteenth century was a convincing explanation given of how evolution takes place. Charles Darwin provided it, but his grandfather Erasmus Darwin, a poet, scientist, and inventor, helped prepare the way. In 1809 French biologist Lamarck gave a comprehensive account of how evolution might occur, but it included a role for spontaneous generation and excluded both extinction and competition for resources by variants in a population as factors. Lamarck, like both Erasmus and Charles Darwin, also believed that use and disuse of organs affected their inheritance.

In 1859 Charles Darwin published an abstract explaining that natural selection is the mechanism behind evolution. The argument can be presented by four statements: (1) organisms vary; (2) more organisms are produced than survive; (3) therefore there is a struggle for existence; (4) more of those variants adapted to their environment will survive to reproduce.

Darwin based part of this on observation. He had seen great variations among members of a species during his voyage on the *Beagle* and in his study of barnacles, and he had observed that these variations were neither absolute distinctions among species nor simply differences among whole organisms. Rather, there is a continuum of variations, which can occur in every part of the organism, not just in the organism as a whole. The fact that fewer organisms survive than reproduce can also be observed; Darwin had been impressed by Thomas Malthus's observations regarding the ability of people to multiply faster than they could produce food. He extended Malthus's argument to other organisms and deduced that there must be competition for resources. And selective breeding of animals and plants, Darwin pointed out, showed that if only those best adapted to the environment reproduced, then, generations later, one would find variants quite different from the ancestral stock.

Darwin's postulates were based on both induction and deduction and, like Mendel's laws, they were statistical statements. He thus contributed to a revolution not only in the content of biology but also in the methodology of science.

The discipline of population genetics today is overwhelmingly statistical. Populations, not individuals, evolve. The genotypes of individuals in a population are combined in an aggregate genotype for the population; "variation" means variation from the population's norm, which is a statistical concept. We say that a population has evolved whenever the frequency of genes within that population changes. These frequencies change, Darwin said, because there is differential survival and differential reproductive rates within a population: more "fit" than "unfit" individuals survive, reproduce, and pass their genes on to the next generation. Differential rates of reproduction are thus at the heart of natural selection, and natural selection is anything that promotes differential reproduction.

If populations were not adapted to their environments, they would not survive, but the environment itself, and thus the pressures of natural selection, changes. Three basic forms of natural selection can be distinguished: (1) Stabilizing selection—in a population adapted to its environment, extreme variants are selected against; they are minimized, and natural selection promotes a stable population. (2) Directional selection the environment may change so that one extreme provides the best adaption; selection will then result in a shift in the frequency of genes away from the norm, toward one extreme, and the population will have evolved. (3) Disruptive selection—environmental change might make two or more variants advantageous, and the frequency of genes will therefore shift in at least two directions.

These shifts in gene frequency do not occur at a single time. The changes usually take place at a slow rate, although rapid evolution can also occur. As a result of changing environments and pressures of natural selection, we find certain recurring phenomena in the evolution of plants and other organisms.

Adaptive radiation is the process of organisms becoming adapted to new niches. If many new and unoccupied niches become available to a population that has a normal range of variation, rapid evolution can result. If different organisms occupy similar niches, they will tend to develop morphological similarities, a process known as convergent evolution. Among plants hybridization has also been important in evolution. Hybrids will often interbreed with the parent population, a process known as introgression. The result can be the exchange of genetic information between populations that do not actually interbreed and the formation of many intermediates that are adapted to newly developing conditions or can exploit new habitats.

Because organisms are interdependent, we also find coevolution. Flowers and bees, for example, began to evolve some 130 million years ago and are adapted to each other. As a result of their coadaptations, flowers and their pollinators have become increasingly specialized, producing greater diversification among both plants and pollinators. Another prominent example of coevolution is the relationship between some species of plants and butterflies: the plants produce chemicals discouraging insects and animals from eating them; larvae of some species of butterflies have evolved immunity to some of these chemicals.

Many mutualistic relationships have evolved between plants and animals, but we cannot ascribe a specific function to all of these or other apparent adaptations. Evolution has no purpose, no preordained direction. Rather, the diversity of organisms results from the varying pressures of natural selection acting on naturally occurring genetic variations within a population. The diversity of organisms thus reflects not a purpose but the diversity of niches, of ways in which organisms can capture energy.

Review Questions

1. Contrast Lamarck's concept of evolution with that of Charles Darwin and his grandfather.
2. How did domesticated plants and animals play an important role in the formulation of the theory of natural selection?
3. Why are organisms past the reproductive age largely expendable in a biological sense?
4. Why would Darwin have been uncomfortable with the statement "plants have evolved to meet their needs"?
5. How did Darwin and Mendel differ in their explanations of organismal variation?
6. Compare adaptive radiation and convergent evolution.
7. How can hybridizaton be a factor in speciation?
8. Honey bees and flowers have a symbiotic relationship. What type is it, and how does it differ from other types of symbiosis?
9. Flowering plants and insects are extremely diverse groups of organisms. How might this diversity be stimulated by the many mutual relationships between the two groups?
10. Discuss energy availability as a selective force in determining diversity.

A Profile of Charles Darwin

Charles Robert Darwin was born on February 12, 1809, a familiar date because on that day we celebrate Lincoln's birthday. These two great emancipators, one of the people's bodies and the other of their intellect, began life on the same day; it was also the year in which Lamarck published his explanation for organic evolution.

Darwin was born in the town of Shrewsbury in the English Midlands about forty miles from the great industrial city of Birmingham. In the Midlands began the industrial revolution, and the Darwin family was close to its origin in the person of the family's first prominent member, Erasmus Darwin (see pages 386–387). An inventor himself, he was a patron of James Watt, who invented the steam engine.

Erasmus Darwin, Watt, Josiah Wedgwood, Joseph Priestly, and other talented men founded the Lunar Society, which met each month during the full moon. Many members were influential in scientific and industrial developments; prominent among them was Josiah Wedgwood, who pioneered mass production in his pottery works. We now think of Wedgwood china as expensive, but his was the first high-quality, mass-produced, and inexpensive kitchenware. Dr. Robert Darwin, son of Erasmus and father of Charles, married a daughter of Josiah Wedgwood.

1 Robert Darwin was a prominent physician and, according to his son Charles, a remarkable man.

Robert Darwin was, in the words of his son, Charles, "a remarkable man," who won great fame and fortune as a doctor. For his day, he was also of remarkable size, six feet two inches tall and 336 pounds at his last weighing. Dr. Robert was a keen observer and a free thinker like his father, so that Charles was raised in an environment where new ideas and ways of doing things were common currency.

Charles exhibited no unusual talents as a child and young man. An indifferent scholar, his chief passion was hunting. He also showed interest in chemistry and in collecting natural objects. Then as now, doctors wanted their sons to follow in their footsteps, so Charles was packed off to the University of Edinburgh to join his brother Erasmus as a medical student. He disliked lectures in general and 8 A.M. lectures in particular, and he could not stand the operating theater (this was before anesthetics and effective painkillers). He did join a group interested in natural history and conducted his first research project on some marine invertebrates.

Not appearing suited to a medical career, after two years at Edinburgh he was sent to Cambridge University to study theology. His own words are the best testament to his formal university experience: "During the three

2 The *H.M.S. Beagle*, a crowded little warship, was Darwin's home off and on over the five years of its voyage around the world. The *Beagle* remains one of history's most famous ships.

years which I spent at Cambridge my time was wasted, as far as the academical studies were concerned, as completely as at Edinburgh and at school." Though he felt his studies were of little value, he thoroughly enjoyed himself in a group of young men with high spirits and sporting interests. Darwin has been characterized as something of a college dropout, but he was a normal and well adjusted college student frequently more interested in extracurricular activities than in formal academic and career training.

He did complete his theological studies and obtained a B.A. degree, but his career was shaped by other experiences at Cambridge. His cousin introduced him to insect collecting, and he soon became an enthusiastic and knowledgeable beetle collector. He went on geological field trips with Adam Sedgwick, but most important, he became a close friend of botanist John Stevens Henslow. Soon after graduation, Darwin was recommended by Henslow to serve as naturalist on *H.M.S. Beagle*, which was about to make a voyage around the world.

Darwin was 22 when he started, two days after Christmas in 1831, on a voyage of almost five years. He would digest the fruits of this voyage for many years, and biology's grand idea would issue from it.

Although the voyage was long, Darwin was ashore for long periods. Indeed, he took every opportunity to leave the ship, because he suffered constantly with seasickness. Conditions on the *Beagle* were difficult. It was 90′ long and 24′6″ wide, 11′ in the hold, and carried 65 men and 9 cannons. It belonged to a class of ships (a brig rigged as a barkentine) known for being difficult to sail and susceptible to swamping in high seas. Crowding was a problem with a 24′ yawl and 5 smaller boats on the main deck. Darwin had little room for his gear and could only sling his hammock with great difficulty. Reading, writing, drawing, collecting, and observing and preparing specimens were difficult enough in these cramped quarters, but consider doing it all while seasick!

The *Beagle* had a remarkable captain. Robert FitzRoy was the second son of the second son of the third Duke of Grafton. The first Duke was the bastard son of Charles II and Barbara Palmer, his long-time mistress, who earned the title Dutchess of Cleveland. London's FitzRoy and Cleveland Squares and Euston Station are all reminders of this charming lady and her descendants (the Graftons were also Earls of Euston).

This was FitzRoy's second voyage in the *Beagle*, the purpose of the trip being to map the coast of South America, to carry a set of accurate clocks to determine longitude, and to keep an eye out for natural resources of possible commercial interest. FitzRoy was only 26 at the time, although he had been in the Royal Navy since the age of 12. An outstanding seaman, he kept the *Beagle* afloat in conditions that had sent many sister ships to the bottom. His charts were so accurate they were used into the twentieth century, and he invented a navigational sextant used for many years. Retiring from the navy soon after the voyage with Darwin, FitzRoy served in Parliament and became governor of New Zealand. He was soon removed from that post, because he supported the rights of the Maori inhabitants over claims by the English colonists. Perhaps his greatest contribution was establishing the British Meteorological Service. FitzRoy was a difficult traveling companion, Darwin said, because of fits of depression and inability to entertain views contrary to his own. These mental problems persisted, and at the age 60 he ended his life by slashing his throat with his razor.

The *Beagle*, though the best-known ship in the history of biology, was sold in 1870 for about $2500, apparently for scrap. It had been taken in 1841 on a surveying trip to Australia and the East Indies and was then anchored near England's east coast as a watch vessel. Darwin, though a wealthy man in 1870, apparently had no interest in preserving even a plank of the ship he had ridden to fame.

Darwin's advisor by mail throughout his voyage on the *Beagle* was Henslow. The letters between the two clearly show Darwin as the initially bewildered novice and Henslow the patient tutor. An accomplished naturalist upon his return, Darwin still looked to Henslow for advice on descriptions and disposition of his collections. This was no easy task, taking ten years to complete, but in doing it he became well acquainted with prominent geologists and biologists.

Two events of great importance took place soon after Darwin's return in 1836. He found that an inheritance from his Wedgwood mother left him financially independent, and he started a notebook in 1837 on "Transmutation of Species." Some time during the following year he formulated the theory of natural selection. A sketch of the theory was written down in 1842, and he wrote a lengthy essay in 1844. Although his first publication on the subject did not appear until 1858, he discussed the theory with his close friends like Sir Joseph Dalton Hooker, the director of Kew Gardens (who married the daughter of Darwin's teacher Henslow), and corresponded on it with Asa Gray, a botanist at Harvard.

By 1840, he was well known for his collections and observations; he was author of a popular book, secretary to the Geological Society, and fellow of the Royal Society. His early scientific reputation came from geological observations on the uplift of the west coast of South America and explanations of the origin and development of coral reefs. Between the geological work and *On the Origin of Species* in 1859, he published a work on barnacles. After the *Origin*, he became something of a writing machine. There were six editions of the latter and many other books, from the pioneering *Expression of the Emotions in Man and Animals* to the *Formation of Vegetable Mould through the Action of Worms*, published six months before he died in 1882. The former book was to have much influence on future research on human behavior; Darwin initiated new trends in psychology as well as in geology, botany, and anthropology. All his contributions to specialized fields were held together by the thread of natural selection.

Emma Wedgwood, Charles' first cousin, was a year older than he. They married in 1839 and moved to a house on Upper Gower Street, a ten-minute walk from Charles Lyell's home. At first, Darwin was socially active and was involved in professional affairs. He was now particularly close to Lyell, frequently exchanging visits. But Darwin began to be ill for long periods, and in 1842 he moved his family to a comfortable eighteen-acre estate near the village of Down about twenty-five miles south of London; there he lived the rest of his life. The Darwins traveled little, and then usually to a health spa.

Darwin's sickness persisted the rest of his life, and has been the source of much speculation. A convincing explanation is that he suffered from Chagas' disease. A carrier for the causative agent of the disease is the insect

3 Darwin is usually pictured as an old man with a striking resemblance to the white-bearded deities of Renaissance painters. This painting of Darwin portrays him at age 31, when he had been married one year, was the father of a new child, and making his mark on the world, as evidenced by his recent election to the Royal Society.

4 Sir Joseph Dalton Hooker (1817–1911) was Darwin's closest friend and advisor during most of the latter's adult life. A world traveler and explorer, he was an early advocate of natural selection and the first to know of Darwin's theory and a pivotal figure in the history of plant taxonomy. He incorporated the theory into his Introduction to the *Tasmanian Flora* which preceded the publication of the *Origin of Species*.

5 Darwin lived in a large, comfortable home near the village of Down in Kent, south of London, from 1842 until his death in 1882. Many of his experiments and all of his writing were done in this study. He wrote in the large leather chair with wheels. A board placed across the chair arms served as a desk. The house is open to the public and this room looks much the same now as it did during his lifetime.

Triatoma infestans; Darwin reported receiving numerous bites from this bug in South America. The disease causes heart damage and various other nasty ills; the symptoms of Darwin's illness, including the heart attack that finally killed him, match those of Chagas' disease.

Like many Victorians, the Darwins had a large family, six boys and four girls. Childbirth was a common occurrence in Down House, and Darwin was active in the birth of his son Francis, who became a distinguished plant physiologist. Even among the affluent, however, mortality in childhood was high then. The Darwins lost an infant girl and a two-year-old boy, and the first-born daughter, to whom Darwin was particularly attached, died at ten. The Darwin children were a distinguished bunch, and the line to this day continues to produce scientists.

Darwin started as a geologist, but he ended his life as a botanist. During his last seven years, he published five books on plants. One of these, *Different Forms of Flowers on Plants of the Same Species,* grew from earlier papers on the implications of the different size combinations of pistils and stamens in the primrose. In spite of his insights in so many fields, he said about his study of this humble plant: "I do not think anything in my scientific life has given me so much satisfaction as making out the meaning of the structure of these plants."

Darwin was frail and in his seventy-third year when he spent part of April 18, 1882, examining a plant brought to his study. He read as usual that evening, but the next afternoon he died. Pallbearers at the funeral included his friends Hooker, Huxley, and Wallace, two dukes, and James Russell Lowell, American minister to Great Britain. He was buried in Westminster Abbey a dozen feet from Newton, who had also changed our view of the universe.

CHAPTER 15

Plants and the Environment

One hundred fifty years ago, and in some places in the United States much more recently, lumber companies were clear-cutting their way through our northern forests, felling trees with abandon and clogging rivers and lakes with logs and debris. Square miles of earth where mature forests had stood were left barren and the environment was drastically altered. While all this was going on it was forgotten that destroying a forest removes the home that belonged not just to trees but to the whole network of forest life. By removing trees and converting entire stands of timber into lumber for buildings or pulpwood for newsprint, birds, squirrels, deer, wolves, moose, bear, insects, and crowds of smaller animals were made homeless. Changes in climate in and around the cleared acres were risked, because forests transpire enormous quantities of water.

ECOSYSTEMS

All those plants and animals, and uncountable microscopic beings, are the forest **ecosystem,** an organization in which a group of organisms interact with one another and with the environment around them. *Ecosystem* derives from **ecology,** a word coined about a century ago from two Greek words meaning "study of the home." The word has been fashionable, if not always understood, just since the environmental movement found its voice in the sixties. People everywhere began looking at the natural world around them

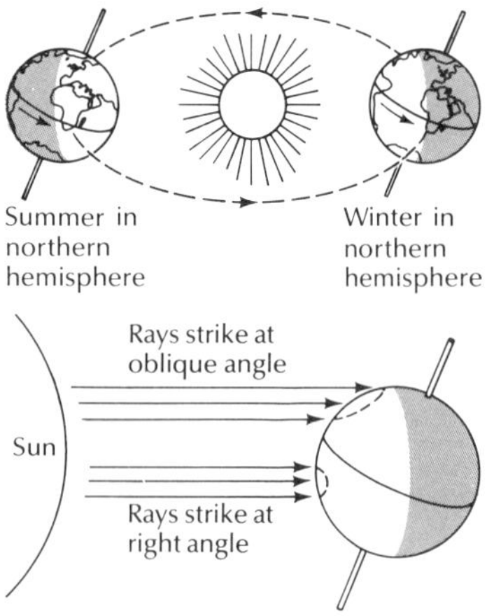

15.1 The earth is tilted at a 23.5-degree angle, so the hemispheres alternate in their distance from the sun as the earth rotates around the sun. The angles of the sun's rays differ and this creates winter and summer in temperate regions. The climate determines the type of vegetation.

and seeing that it is a coexisting community of living and nonliving things, not just a place for one species—humans—to exploit.

"Home" in the ecological sense is not just a place of food and shelter; it is any plant or animal, the place in which it lives, and everything it needs to make its living. Animal populations, plant populations, soil, and climate are all part of it. A mere drop of water may be home for an amoeba; for a timber wolf, home may take in many square miles over which to search for game. Humans are the foremost example of animals that can make a home just about anywhere by adapting the environment to their needs.

Interactions between plants and environment have come into our story many times. We have seen and analyzed the main things plants require for growth—and discussed their ability to adapt to environments in which these are limited. Now we will consider the world's major plant communities and the forces that determine their structure. What adaptations are characteristic of plants in these communities, and how have human actions affected an area's vegetation? In short, we will look into some "homes" and see the plants that live in them.

Communities A front lawn in suburbia may have a **population** (a group of plants of the same kind) of clover plants and another of crabgrass. Together with the other plant populations in that lawn (assorted weeds, and flowers from seeds blown from a nearby garden), these are the lawn **community.** The living (biotic) and nonliving (abiotic) components of the lawn, along with other parts of the environment impinging on or touching it (such as sun, rain, wind, animals) compose an ecosystem. This ecosystem may include things and events well beyond the lawn's edge, perhaps trees and people and garbage and dogs—all the elements that affect a lawn's life.

All these are interdependent and interact with each other, an idea fundamental to the concept of ecology (and one that makes it a difficult subject). If, by introducing new elements or subtracting others, this interaction is inhibited or destroyed, then the ecosystem will be damaged, or it may break down completely. Parts will die; other will move away; in any event, it will not be the same as it was.

Elements of Climate The first and most powerful determinant of an ecosystem is climate, which basically determines what plants live where. Temperature and water are the most important factors characterizing an area's climate.

Three very broad and familiar kinds of climate are tropic, temperate, and arctic. The tropics are strongly and continuously bathed by the sun's rays; temperate regions have large amounts of sunlight during specific parts of the year, alternating with periods of weak

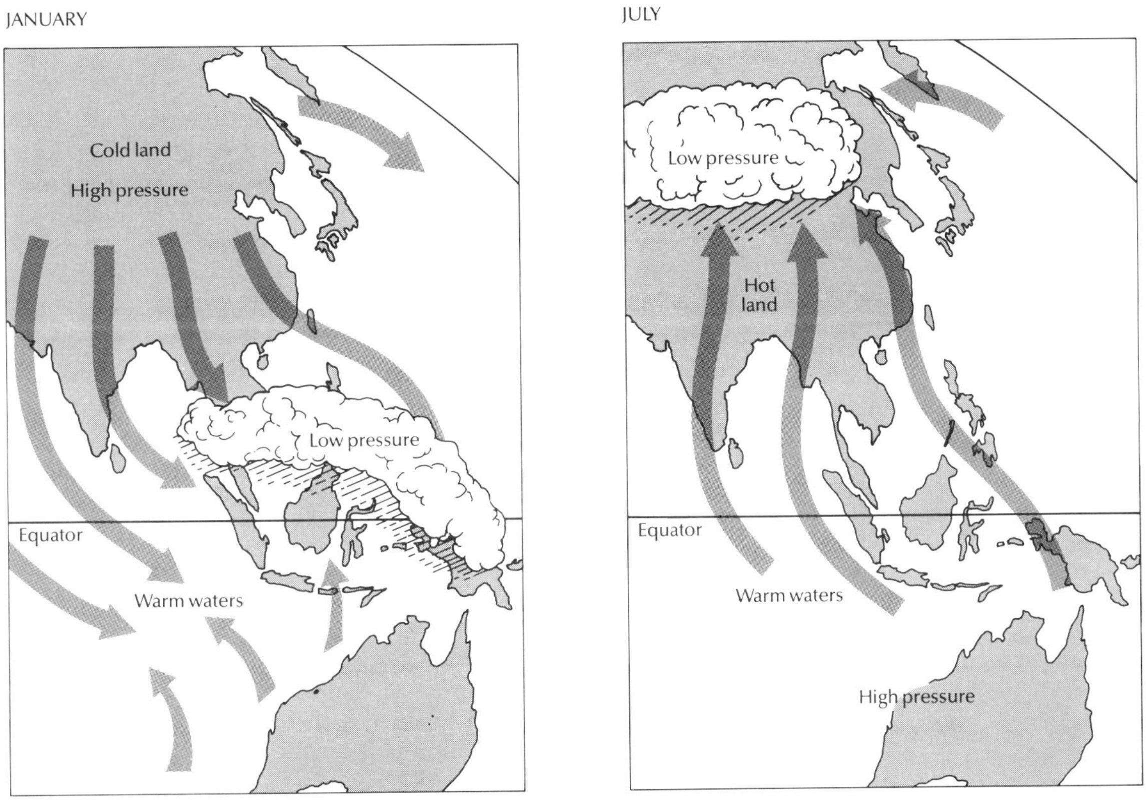

15.2 The monsoon season characteristic of Southeast Asia is the result of a shift from a high-pressure region to a low-pressure region over the land. With the appearance of the low-pressure region, warm, moisture-bearing winds from the ocean sweep over the land and bring abundant rainfall.

light and short days; arctic regions get very little sunlight during part of the year (see Figure 15.1).

If a region has a life-supporting temperature (considerable periods of time when the surface layers of soil are not frozen), then water becomes the primary factor in the construction of a plant community, because water is so crucial in plant growth and transport of materials, as well as for basic metabolism. East Asia, India, and the vast number of islands in the East Indies have patterns of rainfall strikingly different from those in our part of the world: **monsoons.** These are seasonal winds that carry moist air from oceans in one direction, bringing with it a great deal of rainfall, for weeks or even months. Vegetation in these regions depends on the monsoons, which alternate with a dry season that may bring drought. Life in this large region thus depends on seasonal moisture. In culture too we see effects of the monsoons. Failure of the monsoon

15.3 This eighteenth-century miniature painting depicts the wife of Vishnu, a Hindu god, welcoming the arrival of the monsoon. These rains awaken the earth and ensure good crops. Their absence can mean starvation. Weather, plants, and culture are interwoven in monsoon regions.

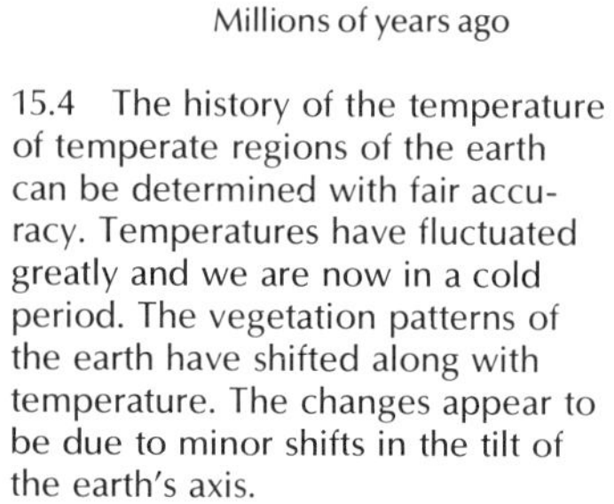

15.4 The history of the temperature of temperate regions of the earth can be determined with fair accuracy. Temperatures have fluctuated greatly and we are now in a cold period. The vegetation patterns of the earth have shifted along with temperature. The changes appear to be due to minor shifts in the tilt of the earth's axis.

means crop failure and starvation. Such a dramatic event always influences religion, art, and customs of a region (see Figure 15.3).

Looking at levels of precipitation across the United States, we see easily the effect of water on plant communities, especially the American grasslands. As rainfall increases from the Rocky Mountains to the Mississippi River, the native grasses are more abundant and larger. This is dramatically illustrated by looking out an airplane window on a clear day as you fly east from the West Coast. The moisture gradient is reflected in the color change as you fly east: the sandpile brown of the dry West gradually changes to light and then dark green as you approach the Mississippi River. The green is a direct reflection of the high precipitation of the eastern United States.

By comparing plants on north- and south-facing slopes, we see the influence of both temperature and moisture. North-facing slopes are cooler and wetter, while south-facing slopes are warmer and drier. The respective slopes often support different plant communities. Moisture in the air forms clouds, greatly affecting earth's climate and therefore ecosystems. When warm moist ocean air is drawn upward into the cooler high altitudes, moisture condenses, forms clouds, and, under certain conditions, falls as rain or snow. At the extreme northwest corner of the United States, the Olympic mountain range deflects the warm moisture-laden oceanic air upward; from the resulting precipitation a rain forest has formed.

By forming clouds, moisture also influences the amount of sunlight reaching the land, and this in turn regulates evaporation. Temperature too is regulated by moisture. Large bodies of water moderate temperature; along the shores of oceans or even large lakes, temperatures tend to be less extreme. A continent's interior is likely to have extremes of heat and cold because no large bodies of water are there to moderate temperatures. Large forests, with their high rates of transpiration, release so much moisture from their leaves that they directly affect the local climate. The amount of moisture in the air influences temperature and air temperature influences cloud formation and precipitation.

Climate and the History of Plant Distribution Although we label climates tropical, temperate, and arctic, names that imply permanence, climate is not constant—it is always changing. Even though the yearly changes may be too slight for us to notice at the time, from the long view of biological history the effects of changes are quite clear. Good evidence now shows that earth has gone through periodic changes in climate and that these are caused by small changes in the earth's orbit (see Figure 15.4).

We can understand how plants have evolved and been distributed in many areas only by realizing that earth's continents are not

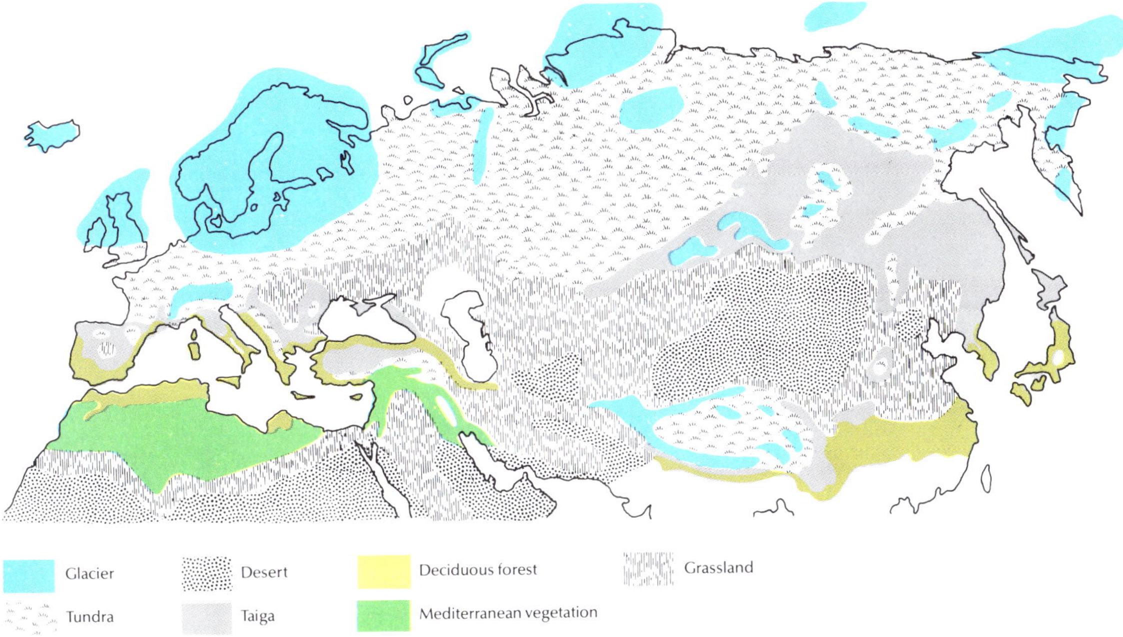

15.5 During the Pleistocene epoch, which began one million years ago, the distribution of plant communities over Eurasia was much different than it is today. Europe was covered by tundra and taiga, and the shores of the Mediterranean were covered with forest.

static (see page 460) and that the world is still recovering from a rather recent Ice Age. During the Pleistocene, the greater part of the temperate zones was covered with glaciers, which shifted current vegetation zones from the higher latitudes of the arctic to lower latitudes. Our temperate vegetation zones are only some ten thousand years old, much younger than the tropical zones that did not go through an Ice Age. Our northern temperate areas are to some degree still in the Ice Age, whose remnants are very long cold winters in such places as Maine, Minnesota, North Dakota, Wyoming, and Montana. Harsh winters and short, almost tropical summers give plants here some very special problems. Much the same thing can be seen in Europe, where, until quite recently, most of the area north of the Alps was tundra (northern parts of Scandinavia and the Soviet Union still are).

The end of the Ice Age coincided with the first attempts to domesticate many of our economic plants. Wild wheat, barley, peas, beans, and olives moved into the region of the Zagros-Taurus mountains of the Middle East (see Chapter 2) about eleven thousand years ago, after glaciers had retreated and a mild Mediterranean climate developed, with winter rains and summer droughts. The development of this characteristic plant community was traced by dating the pollen in sediments from lake bottoms. Glaciers on the move had scoured the land clean; retreating, they left hollows scooped in the bedrock, which filled with water and became lakes.

15.6 RAINFALL

Water is the most important determinant of plant growth. In addition to its direct effect on plant growth and metabolism, water increases the erosion of rocks, releases minerals, and is essential to decomposers which increase the cycling of nutrients. The relationship between mean annual temperature and precipitation and distribution of major biomes is depicted in the graph (below). Water supply, not temperature, determines distribution of desert and grassland, while temperature is the primary limiting factor in distribution of tropical forest. Deciduous forest has a broad tolerance for both temperature and moisture. The effect of water is clearly illustrated by the direct correlation between growth of grasses and rainfall on the midcontinent slope extending from Denver, nestling beneath the Rocky Mountains, to the banks of the Mississippi River at St. Louis (see facing page, bottom). The lower Colorado Basin (facing page, top left) is a desert, receiving no water during most years. Occasional cloudbursts will flood the desert and it will spring to life for a short period. The resultant seeds will lie dormant until the next brief period of rain. The Great Basin of Nevada and Utah (facing page, top right) receives about 10 inches of water per year. It is characterized by sagebrush and other drought-tolerant plants. The east coast of the United States from New Jersey to Florida (facing page, lower left photo) has 35 inches of rain each year, but nutrient-poor sandy soil supports sparse pine forest rather than the luxuriant growth which this amount of rainfall could support. The Olympic Penninsula of western Washington (facing page, lower right photo) receives up to 100 inches of rain each year and is the wettest place in the continent. It produces a temperate rain forest made up of mainly conifers like the Douglas fir. The effects of moisture and temperature on structure of plant communities can often be seen on adjacent north-facing and south-facing slopes. The two banks of the Tuolumne River in the foothills of the Sierra Nevada in California provide an example. The north-facing slope is cooler and wetter and supports lush tree growth (below, right). The south-facing slope a few yards away supports shrubs and small trees (below, left).

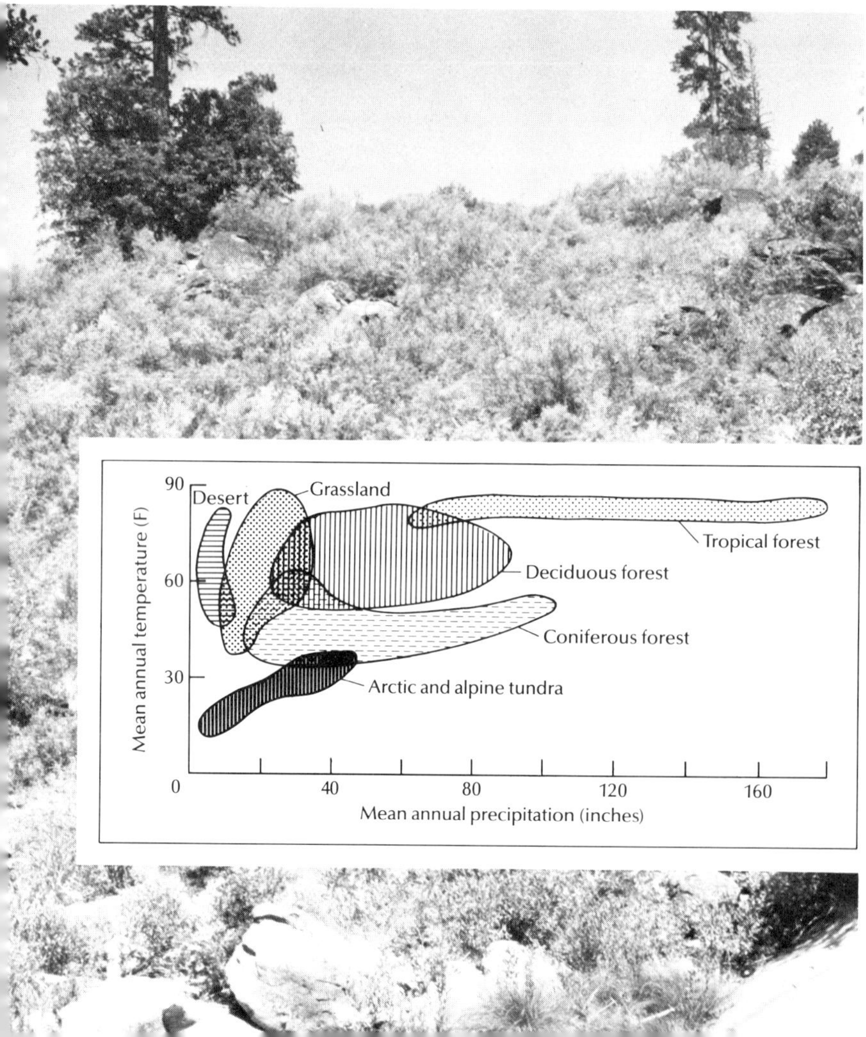

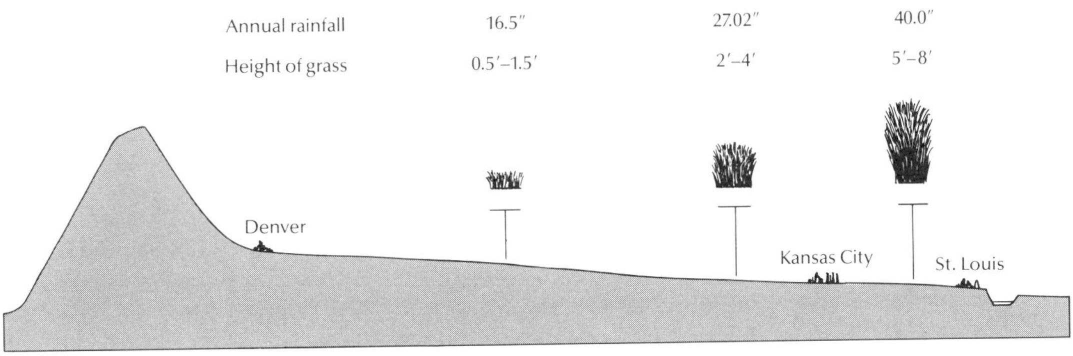
Annual rainfall
16.5″
27.02″
40.0″
Height of grass
0.5′–1.5′
2′–4′
5′–8′
Denver
Kansas City
St. Louis

15.7 BIOMES OF THE WORLD

The biomes of the world (see map below) are defined by major climax plant communities. Their distribution is determined by temperature, moisture, and availability of essential minerals. Tundra (facing page, bottom right) is the biome of arctic regions. A short growing season, shallow soil, and long periods of extremely low temperatures restrict plants to small, shallow-rooted forms. Annuals must have short life cycles and perennials must be resistant to extreme cold. South of the tundra is the taiga (tīga), the belt of mostly coniferous forest of northern temperate climes (see facing page, bottom left). Grasslands occur in semi-arid regions in both temperate and tropic zones. The Great Plains of the United States, in the lee of the Rocky Mountains, are one of the world's major grasslands. Deserts (facing page, top left) occupy about one-fifth of the globe's surface. They are found at sea level and on mountains, in temperate and tropic zones, but always where the surface water supply is low. Desert plant life is sparse but diverse and often spectacular.

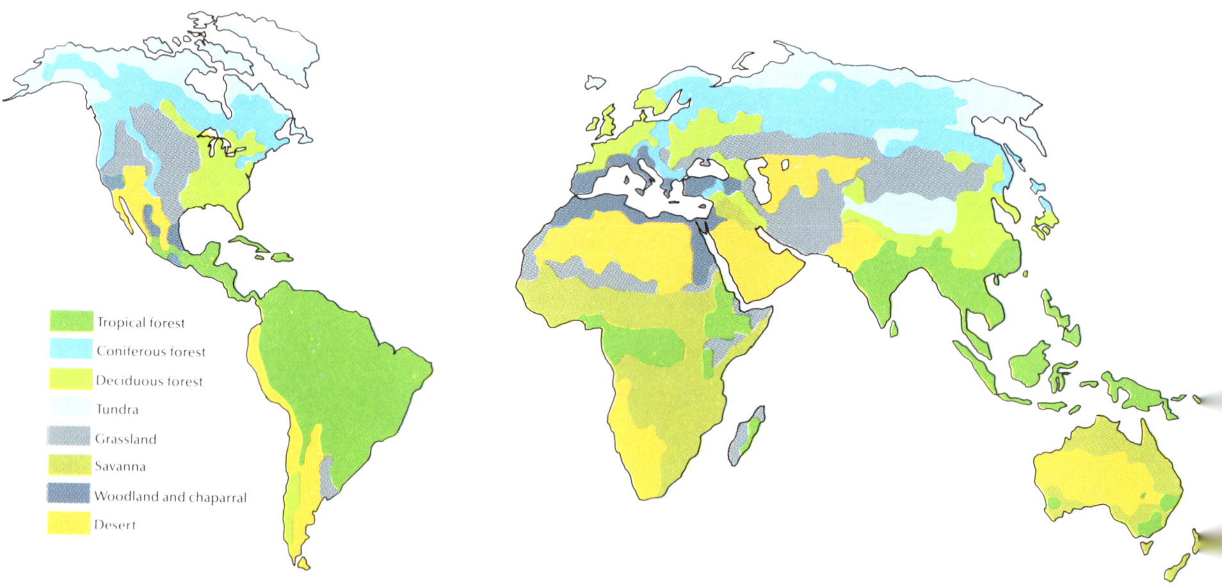

Pollen from plants that took hold in the new soil drifted into the lakes, settled to the bottom, and became fossilized in the mud. Sometimes the lakes filled up completely with sediment and became bogs. A core taken from such a bog shows layers of fossil pollen, a guide for drawing an accurate picture of the plants that grew in that region.

As the climate moderated following the Ice Age, human populations moved into the Zagros-Taurus area. Fossilized remains show that these people domesticated wild grains and olives, taking seeds with them on their seasonal wanderings. Other types of plants, notably vegetables, were cultivated later, and this mountainous area became one of the first, possibly the earliest, densely populated areas.

As glaciers retreated, tundra grew up along their edges, forests followed the tundra, and these were followed by cultivated lands. Within the last ten to fifteen thousand years, all this has been witnessed by human eyes, and cultivated plant communities have been part of the European temperate zone for a very long time. Though change is intrinsic in the biosphere, built in by slowly changing climate and geological events, humanity has hastened change in a good many places by felling trees, clearing fields, and planting crops.

But nothing we have done can compare with the changes wrought by immense natural events—glaciation, movement of earth's crust (see the Supplement to this chapter), and volcanic activity, which builds mountains. Today's distribution of plant communities throughout the world is recent, and it will change. Compress the history of plant distribution into a movie and the present will appear as only one frame of that film; everything we see in the world today is here no longer than the blink of an eye.

Life Zones A particular combination of rainfall, temperature, and soil will produce a specific life zone, a **biome.** Specifically, **biome** refers to the group of organisms restricted by the environment to a particular climatic zone. Climate and soil structure and composition determine which plants can live in an area, and the plants determine which animals live where. Biomes are therefore classified by their dominant plant communities (see Table 15.1). (The distribution patterns of organisms throughout the world can also be classified by the zoological realm, dividing the world into areas according to similarities of animals; classifications of distribution patterns, just like those of diversity, must be considered as relative things.)

No biome is exclusive to one locality; it has its counterparts all over the world. The Soviet Union's coniferous forest biome in Siberia is the same biome as in similar latitudes of North America.

Table 15.1 Major Biomes of the World

Biome	Distinctive characteristics	Representative plant life
AQUATIC		
Ocean	Temperatures range from cold to tropical, ecology varies with climate and depth	Algae
Shore	Subject to a wide variety of environmental changes	Algae, grasses, seaweed
Swamp, bog, marsh	Subject to tidal fluctuations if marine	Algae, cattails, sedges, mosses, mangroves
TERRESTRIAL		
Chaparral	Heavy winter rains, long hot, dry summers	Shrubs, small evergreens and deciduous trees, grasses
Coniferous forest (Taiga)	Subarctic; short cool summers, long cold winters with much snow	Birch trees, evergreens
Deciduous forest	Temperate; marked seasonal cycles, sufficient precipitation	Deciduous trees, oak, maple, beech, flowering herbs
Desert	High temperatures, aridity	Cacti, sagebrush, succulents, shrubs
Grassland	Cool winters, hot dry summers, water supply low	Tall, medium, and short grasses, asters, sunflowers, goldenrod, herbs
Savanna	Tropical; low seasonal rainfall	Grasses, scattered small broad-leaf evergreens and deciduous trees
Tropical forest	High temperatures, abundant rainfall, almost continuous growing season	Epiphytes, vines, broad-leaf trees (evergreen and deciduous), bamboo
Tundra	Arctic; extreme cold, permafrost, short summers, brief growing season	Grasses, lichens, sedges, dwarfed woody plants, shrubs, mosses

Examples of the same biome in widely separated areas, however, are usually made up of different species of plants and animals. In the California foothills there is a community of tough shrubs and small trees and thorny bushes, called **chaparral** (shown in Figure 15.8), and in the hills of Chile and Turkey there is a very similar community. The species in each community are different, but the morphology of the communities is similar: the shrubs and trees have similar shapes and, viewed from a distance, they would look alike, because they are created by similar climates and soils. Widely separated copies of the same biome thus provide a fine example of convergent evolution (page 404): because environmental conditions are the same, plants have faced similar selection pressures and have adapted in similar ways.

15.8 Regions with similar climates tend to have plant communities with a similar morphology. Areas with a Mediterranean climate, with cool wet winters and dry hot summers, have a plant community with many shrubs and small trees. In California and Mexico, it is called *chaparral* (right); and in Turkey and other areas bordering the Mediterranean, it is called *maquis* (left). While these plant communities appear to be similar, they consist of different taxa.

Examples of Biomes Tundra is an arctic biome, an environment in which extreme cold restricts life, and usually grasses, lichens, sedges, and dwarfed woody plants are in the community. The tundra has its seasons, but summers are so short that only the upper layers of soil melt. Beneath is a permanently frozen layer, the **permafrost.** Superficially, tundra is much like bogs, which are found in many parts of the world. But whereas bogs may occur in thick forestlands and form when a shallow lake gradually fills up, true tundra begins at the forest's edge where the climate dictates that trees can no longer grow, as along the shores of Hudson Bay in Canada. A train trip to the town of Churchill in Manitoba takes the traveler hour after hour through thick forest, until at last trees fade out and tundra begins: a vast, rolling landscape of low vegetation that bursts into bloom for only a few short weeks in summer.

Another biome is the **taiga** (tīga), a subarctic coniferous forest. Birch trees (though they are not conifers) are an important subgroup in forests throughout the north woods of the United States and Canada as well as in the European and Siberian taiga. Some native Americans built their birchbark canoes from the ever-present paper birch trees *(Betula papyrifera),* and the French voyageurs who carried on the fur trade throughout the north woods in the seventeenth and eighteenth centuries did the same.

Biomes are distributed not only laterally across the world but also vertically; an increase in latitude is often mirrored by an increase in altitude. Climb a tall mountain in a subarctic region of North or South America, such as the White Mountains of New Hampshire, and the forest will fade out as you get near the top. Then comes the timberline, beyond which no trees grow, and beyond the timberline is tundra, the same sort of tundra biome that lies far to the north near sea level along Hudson Bay and beyond. The birches so plentiful in the cold subarctic coniferous forest are characteristic too of the mountain's coniferous forest.

Other biomes are these: the deciduous forest, found mostly in temperate regions with marked seasonal cycles and familiar to every midwesterner or New Englander, with its flaming autumn colors and leafless winter branches; the tropical forest, found where temperatures are high and rainfall abundant; the grasslands, such as the American great plains, where trees are sparse and the water supply fairly low; the savanna, grasslands where there are scattered trees; the desert, where temperature and aridity are both very high; and finally the boundary areas between biomes. The biomes merge with no clearly defined border, making it difficult to say just where tundra, taiga, or savanna ends and another biome begins. Boundary areas, however, do have distinctive plant communities, and animals from adjacent major biomes wander in and out of them.

FOREST BIOMES

Two of the major biomes are forest and desert, which are also at opposite ends of a moisture gradient. Because of the area of the globe they take up and because we find distinctive plant and animal adaptations in them, forests are worth a close look. Figure 15.9 shows how forests are distributed all over the world today, but biologically speaking, the map is not quite accurate. It actually

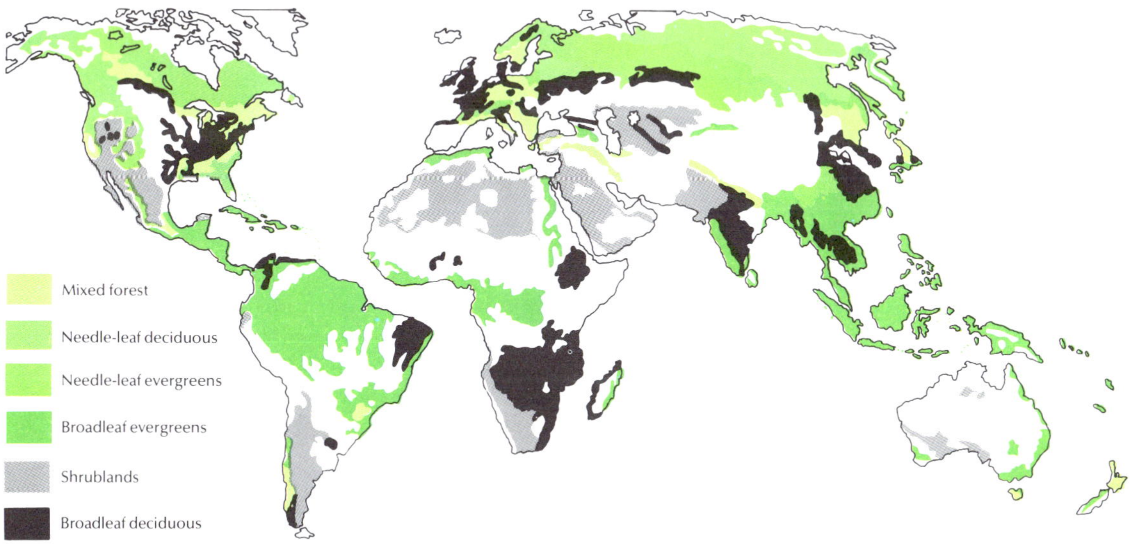

15.9 This map depicts the type of forest that each region of the world is capable of supporting.

shows all the places where, with their climates and soils, forests could be if humans had caused no major disturbances. As it is, many of the great forests of Asia, Europe, and America have been cut down and replaced by croplands. All that remains of the forests in regions of China, India, Europe, and North America is a mere relic of what once lived there.

15.10 A tropical rain forest has several layers of vegetation, each one adapted to a different amount of light. The third layer forms a canopy which screens out much of the light, but a few very tall trees break through this canopy and stand like sentinels above it.

Adaptations in the Forest Tropical forests have the greatest extent and diversity of any terrestrial biome because they have remained undisturbed far longer than forests at higher latitudes and have an almost continuous growing season. The structure of a tropical rain forest is very complex. Many more niches for living are available than in the same space in a temperate zone forest, and these are spread out not only laterally but also vertically, right up to the tree crowns (see Figure 15.10). Each level has its distinctive plant and animal community, making life here varied to a degree never found in the California redwood forest.

A critical limiting factor for growth in the tropical forest is light. Because foliage of tree crowns is so thick and growth of vines and other plants at many levels is so prolific, the light near to and on the forest floor is dim, with sunlight rarely shining through. There are, therefore, many plant adaptations minimizing the effect of the lack of light.

In the lower layers of a tropical forest, plants must be efficient light trappers if they are to receive enough for photosynthesis. Plants in the understories have large thin leaves, which present the maximum surface for light reception, and there is evidence that their photosynthetic pigments are particularly efficient in trapping light.

Some plants have another adaptation for gathering light in the tropical forest: living up closer to the light. You may spot a problem at once: the higher they climb, the farther they are from the soil and water needed for nourishment.

One solution is for plants to divorce themselves entirely from the ground. Such plants living among the upper stories are **epiphytes,** or "air plants," and have no connection with the soil. They trap water that runs off the trees and filters down from leaf to leaf. At the bases of their own leaves, many form cups to store rainwater, like miniponds. Small animals make their homes in these cups, where bits of debris from leaves and the bark of the trees themselves are also captured. As materials decompose in the microaquatic system of the cups, they provide nutrients. The ephiphytes also have specialized roots that attach to the host trees and absorb water and nutrients from cracks in the bark.

There are many different types of epiphytes with their own distinctive ways of making a living high in the trees. The tank bromeliads, relatives of the pineapple, have large flaring "cups" to catch water; these are also homes and breeding places for insects. Another bromeliad is the misnamed but well-known Spanish moss *(Tillandsia usneoides)* of the American deep South; it is neither Spanish nor a moss but a flowering plant with a highly specialized system for obtaining water. Its rootless tendrils have tiny scalelike hairs that form very efficient water traps. Rainwater is caught between the hairs and the outermost cells of the shoot, which absorb it. The plant gets not only moisture but also any mineral nutrients that wash down from the host tree.

Vines are another adaptation for living in the upper layers of the tropical forest. They climb trees by coiling around a trunk or sending out tendrils that give support while the main branch grows upward. Some vines have thorns that dig into the bark of the host tree. Others, like the vine-like tree strangler fig *(Ficus)*, deposit seeds that germinate high on the branches; their stemlike roots grow to the ground and eventually enclose the supporting tree.

Perhaps the most interesting of the jungle vines are the prolific lianas, the kind on which Tarzan swung through the trees in Edgar Rice Burroughs' fictional adventure stories. Lianas reach the top of the forest canopy and keep right on growing, reaching over to another tree and sometimes forming long loops that stretch almost to the ground. A tree that falls and drags a mass of lianas with it makes an almost impenetrable tangle on the forest floor. The liana, like as not, will start to climb again on another tree.

The large trees in the topmost parts of the forest canopy can be thought of as "plant racks," with almost every space in the canopy occupied by some plant. A separate world of life is up there, with humus in which seeds germinate after being excreted by birds, water reservoirs formed by epiphytes, and a great diversity of animal life. By comparison, the forest floor may seem almost deserted.

The canopy atop a mature tropical forest is closed almost like a roof, letting too little light through to support much plant growth on the forest floor. The true topical forest is not a tangled mass of undergrowth, as so many authors of idealized jungle stories would have us believe. There are places where passage is most difficult, but only where a tree has fallen, bringing down masses of leaves, branches, and vines and leaving a window in the canopy through which the sun shines on the plants lower down. A canopy also darkens the forest floor in mature deciduous forests of North America. Vistas at ground level suggest a cathedral's interior. Settlers moving into the Ohio Valley after the Revolutionary War could drive their teams with ease for long distances beneath the closed canopies of the beech-maple forests covering the region like a blanket. But far more familiar today is the forest that grew after logging. The new growth has not yet reached maturity, the canopy is still jagged, underbrush of all kinds clutters the forest floor, and walking through it is difficult.

15.11 Epiphytes, like these bromeliads (left), and vine-like trees such as the strangler fig (right) demonstrate two adaptations for living high above the ground. Epiphytes obtain water and nutrients without having roots in the soil, while vines are supported by trees of the canopy.

Because light is scarce beneath the forest canopy, seedling trees face a tough battle for growth, and so long as the canopy is intact, most lose that battle. Any area opened to the light will have exuberant new growth, but in the severe competition for living space, only a few trees reach maturity and close the canopy once again.

Ample moisture, nutrients, and warm temperatures support growth in forests, but trees do not all make their living in the same manner. Their size gives these often gigantic structures very special problems in engineering and plumbing. Variations in size and shape of leaf and trunk are almost endless; less well known, particularly in temperate regions, is the variation in root structure and function (see Figure 15.12). Among the most interesting examples of the latter are the mangrove *(Rhizophora mangle)* and the swamp cypress *(Taxodium distichum).* These trees stand in water and mud much of the time, where roots can get little oxygen. The trees have erect kneelike root structures that rise above the water or are exposed at low tide, and in these are air channels through which oxygen can diffuse to the submerged roots.

Diversity and Stability Walking from a forest into a recently plowed but unplanted field, you would notice the great contrast in diversity between the two areas. The old forest has many kinds of plants, but the recently disturbed field might be thick with just one or two species of weeds. Given time, the field will become a

15.12 Roots exhibit special adaptations for many environments. The knees of the bald cypress (below, left) are really air ducts for roots growing in soil or water low in oxygen. Roots of epiphytes (left) anchor the plant to supports and absorb water and minerals on the surface of the supports. Tangled masses of adventitious roots grow into every crevice in the rocks and firmly anchor these shoreline trees (below, right).

forest. Think of the great temperate coniferous forests as recently disturbed fields, poor in number of species but rich in numbers within the few species. There are often large areas that come close to resembling a cornfield, with one kind of tree dominant. Like the weedy field, these forests occupy land recently plowed—by glaciers. The trees can be thought of as weeds, but unlike the field, these forests will become more diverse only when the climate again changes.

A tropical forest shows the same contrast in diversity with the temperate forest as did the temperate forest with the field. So many species live in these tropical forests that you can walk a good distance before sighting two of the same. Diversity of species is great, but there are relatively few individuals of each species. Because of this diversity, food webs in these forests are extremely complex, providing the ecosystem with some stability. Although thousands of acres of temperate forests can be destroyed by just one insect pest, the tropical forest is not likely to suffer wholesale destruction, because members of each population are widely dispersed; a population of herbivores will have difficulty finding enough individuals of any plant species to threaten that species' survival. (Pests rarely reach epidemic proportions, because no insect pest can easily escape pressure from its predators.)

Of course, the tropical rain forest is prone to all kinds of disturbances, among them hurricanes and landslides, but it recovers from these fairly quickly, just as temperate forests do. The complexity of the tropical forest provides a great deal of stability, but it can be seriously damaged by major disturbances. There are some indications that clear-cutting large areas of rain forest leaves scars that take a very long time to mend; it can expose the thin soil layer to massive erosion by heavy tropical rains. The loss of soil can permanently alter the ecosystem and prevent regrowth of the forest. Temperate forests, on the other hand, like a field of weeds, are easy to destroy but, like the weeds, they quickly grow back. Many temperate forests that look like pristine wild areas were clear-cut a hundred years ago.

One limitation on forest growth is clearly seen at a mountain timberline, beyond which trees cannot grow. Just below the timberline, trees are dwarfed and twisted by climatic forces. The timberline is a function not of altitude but of frozen soil and freezing temperatures. In the tropics the timberline may be around 4,200 meters (14,000 feet). Beyond that line grows tundralike vegetation: hardy grasses, mosses, and dwarfed shrubs. Still higher are mainly lichens, those remarkable composite plants that are part algae and part fungus and that subsist on bare rock (Chapter 4). Because their water is severely limited at low temperatures, many mountain plants have developed morphological similarities to desert plants, including leaves protected against evaporation by waxy surfaces and far-reaching root systems.

DESERT BIOMES

In the forest, competition for light is so strong that it often is limiting. In deserts, light, temperature, and usually nutrients are not limiting, but water is. Oases in the desert have long been centers of civilization. They differ from the surrounding barrenness only in availability of water. We see this in our new oases in the Southwest, which are centers of population. Approaching these from the air you can see fertile, irrigated areas set off from the dry surroundings as sharply as in an architect's drawing. On one side water is available, and there you see green fields, lawns, trees; on the other side, without water, desert stretches beyond vision (see Figure 15.13). If a way of bringing more water to the vast deserts were found, many of our problems with food would be alleviated.

Most deserts are not devoid of plants, nor are they totally waterless. In fact, the plant life is often rich in species, if sparse in number of plants. Plants of the desert are frequently remarkably similar in morphology and physiology, even though unrelated taxonomically. This is another example of convergent evolution, discussed in the previous chapter: similar conditions of climate and soil tend to select similar plant characteristics. (Figure 15.15 shows the location of today's deserts and factors causing them.)

15.13 The oasis is an island of plant life in the ocean of desert. If provided with water, the desert will support luxuriant plant growth.

15.14 ADAPTATIONS TO ALTITUDE

We are accustomed to think of biomes changing as latitude changes, but the most dramatic survey of biomes is obtained when ascending high mountains like the Rockies. The influence of temperature and moisture is demonstrated over short distances. The diagram of biomes at different altitudes and on north- and south-facing slopes in the Rocky Mountains shows that a particular biome is determined by temperature and moisture rather than altitude. The south-facing slope is hotter and drier, so desert and grassland extend higher and timberline is higher than on north-facing slopes. Changes in plant communities reflected by increases in altitude parallel those due to increases in latitude (below). It follows from this relationship that biomes of higher latitudes are found at higher altitudes in the tropics than in temperate zones. The six-foot lobelia (facing page, right) grows above 12,000 feet on Mount Kenya in East Africa, an altitude where a few plants would grow in temperate zones. This plant looks very much like a desert plant because of its specialized leaves. High altitudes, like deserts, often have little surface water and extreme temperatures. Alpine regions are similar to deserts in possessing a short growing season, so annual plants have a similar morphology. The Himalayas, the world's loftiest mountains, are a laboratory for adaptations of plants to altitude since they range from sea level to over 20,000 feet. A member of the compositae, *Anaphalis* (facing page, top left), lives above 17,000 feet, while gentians (facing page, second photo from top, left) are found from 13,000 to 14,000 feet. A giant member of the arum family grows in the 5000 to 9000 feet zone (facing page, third photo from top, left). Timberline (facing page, bottom left) occurs at different altitudes depending upon latitude. It can range from 6000 feet in the White Mountains of New Hampshire to 12,000 feet in the more southerly Colorado Rockies.

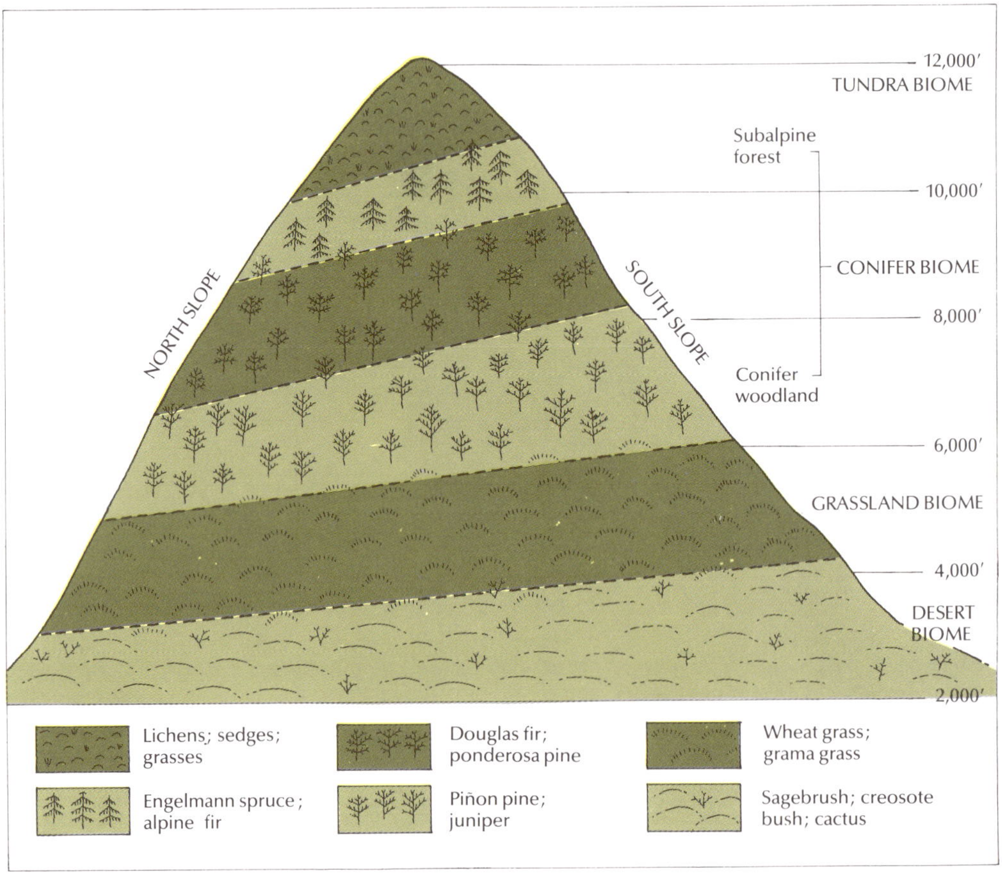

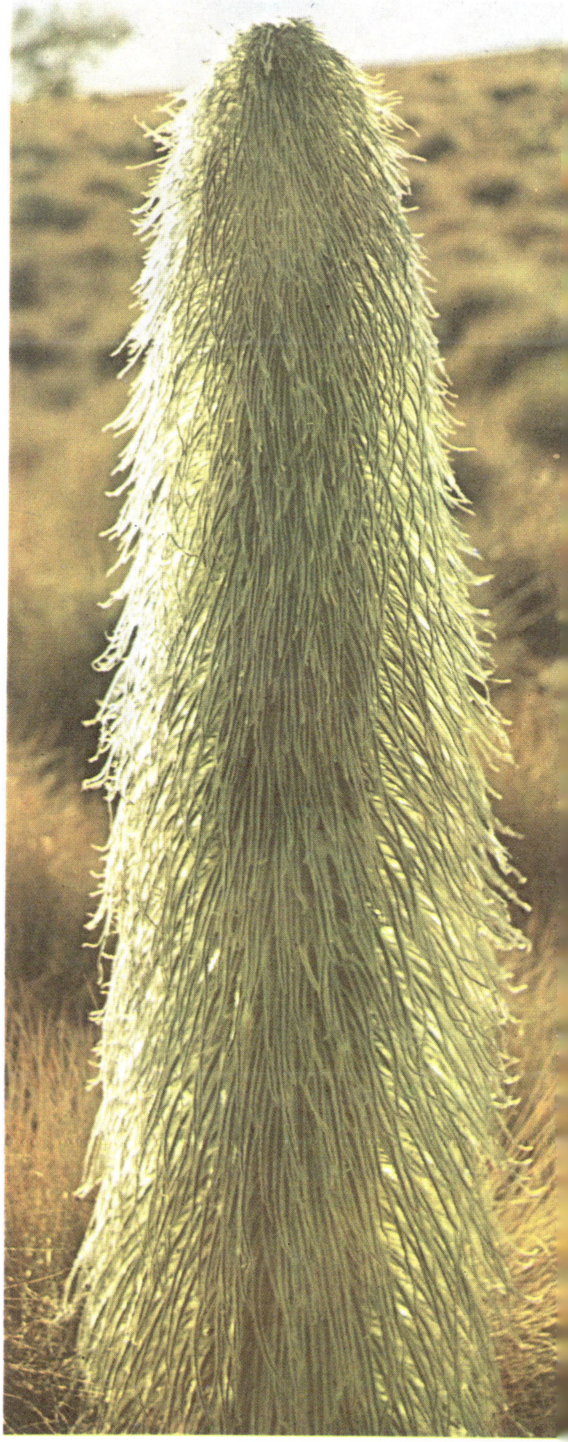

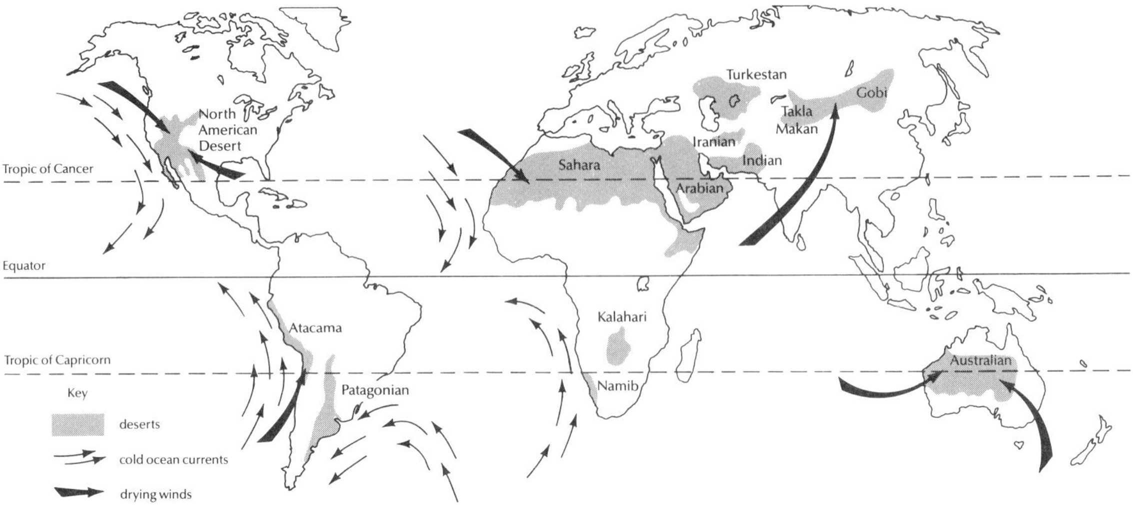

15.15 The world's deserts are usually associated with the Tropic of Cancer and the Tropic of Capricorn. The winds reaching these regions are dry, either because they have been cooled by cold ocean currents or because they have lost their moisture passing over mountains or large continental areas.

Desert-like Environments In regions other than deserts water is also limiting for plant growth, as along seashores. Plenty of water flows in and around a salt marsh washed regularly by the tide, but the water is salty, making the osmotic balance unfavorable. (Remember that osmosis works by diffusion of water through a membrane impermeable to salts until there is a balance of solutes in the two solutions.) Seaside plants washed by salt water must maintain an osmotic balance.

High altitudes or latitudes are also regions where water is often limiting. Cold air holds less moisture than warm air, and this dry air influences the microhabitat of the plant. Think how dry your skin is in the winter. Even when water is available, low temperatures may cause plants to suffer from physiological drought. (Remember that water uptake requires energy for salt intake.) Plants retaining leaves in winter, like conifers, require a good deal of water. But the water is difficult to take up because it is frozen, and the plant's metabolism is slowed and its supply of energy is depleted because of the effect of short, cloudy, and cold days on photosynthesis. Perennial plants may die in the wintertime because they lack water. In a sense, plants live in a desert during the winter.

During part of the year, many semiarid regions have conditions as extreme as any desert. Some places around the Mediterranean have desert conditions for six months a year. (Consider Morocco and Tunisia and, opposite them, the dry coastal areas of Spain, France, and Italy.) This Mediterranean climate is also found in California, southwestern Australia, Chile, and South Africa. At times there is plenty of water, but in summer the climate is almost as dry, if not as hot, as the Sahara. Plants that grow and reproduce in these areas have many of the same problems as desert plants, and so they have evolved many similar adaptations.

Adapting to Aridity The desert is not completely without water. When rain does come, it often comes suddenly and so abundantly that some deserts even have floods. The events that follow this rainfall are astonishing: the earth seems to spring to life and turns into a flower garden. The moisture does not last and the plants do not long outlive it, but their adaptations to drought are fascinating. For example, some of the desert's annual flowering plants have seeds containing chemicals that inhibit germination. These seeds lie dormant in the desert soil for a long time, even through periods of transient rainfall. The amount of water required to wash out the inhibiting chemicals seems nicely balanced with the amount that will sustain the growth of the plant through its short life cycle. The germination inhibitors act like moisture meters and prevent the seed from germinating until the soil has enough water to ensure the plant's survival.

Other morphological adaptations enable plants of the desert to get along with a minimal supply of water. Some have long roots to tap underground water, reaching down to the water table. They are so reliable as indicators of subsurface water that people long ago learned to dig wells where they grow. The mesquite *(Prosopis fuliflora),* the cottonwood tree *(Populus deltoides),* and the California fan palm *(Washington filifera)* are all indicators of water at different depths below the surface. These plants take advantage of an available water supply and are not really adapted to arid conditions.

An adaptation specifically for storing water is succulence; a thick fleshy tissue in which cells absorb and hold water like a sponge. A great many of the cacti are succulents, with thick barrel-shaped trunks and heavy fleshy branches. They can store a great deal of water, and the thick cuticle covering stems and branches prevents the stored water from evaporating. During the desert's infrequent rainfalls, succulents accumulate water which sustains them during extended dry periods.

Desert plants that have neither roots penetrating deep into the soil nor succulent stems and leaves rely on far-ranging root systems to suck up the last bit of moisture. One is the creosote bush *(Larrea divaricata),* familiar in our western desert landscape. One of the most extraordinary plants, however, is the American pygmy cedar *(Peucaphylum),* which can live with no water from the ground at all. It lives entirely on water vapor gathered from the air. At night, when even the tropical desert turns surprisingly cold, lowered air temperature condenses water as dew, which is the water supply for these plants growing on exposed rock. Other plants, for unknown reasons, can recover from extreme desiccation.

Many desert plants have surfaces that reflect sunlight and decrease heat absorption. These reflective surfaces are sometimes produced by elongated epidermal hairs or by a layer of wax produced by epidermal cells. The hairy surfaces have another advan-

15.16 ADAPTATIONS TO DESERTS

When the desert receives water, it blooms. These sand verbenas (the genus *Abronia* in the four o'clock family, Nyctaginaceae) (facing page, right) in the Borrego Desert in southern California are a carpet of color interspersed with the drought-resistant creosote bushes. Cacti (facing page, top left) are associated with deserts of the Western Hemisphere. Their succulent tissues store water and their shape and thick cuticles restrict water loss. The saguaro (facing page, middle left) stores water in its stem. In a drought, its trunk becomes slim but after rain, the trunk is plump. Many arid zone plants in addition to cacti, such as the blue kleinia (facing page, bottom left), are succulents. Their thick leaves provide a low surface-to-volume ratio and lessen evaporation of water. The blue color is due to a coating of wax on the leaves. This waterproofing to keep water in is also characteristic of xerophytes. The root systems of desert plants (below) vary from wide-ranging surface roots, which soak up water from even light showers, to the deep roots of the mesquite. The latter can tap water as deep as 100 feet below the surface.

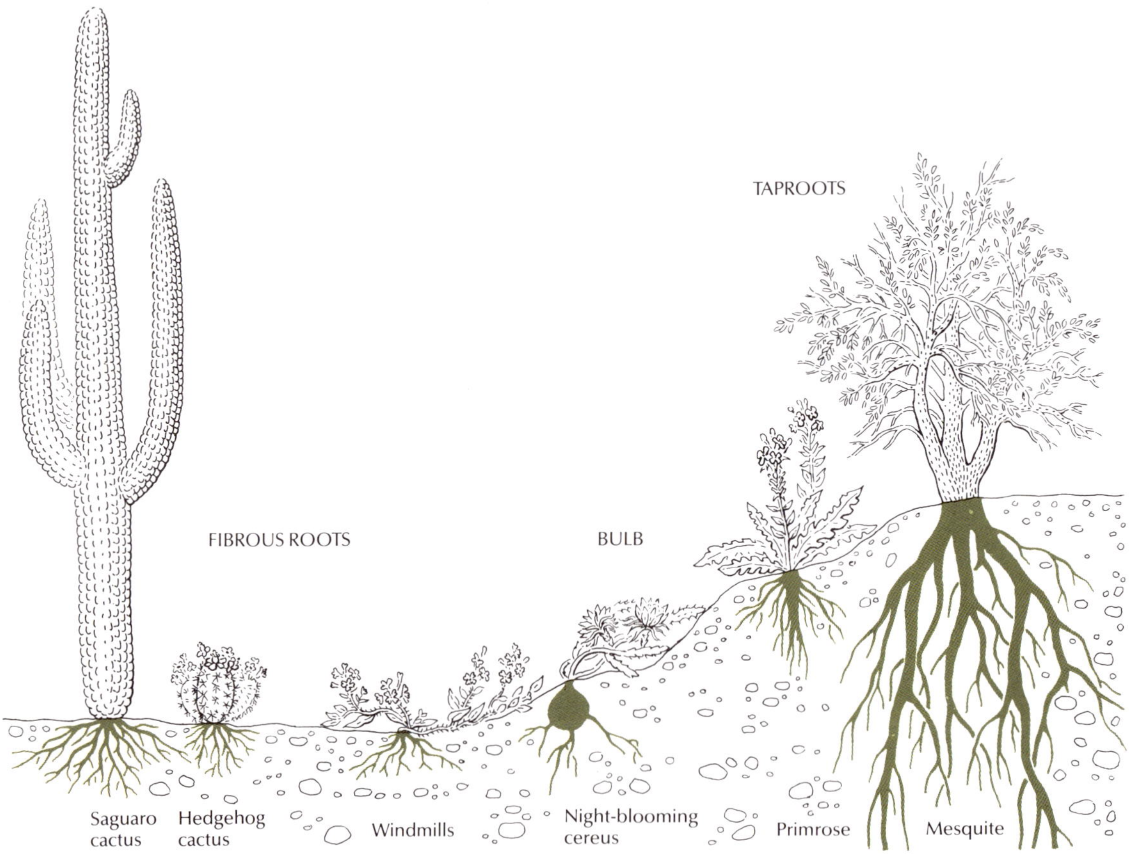

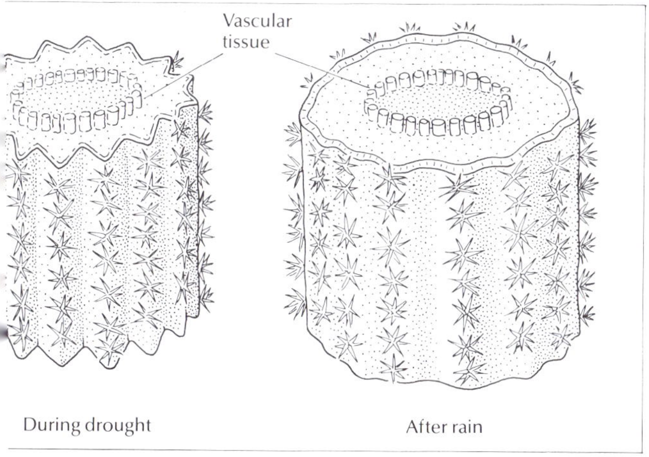

Vascular
tissue
During drought
After rain

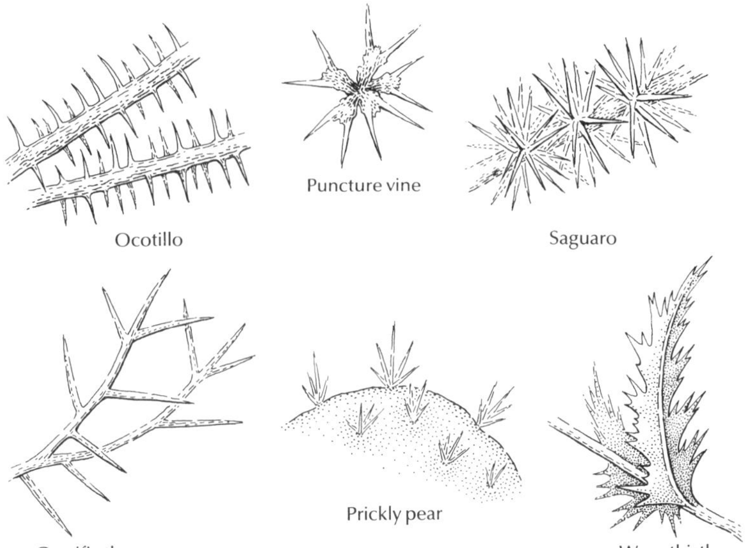

15.17 Many desert plants feature a bristling armament in the form of thorns and spines. These structures can be derived from branches, parts of leaves, or, as in the thistle, be the sharp edges of the leaf itself. Juicy plant tissue is particularly attractive to desert herbivores, so these plants erect the equivalent of their own fences.

tage: they not only decrease evaporation but act as sponges for the dew that collects on the plants at night. They introduce between the plant's surface and the air a boundary with its own microclimate.

Many of the larger cacti have somewhat the same ecological function in the desert as trees do in the forest. They provide homes for desert animals, and their succulent stems and leaves hold water and food for animals. Cacti have sharp spines; these are modified leaves that eliminate leaf surfaces, an adaptation for aridity, and effectively protect against large herbivores. These spines can be quite complex structures and are very diverse in form (see Figure 15.17).

Another protective device is found in the creosote bush and in sagebrush *(Artemisia tridentata),* as well as many other desert plants. Compounds made by the plant coat the leaves and make them unpalatable for animals. Some desert plants, such as members of the Euphorbiaceae, even have sap that is poisonous to animals.

People and Deserts Deserts interest the botanist because of the extreme conditions under which plants live and the adaptations they have made to a dry climate, but deserts are pretty much a loss for the world's supply of food and fiber. Sadly, too, deserts are increasing in some parts of the world, particularly the Sahara Desert in Africa and the Rajasthan Desert in India, both regions that badly need new sources of food. Deforestation and overgrazing cause most of this expansion, just as overcultivation in semiarid regions produced the great American dustbowl during the depression years.

15.18 Crumbled remnants of the ancient city of Palmyra tell the oft-told tale of civilization's temporary conquest of the desert. An important oasis on the caravan route to Damascus, this prosperous city in the Syrian desert fell into ruin after the Roman Emperor Aurelian deposed the famous Queen Zenobia in 273 A.D. Will cities in our own Southwest suffer Palmyra's fate?

In the United States the desert ecosystems (and others) are being damaged by new and unforeseen consumer products of the technological age: the dune buggy and the cross-country motorcycle. These and other off-road vehicles, which can go through almost any kind of terrain at great speed, have made distant desert places all too easily accessible, and the damage they and their riders inflict on the fragile desert ecosystem takes a very long time to repair—longer than the life spans of their riders. A speeding dune buggy can damage fatally a desert shrub that took 2,000 years to grow, about as long as a redwood or bristlecone pine, which, because they are rare and more spectacular, are more easily protected by law.

Even those who love plants can help to destroy the desert flora. Some of the strangest and rarest plants grow in deserts, but these unfortunately have no adaptations to protect them from human collectors. They are great favorites with dealers and plant lovers who seek them out to sell or bring home; these people have been responsible for much destruction in the desert plant communities of the American Southwest and Mexico.

One place in which the desert has been pushed back with some success for human benefit is Israel. This small country has done battle with the desert, building irrigation projects that have created green and fruitful fields where little could grow before. Similar

15.19 SUCCESSION

After an area is disturbed, a series of plant communities will succeed each other until a climax community characteristic of that biome is established. Plant succession on an abandoned farm in the southeastern United States begins with grasses and herbs and progresses to shrubs, pine forest, and finally a mature hardwood forest (see below). Succession of plants is followed by succession of animals. The diversity and density of bird species, for example, multiply many times during this evolution from abandoned field to hardwood forest. During the early stages of succession, part of gross production goes into building extra resources for the ecosystem. A *climax community* may be defined as an energetically efficient ecosystem, because it cycles its resources. It maintains itself—and no more than this (see facing page, bottom). Mountain lakes and meadows (facing page, top left) illustrate the dynamic nature of plant succession. A lake turns into a bog and then a meadow. The meadow will eventually turn into a forest—you can see the red maples beginning to grow (facing page, left, second photo from top). Even a mountain ledge (facing page, top right) can illustrate plant succession. Lichens and moss crumble the rock and provide humus. Grasses and herbs grow and pay their dues to soil construction. After many years, there is enough soil to nurture germination and growth of a pine seedling. Its roots penetrate rock crevices, widening fissures, forcing new cracks, and building soil. In time, and with the aid of wind, water, and ice, plants will wear down and conquer the mountain.

Community type:	grassland	shrubs	pine forest	hardwood forest
Time in years:	1–10	10–25	25–100	100+
Number of common species of birds:	2	8	15	19
Density (pairs per 100 acres):	27	123	113	233

SUCCESSIONAL COMMUNITY

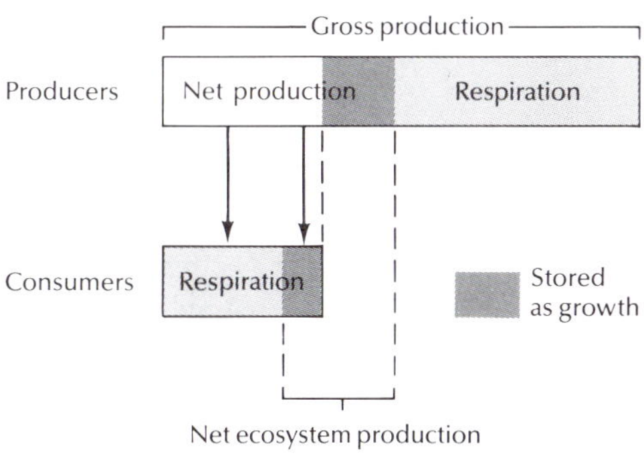

CLIMAX COMMUNITY

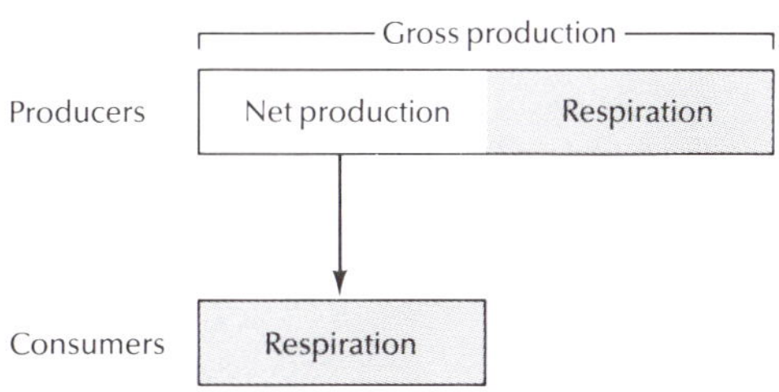

projects are under way elsewhere, as in the Imperial Valley of California, which supplies the country with most of its winter vegetables; before irrigation it was a sandy desert. Whether human technology or the desert will win over the long haul remains to be seen. History tells us that intensive cultivation has rarely succeeded in arid regions for long periods. We can see this problem in Arizona, a rapidly growing state. This growth is being subsidized by pumping water from natural underground reservoirs. Once this water is depleted, the current agriculture and population growth cannot be sustained.

PLANT SUCCESSION

A new volcanic island just risen from the sea tells us much about succession of plants in an ecosystem. Left unmolested, such an island will exhibit one group of plant and animal communities after another. Finally, if the environment is stable, a **climax community** will maintain itself and succession will cease. At this stage, the plants are so well adapted to their environment (which they have helped create) that plants and environment are in balance.

With spores that can travel far on the wind, lichens or mosses often are the first plant forms to appear on a new, barren rock. In time these can break down the rock and create humus. The lichen or moss population is followed by ferns and grasses, their seeds brought by birds. One population succeeds another until the climax plant community establishes itself and the pattern of diversity stabilizes. North America and Europe went through such a succession after the last Ice Age, when retreating glaciers left the land scoured clean of plant life. Stages in plant succession will also take place on a plowed field or in a new or lifeless pond.

Any biome will pass through well-defined stages in succession, each stage with its characteristic plants. Even the biomes themselves can be considered stages in succession regulated by climate. The deciduous forests now in North America went through a tundra and taiga stage as the climate warmed after the Ice Ages. If the climate of the eastern United States continued to warm until it was summer the year around, the deciduous forests would give way to tropical forests.

Succession is particularly clear in mountainous regions. A mountain lake gradually fills with plant debris and silt, turning into a marsh. The marsh slowly accumulates more soil and plants until it is a meadow. In the next stage the meadow becomes a forest. All these stages can be seen in mountainous areas, where today's lake is tomorrow's meadow, and today's meadow tomorrow's forest. The lakes, the marshes, and the meadows may look permanent, but each is a transient stage. The only way to keep a meadow as a meadow is to get rid of invading trees, as a farmer well knows.

The stage from meadow or field to forest, intensively studied by ecologists, is easy to observe in many parts of the United States, where marginal farmland has been abandoned by farmers or is being allowed to revert to forest. Walk through almost any New England woodland and deep in the woods you may stumble on stone walls, some as solid as they were a century ago when they marked off a farmer's pastureland. Cellar holes of crumbled homes are scattered through the woods on lands that early settlers worked so hard to clear. Until this century began, western Pennsylvania was covered with carefully tended farms, where deer had not been seen for many years. Today it is mostly forest, and deer hunting is a popular pastime in season. In Berkeley, California, the hills above the campus of the University of California only fifty years ago were pastureland covered with grass. Today, as a result of plant succession, they are covered with dense chaparral.

Cycling of Energy and Nutrients As succession advances, the trophic levels, the levels at which energy is produced and food consumed (Chapter 1), usually increase, and food webs in the ecosystem grow more complex. In a terrestrial example starting with bare rock, lichens or mosses are the first stage of succession, dissolving rocks and rock particles to form, together with their own organic matter, soil in which the next succession of plants can take hold. In an arctic climate plant succession may go no further. In a temperate climate the lichen-moss stage will be followed by annuals, shrubs, and finally a forest. Each successive stage has more forms of plant and animal life and a more complex food web.

Looked at from the point of view of energy supply, the early stages of succession represent an open system. New plants, herbivorous animals, and predators move in and out of the surrounding areas, bringing energy in with them and leaving some in the form of seeds, dung, and carcasses. The plants take hold. Animals also remove energy by eating new growth and leaving the area, and nutrients are lost as rain and runoff leach them away. But as succession proceeds, the ecosystem is increasingly closed: matter is recycled and energy is transferred more and more within the ecosystem itself.

A good example of this change is nitrogen fixation, which requires a lot of energy (Chapter 9). In the early stages of succession, as plants reestablish themselves, nitrogen fixation is common. There is a large input of energy during these early stages to build up and maintain a store of nitrogen in the soil. But as the climax vegetation is reestablished, less nitrogen is lost to the ecosystem because it is efficiently recycled. Recycling nitrogen takes less energy than fixing it, so the climax community is less energy-intensive and better adapted. Because some nitrogen is always lost, the system cannot be entirely closed, but it is highly efficient. Perhaps this system has a lesson for human society.

15.20 This yellow pine forest in Oregon is recovering from a fire. The burnt trees stand like skeletons among shrubs and young trees. The plant communities in the burned area go through several successional stages, climaxed by the pine trees as the dominant species in the community.

Fire Lightning starts many fires, and early peoples, recognizing the importance of fire in managing ecosystems, regularly burned large sections of countryside, thereby controlling plant succession. By burning back the forest, the Indians extended the prairie, thus winning more grazing ground for buffalo and deer, their principal source of livelihood. In California they burned chaparral to increase grazing land for deer and maintained oak woodland by regular burnings, which helped to ensure a regular supply of acorns, a major source of food. They kept the floor of Yosemite Valley parklike with grass and oaks. Now such management is gone and succession is turning the valley into a coniferous forest.

Chaparral is a fire-adapted plant community: many of its species burn easily and regenerate rapidly by sprouting from the stumps of burned plants. For an arid environment this adaptation is beneficial. Grassland, too, is well adapted to fire, and human efforts to protect chaparral and grassland from fire often do more harm than good.

Whenever plants in climax and advanced successional stages burn, succession starts again and proceeds toward a climax stage. In some communities, however, the climax stage might not be reached. Most often chaparral, for example, is not a climax vegetation. If it were left alone long enough, other plant communities would succeed it, but they rarely have a chance because it catches fire so easily and reestablishes itself so quickly after the fire without going through all the successional stages. You might say that chaparral is a plant community adapted for keeping itself in office.

We worry a great deal about fire, and rightly so, because it can cause property damage and wastes valuable timber. But is fire always bad? Forests often recover quickly from a burn, and the fire itself gives benefits: it kills pests and releases nutrients to the soil. In an area that has no fires for a long time, accumulated fuel stokes a very hot fire when it finally comes. High temperatures damage the soil physically and chemically and kill plants, seeds, and roots. Recovery then takes a very long time, because all the forest plants have to be reintroduced; and when plant cover is gone for a long time, erosion may remove much soil, greatly increasing the time it takes to reestablish the forest.

In a burned-over area, animal succession follows succession of plants. Among the first to arrive are insects and birds, each helping to reestablish plants, bringing in seeds from other areas. Animals like rabbits are followed by their predators, such as foxes. Seeds in animal droppings sprout and grow in the enriched soil. As shrubs and young trees take hold, deer will come, and so, as the forest slowly reclaims the land through the stages of succession, the animal population too will be reestablished.

MANAGING THE LAND

In their pursuit of food for a growing population, people have traditionally taken over management of land in those regions where plants would grow without severe limits. To put it differently, land that is arable, capable of producing crops, has generally been taken over by farmers. Within the last century, as the human population has increased wildly, more and more of the environment has come under their managerial hands and more is to come.

15.21 Germany's Black Forest, home of cuckoo clocks and the scene of hikers getting close to nature, is really a big tree farm. Once a deciduous hardwood forest, it is now composed of conifers. Many of the latter, like the Douglas fir of western North America, are introduced species.

Queen Anne's lace
(wild carrots)

Prickly pear
cactus

15.22 Many of our common and most noxious weeds were introduced from Europe and Asia. A good example is the wild carrot, often called Queen Anne's Lace. In contrast, the prickly pear cactus, native to the Western Hemisphere, has spread to Europe and Asia, where it is often a pest.

For centuries much of the world's best cropland has been cultivated, and flocks of sheep, cattle, and other domesticated animals have grazed on grassland and savannas since the beginning of history. More recently, temperate forests have been turned into "tree farms," and tropical forests that only a century ago covered the equatorial regions have been drastically reduced to make room for more crops.

Without looking far you can see a fast-spreading, new, human-induced successional stage. Expanding populations first stimulated the increase in cultivated land, and the same pressure is now destroying the very cropland upon which it depends for food. Many acres of our best farmland have been transformered into cities and highways. California alone removed 5 to 6 percent of its cultivated land from 1969 to 1974, much of it now replaced by buildings and roads. By 1987, 5 percent of the state's present agricultural land will be lost to urbanization. This loss is significant to the nation, because California produces 25 percent of the table food and 40 percent of the vegetables consumed in the United States. Other states are also cutting deep into their stock of arable land. Farmland near urban areas is often assessed for taxes at its potential (developed) worth, encouraging farmers to sell out to developers rather than pay stiff taxes. The need to keep land in cultivation is raising complex political questions, but people have begun to take into account the fact that society depends on crops that grow in the ground.

Human control over the environment need not be harmful, although it often is. In many parts of the world, land has long been intensively cultivated without degrading it or fouling the waterways. The United States had its dust bowl in the thirties because the land was overused in a thoughtless way, but today much of the area is again fertile soil, producing good crops and forage for animals because of good management. Indians turned forests into prairies with fire, and in these prairies we have substituted wheat and corn for native grasses. There is no intrinsic reason why a field of wheat need be any more or less good than an equal area of grass or trees.

Introduced Plants Human influence has been just as great but less obvious in parts of the world other than our own Midwest and parts of the Southeast, as in Germany's Black Forest. Superficially, it seems just another area of heavily forested hills; in fact, human influence has been predominant here for many centuries. A closer look finds a great many Douglas firs, whose native habitat is the Pacific coast of North America. For all its rugged look, the Black Forest is cultivated almost as closely as a field of corn.

The firs were imported to Germany by lumber companies, as plants are often moved around from one part of the world to another. Weeds as well as crop plants travel with humans. Many of the plants we see every day in lawns, gardens, vacant lots, and sidewalk cracks are weeds, and most of our weeds are immigrants.

The introduced plant can become so much a part of the local landscape that people think it is native. The prickly pear cactus *(Opuntia spp.)* is from the deserts of the New World, but in Europe, Asia, and Australia, where it is often a noxious weed, many people think of it as a native. In the same way, Californians have come to think that the *Eucalyptus* trees covering so many hills have always been there, not realizing that this Australian native has been around for only a century. Even California's golden hills are not truly Californian: the grass that covers many of them is a wild oat grass from the Mediterranean region.

Mismanagement Where human managers have done things badly, people and their animals have created havoc with the environment. China was deforested as Chinese philosophers preached oneness with nature. The teachings of Hindu philosophy have not kept Indian peasants from cutting trees for short-term profit, and long before Christianity, great portions of the Middle East were denuded of plant cover by grazing sheep and goats. Once

15.23 Placer mining near Idaho City, Idaho, destroyed vegetation and severely eroded the soil. Trees quickly became established in pockets of soil, helping to check further erosion. In time, new soil will be formed, and the scars of erosion will be covered by new forest.

the plant cover, whether trees or grass, is stripped away, the soil is bared to wind and rain and is swept away; the American dust bowl of the 1930s, celebrated in John Steinbeck's *Grapes of Wrath*, dramatically demonstrated this sequence. When topsoil is destroyed, semiarid regions can become true deserts.

The most efficient denuders of the land have been goats, axes, and plows. When Julius Caesar invaded Gaul, Europe north of the Alps was covered with forests, and his legions began transforming it. When the Pilgrims landed in America, a deciduous forest stretched from the Atlantic Ocean to the Great Plains. A squirrel, some said, could travel all the way from the Atlantic coast to the Mississippi without touching the ground. The axe swept all away, and when immigration reached the Pacific coast, the same destruction was repeated there. Deforestation was followed by erosion of the soil and silting up of rivers; rapid runoff from storms and meltwater increased floods. Fortunately, this trend has been somewhat reversed and forests are being allowed to grow back in many parts of the country. Soil loss continues, however, and soil managers are increasingly concerned with this problem.

In many parts of the world, however, the lesson is not yet learned. One of the great unpublicized environmental problems is deforestation of hills and mountains in Asia, Latin America, and Africa. Because effective birth control is not practiced, there are tremendous population pressures, and the watershed forests are being recklessly cut down for fuel and to clear land for cultivation. Then eroding soil and unobstructed runoff of water defeat the purpose behind cutting the forests and cause flooding in the valleys. Because such matters are all too often disregarded in short-range planning, humans create deserts where they meant to create cultivable land.

Some blame modern technology as the only current villain in our negative influence on the land, yet the richly successful civilizations of the ancient Middle East and North Africa fell victim to the dust and dryness of human-created deserts. Even peoples close to the soil make the same mistake, like the Masai tribe of Africa, who traditionally subsist on cow's milk mixed with cattle blood. Modern technology has helped their population expand, and now their huge cattle herds are overgrazing their lands, breaking down the natural ecosystem and leaving the Masai increasingly impoverished.

We will manage more land, sometimes effectively and other times disastrously. So it has been through history. But in the long run our disturbances of the ecosystem, like the glaciers that crept down from the poles during the Ice Ages, will be transitory. Plants will outlast people, as they predate them, and they will not be denied. Gradually obliterating human devastation in successional stages like those described here, they will reestablish climax communities. One thing is certain: as long as our earth has life, plants will provide its base.

Summary

Recent years have seen an increasing interest in ecology, the "study of the home." Home here means not just a place but everything an organism needs to make a living. The organization in which organisms live and interact with each other and the environment around them is an ecosystem. Thus an ecosystem includes both biotic and abiotic elements and all factors, even some extending beyond its physical borders, that affect it. A group of organisms of the same kind is known as a population, and various populations will form a community.

Climate is the foremost determinant of an ecosystem, and temperature and water supply are the most important characteristics of climate. A climate may be generally described as tropical, temperate, or arctic.

The effects of moisture are dramatically seen in regions that depend on monsoons—seasonal winds carrying moist air from oceans in one direction, bringing much rain for weeks or months. Vegetation in East Asia, India, and many island of the East Indies depends on these seasonal rains, which alternate with dry periods.

Temperature and moisture affect each other. Large bodies of water moderate temperatures on land, and air temperature influences cloud formation and precipitation. Clouds in turn affect how much sunlight reaches the ground and, hence, the rate of evaporation. The interdependent effects of temperature and moisture can be seen in the formation of rain forests and by comparing vegetation on north-facing slopes with that on warmer and drier south-facing slopes.

There is good evidence that earth has gone through periodic climatic changes caused by small changes in its orbit. Today's temperate zones are only about 10,000 years old and are still recovering from the effects of Pleistocene glaciers. As these glaciers retreated at the end of the last ice age, tundra grew, and then forests. With the spread of domesticated plants, many of these forests were eventually replaced by cultivated fields.

The effects of rainfall, temperature, and soil combine to produce specific life zones, biomes; that is, these environmental conditions together determine which plants can live in an area, and plants determine where animals can live. Biomes are classified by their dominant plant communities. In tundra, for example, one finds extreme cold, short summers, and usually grasses, lichen, sedges, and dwarfed woody plants. Other prominent biomes include taiga, deciduous forests, tropical forests, savannas, grasslands, and deserts.

Examples of the same biome are found in widely separated places, distributed both laterally and vertically. They are usually made up of different species of plants and animals, but these communities will have similar shapes because they have been affected by similar climates and soils. Widely separated examples of the same biome thus will often illustrate convergent evolution.

Tropical forests have more diverse organisms than any other terrestrial biome; they have remained undisturbed longer and have a longer growing season. Many more niches are available here than in a temperate forest, and these are distributed both laterally and vertically. The result is a complex ecosystem.

Because of the tropical forest's closed canopy, light is a crucial limiting factor. At lower levels of the forest, large thin leaves are a frequent adaptation. Other adaptations for obtaining light are illustrated by plants living higher up in the forest; epiphytes ("air plants") and vines are examples.

The ecosystem in a tropical forest can recover rather quickly from most disturbances. There is evidence, however, that clear-cutting large areas can result in long-lasting damage. Temperate forests, in contrast, can be easily destroyed but they grow back quickly. They have less diversity than tropical forests but usually more representatives of each species.

The timberline illustrates one limitation on forest growth; tundra vegetation but not trees grow beyond this line. It is a function of freezing temperatures and frozen soil, which limit the water and minerals that plants can obtain. As a result vegetation beyond the timberline shows adaptations similar to those of desert plants, including waxy surfaces and far-reaching roots.

Light, temperature, and nutrients are usually adequate in the desert, but water, though present, is limiting. Water can be limiting in other regions—seashores and salt marshes for example. Many plants live in a kind of desert in winter, because low temperatures can create the equivalent of a drought, and some places around the Mediterranean and elsewhere have desert like conditions during half of the year.

Adaptations to living with a minimal supply of water include long roots reaching down to the water table, succulence (thick fleshy leaves or stems for maximum absorption and storage of water), far-ranging roots, hairs or a layer of wax reflecting sunlight or decreasing heat absorption, and spines (which are modified leaves with less surface area). Some flowering annuals of the desert have chemicals in their seeds that inhibit germination. The amount of water required to wash out the inhibiting chemicals seems nicely balanced with the amount the plant will need during its brief life.

Although plant life in deserts is diverse, it is sparse and does not contribute significantly to our food supply. Deserts are unfortunately spreading in some parts of the world, usually because of deforestation and overgrazing. In Israel and California deserts have been pushed back, but intensive cultivation of arid regions has rarely succeeded for long. Meanwhile some of the unique plant life of our deserts is being endangered by plant collectors and recreational vehicles.

Plant communities within biomes are not static. Every biome passes through well-defined stages, which have characteristic plants, and biomes themselves can be considered stages in succession. The climax community is the group of populations existing when plants and environment are in balance. One population will succeed another until the climax community is established, unless fire or another disturbance intervenes. As succession advances, tropic levels (the levels at which energy is produced and food consumed) increase, food webs become more complex, and plant and animal life becomes more diverse.

Lichen and moss are often first to appear in a previously barren area; ferns and grasses are likely to come next. In the early stages of succession the systems of energy and matter are open; that is, energy and matter are being both brought into the system and taken out. Later the ecosystem is increasingly closed as matter is recycled and energy transferred more and more within the ecosystem itself.

When plants in advanced stages of succession burn, succession begins again. Some plant communities such as chaparral and grasses are adapted to fire, burning easily and regenerating quickly. Burning is one way, in fact, to manage land. Forests may recover quickly and reap some benefits from a fire. A very hot fire, however, will damage the soil and kill seeds and roots; recovery will then be very slow.

More and more of the environment has come under human management, and even more is likely to. People replaced forests with cultivated fields, and now are replacing fields with highways and shopping centers. People have greatly changed some areas by importing plants (the Douglas firs of Germany's Black Forest and golden grasses of California are examples). In Asia, Latin America, and Africa deforestation is a major problem, creating erosion, silting up rivers, and increasing floods.

Review Questions

1. What biome is characteristic of your locality? How would this biome change if the local climate became warmer or colder and wetter or drier?
2. Contrast selective forces in temperate and tropical forests and explain their effects on species diversity.
3. Fat plants do very well in hot deserts but fat people don't. Explain.
4. Long term intensive cultivation of semi-arid and arid regions has rarely succeeded. What are the reasons for this? Can our increasingly populated arid regions escape a similar fate?
5. What climax plant community is typical of your region? What is a climax community?
6. What is the relationship between efficient energy utilization and plant succession?
7. Prepare a new speech for Smokey the Bear in which safety and conservation are based on an understanding of forest ecology.
8. Population movement back into our cities may have an important impact on future food supplies. Explain.
9. Discuss examples of land mismanagement in the world and present some possible solutions.

SUPPLEMENT

Plant Distribution and Continental Drift

Many features of plant distribution have often baffled scholars. South America, for example, is half a world away from Australia; yet the two continents share several closely related plants and animals (notably marsupials) that are absent or poorly represented on other continents. How can the similarity be explained?

A German meteorologist, Alfred Wegener (1880–1930) proposed that the world's continents were, in effect, adrift on the viscous mass of the earth's interior, and that continents which are separated today might have been joined in the distant past. South America and Africa are an obvious example: the Brazilian bulge fits neatly into the Central African coast. It seems not too far-fetched to say that they might once have fitted together like two pieces in a jigsaw puzzle.

Wegener's theory of continental drift left a good deal to be desired; few scientists took it seriously. How could the continents move? But then in the 1960s, Wegener's theory was buttressed in quite convincing fashion. The concept of **plate tectonics** (dealing with forces involved in movement of the earth's crust) holds that there is indeed motion of the earth's crust. Huge plates of rock are buoyant relative to the denser matter of the earth's interior, and so they float, and are indeed in motion. They are propelled by the enormous forces of molten lava thrust up from deep within the earth's core through vent holes in its mantle. As the lava cools, it builds up *mid-ocean ridges* along the fault lines through which it was pushed upward (see Figure 1). Constant deposition of new lava forces the plates into slow, ponderous movement.

Earthquakes occur, according to plate tectonics, when two continental plates, butting against each other, create mammoth strains; these are relieved when the plates slide over or under one another or buckle upward, creating a force that heaves the crust up. Earthquake zones are the regions where these moving plates are in contact. The incredible forces in these continental shoving matches show in the 40–60 million-year-old collision between the plate bearing the Indian subcontinent and the Eurasian land mass. The displacement caused by the northward migration of India made the upthrust that created the Himalaya Mountains and subsidiary wrinkles

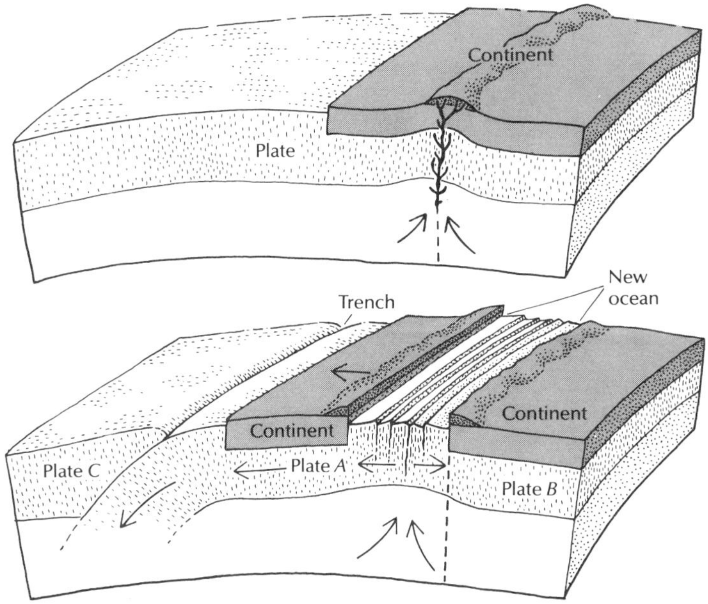

1 Continental drift can be explained by plate tectonics. When a rift appears in the crust or plate upon which continents rest, molten rock beneath the plate is extruded. The formation of a trench pulls plate *A* away from plate *B* at the same time as the extruded molten rock forms a new ocean bottom.

in China that became other high mountain ranges. The movement of these peaks continues.

The theory of plate tectonics has provided new interpretations of earth's history as well as explanations for plant distribution. During the Cretaceous Period about 100 million years ago, Australia, New Zealand, and Antarctica were joined to each other with South America, Africa, Madagascar, and India, all in one enormous land mass known as Gondwanaland. This supercontinent covered a vast portion of the southern hemisphere. Gradually, continental drift split Gondwanaland into the continents and other land masses we know today (see Figure 2).

Related plant species that had evolved before Gondwanaland broke up then drifted apart on their continental rafts and new species evolved from the isolated populations (see page 398). While the continents were still close together, genetic exchange between populations was possible; seeds might be borne across water on driftwood, by wind, or swallowed or stuck to the feet of migrating birds. But migration grew less likely as distance between the continents grew. With the continental land masses floating around on the molten interior at one to twenty centimeters per year, we can visualize some plants and animals being "rafted" from one climatic region to another over the immense periods of geologic time. They simply traveled along with the rest of the continent; if they passed through different climatic zones as a result, populations adapted or perished by natural selection. During the long drift apart, some genera fell into extinction and occasionally members of a species or of related species were fossilized. And so, closely related organisms and fossils that were together on an ancient continent are found, much later, far apart.

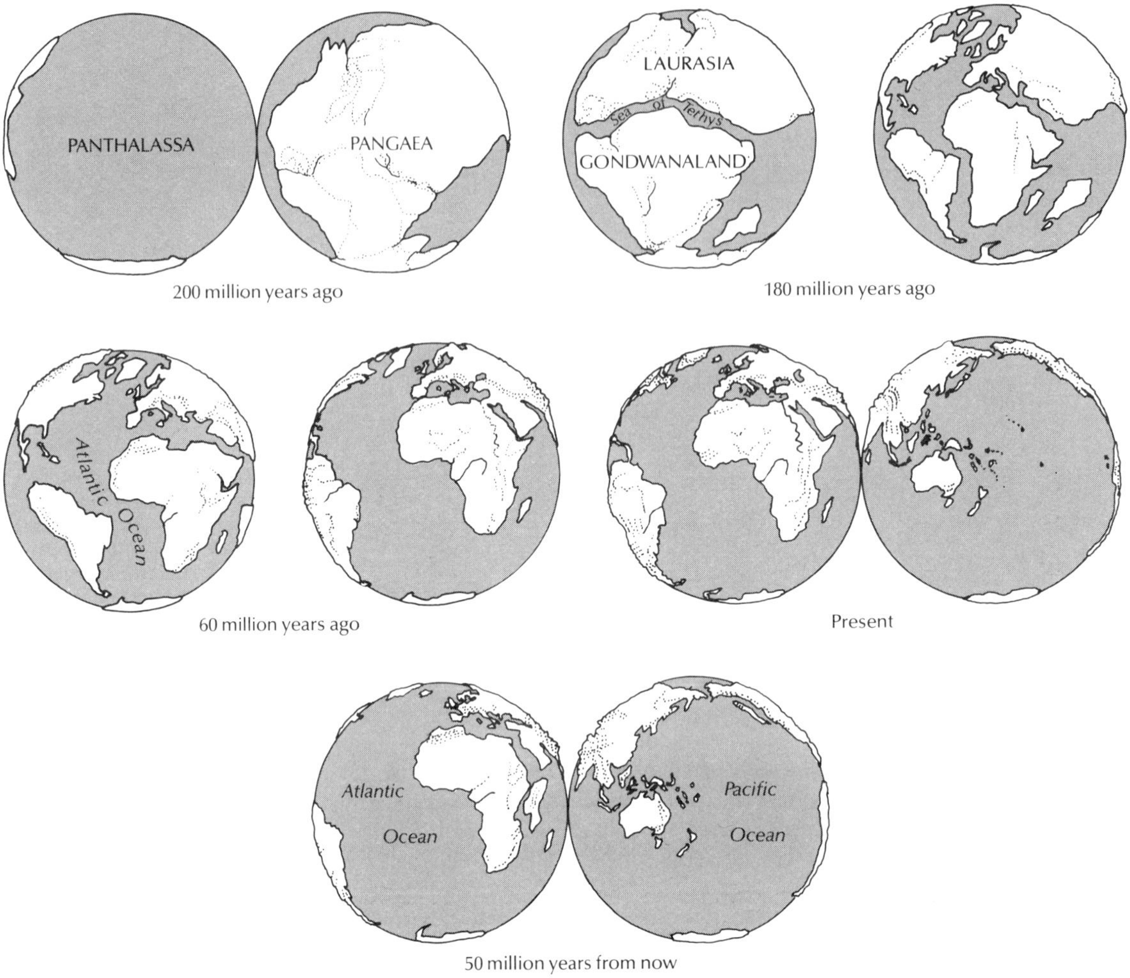

2 The earth's crust has been moving for over 200 million years and it will continue to do so far into the future. At one time the earth had one ocean and one continent, Pangaea. This began the breakup, about 100 million years ago, into continents we recognize today. As recently as 60 million years ago, India began its migration toward Asia. Fifty million years from now, Africa will have split along the great rift valleys and California will have separated along the San Andreas fault and its western part will have drifted well into the Pacific Ocean.

India and the southwestern Pacific islands provide examples of the effects of continental drift. When India joined Eurasia some 50 million years ago, great numbers of plants and animals migrated into India from Eurasia. The fossil record shows no mammals in India before union with Eurasia. Because of this migration from Asia, India is without the uniqueness in flora and fauna that marks Australia, Africa, and South America. The southwestern Pacific islands today are large and small masses scattered over thousands of miles, but they share terrestrial specimens among themselves and with continents not so far away.

Our continents share related organisms, but many of their plants and animals have evolved since becoming isolated from one another. The family Myrtaceae is concentrated in South America and Australia, but a large genus of this family, *Eucalyptus,* is found only in Australia (see the illustrations on page 404). The marsupials of South America and Australia long ago must have roamed areas that if not adjacent, were so similar that biologically similar life forms evolved there. Later movements of the continental plates separated them, but their common origin can still be established.

Continental movement continues; it did not happen just once. Strains still produced by India's continuing thrust into Eurasia are released in the gigantic earthquakes for which the region is known (the 1976 earthquake in China lasted about a minute and killed 600,000 people). California west of the San Andreas fault is migrating seaward because of plate tectonics. California's notorious earthquakes have struck along this fault, including the great one that rocked San Francisco in 1906. Continued slippage along both sides of this fault will move California real estate in one valuable chunk out into the Pacific Ocean in a few million years. Plate tectonics thus brings into clearer focus James Hutton's vision of the dynamic character of the earth's crust.

The development of our knowledge of plate tectonics, and the apparent verification of continental drift, is a major revolution in scientific thought in our time. In addition to providing new insights into the evolution of the earth and its inhabitants, the theory of continental drift provides a vivid lesson in how science works. When I was a graduate student, only twenty years ago, continental drift was thought to be a wild theory, its few adherents eccentric. Today, it is conventional wisdom, and manifestations of its effects are obvious even to the casual observer. The history of this theory should give us pause when wondering at the "primitive" views of our forebears, and it should make us skeptical about current doctrines.

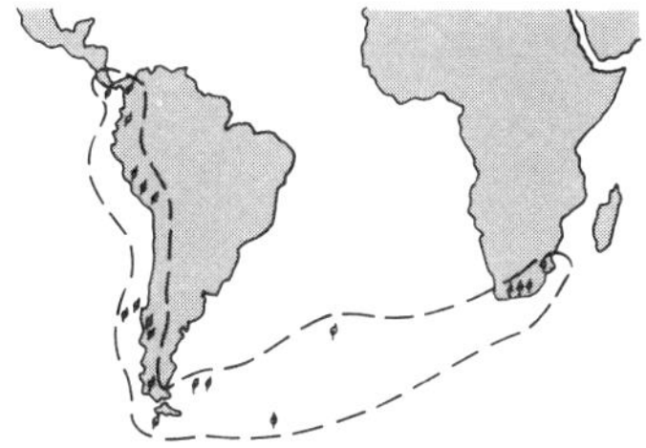

3 *Clasmatocolea vermicularis,* a liverwort, occurs at the southern tip of Africa and at the southern tip and along the west coast of South America. Its dispersal over long distances is unlikely and continental drift presents the best current explanation for its distribution.

Supermarket Taxonomy

This appendix is a summary of the supermarket plants discussed in Chapter 3, and it provides a brief description of the families to which they belong. It can help you to begin to observe the differences in flower structure. (Although leaves, stems, and even roots can provide useful clues to the identity of plants, the flower is the surest and quickest guide to identification.) If you would like further information on a specific plant, consult the many regional and local guides to plant identification.

Family	Examples: Common Name	Examples: Scientific Name
Dicotyledons		
Rosaceae		
Members of this family are found throughout the world. The 3200 species are generally woody and they have alternate leaves. The bisexual flowers have five sepals, five petals, many pistils, and many stamens, which come in whorls of five. Most temperate fruits belong to the rose family. In addition to the familiar roses, spiraea, cotoneaster, hawthorn, and mountain ash are widely planted as ornamentals. The Rosaceae is related to the Leguminosae.	Apple	*Pyrus malus* L. (and other species)
	Blackberry	*Rubus canadensis* L. (and other species)
	Cherry	*Prunus avium* L.
	Peach	*Prunus persica*
	Pear	*Pyrus communis* L. (and other species)
	Plum	*Prunus domestica* L.
	Rose	*Rosa* spp
	Strawberry	*Fragaria virginiana* (and other species)

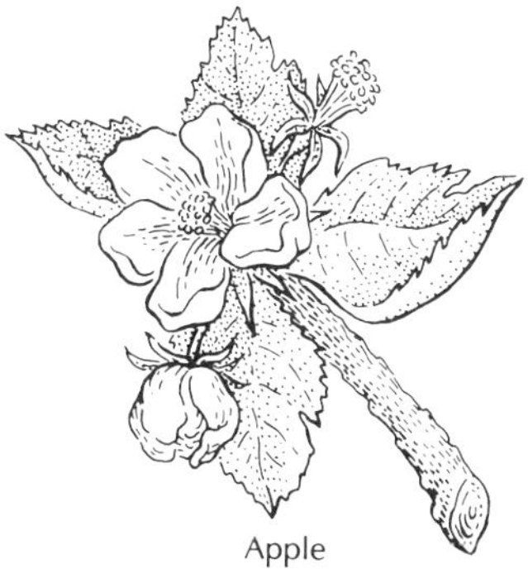

Apple

Family	Examples	
	Common Name	Scientific Name
Cruciferae		
The 2500 species of this family are herbs found throughout the cooler parts of the Northern Hemisphere. Plants have alternate leaves, and the bisexual flowers have four sepals and four petals. The flowers have six stamens in two whorls and one pistil with a single style. This family includes horseradish and it contributes stock, wallflowers, and sweet alyssum to the flower garden.	Broccoli Brussels sprouts Cabbage Cauliflower Mustard Radish Turnip Watercress	*Brassica oleracea* L. *Brassica oleracea* L. *Brassica oleracea* L. *Brassica oleracea* L. *Brassica alba* *Raphanus sativus* L. *Brassica rapa* L. *Nasturtium officinale*

Mustard

Family	Common Name	Scientific Name
Cucurbitaceae		
This rather small family of 850 species consists of climbing or prostrate annual herbs that live in warm regions. Plants are characterized by alternate leaves and coiled tendrils. The unisexual flowers have a five-lobed calyx and either a five-lobed corolla or five petals. There are five stamens and one pistil. The ovary is inferior (below the corolla).	Cantaloupe Cucumber and Gourd Pumpkin Squash Watermelon	*Cucumis melo* L. *Cucumis sativus* L. *Cucurbita pepo* L. (and other species) *Cucurbita pepo* L. *Citrullus vulgaris*

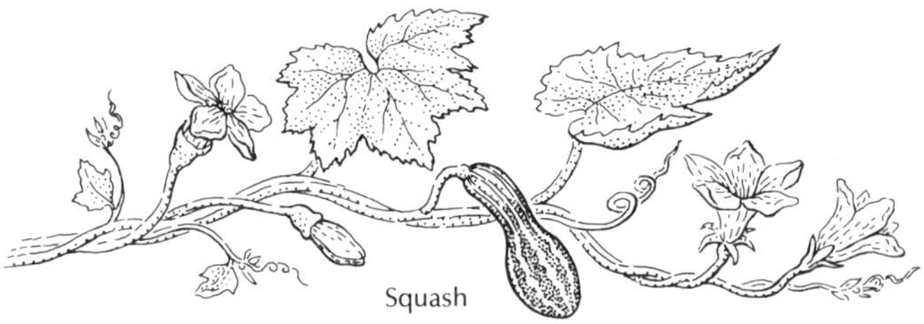

Squash

Family	Common Name	Scientific Name
Chenopodiaceae		
This family of mostly nonwoody plants has a world-wide distribution, but is found particularly in warm regions. The flowers are not showy and have five stamens and a single pistil with three stigmas. Some members of the family are succulent and a large number are xerophytic halophytes—	Beet Lamb's quarters Russian thistle Spinach Sugar beet	*Beta vulgaris* L. *Chenopodium album* L. *Salsola kali* L. *Spinacea oleracea* L. *Beta vulgaris* L.

Family	Examples	
	Common Name	Scientific Name
(Chenopodiaceae continued)		
They are adapted to live in arid regions where the soil has a high salt content. The family is related to the Cactaceae and to a number of families—such as the Amaranthaceae, Portulacaceae, and Molluginaceae—that include many well-known weeds. This group of families is also characterized by having a group of water soluble pigments, called betalains, which range from blue to orange to red. All other plant groups have the blue to red anthocyanin pigments that give most flowers and some leaves their color. Distinguishing between plant families by pigment is a good example of how organic chemistry can be used in plant taxonomy. With the exception of sugar beets, table beets, spinach, and swiss chard, this family contributes little to the economy.		
Compositae		
Many consider this the plant family with the largest number of species (20,000) though some authorities disagree. The species include herbs, shrubs, and trees, and are distributed world-wide. There is a great variety in the shape of leaves and their location on the stem. The flowers are either bisexual or unisexual, and they occur in a whorl to form a disk-shaped inflorescence. The five-lobed corolla is fused, and in bisexual flowers the five stamens are fused to form a cylinder around the style of the single pistil. Except for lettuce and sunflowers (which provide seeds and oil), members of this family are not very important as food, but a large number of species are ornamentals. Many of our most troublesome weeds are also composites.	Artichoke Dandelion Lettuce Sunflower Wormwood	*Cynara scolymus* L. *Taraxacum officinale* *Lactuca sativa* L. *Helianthus annuus* L. *Artemisia absinthium* L.
Leguminosae		
Herbaceous annuals like beans and woody perennials like locust trees are legumes. This family is characterized by compound leaves. The flowers, which have a five-lobed corolla, are bisexual and usually zygomorphic (they can be divided into two symmetrical parts only along one plane). The seeds of legume species usually have no endosperm. The family derives its name from the fruit, which is called a legume. When mature and dried, the fruit splits along one plane and forms two halves. Leguminosae is one of the three largest plant families (the others are Compositae and Orchidaceae) and is spread throughout temperate and tropical regions. The economic importance of this family is manifested by the large number of species used for animal	Alfalfa Bush bean Green pea Kidney bean Lima bean Navy bean Pinto bean Soybean	*Medicago sativa* L. *Phaseolus vulgaris* L. *Pisum sativum* L. *Phaseolus vulgaris* L. *Phaseolus lunatus* L. *Phaseolus vulgaris* L. *Phaseolus vulgaris* L. *Glycine max*

Beet

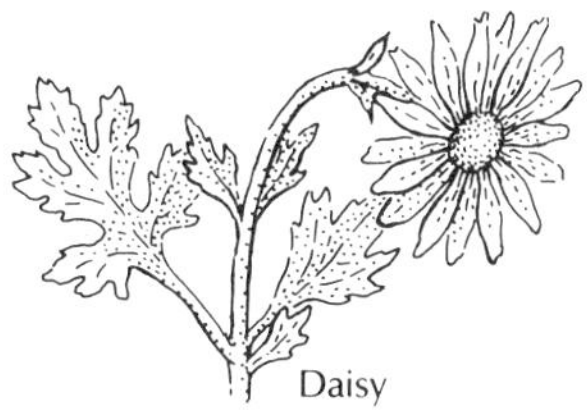

Daisy

Family	Examples: Common Name	Examples: Scientific Name
(Leguminosae continued) food (alfalfa, clover) and the equally large number used in the human diet. Perhaps the most important feature of this family is its ability to fix nitrogen by means of symbiotic bacteria living in root nodules (Chapter 9).	 Green pea	
Solanaceae The nightshade family is large, consisting of more than 2200 species. Approximately half the genera are endemic to tropical parts of the Western Hemisphere and to South America. The family includes herbs, shrubs, and trees. It is best known for its food and drug plants. Common cultivated flowers such as the *Petunia* are also in this family. The flowers are bisexual and the five-lobed corolla is at least fused at the base and often forms a tube. There are commonly five stamens and there is always one pistil.	Eggplant Nightshade Peppers Potato Tobacco Tomato  Nightshade	*Solanum melongena* L. *Solanum nigrum* L. *Capsicum annuum* (and other species) *Solanum tuberosum* L. *Nicotiana tabacum* L. *Lycopersicon esculentum*
Umbelliferae Most species of this family are herbs with compound leaves. The flowers are arranged in small clusters that together provide an umbrella-like appearance. The flowers are generally bisexual and have five petals and five stamens. Leaves and stems are aromatic and the bases of the petioles of leaves sheath the stems to which they are attached. Besides those listed, food plants and herbs in this family include parsnip, anise, caraway, dill, and fennel.	Carrot Celery Parsley Poison hemlock Queen Anne's lace (wild carrot)  Parsley	*Daucus carota* L. *Apium graveolens* L. *Petroselinum sativum* *Conium maculatum* L. *Daucus carota* L.

Family	Examples: Common Name	Examples: Scientific Name
Convolvulaceae		
This family has many tropical representatives, including many vines. The flowers have a calyx with five sepals, a corolla with five fused petals (frequently five-lobed), five stamens, and one pistil. Besides those listed, other members of interest are *Dichondra,* used as ground cover, and the dodder *(Cuscuta),* a parasitic vine.	Morning glory Sweet potato	*Ipomoea purpurea* L. *Ipomoea batatas*

Morning glory

Family	Examples: Common Name	Examples: Scientific Name
Theaceae		
This is a small family of 500 species that live mainly in the tropics and subtropics of Asia. The species are woody plants whose often showy flowers have five sepals, five petals, one pistil, and numerous stamens. Tea is described in Chapter 5.	Camellia Tea	*Camellia japonica* L. *Camellia sinensis* L.

Camellia

Family	Examples: Common Name	Examples: Scientific Name
Euphorbiaceae		
Most species in this family are tropical and are concentrated in Africa and South America. Most species are herbaceous. A milky, frequently toxic latex is characteristic of the family. The flowers are unisexual, and the small calyx and corolla have five lobes. Cassava is the main food plant, but other plants of economic importance provide rubber, castor oil and tung oil.	Cassava Poinsettia Rubber tree	*Manihot esculenta* *Euphorbia pulcherrima* *Hevea brasiliensis*

Poinsettia

Family	Examples: Common Name	Examples: Scientific Name
Monocotyledons		
Gramineae		
Grasses are found almost everywhere plants can grow, and the grass family probably has more individual plants than any other family. Bamboos are woody, but other members of the family are annual or perennial herbs. Fibrous roots are characteristic of the family and many species have rhizomes. Leaves occur singly at nodes and form two ranks.	Corn Crab grass Oats Sorghum Sugar cane Wheat	*Zea mays* L. *Digitaria sanguinalis* *Avena sativa* L. *Sorghum bicolor* *Saccharum officinarum* L. *Triticum aestivum* L.

Family	Examples: Common Name	Examples: Scientific Name

(Gramineae continued)

Leaves are generally strap-shaped and the base of each leaf forms a sheath enclosing the stem. Clusters of flowers are arranged on a spikelike inflorescence. Flower structure of grasses is complex and unique; a special terminology is used to describe grass flower structure. Most species have bisexual flowers, and they all have very small sepals and petals and a feathery stigma. Seeds of grasses are characterized by a high proportion of endosperm. Grasses provide much of the fodder for domestic animals as well as the staples of human diet. Grasses are also widely used for ornamentals and as a playing surface for golf, tennis, and, in some places, baseball and football.

Crab grass

Bromeliaceae

This family grows in tropical and subtropical regions; except for a few species in West Africa, it is found only in America. Most of the species are epiphytes, and the leaves are typically arranged in a rosette. Flowers have a three-lobed calyx and corolla, six stamens, and one pistil. The pineapple is the only species important as food. Dried Spanish moss is used in upholstery, but the main economic interest in other members of the family is for their use as ornamentals. Bromeliads are beginning to rival orchids in popularity as exotic plants to raise.

Common Name	Scientific Name
Pineapple	*Ananas coinosus*
Spanish moss	*Tillandsia usneoides* L.

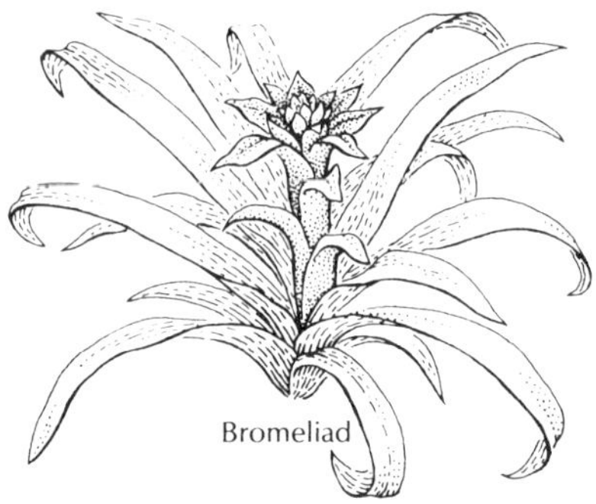

Bromeliad

Liliaceae

Plants with corms, bulbs, tubers, or rhizomes are probably lilies. Perennial herbs with large, showy, generally bisexual flowers, the 4000 or so lily species are found mostly in warm regions. Lily sepals and petals look alike and are called tepals. Flowers have six stamens and one pistil. The lily family is considered a primitive monocot family; its ancestors also developed into other families such as the orchids. Onions and asparagus are important food plants, but the members of this family are most important economically as ornamentals: it is a rare garden without plants of this family.

Common Name	Scientific Name
Asparagus	*Asparagus officinalis* L.
Easter lily	*Lilium longiflorum*
Garlic	*Allium sativum* L.
Onion	*Allium cepa* L.

Easter lily

Family	Examples: Common Name	Examples: Scientific Name
Orchidaceae		
Some authorities consider Orchidaceae the largest plant family. It is most abundant in the tropics, where the species are epiphytes, but species in temperate regions are terrestrial. Some orchids are vines and some are saprophytes (deriving nourishment from dead organic matter). Orchids evolved from lilylike ancestors, and the flowers are characterized by the fusion of floral parts. Flowers are solitary, usually bisexual, and zygomorphic. The style, stigma, and stamens are fused into a complex structure called the column, and one of the fused petals often forms a pouchlike structure. These specialized floral structures contribute to the complex method of pollination characteristic of orchids. Human interest in the orchid family is almost entirely aesthetic.	Lady slipper orchid Vanilla orchid	*Cypridpedium* spp *Vanilla planifolia*

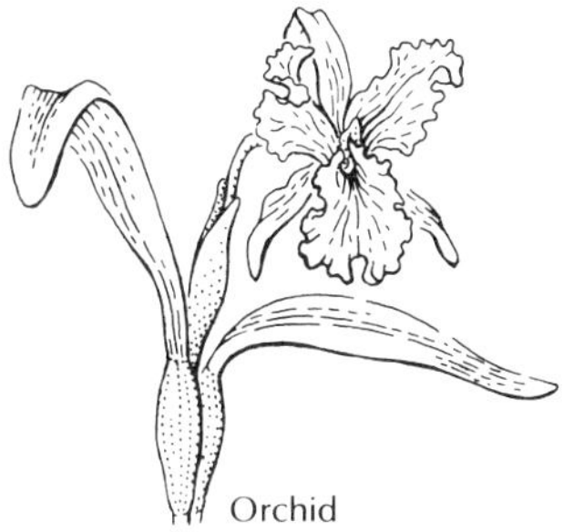

Orchid

Glossary

abscission: the normal falling off the plant of leaves, flowers, fruits, and buds following formation of specialized layers of cells, the abscission zone, near the ends of the petioles.

absorption spectrum: a curve produced by plotting wavelength of light against absorption of light energy by pigments.

action spectrum: a curve produced by plotting wavelength of light against the response by an organism or its parts.

adaptive radiation: splitting of a population into new taxa as a result of natural selection.

adenine: one of a group of chemical compounds, the purines, components of nucleic acids which have two nitrogen-containing cyclic rings.

adhesion: attraction between water molecules and other molecules, especially those in cells along the walls of xylem elements.

adventitious root: root that appears from an uncommon place such as a stem or leaf.

aeon: (ēon): a very long but indefinite time.

aerobic respiration: respiration requiring oxygen.

air layering: a type of asexual reproduction in which roots are induced from a region of the stem while it is attached to the plant.

alkaloid: a nitrogen-containing compound, a base, derived from plants, which produces a physiological reaction in animals.

allele: one of several contrasting forms of a gene at the same location on homologous chromosomes.

alternation of generations: the two basic phases of the plant's life cycle. The gametophyte phase has half as many chromosomes as the sporophyte generation. The gametophyte produces gametes and the sporophyte produces spores, which are the products of meiosis.

amino acid: organic compounds containing an amino group (NH_2) and a carboxyl group (COOH). Proteins are formed of amino acids that have been linked together.

anaerobic organism: an organism that does not require oxygen to live.

anaerobic respiration: respiration without oxygen.

anaphase: the phase of mitosis in which the chromosomes move apart to opposite poles of the spindle.

angiosperms: plants with flowers; their seeds are enclosed by an ovary, a fruit.

annuals: plants that live for only one growing season.

anther: the floral organ, part of the stamen, which produces pollen.

antheridia: structures on gametophytes of non-seed-bearing plants, which produce male gametes or sex cells.

anthropomorphism: assigning human characters to other organisms and objects.

apex: the tip or pointed end.

apical bud: the meristem and young leaves at the end of a branch.

apical dominance: inhibition of lateral or axillary shoot growth by the apical bud of a stem.

apical meristem: the group of embryonic dividing cells at the tip of a plant axis.
apomixis: reproduction from seeds that developed without fertilization.
arable land: soil fit for cultivation.
archegonia: structures of the plant gametophyte that produce female gametes; female sex organs.
asexual reproduction: reproduction, such as fission, without fertilization.
autotroph: an organism that does not depend on other organisms for energy; its nutrients are inorganic.
auxin: a group of hormones that control plant growth; characterized by their activity in the oat coleoptile curvature assay.
axil: the angle formed by the upper side of a petiole or leaf stalk base and the plant axis.
axillary buds: buds formed in axils.

bacilli: rod-shaped bacteria.
basal meristem: the meristem, source of new cells at the base of leaves such as those of grasses.
biennial: plant that lives for two growing seasons.
binary fission: asexual reproduction, as in bacteria, by an individual dividing in two.
binomial system: a system of classification in which every organism is assigned a generic (genus) and a specific (species) name.
bioassay: determining a compound's activity by testing it with a living system, a plant, animal, or microbe.
biological clock: the built-in means of directing an organism's periodic responses.
biological control: control of pests by applying other organisms, especially natural enemies.
biomass: amount of living material in a specified habitat.
biome: a major climatic region, characterized by its dominant vegetation.
biosphere: the fraction of earth's surface including atmosphere, land, and sea, in which organisms live.
bisexual: containing male and female organs; flowers with both pistil and stamens.
bolting: elongation of a stem before flowering.
bracts: modified leaves.
bulb: an underground stem modified for storage with short internodes and bearing tightly pressed, fleshy leaves.

callose: a complex carbohydrate formed in phloem cells. It plugs up aged and injured sieve plates.
calyx: the collective name for sepals, the ring of leaflike structures below flower petals that protect the bud.
capillary action: the movement of fluids in small spaces by attraction between molecules.
capillary water: moisture held between soil particles or adhering to them by capillarity; main source of water for plants.
carbon cycle: the flow of carbon between living and nonliving elements of the biosphere.
carotenoids: yellow to orange-red pigments found in cells. A carrot's color is due to carotenoids.
casparian strip: a band of waxy material in the walls of endodermis cells.
cauliflory: formation of flowers from older regions of the stem.
cell elongation area: the part of the plant axis with elongating cells, responsible for growth of the axis.
cellular respiration: the burning of hydrogen from food in cells and production of chemical energy.
cellulose: a carbohydrate composed of strings of glucose molecules; a polymer of which plant cell walls are made.
center-pivot irrigation: water from a central source is applied to the crop inside a circle up to a mile wide from a rotating boom equipped with nozzles.
chaparral: a fire-adapted plant community found in parts of the Western Hemisphere with hot dry summers and mild wet winters: a thicket of shrubs and small trees.
chemosynthesis: deriving of energy by autotrophic bacteria from chemical reactions with inorganic substances.
chernozems: very black, highly fertile soils of temperate regions.
chert: a type of flint-hard quartz rock.
chitin: a carbohydrate-like compound containing nitrogen. It forms much of the exoskeleton of insects and other arthropods and is a component of fungal cell walls.
chloroplast: organelle in plant cells containing chlorophyll, in which photosynthesis goes on.
chromatid: the half-chromosomes observed from prophase to metaphase in mitosis.
clay: finest-grained mineral soil, with particles 0.002 mm in diameter or less.
climax community: the plant community at the final or mature phase of succession. It will persist unless it is disturbed or the climate changes.
clones: all the offspring produced by asexual reproduction from an individual.
closed system: a set of interacting elements from which nothing leaves and to which nothing is added, such as earth's biosphere.
coadapted: organisms adapted to each other and to the same environment, like bees and flowers.
cocci: sphere-shaped eubacteria (true bacteria).
cohesion: mutual attraction of molecules; the weak bonding between water molecules.
coleoptile: a cylindrical structure of grass embryos enclosing the apex and the first leaf.
coleorhiza: a caplike structure around the embryonic root of grasses.

commensalism: a symbiotic relationship in which one or more organisms benefits and the others are not harmed.

community: a group of different organisms living together.

convection: movement of an unevenly heated fluid, such as gas, transferring heat.

convergent evolution: morphological similarity between unrelated organisms adapting to similar selection pressures.

coprolite: fossilized dung.

cork cambium: the meristematic region in stems and roots that produces cork cells.

corolla: the collective name for petals of a flower.

cortex: primary tissue of the root or stem axis between the epidermis and the vascular tissue.

cotyledon: food-storage organ of many plant embryos, inside the seed. It often functions as a photosynthetic organ on dicot seedlings.

cross: mating of two organisms.

crossing over: entangling and exchange of segments by chromosomes during metaphase of the first meiotic division, mixing the genes.

cross-pollination: transferring pollen from anther of one flower to stigma of another.

cuticle: waxy layer on outer skin of plant organs that slows evaporation of water.

cytoplasm: the proteinaceous material of cells, outside the nucleus, in which the organelles circulate.

cytosine: a nitrogenous base, component of the nucleic acids RNA and DNA.

dark reaction: fixing and reduction of CO_2 during the part of photosynthesis that does not require light.

day-neutral plants: plants that flower regardless of photoperiod or duration of daylight.

deciduous plants: plants that lose their leaves following the growing season each year.

decomposers: scavenger organisms like bacteria and fungi that feed upon and break down complex organic matter, recycling it.

deoxyribonucleic acid (DNA): a compound carrying genetic information and consisting of two strands of nucleotides bound in a spiral. Genes are composed of DNA.

dendrochronology: the science of determining time by studying growth rings in trees.

denitrification: the release of nitrogen (N_2) to the atmosphere from its bound form in the soil (NO_3) by bacterial action.

detritus: bits and pieces of dead organic matter as on the forest floor, mostly remnants of plants, which are food for decomposers.

diatom: a division of algae. Many diatoms have beautifully symmetrical silica cell walls.

dicotyledonae: one of the two subclasses of flowering plants. Their embryos have two cotyledons or seed leaves.

differential survival: more individuals are produced than survive; therefore some will have shorter lives.

differentiation: increasing specialization in form or function of cells, organs, or the organism during development.

diploid: having two sets of homologous chromosomes.

directional selection: natural selection producing a shift in gene frequency of a population so that the whole population evolves.

disruptive selection: natural selection changing the gene frequency and splitting the population into two or more units.

division: the major taxonomic category of the plant kingdom, equivalent to the phylum of the animal kingdom.

dominant: the phenotypic expression of one allele in a heterozygous combination.

dormant: having reduced metabolism, which causes inactivity and halted growth.

ecology: the science of interactions among organisms and between organisms and their environment.

economic plants: plants used by human beings.

ecosystem: a system in which organisms and environment interact.

embryo: in a seed plant, early, heterotrophic stages in the development of the sporophyte.

endemism: native to a particular region.

endodermis: a layer of water-channeling cells in young regions of roots separating the cortex from the vascular region.

endosperm: storage tissue that nourishes the embryo in seeds of gymnosperms and angiosperms; has the same function in the two groups, but different origins.

enzyme: a protein that catalyzes (accelerates) chemical reactions in cells; it is not used up in the reaction.

epidermis: the outer cell layer of leaves and young plant organs.

epiphyte: a plant that lives on another, but is nutritionally independent of it.

epoch: a subdivision in the Quaternary and Tertiary Periods of the Cenozoic Era.

era: the longest category in the classification of geologic time.

erosion: wearing away of soil or rock on the earth's surface by water, wind, or ice.

ethylene: a gas (C_2H_4) produced by plants that acts as a hormone

etiolated: the bleached and spindly appearance of a plant grown in the dark.

eukaryotic: describes an organism with a defined nucleus.

eutrophic: describes a nutrient-rich aquatic environment.

evaporation: transformation of a substance to a gas, as water to water vapor.

family: a taxonomic category between order and genus.

fermentation: cellular respiration not requiring oxygen; alcohol is a common product.

fertilization: union of male and female gametes; the zygote, a fertilized egg, is its product.

fiber: long, thick-walled cells.

fibril: strands of cellulose composing cell walls.

fibrous root system: thin, shallow growing roots characteristic of grasses and many other plants.

fission: cleavage of a cell into two parts.

fixation of nitrogen: the reduction by microorganisms of atmospheric nitrogen (N_2) to form ammonia (NH_3), which is usable by plants.

flagella: whiplike appendages that propel many cells.

flavins: a class of yellow pigments involved in many aspects of cell metabolism.

florigen: the hypothetical chemical messenger stimulating flowering.

flushes: periodic stem growth and leaf production common in woody tropical plants.

food chain: the predator-prey pathway: each organism eats the next link below and is eaten by the next above.

food pyramid: a way of describing a food chain. The top of the food chain has less biomass than the bottom.

food web: a group of interconnecting food chains.

fragmentation: a method of asexual reproduction. A multicellular or colonial organism is subdivided by fragmentation.

fruit: a structure containing seeds—the ovary—derived at least in part from one or more flowers.

gamete: a sex cell or germ cell.

gametophyte: the haploid phase of the plant life cycle, which produces gametes.

gene: the hereditary unit, consisting of DNA and occurring on the chromosome.

gene pool: the collective genotype of a population.

genotype: the genetic composition or makeup of an organism: all its hereditable characteristics.

genus: the taxonomic category between family and species.

geotropism: movement of an organism in response to gravity.

germinate: breaking of dormancy by a seed or spore and commencement of growth.

girdling: cutting away phloem and cambium completely around a stem to kill it.

grafting: the joining of plant parts, often from different individuals, to form a union.

gram negative: an organism that does not hold dye applied by Gram's technique of staining specimens for microscopical examination.

gram positive: an organism that does hold Gram's purple stain.

gravitational water: water that drains quickly away in loose soils, depriving roots of nourishment, or saturates the soil, displacing needed oxygen.

gross productivity (of plants): total amount of carbon incorporated.

growth: an irreversible increase in size.

guanine: a purine base and component of the nucleic acids DNA and RNA.

guard cell: specialized epidermal cells occurring in pairs that form a stoma when turgid.

halophyte: salt-tolerant plant.

haploid: having one set of chromosomes following meiosis.

haustoria: specialized structures with which parasitic plants obtain nutrients from a host plant by penetration.

heartwood: the dead, dark wood in the center of a tree trunk.

herbaceous plant: a nonwoody plant.

herbicide: a chemical used to kill plants, particularly weeds.

heterosis: increased vigor, possessed by many hybrids, which shows in size (offspring larger than parents) and other characteristics.

heterotroph: an organism that, unable to make its own food, depends on other organisms for energy.

heterozygous: a mixed genotype, consisting of different alleles.

holdfasts: specialized structures attaching stalks of algae to underwater rocks.

homologous: having the same origin.

horizon: distinct layer of soil.

hormone: a compound produced in cells of one part of an organism, very small amounts of which can influence cells in another part.

humus: partially decomposed organic matter in soil that nourishes plants and helps the soil hold water.

hybrid plant: a cross-bred plant. The parents are of different varieties or species.

hybrid swarm: a population of interbreeding hybrids.

hydrophyte: a water-tolerant plant.

hydroponics: growing plants with roots immersed in a nutrient solution instead of soil.

hypha: a filamentous element of a fungal mycelium, the body of the fungus.

hypocotyl: in seedlings like beans, the stem between cotyledons and the root.

inflorescence: a flower cluster or stalk.

insectivorous plants: plants with leaves specialized for capturing and digesting insects.
integrated pest control: a combination of chemical and biological methods.
intergeneric crosses: mating species in two genera.
internode: parts of stems between nodes.
interspecific crosses: mating between species.
introgression: interbreeding of hybrids with the parent population, transferring genetic information between species or populations adapted to different habitats.
ion: an atom or group of atoms with either more or fewer electrons than normal; it therefore has a positive or negative charge.

kingdom: the most comprehensive taxonomic category.

lamina: the blade or broad portion of a leaf.
lateral branch: a branch arising from an axillary bud at the base of a leaf.
latex: a milky, sticky hydrocarbon sap that coagulates in air; produced by many plants.
latosols: a type of red or yellow tropical soil, heavily weathered.
law of the minimum: plant growth in an ecosystem is limited by the least available required element.
leaflet: segment of a compound leaf.
lenticels: openings in the corky layer of stems through which gases diffuse between the atmosphere and interior of the plant.
light reactions: that part of photosynthesis where light is absorbed by chlorophyll, and water molecules are split into hydrogen and gaseous oxygen, and ATP is formed.
lignin: a complex compound in the cell wall; a major constituent of wood, providing its durability.
linked genes: genes inherited together because they are together on one part of a chromosome.
long-day plants: photoperiodic plants that flower only when daylight exceeds a critical length.

macroherbivore: large plant eaters such as vertebrate animals.
mass flow: a model to explain translocation of food in the phloem: downward movement of sap to the roots.
maternal inheritance: genes passed on only through the egg cell, so that Mendel's laws of inheritance are not followed.
meiosis: distribution of chromosomes in cell division, reducing the number of chromosomes in a cell by half. In plants, meiosis produces haploid spores, which produce gametophytes.
meristems: groups of dividing cells; source of new cells.
mesophyll: chloroplast-containing parenchyma cells in leaves between epidermal layers.
messenger RNA: RNA, a nucleic acid that copies the genetic code from DNA in the nucleus and travels to the cytoplasm, where it serves as template for assembly of proteins.
metabolism: the physical and chemical activities in living cells, including transformation of food into energy.
metaphase: the stage in mitosis when chromosomes are aligned on the spindle along the middle of the cell.
microbe: a microscopic organism.
microfossil: a microscopic fossil.
microherbivore: a small, often microscopic plant eater.
mimicry: imitation, as in animals that have evolved to look like like plants.
mitochondria: cell organelles in which aerobic respiration and assembly of ATP goes on.
mitosis: division and segregation of chromosomes to form an equal number in the two new nuclei.
monocotyledoneae: one of two subclasses of flowering plants; embryos have one cotyledon.
monoecy: male and female flowers on the same plant.
monsoon: seasonal wind bringing heavy rainfall from June to October in Southeast Asia.
morphogenesis: change in form of cells, tissue, and organs during the plant's growth.
morphology: the science of form or structure.
mutagenic agent: anything causing a change in genes.
mutation: new heritable feature, caused by change in genes and chromosomes.
mutualism: a symbiotic relationship in which all species benefit.
mycelium: a mass of hyphae or filaments; vegetative body of a fungus.
mycorrhizae: fungi forming a close symbiotic association with plant roots.

natural selection: the selection in a population by natural forces of a genetic variant that is better adapted and hence leaves more progeny; the driving force for evolution.
negative phototropism: growing away from a source of light.
neptunism: the theory that all geological features of the earth were molded by the rise and fall of seas.
net productivity: the amount of reduced carbon available for primary consumers.
niche: all the factors determining an organism's position in its environment, including how it obtains energy and reproduces.

node: a place where a leaf is attached to a stem.
nodule: growths on roots containing bacteria that reduce atmospheric nitrogen, making it available to plants.
nucleic acid: class of complex compounds in all cells, made up of double or single structural strands of nucleotides; DNA and RNA are included.
nucleotide: a class of compounds made up of a nitrogenous base, a 5-carbon sugar, and phosphoric acid; the structural part of a nucleic acid.
nucleus: the cell organelle containing the chromosomes and taking part in growth and reproduction.

oligotrophic: a body of water with few of the elements essential for plant growth.
organs: differentiated parts of an organism, such as roots or leaves, with well-defined structure and function.
osmosis: diffusion of water across a semipermeable membrane, moving from a solution with a low concentration of solute to the more concentrated side.
osmotic gradient: the movement of water from a solution with low concentration of solute to one with higher concentration.
outcrossing: pollination between flowers on the same plant or, particularly, between flowers on different plants.
ovary: base of the angiosperm pistil containing the ovules, in which the female gametophytes are; it becomes the fruit.
ovule: the structure(s) within the ovary containing the female gametophyte; after fertilization, it will become a seed.
oxidize: the removal of electrons from atoms; in cells, it usually removes hydrogen and adds oxygen.

palisade cells: elongated, chloroplast-containing cells forming the top layer of mesophyll in many leaves.
parasite: an organism that lives and feeds on or in another living organism; the parasite benefits but the host is harmed.
parenchyma: thin-walled unspecialized plant cells, as in pith, cortex, and mesophyll.
parthenogenic: development of an embryo from an unfertilized egg.
pathogenic: disease-causing.
pectin: complex carbohydrate in primary cell walls; it helps hold cells together.
peds: natural aggregates formed by particles of soil.
perennials: plants that live for many growing seasons.
pericycle: a layer of root cells inside the endodermis; the meristem from which lateral roots originate.
period: subdivision of an era in the geologic time table.
permafrost: the permanently frozen subsurface layer of soil in arctic regions.
permanent embryo: apical meristems of shoots that constantly produce new organs.
petiole: stalk of the leaf.
phenotype: the outward appearance of an organism generated by its genotype and environment.
phloem: tissue of vascular system in higher plants, composed of sieve cells, sieve tubes, parenchyma cells, and fibers; transports most of the plant's organic compounds, especially its food.
photochemical reaction: a chemical reaction initiated by light.
photomorphogenesis: change in form induced by visible light.
photomorphogenic pigment (phytochrome): a pigment that absorbs light, causing a change in form.
photoperiodism: control of flowering and other physiological responses by length and timing of light and dark periods.
photosynthesis: transformation of light energy to chemical energy by green plants.
phototropism: effect of a unilateral source of light on growth movement of a plant axis.
phytochrome: a photoreversible pigment that controls many phases of plant development.
pigment: a chemical compound that absorbs selected wavelengths of visible light and has a characteristic color because it reflects wavelengths not absorbed.
pistil: a reproductive organ, ovary, style, and stigma, of the flower; usually called the female structure because it will produce the gametophyte that has an egg cell.
pith: the soft primary parenchyma tissue in the center of young vascular stems.
pits: thin-walled regions of the lignified wall of vessel elements and tracheids.
plasmalemma (plasma membrane): the membrane next to the cell wall surrounding the cytoplasm.
plasmodesmata: strands of cytoplasm bounded by plasmalemma connecting cells through holes in their walls.
plasmolysis: loss of water by osmosis from the cell vacuole because the surrounding medium has more solute than the vacuole; the cytoplasm pulls away from the cell wall.
plastids: organelles with two envelope membranes and a variety of internal membranes. Chloroplasts are one type of plastids.
plate tectonics: theory that the continents of earth's crust, because they are lighter than the fluid mantle beneath, slowly drift apart and together on a dozen or so plates.
plowless planting: a new method of planting and tilling without extensive turning of the soil, cutting erosion and evaporation.

podzols: leached but fertile soils of temperate regions.

pollen: small, thick-walled male gametophyte formed by stamens of gymnosperms and angiosperms.

pollen tube: a tube that grows from a pore in the pollen wall through the style, transporting sperm to the female gametophyte.

pollination: transfer of pollen from anther to stigma in angiosperms and to ovules in gymnosperms.

polymer: a complex molecule formed of smaller molecules (monomers) in a chain of repeating units.

population: a group of individuals of one species, capable of interbreeding.

population genetics: analysis of inheritance of characteristics in populations instead of among individuals.

preadaptation: ability of some genetic variants in a population to adapt to changed conditions.

primary consumer: an animal that eats only plants; a herbivore.

primary growth: growth of a plant axis by elongation of cells derived from the apical meristem.

primordia: embryonic cells beginning an organ's development.

prokaryotic: describes an undifferentiated cell, such as bacterium, without defined nucleus or other organelles.

prophase: the first phase of mitosis or cell division, during which the chromosomes take shape, thicken, and divide into pairs and the nuclear membrane disappears.

protoplasm: the living, complex, colloidal jelly of proteins and water inside the cell wall.

pyruvic acid: $C_3H_4O_3$, the raw material for anaerobic respiration in cells.

radioactivity: spontaneous emission of waves or particles of radiation by some elements as their atomic nuclei disintegrate into other isotopes.

recessive: said of an allele (form of a gene) that is not expressed in the phenotype when combined with an allele for a different character.

reduction (carbon): combining of carbon with hydrogen in photosynthesis in the formation of carbohydrates.

reduction division: meiosis, the first division of which produces two cells with half the original number of chromosomes.

reproductive organ: an organ producing gametes.

respiration (aerobic): oxidation of food (glucose) in cells, which yields energy, water, and carbon dioxide.

rhizomes: underground stems.

rhizosphere: the part of the soil penetrated by roots.

ribonucleic acid (RNA): RNA differs from DNA in having nucleotides with the base uracil rather than thymine and a 5-carbon sugar with an additional oxygen atom; it functions in protein synthesis.

ribosome: cell particles of RNA and protein, incredibly numerous and the site of protein synthesis.

root cap: the mound of parenchyma tissue covering and derived from the root apical meristem.

root-crop agriculture: many of the human-nourishing calories of the humid tropics are provided by root crops such as cassava and yams.

root hairs: epidermal cells extending from young regions of roots; they absorb water and minerals.

runners: horizontal stems, which grow along the surface of the ground, sometimes forming adventitious roots.

sap: sugar-rich solution moving through phloem cells.

saprophyte: an organism that grows on and obtains nourishment from dead organic material; includes most fungi.

sapwood: outer, younger part of the xylem of a tree, which conducts water.

secondary consumer: an organism that eats organisms that consume plants.

secondary tissue: tissue derived from lateral meristems such as cambium.

seed coat: the outer layer of a seed, derived from outer layers of the ovule.

self-pollination: transfer of pollen from a flower's anthers to its own stigma.

senesce: to undergo the physiological changes caused by aging.

sepal: leaflike structure beneath the petals of a flower; part of the calyx.

sexual reproduction: the union of gametes to produce a zygote.

shoot: a stem and associated organs; the part of a plant above the soil.

short-day plants: plants that flower when the light period does not exceed a critical length.

sierozems: grayish infertile desert soils.

sieve tube: the food-transporting system of phloem.

silt: mineral particles of soil larger than clay, smaller than sand, deposited by water.

soil profile: succession of horizons visible in a sample of soil.

solvent: a liquid such as water that dissolves another substance; the largest part of a solution.

somatic cell hybridization: the fusion of somatic cells (body cells, not gametes) to form a single hybrid cell.

sori: an aggregation of sporangia, spore-producing organs, on fern leaves.

species: the taxonomic category subordinate to genus, describing populations of similar organisms.

spirilla: flagellated bacteria with spiral form.

sporangia: a reproductive structure producing spores.

spores: generally a thick-walled, tough reproductive cell; in plants, the product of meiosis, they produce the gametophyte.
sporophyte: the phase of the alternating-generation cycle of plant life that begins with the zygote and ends with meiosis; the spore-bearing generation.
stabilizing selection: natural selection eliminating from a population genotypes that deviate significantly from the norm.
stamen: the male reproductive structure in flowers that has a stalk and an anther that produces pollen.
staple foods: foods that provide populations with a large percentage of their necessary calories.
stigma: tip of the pistil that receives pollen, which germinates there.
stomata: the openings between guard cells in the epidermal layers of leaves that open the leaf's interior for exchange of gases with the atmosphere.
style: the stalk of the pistil that connects stigma and ovary; pathway for pollen tubes.
stylet: hollow, needlelike mouth part of an aphid.
suberin: a waxy substance that fills cells produced by the cork cambium.
succulent: fleshy and juicy.
sucker: secondary shoot arising from older portions of stems, as from the base of stumps or from roots.
sucrose: water-soluble sugar ($C_{12}H_{22}O_{11}$) composed of one molecule of glucose and one of fructose.
symbiosis: the close relationship of two or more kinds of organisms, at least one of which depends on the relationship.
symbiotic bacteria: organisms such as nitrogen-fixing bacteria that live in root nodules and derive their nutrition from plants.

taiga: a biome of subarctic regions with mixed coniferous and deciduous forest.
taproot: a vertical, thick primary root with small lateral roots.
taxa: groups of organisms forming taxonomic categories.
taxonomy: the science of classification.
teleology: philosophical study assigning purpose or design to natural phenomena.
telophase: The last phase of mitosis during which chromosomes reform into nuclei and a new cell wall is initiated.
thymine: a single-ringed nitrogenous base; component of nucleotides from which DNA is synthesized.
tiller: grass shoots that grow from buds at the base of stems.
tissue: in plants, many cells of one type, or cells related to a function or region.
tissue culture: growth of plant cells and organs in sterile conditions on a synthetic medium.
tonoplast: the membrane around a vacuole of a plant cell.
trace element: a chemical element required by plants but in very small amounts.
tracheid: a long, thick-walled conducting and supporting xylem cell.
traits: genetic characteristics; as expressed in Mendel's laws of genetics, the phenotypic expression of a gene's alleles.
transfer RNA: RNA that transfers specific amino acids from the cytoplasm to the surface of ribosomes where they are linked to form proteins.
transpiration: movement of water through the plant and its evaporation as water vapor from the leaf surface through stomata.
trichomes: hairlike or bristlelike structures on a plant's epidermis; they can be unicellular or complex multicellular outgrowths.
trophic: nourishment, specifically referring to feeding levels in food chains.
true-breeding plants: reappearance of the same characters generation after generation because of a homozygous genotype for specific characters.
tuber: a thickened portion of a rhizome due to accumulated food.
turgid, turgor: when a cell takes in so much water that the pressure stretches the cell walls, it is said to be turgid (swollen) and to exhibit turgor.

uniformitarianism: a theory proposed by James Hutton to explain past geological events: the past could be interpreted by assuming the earth then was subject to the same forces observable in his time.
unisexual: flowers with only male or only female reproductive organs.
uracil: a nitrogenous base with a single ring structure; a component of nucleotides in RNA, but not DNA.

vacuole: the usually large cavity in a plant cell filled with water and solutes, separated from cytoplasm by a membrane, the tonoplast.
vascular bundles: the strands of mostly primary xylem and phloem in herbaceous stems and young portions of woody stems.
vascular cambium: the ring of meristematic cells that produces secondary xylem and phloem.
vascular plants: plants with specialized tissues, xylem and phloem, for support and for conducting water, minerals, and organic compounds; the tracheophyta.
vector: a carrier.
vegetative organ: an organ involved in growth or nutrition, but not reproduction.
vegetative reproduction: reproduction not involving sex (the union of gametes).
vein: strands of branching vascular tissue forming the framework of a leaf.
vessel elements: specialized xylem cells with perforated end walls.

vitamins: organic molecules essential in small quantities for regulating metabolic processes in plants and animals.

weathering: chemical and mechanical breaking down of surface rocks into small mineral particles by the atmosphere.

woody plants: plants with secondary xylem.

xerophyte: plants tolerant of arid conditions.

xylem: water-conducting cells and associated cells of higher plants.

zooplankton: microscopic animals that feed on the phytoplankton in aquatic ecosystems.

zygote: a fertilized egg, formed by union of two sex cells.

Suggested Readings

Chapter 1

Batzli, G. O. 1974. Production, assimilation and accumulation of organic matter in ecosystems. *Journal of Theoretical Biology* 45:205–217.

Brink, R. A., Desmore, J. W., and Hill; G. A. 1977. Soil deterioration and the growing world demand for food. *Science* 197: 625–630.

Carter, L. C. 1973. *Guide to Cellular Energetics.* San Francisco: W. H. Freeman.

Duchham, A. N., Jones, J. G. W., and Roberts, E. H., eds. 1976. *Food Production and Consumption; The Efficiency of Human Food Chains and Nutrient Cycles.* Amsterdam: Elsevier/North-Holland Biomedical Press.

Gressit, J. L. 1977. Symbiosis runs wild in a Lilliputian "forest" growing on the backs of high-living weevils in New Guinea. *Smithsonian* 7:135–140.

Hutchinson, G. E. 1959. Homage to Santa Rosalia or why are there so many kinds of animals. *American Naturalist* 93:145–159.

Lugo, A. E., Farnworth, E. G., Pool, D., Jerez, P., and Kaufman, G. 1973. The impact of the leaf cutter ant *Atta colombica* on the energy flow of a tropical wet forest. *Ecology* 54:1292–1301.

Mattson, W. J., and Addy, N. D. 1975. Phytophagous insects as regulators of forest primary production. *Science* 190:515 522.

Morgan, J. M., Morgan, M.D., and Wiersma, J. H. 1973. *An Introduction to Environmental Sciences.* Boston: Little, Brown.

Odum, E. P. 1975. *Ecology.* New York: Holt, Rinehart & Winston.

Porter, K. G. 1977. The plant-animal interface in freshwater ecosystems. *American Scientist* 65:159–170.

Scientific American Editions. 1970. *The Biosphere.* San Francisco: W. H. Freeman (Scientific American Books).

Soucie, G. A. 1975. We can still save salt marshes of Georgia, Carolina. *Smithsonian* 5:82–89.

Van Horn, H. H., Cunha, T. J., and Harms, R. H. 1972. The role of livestock in meeting human food needs. *Bioscience* 22:710–714.

Wiegert, R. G., and McGinnis, J. T. 1975. Annual production and disappearance of detritus of three South Carolina old fields. *Ecology* 56:129–140.

Chapter 2

Borgstrom, G. 1976. Never before has humankind had to face the problem of feeding so many people with so little food. *Smithsonian* 7:70–77.

Brown, L. R. 1975. The world food prospect. *Science* 190:1053–1059.

Campbell, B. G., ed. 1976. *Humankind Emerging.* Boston: Little, Brown.

Carrier, L. 1957. *Agriculture in Virginia 1607–1699.* Charlottesville: University Press of Virginia.

Chiarappa, L., Chian, H. C., and Smith, R. F. 1972. Plant pests and diseases: Assessment of crop losses. *Science* 176:769–773.

Dovring, F. 1974. Soybeans. *Scientific American* 230:14–21.

Fagan, B. M. 1974. *Men of the Earth: An Introduction to World Prehistory:* Boston: Little, Brown.

Fussell, G. E. 1965. *Farming Technique from Prehistoric to Modern Times.* New York: Pergamon Press.

Giese, R. L., Peart, R. M., and Huber, R. T. 1975. Pest management. *Science* 187:1045–1052.

Hardy, R. W. F., and Havelka, U. D. 1975. Nitrogen fixation research: A key to world food? *Science* 188:633–643.

Heichel, G. H. 1976. Agricultural production and energy resources. *American Scientist* 64:64–72.

Hodgson, H. J. 1976. Forage crops. *Scientific American* 234:60–75.

Janick, J. 1972. *Horticultural Science.* San Francisco: W. H. Freeman.

Jensen, N. F. 1978. Limits to growth in world food production. *Science* 201: 317–320.

Leonard, J. N. 1973. *The First Farmers.* New York: Time-Life Books.

Murdoch, W. W., and Oaten, A. 1975. Population and food: Metaphors and the reality. *Bioscience* 25:561–567.

Poleman, T. T. 1975. World food: A perspective. *Science* 188:510–518.

Renfrew, J. M. 1973. *Palaeoethnobotany.* New York: Columbia University Press.

Schlebeecker, J. T. 1975. *Whereby We Strive: A History of American Farming, 1607–1972.* Ames: Iowa State University Press.

Seigler, D. S., ed. 1977. *Crop Resources.* New York: Academic Press.

Steinhart, J. S., and Steinhart, C. E. 1974. Energy use in the U.S. food system. *Science* 184:307–316.

Wade, N. 1974. Green revolution (1): A just technology, often unjust in use. *Science* 186:1093–1096.

Wiley, J. P., Jr. 1975. Maine and eastern Canada are trying bacterial warfare against the spruce budword, an enemy of the forests. *Smithsonian* 6:84–89.

Zohary, D., and Hopf, M. 1973. Domestication of pulses in the old world. *Science* 182:887–894.

The following publications are devoted to issues related to food and agriculture:

The World Food Situation and Prospects to 1985, Economic Research Service, U.S. Department of Agriculture, Foreign Agricultural Economic Report No. 98, 1974.

A Mosaic Special; Food, Mosaic 6: No. 3, 1975, National Science Foundation.

Food. A special issue of *Science* 188: No. 4188, 1975.

Food and Agriculture. A special issue of *Scientific American* 235: No. 3, 1976.

Chapter 3

Arber, A. 1970. *Herbals: Their Origin and Evolution.* Darien, Conn.: Hafner Publishing Co.

Benson, L. 1957. *Plant Classification.* Boston: D. C. Heath.

Constance, L. 1955. The systematics of the angiosperms. *A Century of Progress in the Natural Sciences, 1853–1953.* California Academy of Sciences, pp. 405–483.

Corner, E. J. H. 1964. *The Life of Plants.* New York: Mentor Books.

Cronquist, A. 1968. *The Evolution and Classification of Flowering Plants.* Boston: Houghton Mifflin.

Dodson, E. O. 1971. The kingdoms of organisms. *Systematic Zoology* 20:265–281.

Eames, A. J. 1961. *A Morphology of the Angiosperms.* New York: McGraw-Hill.

Fernald, M. L. 1950. *Gray's Manual of Botany,* 8th ed. New York: American Book Co.

Foster, A. S., and Gifford, E. M., Jr. 1974. *Comparative Morphology of Vascular Plants,* 2d ed. San Francisco: W. H. Freeman.

Hutchinson, J. 1969. *Evolution and Phylogeny of Flowering Plants.* London and New York: Academic Press.

Jensen, W. A. 1973. Fertilization in flowering plants. *Bioscience* 23:21–27.

Lawrence, G. H. M. 1951. *Taxonomy of Vascular Plants.* New York: Macmillan.

Leedale, G. F. 1974. How many are the kingdoms of organisms? *Taxon* 23:261–270.

Li, Hui-Lin. 1974. Plant taxonomy and the origin of cultivated plants. *Taxon* 23:715–724.

Maheshwari, P. 1950. *An Introduction to the Embryology of Angiosperms.* New York: McGraw-Hill.

Stafleu, F. A. 1971. *Linnaeus and the Linnaeans.* Utrecht: A. Oosthoek's Uitgeversmaatschappij H. V.

Van der Pijl, L. 1969. *Principles of Dispersal in Higher Plants.* New York: Springer-Verlag.

Went, F. W. 1963. *The Plants.* New York: Time-Life Books.

Chapter 4

Alexopoulos, C. J. 1962. *Introductory Mycology.* New York: Wiley.

Alexopoulos, C. J., and Bold, H. C. 1967. *Algae and Fungi.* New York: Macmillan.

Bell, P. R., and Woodcock, C. L. F. 1968. *The Diversity of Green Plants.* Reading: Addison-Wesley.

Bierhorst, D. W. 1971. *Morphology of Vascular Plants.* New York: Macmillan.

Bold, H. C. 1957. *Morphology of Plants.* New York: Harper.

Bower, F. O. 1935. *Primitive Land Plants.* London: Macmillan.

Cambell, D. H. 1940. *The Evolution of the Land Plants.* Stanford: Stanford University Press.

Chamberlain, C. J. 1935. *Gymnosperms: Structure and Evolution.* Chicago: University of Chicago Press.

Christensen, C. M. 1961. *The Molds and Man,* 2d. ed. Minneapolis: University of Minnesota Press.

Dawson, E. Y. 1966. *Marine Botany.* New York: Holt, Rinehart & Winston.

Esau, K. 1961. *Plants, Viruses, and Insects.* Cambridge: Harvard University Press.

Foster, A. S., and Gifford, E. M., Jr. 1974. *Comparative Morphology of Vascular Plants,* 2d ed. San Francisco: W. H. Freeman.

Jackson, D. F. 1964. *Algae and Man.* New York: Plenum Press.

Scagel, R. F., Bandoni, R. J., Rouse, G. E., Schofield, W. B., Stein, J. R., and Taylor, T. M. C. 1965. *An Evolutionary Survey of the Plant Kingdom.* Belmont, Calif.: Wadsworth.

Smith, D. C. 1973. *The Lichen Symbiosis.* Oxford: Oxford University Press.

Smith, G. M. 1955. *Cryptogamic Botany* vols. 1 and 2. New York: McGraw-Hill.

______. 1944. *Marine Algae of the Monterey Peninsula.* Stanford: Stanford University Press.

Sporne, K. R. 1965. *The Morphology of Gymnosperms.* London: Hutchinson.

Stanier, R. Y., and Doudoroff, M. 1963. *The Microbial World,* 3d ed. Englewood Cliffs: Prentice-Hall.

Taylor, W. R. 1957. *Marine Algae of the Northeastern Coast of North America,* 2d ed. Ann Arbor: University of Michigan Press.

Watson, E. V. 1972. *Mosses.* Oxford: Oxford University Press.

Wheeler, B. E. J. 1975. *The Control of Plant Disease.* Oxford: Oxford University Press.

Wood, R. K. S. 1974. *Disease in Higher Plants.* Oxford: Oxford University Press.

Zentmyer, G. A. 1976. *Phytophthora*—plant destroyer. *Bioscience* 26:686–689.

Chapter 5

Abel, E. L. 1977. When tobacco was king. *American History Illustrated* 11:11–19.

Anderson, E. 1969. *Plants, Man and Life.* Berkeley: University of California Press.

Baker, H. G. 1965. *Plants and Civilization.* Belmont, Calif.: Wadsworth Publishing Co.

Dodge, B. S. 1959. *Plants that Changed the World.* Boston: Little, Brown.

Hedrick, U. P. 1972. *Sturtevant's Edible Plants of the World.* New York: Dover.

Hill, A. F. 1952. *Economic Botany.* New York: McGraw-Hill.

Hyams, E. 1971. *Plants in the Service of Man.* New York: Lippincott.

Jaques, H. E. 1943. *Plants We Eat and Wear.* New York: Dover.

Jennings, P. R. 1974. Rice breeding and world food production. *Science* 186: 1085–1088.

Kingsbury, J. M. 1964. *Poisonous Plants of the United States and Canada.* Englewood Cliffs, N.J.: Prentice-Hall.

Lehner, R., and Lehner, J. 1962. *Folklore and Odysseys of Food and Medicinal Plants.* New York: Tudor.

National Academy of Sciences. 1975. *Underexploited Tropical Plants with Promising Economic Value.*

Purseglove, J. W. 1968. *Tropical Crops, Dicotyledons* (2 vols.) New York: Wiley.

______. 1972. *Tropical Crops, Monocotyledons* (2 vols.). New York: Wiley.

Rick, C. M. 1978. The tomato. *Scientific American* 239:76–87.

Robinson, T. 1974. Metabolism and function of alkaloids in plants. *Science* 184:430–435.

Schery, R. W. 1972. *Plants for Man.* Englewood Cliffs: Prentice-Hall.

Schultes, R. E. 1969. Hallucinogens of plant origin. *Science* 163:245–254.

Schwanitz, F. 1966. *The Origin of Cultivated Plants.* Cambridge: Harvard University Press.

Taylor, N. 1965. *Plant Drugs that Changed the World.* New York: Dodd, Mead.

Uphof, J. C. 1968. *Dictionary of Economic Plants.* New York: J. Cramer.

U.S. Department of Agriculture. 1971. *Common Weeds of the United States.* New York: Dover.

Wheelwright, E. G. 1974. *Medicinal Plants and Their History.* New York: Dover.

Chapter 6

Clowes, F. A. L. 1972. *Morphogenesis of the Shoot Apex.* Oxford: Oxford University Press.

______. *Apical Meristems.* 1961. Oxford: Blackwell.

Clowes, F. A. L., and Juniper, B. E. 1968. *Plant Cells.* Oxford: Blackwell.

Cutter, E. G. 1978. *Plant Anatomy, Part I, Cells and Tissues,* 2d ed. London: Arnold.

______. 1971. *Plant Anatomy, Part II, Organs.* London: Arnold.

______. 1969. *Plant Anatomy: Experiment and Interpretation.* Reading, Mass.: Addison-Wesley.

______. ed. 1966. *Trends in Plant Morphogenesis,* London: Arnold.

Esau, K. 1965. *Plant Anatomy,* 2d ed. New York: Wiley.

______. 1977. *Anatomy of Seed Plants.* 2d ed. New York: Wiley.

Fahn, A. 1974. *Plant Anatomy,* 2d ed. New York: Pergamon.

Goebel, K. E. 1969. *Organography of Plants* (facsimile of 1905 edition). New York: Hafner.

Hayward, H. E. 1938. *The Structure of Economic Plants.* New York: Macmillan.

Milthorpe, F. L., ed. *The Growth of Leaves.* London: Butterworth.

Romberger, J. A. 1963. *Meristems, Growth, and Development in Woody Plants.* U.S. Department of Agriculture Forest Service Tech. Bull. No. 1293.

Salisbury, F. B., and Parke, R. V. 1964. *Vascular Plants: Form and Function.* Belmont, Calif.: Wadsworth Publishing Co.

Sinnot, W. W. 1960. *Plant Morphogenesis.* New York: McGraw-Hill.

Steeves, T. A., and Sussex, I. M. 1972. *Patterns in Plant Development.* Englewood Cliffs, N.J.: Prentice-Hall.

Thompson, D. W. 1961. *On Growth and Form.* Cambridge: Cambridge University Press.

Torrey, J. G. 1967. *Development in Flowering Plants.* New York: Macmillan.

Torrey, J. G., and Clarkson, D. T., eds. 1975. *The Development and Function of Roots.* New York: Academic Press.

Wardlaw, C. W. 1952. *Morphogenesis in Plants.* London: Methuen.

———. 1955. *Embryogenesis in Plants.* London: Methuen.

Chapter 7

Annual Review of Plant Physiology. Palo Alto: Annual Reviews Inc. Topics covered in this chapter, as well as many others, are reviewed regularly in this series which began in 1950.

Albersheim, P. 1975. The walls of growing plant cells. *Scientific American* 232:80–95.

Black, M. 1972. *Control Process in Germination and Dormancy.* Oxford: Oxford University Press.

Cleland, C. F. 1978. The flowering enigma. *Bioscience* 28:265–269.

Evans, L. T., ed. 1969. *The Induction of Flowering.* Ithaca: Cornell University Press.

Galston, W. W., and Davies, P. J. 1970. *Control Mechanisms in Plant Development.* Englewood Cliffs: Prentice-Hall.

Hillman, W. S. 1962. *The Physiology of Flowering.* New York: Holt, Rinehart & Winston.

Khan, A. A., 1977. *The Physiology and Biochemistry of Seed Dormancy and Germination.* Amsterdam: Elsevier/North-Holland Biomedical Press.

Khudairi, A. K. 1972. The ripening of tomatoes. *American Scientist* 60: 696–707.

Laetsch, W. M., and Cleland, R. E., eds., 1967. *Papers on Plant Growth and Development.* Boston: Little, Brown.

Leopold, A. C. 1964. *Plant Growth and Development.* New York: McGraw-Hill.

Machlis, L., and Torrey, J. G. *Plants in Action.* San Francisco: W. H. Freeman.

Ray, P. M. 1963. *The Living Plant.* New York: Holt, Rinehart & Winston.

Salisbury, F. B., and Ross, C. W. 1978. *Plant Physiology,* 2d ed. Belmont, Calif.: Wadsworth Publishing Co.

Steward, F. C. 1968. *Growth and Organization in Plants.* Reading, Ma.: Addison-Wesley.

Steward, F. C., and Krikorian, A. D. 1971. *Plants, Chemicals and Growth.* New York: Academic Press.

Steward, F. C., Krikorian, A. D., and Holsten, R. D. 1966. *About Plants: Topics in Plant Biology.* Reading, Ma.: Addison-Wesley.

Van der Veen, R., and Meijer, G. 1959. *Light and Plant Growth.* Eindhoven: Philip's Technical Library.

Wareing, P. F., and Phillips, I. D. J. 1970. *The Control of Growth and Differentiation in Plants.* New York: Pergamon.

Went, F. W. 1963. *The Plants.* New York: Time-Life Books.

Wilkins, B. B. 1969. *The Physiology of Plant Growth and Development.* New York: McGraw-Hill.

Woolhouse, H. W. 1972. *Ageing Processes in Higher Plants.* Oxford: Oxford University Press.

Chapter 8

Arnon, D. I. Photosynthesis 1950–1975: Changing concepts and perspectives. In A. Trebst and M. Avron, eds. *Encyclopedia of Plant Physiology,* vol. 5, pp. 7–56.

Baker, J. J. W., and Allen, G. E. 1974. *Matter, Energy, and Life,* 3d ed. Reading, Ma.: Addison-Wesley.

Bassham, J. 1977. Increasing crop production through more controlled photosynthesis. *Science* 197:630–638.

Bolin, B. 1970. The carbon cycle. *Scientific American* 223: 124–132.

Chappell, J. B. 1977. *ATP.* Burlington, N.C.: Carolina Biological Supply Company.

Conn, E. E., and Stumpf, P. K. 1976. *Outline of Biochemistry,* 4th ed. New York: Wiley.

Cheldelin, W. H., and Newburgh, R. W. 1964. *The Chemistry of Some Life Processes.* New York: Reinhold.

Dagley, S. 1975. Microbial degradation of organic compounds in the biosphere. *American Scientist* 63:681–689.

Garrels, R. M., Lerman, A., and MacKenzie, F. T. 1976. Controls of atmospheric O_2 and CO_2: Past, present, and future. *American Scientist* 64:306–315.

Gibbs, M., ed. 1971. *Structure and Function of Chloroplasts.* New York: Springer-Verlag.

Jensen, W. A. 1970. *The Plant Cell,* 2d ed. Belmont, Calif.: Wadsworth

Ledbetter, M C., and Porter, K. R. 1970. *Introduction to the Fine Structure of Plant Cells.* New York: Springer-Verlag.

Lehninger, A. L. 1971. *Bioenergetics.* Menlo Park, Calif.: W. A. Benjamin.

Whittingham, C. P. 1977. *Photosynthesis.* Burlington, N.C.: Carolina Biological Supply Co.

Woodwell, G. M. 1978. The carbon dioxide question. *Scientific American* 238:34–43.

Zelitch, I. 1971. *Photosynthesis, Photorespiration, and Plant Productivity.* New York: Academic Press.

Chapter 9

Bormann, F. H., Likens, G. E., and Melillo, J. M. 1977. Nitrogen budget for an aggrading northern hardwood forest ecosystem. *Science* 196: 981–983.

Brill, W. J. 1977. Biological nitrogen fixation. *Scientific American* 236: 68–81.

Epstein, E. 1972. *Mineral Nutrition of Plants: Principles and Perspectives.* New York: Wiley.

Garrels, R. M., Mackenzie, F. T., and Hunt, C. 1975.

Chemical Cycles and Global Environment. Los Altos, Calif.: William Kaufmann.

Goldman, J. C., Tenore, K. R., and Stanley, H. I. 1973. Inorganic nitrogen removal from wastewater: Effect on phytoplankton growth in coastal marine waters. *Science* 180:955–956.

Gosz, J. R., Kilens, G. E., and Bormann, F. H. 1972. Nutrient content of litter fall on the Hubbard Brook experimental forest, New Hampshire. *Ecology* 53:769–784.

Harley, J. L. 1971. *Mycorrhiza.* Burlington, N.C.: Carolina Biological Supply Co.

Hutchinson, G. E. 1973. Eutrophication. *American Scientist* 61:269–279.

Jackson, R. M., and Raw, R. 1966. *Life in the Soil.* New York: St. Martins's Press.

Kellogg, W. W., Cadle, R. D., Allen, E. R., Lazrus, A. L., and Martell, E. A. 1972. The sulfur cycle. *Science* 175:587–596

Parkinson, D., and Waid, J. S. 1960. *The Ecology of Soil Fungi.* Liverpool: Liverpool University Press.

Postgate, J. R. 1972. *Biological Nitrogen Fixation.* Watford, Herts: Merrow

Regier, H. A., and Hartman, W. L. 1973. Lake Erie's fish community: 150 years of cultural stresses. *Science* 180:1248–1255.

Russell, E. W. 1961. *Soil Conditions and Plant Growth,* 9th ed. New York: Longmans, Green.

Ryther, J. H., and Dunstan, W. M. 1971. Nitrogen, phosphorus, and eutrophication in the coastal marine environment. *Science* 171.

Safrany, D. R. 1974. Nitrogen fixation. *Scientific American* 231:64–80.

Sanchez, P. A., and Buol, S. W. 1975. Soils of the tropics and the world food crisis. *Science* 188: 598–603.

Schindler, D. W., and Lean, D. R. S. 1974. Biological and chemical mechanisms in eutrophication of freshwater lakes. *Annals New York Academy of Science* 250:129–135.

Silver, W. S., and Postgate, J. R. 1973. Evolution of asymbiotic nitrogen fixation. *Journal of Theoretical Biology* 40:1–10.

Stevenson, F. J. 1972. Role and function of humus in soil with emphasis on adsorption of herbicides and chelation of micronutrients. *Bioscience* 22:643–650.

Stewart, W. D. P. 1973. Nitrogen fixation by photosynthetic microorganisms. *Annual Review of Microbiology* 27:283–316.

———. 1969. Biological and ecological aspects of nitrogen fixation by free-living micro-organisms. *Proceedings of the Royal Society of Biology* 172: 367–388.

Tisdale, S. L., and Nelson, W. L. 1966. *Soil Fertility and Fertilizers,* 2d ed. New York: Macmillan.

Chapter 10

Epstein, E. 1973. Roots. *Scientific American* 228:48–58.

Esau, K. 1965. *Plant Anatomy,* 2d ed. New York: Wiley.

———. 1969. *The Phloem.* Berlin: Gebrüder Borntraeger.

Heath, O. V. S. 1975. *Stomata.* Oxford: Oxford University Press.

Kramer, P. J. 1969. *Plant and Soil Water Relationships.* New York: McGraw-Hill.

Kozlowski, T. T. 1964. *Water Metabolism in Plants.* New York: Harper and Row.

Meidner, M., and Mansfield, T. A. 1968. *The Physiology of Stomata.* New York: McGraw-Hill.

Rutter, A. J. 1972. *Transpiration.* Oxford: Oxford University Press.

Slatyer, R. O. 1967. *Plant-Water Relationships.* New York: Academic Press.

Sutcliffe, J. F. 1968. *Plants and Water.* London: Arnold.

Wardlaw, I. F. 1974. Phloem transport: Physical chemical or impossible. *Annual Review of Plant Physiology* 25:515–539.

Zelitch, I. 1963. *Stomata and Water Relations in Plants.* Bulletin 664. New Haven: Conn. Agricultural Experiment Station.

Zimmerman, M. H., ed. 1963. How sap moves in trees. *Scientific American* 208:132–142.

Chapter 11

Allard, R. W. 1960. *Principals of Plant Breeding.* New York: Wiley.

Beadle, G., and Beadle, M. 1977. *The Language of Life.* Garden City: Anchor Books.

Briggs, D., and Walters, S. M. 1969. *Plant Variation and Evolution.* New York: McGraw-Hill.

Dobzhansky, T. 1955. *Evolution, Genetics, and Man.* New York: Wiley

Hanawalt, P. C., and Haynes, R. H. 1973. *The Chemical Basis of Life.* San Francisco: W.H. Freeman.

Kendrew, J. C. 1966. *The Thread of Life.* Cambridge: Harvard University Press.

Lerner, I. M. 1968. *Heredity, Evolution and Society.* San Francisco: W. H. Freeman

Raff, R. A., and Mahler, H. R. 1972. The non-symbiotic origin of mitochondria. *Science* 177: 575–582.

Raven, P. H. 1970. A multiple origin for plastids and mitochondria. *Science* 169: 641–646.

Sinsheimer, R. L. 1967. *The Book of Life.* Reading, Ma.: Addison-Wesley.

Srb, A. M., Owen, R. D., and Edgar, R. S. 1965. *General Genetics.* 2d. ed. San Francisco: W.H. Freeman.

Stebbins, G. L. 1959. The role of hybridization in evolution. *Proceedings American Phil. Soc.* 103: 231: 251.

Sturtevant, A. H. 1965. *A History of Genetics.* New York: Harper & Row.

Wallace, B. 1966. *Chromosomes, Giant Molecules, and Evolution.* New York: W. W. Norton.

Watson, J. D. 1965. *Molecular Biology of the Gene,* New York: W.A. Benjamin

———. 1968. *The Double Helix.* New York: Atheneum.

Chapter 12

Crockett, J. V. 1971. *Bulbs.* New York: Time-Life Books.

———.1971. *Flowering House Plants.* New York: Time-Life Books.

Hartmann, H. T., and Kester, D. E. 1968. *Plant Propagation,* 2d ed. Englewood Cliffs, N. J.: Prentice-Hall.

Murashige, T. 1974. Plant propagation through tissue cultures. *Annual Rev. Plant Physiol.* 25:135–166.

Nickell, L. G. 1977. Crop improvement in sugarcane: Studies using in vitro methods. *Crop Science* 17:717–719.

Street, H. E., ed. 1977. *Plant Tissue and Cell Culture.* Berkeley: University of California Press.

Thomas, E., and Davey, M. R. 1975. *From Single Cells to Plants.* London: Wykeham Publications.

Chapter 13

Andrews, H. N. 1961. *Studies in Paleobotany.* New York: Wiley.

Banks, H. P. 1975. Early vascular land plants: Proof and conjecture. *Bioscience* 25:730–737.

Beiser, A. 1963. *The Earth.* New York: Time-Life Books.

Berry. W. B. N. 1968. *Growth of a Prehistoric Time Scale.* San Francisco: W. H. Freeman.

———. 1966. *Plant Diversification.* New York: Holt, Rinehart & Winston.

Delevoryas, T. 1962. *Morphology and Evolution of Fossil Plants.* New York: Holt, Rinehart & Winston.

Eicher, D. L. 1968. *Geologic Time.* Englewood Cliffs, N.J.: Prentice-Hall.

Faul, H. 1978. A history of geologic time. *American Scientist* 66:159–165.

Newell, N.D. 1959. The nature of the fossil record. *Proceedings American Phil. Society* 103:264–285.

Palmer, A. R. 1974. Search for the cambrian world. *American Scientist* 62:216–224.

Rudwick, M. J. S. 1972. *The Meaning of Fossils.* London and New York: MacDonald and American Elsevier Inc.

Schopf, J. W., and Oehler, D. Z. 1976. How old are the eukaryotes? *Science* 193:47–49.

Sporne, K. R. 1971. *The Mysterious Origin of Flowering Plants.* Burlington, N.C.: Carolina Biological Supply Co.

———. 1973. The survival of archaic dicotyledons in tropical rain forests. *New Phytologist* 72:1175–1184.

Sutton, J., and Windley, B. F. The precambrian. *Science Progress* 61:401–420.

Takhtajan, A. 1969. *Flowering Plants, Origin and Dispersal.* Washington, D.C.: Smithsonian Institution Press.

Terborgh, J. 1974. Preservation of natural diversity: The problem of extinction prone species. *Bioscience* 24:715–722.

Van Valen, L. 1971. The history and stability of atmospheric oxygen. *Science* 171:439–443.

West, R. G. 1971. *Studying the Past by Pollen Analysis.* Oxford: Oxford University Press.

Chapter 14

Anderson, E. 1949. *Introgressive Hybridization.* New York: Wiley.

Baker, H. G. 1963. Evolutionary mechanisms in pollination biology. *Science* 139:877–883.

Barlow, N., ed. 1976. *Darwin and Henslow: The Growth of an Idea.* London: Bentham-Moxon and Murray.

———. ed. 1958. *The Autobiography of Charles Darwin, 1809–1882.* New York: W. W. Norton.

Bowman, R. I., ed. 1966. *The Galapagos.* Berkeley: University of California Press.

Carlquist, S. 1965. *Island Life. A Natural History of the Islands of the World.* New York: Natural History Press.

Darwin, C. 1966. *On the Origin of Species.* A facsimile of the first edition. Cambridge: Harvard Univerisity Press.

———. 1962 *The Origin of Species.* New York: Collier Books.

———. 1958. *The Voyage of the Beagle.* New York: Bantam Books.

De Beer, G. 1965. *Charles Darwin, A Scientific Biography.* New York: Anchor Books.

Dobzhansky, T. 1951. *Genetics and the Origin of Species,* 3d ed. New York: Columbia University Press.

Dowdeswell, W. H. 1958. *The Mechanism of Evolution,* 2d ed. New York: Harper.

Ehrlich, P. R., Holm, R. W., and Raven, P. H., eds. 1969. *Papers on Evolution.* Boston: Little, Brown.

Eiseley, L. 1961. *Darwin's Century.* New York: Anchor Books.

Fisher, R. A. 1958. *The Genetical Theory of Natural Selection,* 2d ed. New York: Dover.

Gilbert, L. E., and Raven, P. H. 1975. *Coevolution of Animals and Plants.* Austin: University of Texas Press.

Grant, V. 1963. *The Origin of Adaptations.* New York: Columbia University Press.

Hardin, G. 1959. *Nature and Man's Fate.* New York: Mentor Books.

Heinrich, B. 1973. Energetics of the bumblebee. *Scientific American* 228: 96–102.

Herbert, S. 1971. Darwin, Malthus, and selection. *Journal of the History of Biology* 4:209–217.

Hubbell, T. H. 1968. The biology of islands. *Proceedings National Academy of Science* 60:22–32.

Huxley, J. 1953. *Evolution in Action.* New York: Mentor Books.

Jansen, D. H. 1966. Coevolution of mutualism between ants and acacias in Central America. *Evolution* 20:249–275.

Joravsky, D. 1970. *The Lysenko Affair.* Cambridge: Harvard University Press.

King-Hele, D. 1974. Erasmus Darwin, master of many crafts. *Nature* 247: 87–91.

Lerner, I. M. 1959. The concept of natural selection: A centennial view. *Proceedings of the American Philosophical Society.* 103:173–182.

Mayr, E. 1970. *Populations, Species, and Evolution.* Cambridge. Harvard University Press.

Moore, R. 1969. *Evolution.* New York: Time-Life Books.

Moorehead, A. 1969. *Darwin and the Beagle.* New York: Harper & Row.

Ornduff, R., ed. 1967. *Papers on Plant Systematics.* Boston: Little, Brown.

Sheppard, P. M. 1959. *Natural Selection and Heredity.* London: Hutchinson.

Simpson, G. L. 1949. *The Meaning of Evolution.* New Haven: Yale University Press.

———. 1953. *The Major Features of Evolution.* New York: Columbia University Press.

Stebbins, G. L. 1950. *Variation and Evolution in Plants.* New York: Columbia University Press.

———. 1966. *Processes of Organic Evolution.* Englewood Cliffs, N.J.: Prentice-Hall.

Thoday, J. M. 1972. Disruptive selection. *Proceedings Royal Society* 182: 109–143.

Wallace, A. R. 1962. *The Malay Archipelago.* New York: Dover.

Wilson, L. G. 1971. Sir Charles Lyell and the species question. *American Scientist* 59:39–55.

———. 1972. *Charles Lyell, The Years to 1841: The Revolution in Geology.* New Haven: Yale University Press.

Chapter 15

Alexander, T. A revolution called plate tectonics has given us a whole new earth. *Smithsonian* 5:30–41.

Ayala, F. J. 1972. Competition between species. *American Scientist* 60: 348–357.

Billings, W. D. 1968. *Plants and the Ecosystem.* Belmont, California: Wadsworth Publishing Co.

Calder, N. 1978. Head south with all deliberate speed: Ice may return in a few thousand years. *Smithsonian* 8:32–41.

Connell, J. H. 1978. Diversity in tropical rain forests and coral reefs. *Science* 199:1302–1310.

Cracraft, J. 1973. Continental drift, paleoclimatology, and the evolution and biogeography of birds. *Journal of Zoology, London* 169:455–545.

Ehrlich, P. R., and Ehrlich, A. H. 1970. *Population Resources Environment.* San Francisco: W. H. Freeman.

Ehrlich, P. R., Holdren, J. P., and Holm, R. W., eds. 1971. *Man and the Ecosphere.* San Francisco: W. H. Freeman.

El-Baz, F. 1977. Expanding desert creates grim beauty but also threatens crucial cropland. *Smithsonian* 8:34–41.

Emmel, T. C. 1973. *An Introduction to Ecology and Population Biology.* New York: W. W. Norton.

Farb, P. 1967. *Ecology.* New York: Time-Life Books.

———. 1969. *The Forest.* New York: Time-Life Books.

Gomez-Pompa, A., Vazquez-Yanes, C., and Guevara, S. 1972. The tropical rain forest: A nonrenewable resource. *Science* 177: 762–765.

Hadley, N. F. 1972. Desert species and adaptation. *American Scientist* 60:338–347.

Hardin, G. 1969. *Population, Evolution, and Birth Control,* 2d ed. San Francisco: W. H. Freeman.

Heinrich, B., and Raven, R. H. 1972. Energetics and pollination ecology. *Science* 176: 597–602.

Levitt, J. 1972. *Responses of Plants to Environmental Stresses.* New York: Academic Press.

McGinnies, W. G., Goldman, B. J., and Paylore, P., eds. 1971. *Food, Fiber and the Arid Lands.* Tucson: University of Arizona Press.

McKenzie, D. P. 1977. Plate tectonics and its relationship to the evolution of ideas in the geological sciences. *Daedalus* 1:97–124.

National Academy of Sciences-National Research Council. 1969. *Resources and Man.* San Francisco: W. H. Freeman.

Odum, E. P. 1963. *Ecology.* New York: Holt, Rinehart & Winston.

———. The strategy of ecosystem development. *Science* 164: 262–270.

Ormerod, W. E. 1976. Ecological effect of control of African trypanosomiasis. *Science* 191:815–821.

Pimentel, D., Terhune, E. C., Dyson-Hudson, R., Rochereau, S., Samis, R., Smith, E. A., Denman, D., Reifschneider, D., and Shepard, M. 1976. Land degradation: Effects on food and energy resources. *Science* 194:149–155.

Raven, P., and Axelrod, D. I. 1974. Angiosperm biogeography and past continental movement. *Annual of Missouri Botanical Garden* 61:539-673.

Reichle, D. E., ed. 1970. *Analysis of Temperate Forest Ecosystems.* New York: Springer-Verlag.

Riclefs, R. E. 1973. *Ecology.* New York: Chiron Press.

Smith, A. P. 1973. Stratification of temperate and tropical forests. *American Naturalist* 107: 671–683.

Waisel, Y. 1972. *Biology of Halophytes.* New York: Academic Press.

Watt, K. E. F. 1973. *Principles of Envioronmental Science.* New York: McGraw-Hill.

Whittaker, R. H. Evolution of diversity in plant communities. In *Diversity and Stability in Ecological Systems,* Brookhaven Symposia in Biology, No. 22, Upton, N.Y.

Woodell, S. R. J. 1973. *Xerophytes.* Oxford: Oxford University Press.

Wright, H. E., Jr. 1976. The environmental setting for plant domestication in the near east. *Science* 194:385–389.

Credits

The author gratefully acknowledges permission to use material from the following sources.

Figure 1.1: NASA. *Figure 1.3:* page 6: NASA; page 7: (left) Brown Brothers; (right) Bureau of Reclamation, U.S. Dept. of the Interior, Lower Colorado Division, Boulder City, Nevada. *Figure 1.9:* adapted from *Living in the Environment* by G. Tyler Miller, © 1975 by Wadsworth Publishing Company, Inc., Belmont, Calif. 94002. Reprinted by permission of the publisher. *Figure 1.10:* page 14: Harvey Lloyd, Nancy Palmer Photo Agency; page 15: (photos) (top) Bruce Coleman Inc.; (middle) M. Freeman, Bruce Coleman Inc.; (bottom) Francois Gohier, Photo Researchers, Inc. *Figure 1.13:* adapted from Victor A. Greulach and J. Edison Adams, *Plants: An Introduction to Modern Botany,* 3rd edition, p. 395. Copyright © 1962, 1967, 1976 by John Wiley & Sons, Inc. Reprinted by permission. *Supplement Figure 6:* adapted from Life Science Library/*The Cell,* p. 37, published by Time-Life Books Inc. Reprinted by permission.

Figure 2.1: Anthony Howarth, *Daily Telegraph* Magazine, Woodfin Camp & Assoc. *Figure 2.2:* adapted from The Emergence of Man/*The First Farmers,* p. 11, published by Time-Life Books Inc., and from *People of the Earth,* 2nd edition, by Brian M. Fagan, pp. 212, 352. Copyright © 1977 by Little, Brown and Company (Inc.) Reprinted by permission. *Figure 2.5:* from Jack R. Harlan, "The Plants and Animals That Nourish Man," *Scientific American,* September, 1976, pp. 94–95. Copyright © 1976 by Scientific American, Inc. All rights reserved. Reprinted by permission of W. H. Freeman and Company. *Figure 2.6:* courtesy of the Oriental Institute, University of Chicago. *Figure 2.7:* page 36: (left) Staatliche Museum, Berlin; (right) The University Museum, University of Pennsylvania; page 37: (top) Paulus Leeser, courtesy American History Division, The New York Public Library, Astor, Lenox and Tilden Foundations; (bottom) rendering by James Alexander based on a photo courtesy Giraudon. *Figure 2.8:* adapted from Jim Flore in Life Nature Library/*The Plants,* p. 162, published by Time-Life Books Inc. Reprinted by permission. *Figure 2.9:* (top left) adapted from The Emergence of Man/*The First Farmers,* artist, Leslie Martin, Time-Life Books Inc., publisher; (top right) courtesy of American Museum of Natural History; (middle) from Great Ages of Man/*Ancient Egypt,* photographer, Erich Lessing, Time-Life Books Inc., publisher; (bottom) Historical Pictures Service, Chicago. *Figure 2.10:* The Granger Collection, New York. *Figure 2.11:* (photo) Earl Roberge, Photo Researchers, Inc. Adapted from C. W. Thornthwaite, *The Geographical Review,* January 1948, and Life Science Library/*Food and Nutrition,* p. 46, published by Time-Life Books Inc. Reprinted by permission. Redrawn from Life Science Library/*Food and Nutrition,* p. 171, published by Time-Life Books Inc. Reprinted by permission. *Figure 2.12.:* Terence Spencer, © Time, Inc. *Figure 2.14:* from Gary H. Heichel, "Energy Needs and Food Yields," *Technology Review* 76, 8 (1974). Reprinted by permission. *Figure 2.15:* (top) Allis-Chalmers Corporation; (bottom) Derek Bayes. *Figure 2.16:* from Carol Steinhart and John Steinhart, *Energy: Sources, Use, and Role in Human Affairs,* © 1974 by Wadsworth Publishing Co., Inc., Belmont, Calif. 94002. Reprinted by permission of the publisher, Duxbury Press. *Figure 2.17:* from Gary H. Heichel, "Agricultural Production and Energy Resources," *American Scientist* 64 (1976). Reprinted by permission. *Figure 2.18:* data from Food and Agriculture Organization and U.S. Department of Agriculture. *Supplement Figure 3:* data courtesy of U.S. Senate Select Committee on Nutrition and Human Needs, 1977.

Figure 3.1: Dmitri Kessel, © Time, Inc. Reprinted by permission. *Figure 3.2:* courtesy of the Hunt Institute for Botanical Documentation, Carnegie-Mellon University. *Figure 3.3:* redrawn from G. Ledyard Stebbins, *Processes of Organic Evolution,* 2nd edition, © 1971. Reprinted by permission of Prentice-Hall, Inc., Englewood Cliffs, New Jersey. *Figure 3.4:* Jeanne White from National Audubon Scoiety, Photo Researchers, Inc. *Figure 3.5:* Stephen Dalton. *Figure 3.6:* courtesy Missouri Botanical Garden. *Figure 3.9:* Grant Heilman Photography. *Figure 3.10:* page 86: (left) courtesy Missouri Botanical Garden; (bottom left) Al Bussewitz; (bottom right) Al Bussewitz; page 87: (left) David M. Stone; (top right) A. W. Ambler from National Audubon Society, Photo Researchers, Inc., (bottom) William M. Harlow. *Figure 3.11:* (left) W. M. Laetsch, (right) Dr. John H. Miller, Syracuse University. *Figure 3.12:* Dr. William C. Steere, The New York Botanical Garden. *Figure 3.13:* redrawn from Life Nature Library/*The Plants,* published by Time-Life Books Inc. *Figure 3.15:* redrawn from illustration by Anthony D'Attilio from Life Nature Library/*The Plants,* published by Time-Life Books Inc. *Supplement Figure 1:* (left) Dr. Norma J. Lang, University of California, Davis; (middle) Harold E. Sweetman; (right) W. M. Laetsch. *Supplement Figure 2:* Dr. Myron C. Ledbetter, Brookhaven National Laboratory.

Figure 4.1: William M. Harlow. *Figure 4.2:* (photos) The Time-Life Encyclopedia of Gardening, *Evergreens,* Time-Life Books Inc. publisher. Reprinted by permission. *Figure 4.3:* Grant Heilman Photography. *Figure 4.4:* redrawn from Harold C. Bold, *The Plant Kingdom,* 3rd edition, p. 136 (Figure 8.14), © 1961, 1964, 1970. Reprinted by permission of Prentice-Hall, Inc., Englewood Cliffs, N.J. *Figures 4.6 and 4.7:* Courtesy of Carolina Biological Supply Company. *Figure 4.8:* page 111: Douglas Faulkner; inset, courtesy Verne N. Rockcastle, Cornell University; *Figure 4.9:* page 113: (top) Grant Heilman Photography; (bottom) courtesy Verne N. Rockcastle, Cornell University. *Figure 4.10:* (top left) Dr. Jack A. Robertson, Grinnell College; (top right) Dr. Albert Loeblich III, University of Houston; (middle) Dr. Ralph A. Lewin, Scripps Institution of Oceanography; (bottom) Ewing Krainin, Photo Researchers, Inc. *Figure 4.11:* Walter H. Hodge. *Figure 4.12:* page 118: (top left) Steven C. Wilson; (right) Grant Heilman Photography; (bottom left) Grant Heilman Photography; (right) Grant Heilman Photography; page 119; (right) Grant Heilman Photography; (left) from *The Plant World,* 5th edition, p. 371, by Harry J. Fuller, Zane B. Carothers, Willard W. Payne, and Margaret K. Balbach, Copyright 1941, 1951, © 1955, 1963, 1972 by Holt, Rinehart and Winston, Inc. Reprinted by permission of Holt, Rinehart and Winston. *Figure 4.13:* redrawn from *The Plant World,* 5th edition, by Harry J. Fuller, Zane B. Carothers, Willard W. Payne, Margaret K. Balbach, p. 378. Copyright 1941, 1951, © 1955, 1963, 1972 by Holt, Rinehart and Winston, Inc. Reprinted by permission of Holt, Rinehart and Winston. *Figure 4.14:* (both) Dr. David Pramer, Rutgers University. *Figure 4.15:* from *Plant Biology,* Figure 3.15, p. 76, by Drs. Norstog and Long, © 1976 by W. B. Saunders Company, Philadelphia, Pa. *Figure 4.16:* adapted from Life Nature Library/*The Forest,* p. 48, published by Time-Life Books Inc. Reprinted by permission. *Supplement Figure 1:* (top left) Lee D. Simon, Wakeman Institute of Microbiology, Rutgers University, New Brunswick, N.J. *Supplement Figure 4:* redrawn from *The Plant World,* 5th edition, p. 349, by Harry J. Fuller, Zane B. Carothers, Willard W. Payne, Margaret K. Balbach. Copyright 1941, 1951, © 1955, 1963, 1974 by Holt, Rinehart and Winston, Inc. Reprinted by permission of Holt, Rinehart, and Winston. *Supplement Figure 5:* Dr. C. A. Knight, University of California, Berkeley.

Figure 5.1: Edmund Gerard, © Time, Inc. *Figure 5.2:* redrawn from Harold C. Bold, *The Plant Kingdom,* 3rd edition, p. 145 (Figure 9.3), © 1961, 1964, 1970. Reprinted by permission of Prentice-Hall, Inc., Englewood Cliffs, N.J. *Figure 5.3:* Cary Wolinsky, Stock, Boston. *Figures 5.5 and 5.6:* W. M. Laetsch. *Figure 5.8:* (left) Dr. Lawrence Rappaport, University of California, Davis; (right) Mansell Collection. *Figure 5.9:* W. M. Laetsch. *Figure 5.10:* Walter H. Hodge. *Figure 5.11:* W. M. Laetsch. *Figure 5.12:* Walter H. Hodge. *Figure 5.13:* Helen and Frank Schreider. *Figure 5.15:* Royal Tropical Institute, Photograph Archives, Amsterdam. *Figure 5.16:* (left) W. M. Laetsch; (right) USDA, photo by Southern Regional Research Center. *Supplement Figure 1:* courtesy of the Hunt Institute for Botanical Documentation, Carnegie-Mellon University.

Figure 6.3: (photos) Lewis Feldman, University of California, Berkeley; page 169: redrawn from T. Elliott Weier *et al., Botany: An Introduction to Plant Biology,* 5th edition, pp. 309, 311. Reprinted by permission of John Wiley & Sons, Inc. *Figure 6.4:* (top) W. M. Laetsch; (bottom) Dennis Greathouse, University of California, Berkeley. *Figure 6.7:* (photos) W. M. Laetsch; (right) redrawn from Life Nature Library/*The Plants,* published by Time-Life Books Inc. *Figure 6.8:* W. M. Laetsch. *Figure 6.9:* (left) Andreas Feininger; (middle and right) W. M. Laetsch. *Figure 6.10:* W. M. Laetsch. *Figure 6.11:* Sanford S. Tepfer, University of Oregon, Eugene. *Figure 6.12:* (left) W. M. Laetsch; (right) Al Bussewitz. *Figure 6.13:* from David Rayle and Lee Wedberg, *Botany: A Human Concern,* Figure 6-25, p. 185. Copyright © 1975 by Houghton Mifflin Company. Reprinted by permission. *Figure 6.14:* (photos) page 183: Jerome Wexler, Photo Researchers, Inc. *Figure 6.15:* courtesy of Carolina Biological Supply Company. *Figure 6.16:* derived from *The Plant World,* 5th edition, by Harry J. Fuller, Zane B. Carothers, Willard W. Payne, Margaret K. Balbach, p. 147. Copyright 1941, 1951, © 1955, 1963, 1972 by Holt, Rinehart and Winston, Inc. Reprinted by permission of Holt, Rinehart and Winston. *Figure 6.17:* page 186: W. M. Laetsch; page 187: (bottom) courtesy Carolina Biological Supply Company; (top) adapted from T. Elliot Weier *et al., Botany: An Introduction to Plant Biology,* 5th edition, p. 141. Reprinted by permission of John Wiley & Sons. *Figure 6.18:* W. M. Laetsch. *Figure 6.19:* page 192: W. M. Laetsch; page 193: Lewis Feldman, University of California, Berkeley; adapted from Walter H. Muller, *Botany: A Functional Approach,* 3rd edition, pp. 116, 119. Copyright © 1974. Reprinted by permission of Macmillan Publishing Co., Inc.

Figure 7.1: Edmund Gerard, © Time, Inc. *Figure 7.4:* (photos) Ralph W. G. Wyckoff, University of Arizona, Tucson; redrawn and adapted from Life Nature Library/*The Plants,* published by Time-Life Books Inc. *Figure 7.5:* adapted from *The Living Plant* by Peter Ray, p. 181. Copyright © 1963, 1972 by Holt, Rinehart and Winston, reprinted by permission; page 206: Albert Fenn © Time, Inc.; page 207: (bottom) Jerome Wexler, Photo Researchers, Inc.; (top) Lewis Feldman, University of California, Berkeley. *Figure 7.6:* adapted from David Rayle and Lee Wedberg, *Botany: A Human Concern,* Figure 8-12, p. 256, Copyright © 1975 by Houghton Mifflin

Company. Reprinted by permission. *Figure 7.8:* W. M. Laetsch. *Figure 7.10:* William Vandivert; from "Light and Plant Development," by W. L. Butler and R. J. Downs, *Scientific American,* December 1960, p. 57. *Figure 7.12:* Edmund Gerard, © Time, Inc. *Figure 7.13:* Grant Heilman Photography.

Figure 8.1: redrawn from Life Science Library/*The Cell,* p. 36, published by Time-Life Books Inc. *Figure 8.2:* W. M. Laetsch. *Figure 8.3:* adapted from David Rayle and Lee Wedberg, *Botany: A Human Concern,* Figures 2-8 and 2-11, pp. 44, 46. Copyright © 1975 by Houghton Mifflin Company. Reprinted by permission. *Figure 8.6:* from Life Science Library/*Water* pp. 112-113, published by Time-Life Books Inc.; from Walter H. Muller, *Botany: A Functional Approach,* 3rd edition, p. 171. Copyright © 1974. Reprinted by permission of Macmillan Publishing Co., Inc. *Figure 8.9:* adapted from A. L. Lehniager, "How Cells Transform Energy," *Scientific American,* September 1961, Figure 3.9. *Figure 8.12:* Courtesy, Field Museum of Natural History, Chicago. *Supplement Figure 1:* redrawn from Diane Robertson and W. M. Laetsch, "Structure and Function of Developing Barley Plastids," *Plant Physiology,* Volume 54 (1974), p. 158, Figure 17. Used by permission. *Supplement Figures 2, 3, 4:* W. M. Laetsch.

Figure 9.4: (top photo) D. Goodchild; (bottom photo) courtesy J. C. Burton, The Nitragin Co., Inc. Milwaukee, Wis.; adapted from Frank B. Salisbury and Cleon Ross, *Plant Physiology,* © 1969 by Wadsworth Publishing Company, Inc., Belmont, Calif. 94002. Reprinted by permission of the publisher. *Figure 9.5:* W. M. Laetsch. *Figure 9.6:* (left) Al Bussewitz; (right and bottom) Ruth Strand, University of California, Berkeley. *Figure 9.7:* (left and middle) David M. Stone; (right) George S. Sheng; (bottom) David M. Stone. *Figure 9.9:* W. M. Laetsch. *Figure 9.11:* from Robert H. Whittaker, *Communities and Ecosystems,* p. 122. Copyright © 1970 by Robert H. Whittaker. Reprinted by permission of Macmillan Publishing Company. *Figure 9.12:* Carl H. Strandberg. *Figure 9.13:* redrawn from Life Science Library/*Water,* p. 174, published by Time-Life Books Inc. *Figure 9.14:* The Sacramento Bee. *Supplement Figure 1:* Redrawn from Jules Janick, Carl H. Noller, and Charles L. Rhkerd, "The Cycle of Plant and Animal Nutrition," *Scientific American,* September 1976. Reprinted by permission. *Supplement Figure 2:* Dur Morton from National Audubon Society, Photo Researchers, Inc. *Supplement Figure 3:* adapted from Maarten Chrispeels, *The Green Revolution* (unpublished manuscript). *Supplement Figure 4:* redrawn from *Living in the Environment* by G. Tyler Miller, © 1975 by Wadsworth Publishing Company, Inc. Belmont, California 94002. Reprinted by permission of the publisher. *Supplement Figure 5:* courtesy of Soil Conservation Service, USDA.

Figure 10.1: J. Arthur Herrick, Kent State University. *Figure 10.2:* from Life Nature Library/*The Plants,* p. 75, published by Time-Life Books, Inc. Reprinted by permission. *Figure 10.3:* scanning electron micrograph by J. Heslop-Harrison, University of Wales, Welsh Plant Breeding Station. *Figure 10.5:* page 281: (right) Dr. Vincent J. Schaefer. *Figure 10.6:* Richard Weymouth Brooks. *Figure 10.8:* redrawn from Life Science Library/*Water,* p. 23, published by Time-Life Books Inc. *Figure 10.9:* page 286: redrawn from Martin Zimmerman, "How Sap Moves in Trees," *Scientific American,* March 1963. Reprinted by permission; page 287: (bottom) from *Humanistic Botany* by Tippo and Stern; courtesy Dr. Brian G. Butterfield; page 287: (top and middle) W. M. Laetsch. *Figure 10.10:* W. Thomson. *Figure 10.11:* Lewis J. Feldman, University of California, Berkeley. *Figure 10.12:* redrawn from Emanuel Epstein, "Roots," *Scientific American,* vol. 228, no. 5 (May 1973). *Figure 10.14:* Dr. Martin H. Zimmerman. *Figure 10.15:* page 295: R. Evert and S. Eichorn, University of Wisconsin. *Figure 10.16:* USDA. *Figure 10.17:* courtesy Carolina Biological Supply Company. *Figure 10.18:* W. M. Laetsch. *Figure 10.19:* redrawn from Katherine Esau, *Plant Anatomy,* 2nd edition, page 332. Reprinted by permission of John Wiley & Sons.

Figure 11.1: Paul Slaughter, Photo Researchers, Inc. *Figure 11.2:* Mansell Collection. *Figure 11.7:* (top left) The Bettmann Archive Inc.; (middle and right) The Granger Collection, New York; (bottom) Philips Electronic Instruments, Inc.; (inset) A. J. Belling Photo Researchers, Inc. *Figure 11.8:* David A Stetler, Virginia Polytechnic Institute and State University. *Figure 11.9:* from Dr. William Jackson, Dartmouth College. *Figure 11.12:* redrawn from Victor A. Greulach and J. Edison Adams, *An Introduction to Modern Botany,* 3rd edition, page 492. Copyright © 1962, 1967, 1976 by John Wiley & Sons, Inc. Reprinted by permission. *Figure 11.15:* from George and Muriel Beadle, *The Language of Life,* Figure 102. Copyright © 1966 by George W. Beadle and Muriel Beadle. Reproduced by permission of Doubleday & Company, Inc. *Figure 11.16:* Dr. Myron C. Ledbetter, Brookhaven National Laboratory. *Figure 11.17:* Dr. Arnold H. Sparrow and Robert F. Smith (photographer), Brookhaven National Laboratory. *Figure 11.18:* C. S. Webber, Jepson Herbarium, University of California, Berkeley. *Figure 11.20:* (top left) Al Bussewitz; (top right) Bruce Coleman, Inc.; (inset) Richard Gross, Motlow State Community College; (bottom) Bruce Coleman, Inc. *Figure 11.23:* John Dominis © Time, Inc.

Figure 12.1: redrawn from Victor A. Greulach and J. Edison Adams, *An Introduction to Modern Botany,* 3rd edition, p. 429. Copyright © 1962, 1967, 1976 by John Wiley & Sons, Inc. Reprinted by permission. *Figure 12.2:* page 351: (bottom photo) W. M. Laetsch; (top photo) Grant Heilman Photography; (top left) adapted from The Time-Life Encyclopedia of Gardening/*Bulbs,* p. 10, published by Time-Life Books Inc. Reprinted by permission; page 350: (middle) adapted from The Time Life Encyclopedia of Gardening/*Bulbs,* pp. 10, 11, published by Time-Life Books Inc. Reprinted by permission. Page 350: (bottom) adapted from The Time-Life Encyclopedia of Gardening/*Bulbs,* p. 13, published by Time-Life Books, Inc. Reprinted by permission. *Figure 12.3:* W. M. Laetsch. *Figure 12.6:* Eldon S. Banta, Ohio State University. *Figure 12.7:* W. M. Laetsch. *Figure 12.8:* redrawn and adapted from Knut Norstog and Robert W. Long, *Plant Biology.* Copyright © 1976 by W. B. Saunders Company. Reprinted by permission. *Figure 12.9:* W. M. Laetsch.

Figure 13.1: Editorial Photocolor Archives. *Figure 13.2:* The Granger Collection, New York. *Figure 13.3:* page 368: (left) from Bernard G. Campbell, *Humankind Emerging,* p. 74. Copyright © 1976 by Little Brown and Company (Inc.). Reprinted by permission; page 368: (right) adapted from Life Nature Library/*Eurasia,* p. 110, published by Time-Life Books Inc. Reprinted by permission; page 369: (top) Time-Life, The Emergence of Man/*Life Before Man,* p. 61; photographed by Fritz Goro; Botanical Museum, Harvard University, Time-Life Books Inc., publisher. Reprinted with permission; page 369: (middle) Time-Life, The Emergence of Man/*Life Before Man* p. 59, photographed by Fritz Goro, Peabody Museum of Natural History, Yale University, Time-Life Books Inc., publisher. page 369: (bottom left) Frank M. Carpenter, Harvard University; page 369: (bottom right) Time-Life, The Emergence of Man/*Life Before Man,* p. 50, photographed by Fritz Goro, Dr. T. Delevoryas, Yale University, Time-Life Books Inc., publisher. *Figure 13.5:* redrawn from Life Nature Library/*The Earth,* pages 138-139, published by Time-Life Books, Inc. *Figure 13.7:* Dr. E. S. Barghoorn, Harvard University. *Figure 13.8:* adapted from *Life: An Introduction to Biology,* by George Gaylord Simpson and William S. Beck. © 1965 by Harcourt Brace Jovanovich, Inc. Reproduced by permission of the publisher. *Figure 13.9:* (top left) Grant Heilman Photography; (top right) W. M. Laetsch; (bottom) Richard Gross, Motlow State Community College.

Figure 14.1: The Granger Collection, New York. *Figure 14.2:* Mansell Collection. *Figure 14.3:* Mansell Collection. *Figure 14.4:* Radio Times Hulton Picture Library. *Figure 14.7:* from Sherwin Carlquist, *Island Life* (The Natural History Press). © 1965 by Sherwin Carlquist. *Figure 14.8:* adapted by permission from *Encyclopaedia Britannica,* 15th edition (1974), 11:542. *Figure 14.10:* redrawn from Sherwin Carlquist, *Island Life* (The Natural History Press). © 1965 by Sherwin Carlquist. Reprinted by permission. *Figure 14.12:* (top) W. M. Laetsch; (bottom) Lewis Feldman. *Figure 14.15:* Page 408: (top left) redrawn from Life Nature Library/*Africa,* p. 65, published by Time-Life Books Inc. Reprinted by permission; Page 408: (right) redrawn from Alfred D. Romer, *The Procession of Life,* p. 253, © 1968. Reprinted by permission of William Collins and World Publishing Co., Inc. Page 409: (top photo) Stephen Collins, Photo Researchers, Inc.; (middle photo) W. M. Laetsch; (bottom photo) H. Eisenbeiss, Photo Researchers, Inc.; (middle) redrawn from Life Nature Library/*Australia,* p. 103 published by Time-Life Books Inc., reprinted by permission. *Figure 14.16:* (top left) W. M. Laetsch; (top right) Lilo Hess; (bottom) Dr. John R. Hendrikson. *Supplement Figures 1, 2, 3, 4:* The Granger Collection, New York. *Supplement Figure 5:* Mansell Collection.

Figure 15.1: adapted from Laurence H. Elliot, *Climate and Man,* pp. 8, 9, © 1969 McGraw-Hill Publishing Company Ltd. Reprinted by permission. *Figure 15.2:* adapted from Laurence H. Elliot, *Climate and Man,* p. 19. Copyright © 1969 McGraw-Hill Publishing Company Ltd. Reprinted by permission. *Figure 15.4:* Courtesy National Science Foundation. *Figure 15.6:* Page 426: (both photos) W. M. Laetsch; (graph) courtesy of National Science Foundation; Page 427: (photos) (top left) W. D. Billings, Duke University; (top right) Nevada State Highway Dept.; (bottom left) Eliot Eliofson, © Time, Inc.; (bottom right) Ansel Adams, Magnum Photos. Page 427: (bottom) adapted from Life Nature Library/*The Land and Wildlife of North America,* p. 119, published by Time-Life Books Inc. *Figure 15.7:* page 429: (photos) (top right) Karl Gullers, Photo Researchers, Inc.; (bottom left) Eliot Eliofson, © Time, Inc.; (top left) Dmitri Kessel © Time, Inc.; (bottom right) Ernst Haas. *Figure 15.8:* (left) W. M. Laetsch; (right) Spencer Barnet, University of Toronto. *Figure 15.9:* redrawn from Life Nature Library/*The Forest,* pp. 62–63, published by Time-Life Books Inc. *Figure 15.10:* redrawn from Life Nature Library/*The Forest,* p. 74, published by Time-Life Books Inc. *Figure 15.11:* (left) W. M. Laetsch; (right) Ross E. Hutchins. *Figure 15.12:* (top) Ross E. Hutchins; (bottom left) Walter H. Hodge; (bottom right) W. M. Laetsch. *Figure 15.13:* Bruce Coleman Inc. *Figure 15.14:* page 440: adapted from Clarence J. Hylander, *Wildlife Communities from the Tundra to the Tropics in North America,* p. 103. Copyright © 1966 by Doris S. Hylander. Reprinted by permission of Houghton Mifflin Company. Page 441: (right) from H. von Meiss Teuffen, Photo Researchers, Inc.; (left) World Book Photo. © 1964 Field Enterprises Educational Corporation; (bottom) U.S. Forest Service. *Figure 15.16:* page 445: (bottom left): from Horace Bristol, Photo Researchers, Inc.; (right) Andreas Feininger, © Time, Inc.; (top left) Richard Weymouth Brooks; drawings adapted from Life Nature Library/*The Desert,* pp. 54–55, 57, 59, published by Time-Life Books Inc. *Figure 15.18:* Inge Morath, Magnum Photos. *Figure 15.19:* page 449: (photos) (right) W. M. Laetsch; (left) Harald Sund; (bottom) Bruce Coleman Inc. Page 448: adapted from George M. Woodwell, "The Energy Cycle of the Biosphere," *Scientific American,* September 1970 (223), p. 69. Reprinted by permission. Page 449: (bottom) adapted from D. W. Johnston and E. P. Odum, *Ecology,* vol. 37, 1956. Reprinted by permission. *Figure 15.20:* J. Arthur Herrick, Kent State University. *Figure 15.21:* Lee Talbot. *Figure 15.23:* U.S. Forest Service. *Supplement Figure 1:* adapted from Robert Dietz and John C. Holden, "The Breakup of Panagaea," *Scientific American,* October 1970. Copyright © 1970 by Scientific American, Inc. All rights reserved. Reprinted by permission of W. H. Freeman Company. *Supplement Figure 2:* adapted from Robert Dietz and John C. Holden, "The Breakup of Panagaea," *Scientific American,* October 1970.

Index